G26 INT

CD ROM
AT ISSUE
DESK

WITHDRAWN

For sale

G26 INT

INTEGRATED HUMAN BRAIN SCIENCE:
Theory, Method, Application (Music)

Front Cover:
Simulated brain. (See Chapter I : Vortex Model of the Brain.)

INTEGRATED HUMAN BRAIN SCIENCE:
Theory, Method, Application (Music)

Edited by

Tsutomu Nakada

Department of Integrated Neuroscience
Brain Research Institute
University of Niigata
1 Asahimachi
Niigata 951-8585, Japan

2000

ELSEVIER

Amsterdam - Lausanne - New York - Oxford - Shannon - Singapore - Tokyo

ELSEVIER SCIENCE B.V.
Sara Burgerhartstraat 25
P.O. Box 211, 1000 AE Amsterdam, The Netherlands

Co-Published by Elsevier Science B.V. and Nishimura Co. Ltd.
Exclusive sales rights for Japan: Nishimura Co., Ltd.

© 2000 Elsevier Science B.V. All rights reserved.

This work is protected under copyright by Elsevier Science, and the following terms and conditions apply to its use:

Photocopying
Single photocopies of single chapters may be made for personal use as allowed by national copyright laws. Permission of the Publisher and payment of a fee is required for all other photocopying, including multiple or systematic copying, copying for advertising or promotional purposes, resale, and all forms of document delivery. Special rates are available for educational institutions that wish to make photocopies for non-profit educational classroom use.

Permissions may be sought directly from Elsevier Science Global Rights Department, PO Box 800, Oxford OX5 1DX, UK; phone: (+44) 1865 843830, fax: (+44) 1865 853333, e-mail: permissions@elsevier.co.uk. You may also contact Global Rights directly through Elsevier's home page (http://www.elsevier.nl), by selecting 'Obtaining Permissions'.

In the USA, users may clear permissions and make payments through the Copyright Clearance Center, Inc., 222 Rosewood Drive, Danvers, MA 01923, USA; phone: (978) 7508400, fax: (978) 7504744, and in the UK through the Copyright Licensing Agency Rapid Clearance Service (CLARCS), 90 Tottenham Court Road, London W1P 0LP, UK; phone: (+44) 207 631 5555; fax: (+44) 207 631 5500. Other countries may have a local reprographic rights agency for payments.

Derivative Works
Tables of contents may be reproduced for internal circulation, but permission of Elsevier Science is required for external resale or distribution of such material. Permission of the Publisher is required for all other derivative works, including compilations and translations.

Electronic Storage or Usage
Permission of the Publisher is required to store or use electronically any material contained in this work, including any chapter or part of a chapter.

Except as outlined above, no part of this work may be reproduced, stored in a retrieval system or transmitted in any form or by any means, electronic, mechanical, photocopying, recording or otherwise, without prior written permission of the Publisher.
Address permissions requests to: Elsevier Global Rights Department, at the mail, fax and e-mail addresses noted above.

Notice
No responsibility is assumed by the Publisher for any injury and/or damage to persons or property as a matter of products liability, negligence or otherwise, or from any use or operation of any methods, products, instructions or ideas contained in the material herein. Because of rapid advances in the medical sciences, in particular, independent verification of diagnoses and drug dosages should be made.

First Edition 2000

Library of Congress Cataloging in Publication Data
A catalog record from the Library of Congress has been applied for.

ISBN: 0-444-50629-2

♾ The paper used in this publication meets the requirements of ANSI/NISO Z39.48-1992 (Permanence of Paper).

Printed in The Netherlands.

Preface

In the early 1990's, the Japanese Ministry of Education, Science, Sports, and Culture (Monbusho) launched a revolutionary plan for funding academic science in Japan. The project provides focused support to super select research groups deemed capable of developing into world class "centers of excellence (COE)" in fields ranging from cosmology to economics. Monbusho aimed to select approximately 25 groups on the cutting edge of their fields in five years. The first five groups were named in 1995 (Science 269:474, 1995). In the third year of this project, the research team I directed was chosen to be a representative in the field of human brain science. The project is formally titled the "Neuroscience of Music" to reflect the final goal of our team, i.e. the elucidation of how the human brain works. This book represents our efforts in this venue.

The book is composed of three chapters. The contributors of the first two chapters were staff and guest scientists at the University of Niigata who have closely collaborated with me to build the newly established COE in human brain research. The third chapter represents the proceeding of the first International Symposium on the Neuroscience of Music held October 15-17, 1999, in Niigata. The symposium represented the first international gathering of experts form diverse neuro-scientific fields for in depth discussions regarding the human brain and music. We were honored to welcome His Imperial Highness Prince Yoshihito Katsura as Honorary Chairperson and His Imperial Highnesses Prince and Princess Takamado as Guests of Honor.

The 21st century is said to be "the Century of the Brain." This book is dedicated to the opening century of the new millennium.

March 14, 2000, in Niigata
Tsutomu Nakada

Contents

Preface v

Section I: Brain Theory

1. Vortex model of the brain: the missing link in brain science?
   T. Nakada (Japan/USA) 3

2. Interactions that coordinate cortical activity
   W.A. Phillips and M.E. Pflieger (UK/USA/Japan) 23

Section II: Advanced Methodology

1. Myths and truths in fMRI
   T. Nakada (Japan) 43

2. Development of 3.0 Tesla vertical MRI system for advanced fMRI applications
   T. Nakada and V3T Team (Japan/USA/UK) 71

3. Using adiabatic rapid passage to minimize RF power requirements in NMR
   M. Garwood (USA) 79

4. An efficient method for ICS analysis of functional magnetic resonance imaging
    K. Suzuki, T. Kiryu and T. Nakada (Japan/USA)............ 91

5. Recording the EEG during fMRI: advantages and disadvantages
    J.R. Ives (USA)............ 101

6. A framework for the integration of fMRI, sMRI, EEG, and MEG
    M. Wagner and M. Fuchs (Germany)............ 109

7. Independent component analysis of simulated ERP data
    S. Makeig, T.-P. Jung, D. Ghahremani and T.J. Sejnowski (USA) .. 123

8. The spatial resolving power of high-density EEG: an assessment of limits
    M.E. Pflieger and T. Nakada (Japan/USA)............ 147

9. All phrase event-related potentials during listening to music
    S. Suwazono, M.E. Pflieger, R.E. Jacobs and T. Nakada (Japan/USA)............ 193

10. A new technique in magnetoencephalography and its application to visual neuroscience
    K. Toyama, K. Yoshikawa, S. Tomita and S. Kajihara (Japan) 201

11. Lambda chart analysis and eigenvalue imaging
    H. Matsuzawa and T. Nakada (Japan/USA)............ 219

12. Single ellipsoid diffusion tensor analysis
    N. Nakayama, Y. Fujii and T. Nakada (Japan/USA)............ 227

13. Principal eigenvector imaging ($PE_{vec}I$): comparison of deterministic vs. optical methods
    H. Matsuzawa, Y. Suzuki, H. Saito and T. Nakada (Japan/USA) .. 233

14. NIRS imaging
    Y. Hoshi (Japan)............ 239

15. Clinical application of near-infrared spectroscopy (NIRS) to cerebrovascular disorders
    S. Kuroda and K. Houkin (Japan)............ 249

16. fMRI guided near-infrared spectroscopy: an example for Parkinson's disease
    I.L. Kwee and T. Nakada (USA/Japan)............ 267

Section III: Neuroscience of Music

Poster and Welcome Statement for the Symposium 277

1. Perception of musical sound: simulacra and illusions
   J.-C. Risset (France) 279

2. Primitive intelligence at the sensory level in audition
   R. Näätänen and M. Tervaniemi (Finland) 291

3. Grouping and differentiation: two main principles in the performance of music
   J. Sundberg (Sweden) 299

4. Horizontal and vertical programming in musical performance
   P.L. Divenyi (USA) 315

5. Automatic processing of musical information as evidenced by EEG and MEG recordings
   M. Tervaniemi (Finland) 325

6. Music as a second language
   T. Nakada (Japan/USA) 337

7. The neural basis of musical processes
   R.J. Zatorre (Canada) 345

8. The relationship between musical pitch and temporal responses of the auditory nerve fibers
   K. Ohgushi and Y. Ano (Japan) 357

9. What can the brain tell us about the specificity of language and music processing?
   M. Besson (France) 365

10. Attention to pitch in musicians and non-musicians: an event-related brain potential study
    T.F. Münte, W. Nager, O. Rosenthal, S. Johannes and E. Altenmüller (Germany) 389

11. Cortical activation patterns during perception and imagination of rhythm in professional musicians: a DC-potential study
    E. Altenmüller, R. Beisteiner, W. Lang, G. Lindinger and L. Deecke (Germany and Austria) 399

12. Interaction in musical-pitch naming and syllable naming: an experiment on a Stroop-like effect in hearing
   K. Miyazaki (Japan)... 415

13. Organization of two-tone facilitation in neurons of the primary auditory cortex in rats
   T. Donishi, Y. Tamai and A. Kimura (Japan) 425

14. Music therapy in Parkinson's disease: improvement of parkinsonian gait and depression with rhythmic auditory stimulation
   N. Ito, A. Hayashi, W. Lin, N. Ohkoshi, M. Watanabe and S. Shoji (Japan)... 435

15. A beat tracking model by recurrent neural network
   K. Ohya (Japan) 445

16. Increased activation of supplementary motor area during a hand motor task tuned to auditory rhythm
   M. Saito, T. Kujirai, N. Saito, K. Kujirai, G. Izuta and T. Yamaguchi (Japan)... 449

17. Somatosensory gating during a hand motor task tuned to auditory rhythm
   H. Watanabe, T. Kujirai, M. Saito, N. Saito, K. Kujirai and S. Ueno (Japan) 457

18. Motor programming as a hand motor task tuned to auditory rhythm
   T. Kujirai, H. Watanabe, K. Kujirai, G. Izuta, T. Yamaguchi and S. Ueno (Japan) 463

19. Motor imagery tuned to auditory rhythm activates the motor cortex
   K. Kujirai, T. Kujirai, T. Kato and M. Tominaga (Japan)....... 471

20. Effect of different music contents on EEG activity and mood states
   K. Naruse and H. Sakuma (Japan) 479

21. Strong rightward asymmetry of the planum parietale associated with the ability of absolute pitch
   K. Katanoda, K. Yoshikawa and M. Sugishita (Japan)......... 487

Subject Index 493

Section I

Brain Theory

T. Nakada (Ed.)
Integrated Human Brain Science: Theory, Method Application (Music)
© 2000 Elsevier Science B.V. All rights reserved

Chapter I.1

Vortex Model of the Brain:
- The Missing Link in Brain Science? -

Tsutomu Nakada[1]

Department of Integrated Neuroscience, Brain Research Institute, University of Niigata

Introduction

There is no doubt that discrete neuronal networks play an essential role in brain function. Nevertheless, many fundamental questions regarding how the brain works remain unanswered. What is the neuronal substrate of consciousness? Why do anesthetic effects diminish at higher atmospheric pressure? How can purely endogenous processes be initiated? These are some examples. In spite of concerted effort by the world's preeminent neuroscientists, no complete theory of brain function has thus far been offered. This void strongly suggests that there must be a *missing link* in the current fundamental concept of how the brain works.

All modern theories eventually require confirmatory proof based on hypothesis-driven experimental results. However, it is often difficult to formulate such a hypothesis in a science which is already highly developed. The successful formulation of the best hypothesis itself should reflect the final accomplishment in the field. In contrast to other fields of science, biological science still remains heavily phenomenology oriented. This tendency toward a description based science has likely served as a major obstacle in the derivation of a global hypothesis of brain function. On the other hand, purely mathematical hypotheses introduced by scientists in non-biological fields, such as mathematics, physics, and engineering, have often failed to provide biological realization. It is clear that the world of neuroscience is in desperate need for a truly multi-disciplinary approach to solving its challenges.

Any model of the brain has to be: (1) completely compatible with all phenomenology described; (2) consistent with principal rules of the universe and phylogeny; and (3) preferably in the form of simple mathematical equations. I present here such a model, the *Vortex Model* of the brain.

---

[1] Correspondence: Tsutomu Nakada, M.D., Ph.D., Department of Integrated Neuroscience, Brain Research Institute, University of Niigata, Niigata 951-8585, Japan, Tel: (81)-25-227-0677, Fax: (81)-25-227-0821, e-mail tnakada@bri.niigata-u.ac.jp

Axiomatic Bases

Axiomatic Basis I: Brain self-organizes based on the rule of free convection.

This axiomatic basis is the starting point of the *Vortex Model*. I was first inspired by the possibility of utilizing a specific differential equation to model the shape of the brain in 1989 when I encountered the pictorial display of the fate of a plume published in Science as an example of a new simulation algorithm, the piecewise parabolic method (PPM) (Cipra, 1989). It was not until ten years later that I got the opportunity to confirm the concept using an Origin 2000 for running the simulations especially aimed at the shape of the brain (Nakada and Suzuki, 2000). The detailed description of the study is presented elsewhere[2]. The following is a brief presentation.

Suppose that a fluid is at temperature $T+\theta$, where T is uniform but θ is not. Free convection associated with a thermal core of steady temperature is given by the Boussinesq equations:

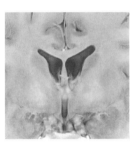

Simulation   **MRI**

Figure 1

A representative example of the results of the simulation for free convection (see Appendix). Due to imperfection of the selected initial conditions, the simulated image (Simulation) is not "perfectly identical" to the actual brain shown by magnetic resonance imaging (MRI). Nevertheless, the surprising similarities in detail is demonstrated, clearly indicating that brain organization indeed follows the self-organization schema of free convection.

$$(\mathbf{v} \cdot \nabla)\mathbf{v} = -\nabla \frac{p^*}{\rho} - \beta\theta \, \mathbf{g} + \nu \Delta \mathbf{v}$$

$$\mathbf{v} \cdot \nabla \theta = \chi \Delta \theta$$

$$\nabla \cdot \mathbf{v} = 0$$

where $\chi = \kappa/\rho c_p$ represents thermometric diffusivity, β, thermo-expansion coefficient, and ν, kinetic viscosity. The equations stand to an effectively incompressible fluid of uniform density ρ, and the gradient of excess pressure is given by $\nabla p^* = \nabla p + \rho \mathbf{g}$ (Faber, 1995, Landau & Lifshitz, 1987). With proper initial conditions, "the fate of a plume" based on the above differential equation effectively outlines the shape of the brain (Figure 1) (see Appendix). The result is not totally surprising since it has been believed that the process of self-organization based on certain rules should be the key element to defining the highly complex brain networks (see Corollaries below). Additionally, this fact leads us to the important conclusion: the shape of the human brain is *not accidental*. Rather, it exhibits specific physical significance of the creation of a *virtual sphere in accordance to convective flow* (Figure 2).

Figure 2

---

[2] Nakada T. and Suzuki K.: Human brain and self-organization. (Submitted)

An important consequence of the fact that the brain is in actuality a virtual sphere is that it ensures equivalence within the entire cortex of any of the columns, the basic units of brain organization, in an orientation perpendicular to the surface of the brain (Figure 3). This significance will be further elaborated later.

Axiomatic Basis II: Astrocytes have all the structural significance necessary for anatomical realization.

The most important factor in a biological hypothesis is firm biological realization. The human species has created various state-of-the-art machines. The principle of any human created technology can, however, be found in nature. Even though the available materials in the biological world are significantly limited, Mother Nature has nevertheless succeeded in creating virtually all conceivable functional units. She has accomplished this intuitively impossible job so well by perfecting two basic rules, namely, sophistication of the structure/shape and sophistication of the steady state condition (homeostasis). Therefore, in the biological world, anatomical realization for a given functionality represents not only the necessary condition but also virtually the sufficient condition to prove the hypothesis.

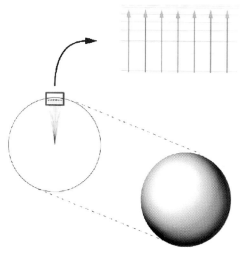

Figure 3

Astrocytes possess all the structural significance necessary for anatomical realization of the *Vortex Model* presented below. These properties are schematically summarized in Figure 4. The functional significance of these structures will be elaborated further below.

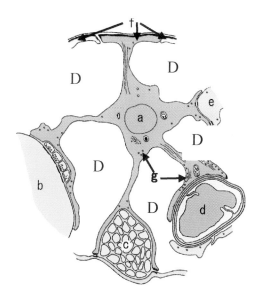

Figure 4

Schematic presentation of an astrocyte and its principal processes. Modified and redrawn based on the figures from Hirano, 1999 and Sasaki, 1999. Astrocytes possess all the structural significance necessary for anatomical realization of the hypothesis. The key elements include: (1) electron rich layer formed by the principal processes just under the pia mater; (2) two compartments of extracellular spaces segregated by principal processes; and (3) assemblies. "D" indicates "dry area", the second extracellular space. See text for details. a: astrocyte, b: neuron, c: axon bundle, d: vessel, e: neighboring astrocyte, f: electron rich layer, g: assemblies.

Elaboration of the Hypothesis

Entropy-Vortex Wave and Cortical Information Processing

Axiomatic Basis I indicates that the shape of the human brain is *not accidental*. Rather, it possesses specific physical meaning in the creation of a *virtual sphere in accordance to convective flow*. An important consequence of this fact is that it ensures equivalence of any of the columns within the entire cortex in an orientation perpendicular to the surface of the brain and a steady flow in the direction from the core to the surface (Layer VI to I) in all the columns (Figure 3).

To demonstrate the functional significance of such an organization, here, the effects of a minor perturbation on such a steady flow will be elaborated (Faber, 1995, Landau & Lifshitz, 1987).

Euler's equation is given by:

$$\frac{\partial \delta \mathbf{v}}{\partial t} + (\mathbf{v} \cdot \nabla)\delta \mathbf{v} + \frac{1}{\rho}\nabla \delta p = 0$$

where δv and δp represent small perturbations in velocity and pressure, respectively.

Similarly, conservation of entropy and the equation of continuity give:

$$\frac{\partial \delta s}{\partial t} + \mathbf{v} \cdot \nabla \delta s = 0$$

$$\frac{\partial \delta p}{\partial t} + \mathbf{v} \cdot \nabla \delta p + \rho c^2 \nabla \cdot \delta \mathbf{v} = 0$$

where $\delta \rho = \frac{\delta p}{c^2} + (\frac{\partial \rho}{\partial s})_p \delta s$ and c represent sound velocity.

For a perturbation having the form $\exp[i\mathbf{k} \cdot \mathbf{r} - i\omega t]$, one gets:

$$(\mathbf{v} \cdot \mathbf{k} - \omega)\delta s = 0$$

$$(\mathbf{v} \cdot \mathbf{k} - \omega)\delta \mathbf{v} + \mathbf{k}\frac{\delta p}{\rho} = 0$$

$$(\mathbf{v} \cdot \mathbf{k} - \omega)\delta p + \rho c^2 \mathbf{k} \cdot \delta \mathbf{v} = 0$$

This result prescribes that there will be two types of perturbations, namely, entropy-vortex wave and sound wave as defined below.

Entropy-Vortex Wave

$$\omega = \mathbf{v} \cdot \mathbf{k}$$
$$\delta s \neq 0$$
$$\delta p = 0$$
$$\delta \rho = \left(\frac{\delta \rho}{\delta s}\right)_p \delta s$$
$$\mathbf{k} \cdot \delta \mathbf{v} = 0$$
$$\nabla \times \delta \mathbf{v} = i\mathbf{k} \times \delta \mathbf{v} \neq 0$$

Sound Wave

$$(\omega - \mathbf{v} \cdot \mathbf{k})^2 = c^2 k^2$$
$$\delta s = 0$$
$$\delta p = c^2 \delta \rho$$
$$(\omega - \mathbf{v} \cdot \mathbf{k})\delta p = \rho c^2 \mathbf{k} \cdot \delta \mathbf{v}$$
$$\mathbf{k} \times \delta \mathbf{v} = 0$$

For the purpose of brain modeling, the following points should be emphasized: (1) for entropy-vortex wave, $\delta s \neq 0$ and $\omega = \mathbf{v} \cdot \mathbf{k}$; and (2) for sound wave, $\delta s = 0$. These conditions predict that a perturbation can produce entropy changes ($\delta s \neq 0$) and, hence, information processing (Cover & Thomas, 1991, Arbib, 1995) and create entropy-vortex waves. The fact $\omega = \mathbf{v} \cdot \mathbf{k}$ ensures that an entropy-vortex wave, which carries newly processed information, travels only in the direction identical to the original flow. In the case of the brain where the original flow is always perpendicular to the surface of the brain, this flow is in the direction parallel to the columnar arrangement from layer VI to I (Figure 5). The velocity of the sound wave is subsonic and, hence, will travel in all directions[3]. However, a sound wave does not carry any new information ($\delta s = 0$).

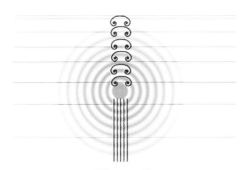

Figure 5

Minor perturbation results in formation of entropy-vortex wave and sound wave. The former travels in the direction identical to the original flow, whereas the latter in all directions. An entropy-vortex wave carries newly processed information ($\delta s \neq 0$), while s sound wave does not ($\delta s = 0$).

---

[3] Sound waves travel faster in water than in air. So a sound wave perturbation is likely to enter the "wet" area immediately.

The Astrocyte Matrix Creates a "Dry" Compartment

In spite of intensive investigation, the precise functions of astrocytes are thus far not entirely understood. Astrocytes are a cousin of neurons. Both astrocytes and neurons evolve from the identical stem cells. It is hardly conceivable that the sophisticated architectural structures of astrocytes are simply supportive in nature. It is credible, although unorthodox, to advance the notion that astrocytes play an active role with respect to the function of information processing.

Astrocyte networks are generally believed to play a key role in the formation of the blood brain barrier (BBB). Nevertheless, the astrocyte network *per se* is not the BBB itself. The principal elements of the blood brain barrier (BBB), the structure involved in gating substrate transport, are believed to be endothelial cells of the small blood vessels and not astrocytes[4]. It is clear that astrocyte networks do not represent the principal structural components of the BBB (Kettenmann & Ransom, 1995). Rather, astrocytes influence endothelial cell specialization in establishing the BBB. Such a regulatory role cannot justify the presence of the well described dense astrocyte networks.

Because of the scarcity of extracellular space in the brain, the astrocyte cell body *per se* was once thought to play the role of "extracellular" space. Since the intracellular environment of astrocytes is cytoplasmic in nature, the astrocyte cell body cannot replace the extracellular space of the brain. However, astrocyte processes are obviously involved in establishing extracellular fluid compartments and maintaining these environments, e.g. the surroundings of the synaptic areas and nodes of Ranvier[5] (Figure 4). These extracellular spaces, containing extracellular fluid, are created by the primary processes of astrocytes which effectively serve as a seal. Astrocytes therefore appear to be actively involved in compartmentalizing the extracellular space in the brain (Kettenmann & Ransom, 1995).

Why is extracellular space compartmentalization necessary? It is highly unlikely that the brain requires *two* independent extracellular spaces of identical composition. Given that astrocytes create a specific extracellular compartment using their primary processes, how does this compartment differ from conventional extracellular space? There must be a rather drastic difference between the astrocyte established extracellular compartment and the conventional extracellular space.

As is commonly encountered in biology, the astrocyte matrix may play a dual role, namely, a physical-structural as well as a functional role. Let's first consider the physical role, namely, the role of astrocytes as supportive structure. This exercise may yield clues about its functional role. What is the advantage of having a matrix-style support structure? A well known biological example of a matrix support system is bone. Compared to solid bone, bone which consists of matrix formation may be lighter in weight when filled with low specific gravity material such as air. The structure also is much more efficient in preserving structural integrity. Could the astrocyte matrix also be effecting a lighter brain by filling the space with lower specific gravity material? Such a compartment would be drastically different from the compartment comprising conventional extracellular space. I refer to this compartment as the "dry" space (relatively speaking) in comparison to the conventional composition of extracellular fluid (Figure 4). Is there any anatomical structure within the astrocyte to support this concept? The answer is yes. The astrocyte assembly fulfills this requirement.

Astrocytes contain a curious structure termed assembly. Recent molecular genetics have

---

[4] The primary processes of astrocytes cover 80-90%, not 100 % of blood vessels.

[5] Secondary processes of astrocytes form broad sheet-like structures which do not attach to any specific structures.

shed light as to the functional identity of this peculiar structure. There is substantial evidence indicating that the astrocyte assembly is identical to aquaporin 4, a water channel found in great abundance in the collecting tubules of the kidney (Yang et al., 1996, Sasaki, 1999). This function strongly suggests that one of the specific functions of astrocytes is related to the control of water contents.

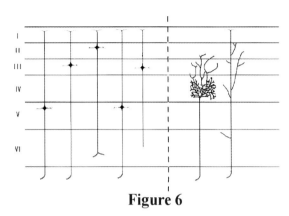

Figure 6

Schematic representation of cortical outputs (left) and inputs (right). Modified and redrawn based on figures presented by Parent, 1996.

If indeed astrocytes regulate water transport, astrocytes can be expected to be involved in establishing compartments which have substantial differences in water contents. Here, I propose that one of the compartments created by astrocytes is a "dry" area in contrast to the wet area of conventional extracellular fluid[6]. It should be clearly emphasized that this dry area is by no means totally dry in the physical sense. Rather, it implies the condition where water contents are substantially lower than in ordinary extracellular space. To be compatible with the hypothesis, the degree of dryness should be at the level where the Reynolds number of the contents (fluid) should be sufficiently high to produce steady flow as convection[7].

Electron-dense Layer and Dendritic Ramification (ELDER) forms a Synapse-like Unit

Another intriguing anatomically well described but functionally poorly understood structural aspect of astrocytes is the electron dense layer of the principal processes[8]. This layer is formed by those astrocyte processes lining up immediately underneath the subpial basement membrane, just at the surface of the brain (Figure 4). What indeed could be their functional significance?

In daily life, a common example of an electron rich plate in an electrical

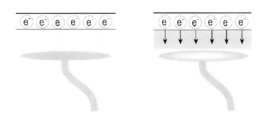

Figure 7

Schematic presentation of the ELDER system. With the permittivity lower than threshold (left), current does not occur (rest). The higher water density increases permittivity. Once threshold is reached (right), current will be introduced (excitation).

---

[6] The presence of a "third" space has been implicated in various contexts including interstitial fluid flow (Cserr & Ostrach, 1974, Rosenberg et al., 1980). However, it has never been formally proposed. It is extremely difficult to confirm this dry compartment which exists only as a micro-environment in the live, functioning brain. This difficulty itself is supportive evidence of the existence of this compartment.

[7] There are several reasons to believe that CO_2 would be the primary gas within the dry area. High concentrations of carbonic anhydrase in oligodendroglia may indeed reflect this functionality.

[8] Basic physics tells us that the presence of a static electric field implies the presence of dielectric material. The implication of such a material represents another piece of supportive evidence for the presence of a dry area, where, this time, dry also refers to low electric permittivity.

device is the source of electrons in the heated filament of the vacuum tube. It appears intuitive that the electron dense layer lining the entire inner surface of the pia matter (and, hence, the surface of the brain) is part of an apparatus for current generation which emits electrons inwards[9].

Layer I of the cortex (molecular layer) contains the terminal dendritic ramifications of the pyramidal and fusiform cells from the deeper layers (Figure 6). Accordingly, I propose that the electron dense layer formed by the primary processes of astrocytes and dendritic ramifications of pyramidal and fusiform neurons constitute a synapse-like unit, I term ELDER.[10] The electron dense layer plays the role of synaptic button, while the dendritic ramifications that of receptor (Figure 7). How does transmission occur? The answer is by turbulence created by entropy-vortex waves detailed below.

The Hypothesis: the Vortex Model of the brain

According to the hypothesis, in addition to conventional neuronal networks, the brain has another information processing system, namely, entropy-vortex wave based cortical processing. The system is based on steady fluid (gas) flow in the form of a free convection pattern.

The convection schema has two possible discrete patterns for *self-organization* (Figure 8). The first pattern is represented by the case of a heated solid body immersed in fluid. Such a system shows a relatively large scale free convection. The second pattern is represented by the case of fluid within a shallow space between two parallel plates of different temperatures. Such a system shows a Bénard type of convection (see Appendix).

In the context of the free convective self-organization schema, the reticular activation system (RAS or reticular formation) and its connectivity within thalamic regions are likely to play the primary role of a solid body with steady temperature for generating convective flow. Neuronal impulses arriving at the mesencephalic reticular formation have eventually to be converted into heat[11]. This

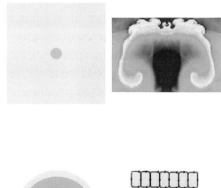

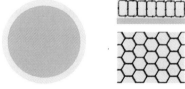

Figure 8

Two types of self-organization based on convection schema: free convection (above) and Bénard type convection (below). The brain appears to utilizes both schemas. Strictly speaking, the primary force which drives a Bénard type of convection in biological systems is likely to be the release of surface free energy. Therefore, this type of convection should be referred to as Marangoni convection, the significance of which is a hexagonal shape of the convection cells as shown here.

---

[9] The direction from the subpial area (superficial cortex) toward the deep cortex.

[10] *E*lectron-dense *L*ayer and *DE*ndritic *R*amification.

[11] The RF receives a large quantity of neuronal impulses (energy) while having little output. For example, more than 40% of the pain fibers arising from the spinal cord terminates within the RF. The abundance of pain fibers within the RF also illustrates how pain is a most effective stimulation for arousal.

heat is the likely energy source for the heated solid body. The convective flow ensures a steady fluid (gas) flow in the brain. The route of fluid (gas) flow in the cortex is secured by the astrocyte network forming the dry extracellular compartment. Brain shape, which follows a free convection pattern, ensures steady flow in all cortical columns to be equivalent qualitatively as well as quantitatively over the entire cortex. Neuronal input (energy in electrical form) to the deep layers of the cortex introduces a small perturbation within the steady flow of the column as illustrated in Figure 5. The resultant vortex-entropy wave travels together with the steady flow, namely, identical columns, and reaches the molecular layer.

The second convective self-organization described previously is represented by the most superficial layer of the cortex, which forms a shallow shell (dry area) filled with fluid (gas) between two parallel plates. In this schema, the electron dense layer of astrocyte processes constitutes the plate with the lower temperature, whereas the remainder of the brain the higher temperature plate. This convection system shows a Bénard type of convection which creates a well known *self-organization* pattern, namely, *convection cells* (see Appendix). This multiple-cell oriented self-organization pattern is highly consistent with cellular automata (see below). The self-organized dry area defines the baseline electric permittivity of the ELDER unit (Figure 7).

The turbulence generated by the vortex reaches the surface area and alters the condition of convection cells and hence the electric permittivity of the extracellular dry area between the electron dense layer and dendritic ramifications of the ELDER unit. At a certain level of water density, electric permittivity reaches the threshold and electrons are released. The release of electrons results in formation of a dendritic impulse. The turbulence may facilitate or suppress the baseline ELDER discharge.

Corollaries

Thermal Convection as a Prototype of Self-Organization Phenomena

Genes represent the principal blue print of all biological organisms on this planet coding for both structural and functional proteins[12]. Because of this, it is intuitively believed that all the detailed structural traits of an organism are genetically determined. Such is not the case. Consider the fact that the human cortex contains more than 10^{14} synapses. Even without regard to the size of the genome, one can easily deduce that a deterministic blueprint for the connectivity of such an enormous number of networks is totally unrealistic, if not ridiculous. As is the case for the physics world, determinism should be abandoned when one considers the principles of how the brain works. The conclusion naturally derived from the accumulated knowledge in science is that Mother Nature is likely to utilize principal rules instead of complete deterministic descriptions to create a desired structure: the principle rule of *self-organization*.

Thermal convection is often treated as a prototype of self-organization phenomena (Davies, 1989, Arbib, 1995). As shown in Axiomatic Basis I, the first type of *self-organization* seen in the brain is free convection, which defines the general configuration of the brain. According to this schema, the heated solid body generating the force of convection flow is located deep

---

[12] Except in some rare exceptions such as prions.

within the *virtual sphere*. For the brain, the mesencephalic reticular formation and its connections within the thalamic area perform this role (Figure 8).

Another well known example of convective *self-organization* that appears to occur in the brain is in the form of Bénard convection (see Appendix). The surface of the superficial layer (Layer I) of the brain where ELDER is found can be regarded as a thin fluid containing shell covering a large spherical solid (the brain itself). It is immediately apparent that the core would have a temperature higher than the superficial layer. This temperature differential fulfills the condition to create a *Bénard convection*. *Self-organization* of the fluid (gas) within this thin shell area, here referred to as *lattice-gas shell*, results in the formation of convection cells as illustrated in Appendix. In biological systems, the primary force which drives this type of convection is likely to be the release of surface free energy rather than gravitational potential energy as is the case for the shallow layer of water covering the earth. This type of *Bénard convection* is often referred to as *Marangoni convection*. The significance of the latter is that the shape of the self-organized convection cells is invariably *hexagonal* (Figure 8).

Cellular Automata to Neuronal Network: Dual Shell Processing

The functional aspects of the *Vortex Model* described so far can be further simplified as illustrated in Figure 9. In this schema, the cortex has dual processing shells, namely, the *lattice-gas shell* formed by the ELDER component with *Marangoni hexagonal convection cells* and the *neuronal network shell* formed by conventional cortical networks.

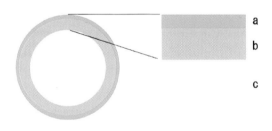

Figure 9

Schematic presentation of the dual processing shell system. a: lattice-gas shell, b: neuronal network shell, c: white matter and other deep structures.

The thickness of convection cells can be conceptually treated as the axis representing the state of the cell (up vs. down). In turn, the *lattice-gas shell* can be considered as a two-dimensional sheet composed of an astronomical number of identical Ising neurons (cells) physically connected to neighboring cells without the use of axons. This structural configuration is highly consistent with cellular automata (Arbib, 1995, Copard & Droz, 1998). The output of this *lattice-gas shell* is passed onto the neuronal network shell where deterministic connectivity (neuronal network) takes over the main role in information processing in the brain. In other words, information is first processed using programmable "matter" (cellular automata) which has a significantly higher degree of freedom. Subsequently results are passed onto the programmable "machine" (neuronal network) which has a much lower degree of freedom but a better reliability in executing discrete tasks. In this context, the ELDER system plays the role of connecting the non-deterministic information processing system to the deterministic system.

Boldly speaking, I conclude that the fundamental unit which makes the human, and perhaps other mammalian, brain cortex unique is the *lattice-gas shell* part of the brain.

Critical Point Phenomena: Universality and Re-normalization

In the early 1970's, Kenneth Wilson (1982 Nobel prize laureate for physics) published a series of papers opening up a new field of physics now generally referred to as critical point physics. It is far beyond the scope of this short paper to describe this highly innovative concept in detail. Nevertheless, it cannot be overemphasized that the conceptual revolution that Wilson initiated indeed represents the beginning of a new era reflecting the final stage of science: *principal rules for everything.*

The key words of the concept are *universality* and *re-normalization*. The *universality* concept provides strong motivation for studying physical phenomena which possess (a) critical point(s) at which apparently greatly disparate physical systems actually demonstrate similar essentials. There are generic problems which are universal to a whole class of problems in different systems in physics, from turbulent flow to gauge theory. The *re-normalization group theory* provides technical tools with which to solve actual problems in the tangible world.

The concept of the *re-normalization group* provides scaling of configurations (patterns) based on *correlation length*, and, therefore, "what you can take with you and what you must leave behind as you gradually move upscale from a microscopic description to a macroscopic one" (Davies, 1989). A macroscopic configuration actually represents the step-wise repeats of microscopic configurations, a relationship exemplified by well-formed turbulence and its eddies, respectively.

There is no doubt that the nervous system has evolved to deal with extra-personal space and the non-self environment by generating the most appropriate behavior or response. The brain accomplishes these highly complex tasks by creating effective replicas or environmental maps of the world with which it interacts. Should the brain follow the basic principle of *re-normalization*, an effective replica of the world can be easily constructed in a scale where brain neuronal networks can effectively handle all the essential information for generating a timely and appropriate behavior or response. The presented model of the brain is highly consistent with such an "ultimate" machine.

Re-Normalization and Multiple Resolution Representation

It is well known that identical feature/information representations appear repeatedly in lower degrees of resolution in the brain (Kandel et al., 1991, Arbib, 1995). This apparent redundancy is a key characteristic of the brain and distinguishes it from modern engineering. The essentials of this redundancy are readily explained by the presented model and the *re-normalization* concept described above.

Let's consider visual information processing as example. The human retina contains over 100 million neurons comprising at least 30 different cell types utilizing 20 different neurotransmitters (functionality). In a highly simplified view, a single instant parallel two-dimensional set of visual information can be regarded to be handled by 1 million neurons or approximately a 1024 x 1024 matrix. Supposing the brain indeed uses the re-normalization schema in the visual cortex, six consecutive repetitions of a three to one reduction of the matrix size by hexagonal decimation (see Appendix) will reduce the matrix size to less than 32 x 32. This matrix size is highly compatible with a group of 1000 neurons, the approximate size with which a single feature/information is believed to be processed in the cortex.

The re-normalization process described above is generally referred to as *coarse graining*. Each re-normalization process, such as the decimation process, provides the identical features in lower resolution, the reduction rate of which is determined by the *correlation length* (Davies, 1989). Each re-normalization step produces an effective replica of the feature/information in a smaller matrix (lower spatial resolution). The successive repetition of this re-normalization process eventually yields a replica which is small enough to be handled as *principal quantum* of the brain (see below). The series of compression processes require the identical feature/information to re-appear in successively lower resolutions.

The universal rule of the *re-normalization* schema represented by *coarse-graining* is not only highly consistent with the phenomenology of actual brain processing, but also provides the basic essential reason for having multiple representations.

Two Dimensional Hexagonal Matrix and "Principal Quantum"

In general life, a two-dimensional matrix is often treated as a square matrix primarily due to its simplicity. Theoretically, however, a hexagonal shaped matrix is ideal (Rosenfeld & Kak, 1976). As illustrated by the *Marangoni convection* cells mentioned before, many macroscopic phenomena often encountered in nature provide observable examples of hexagonal matrices. In the case of lattice-gas cellular automata in fluid mechanics, it is known that in order to obtain the Navier-Stocks equation exactly, the lattice has to possess sufficient symmetry to guarantee isotropy of the fourth order tensor. It is well known that the hexagonal Frish-Hasslacher-Pomeau (FHP) lattice in two-dimension meets this condition (Doolen, 1991). The two-dimensional hexagonal matrix appears in various structures of the brain as well. The primary visual cortex and the cerebellum represent typical examples (Eccles & Szentágothai, 1967, Lund et al., 1993).

A two-dimensional hexagonal matrix, which possesses a totally isotropic degree of freedom of six, appears to be the principal matrix in nature when *self-organization* plays a main role. This type of matrix appears to be applicable to brain organization, especially in information processing. The ripple model of auditory processing provides strong support to the notion that feature/information in the brain is processed and/or travels in a fashion similar to the visual modality (Shamma et al., 1993, Arbib 1995). As discussed previously, a hexagonal matrix equivalent to approximately a 32 x 32 square matrix is a good candidate for archived visual

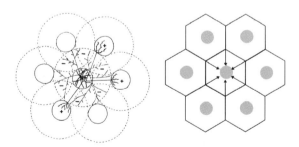

Figure 10

Schematic presentation of the primary visual cortex (left: redrawn from Lund et al., 1993) and equivalent hexagonal matrix (right). Arrows indicate "decimation" (see Appendix).

information created by successive re-normalization processes. In order to be able to accomplish multi-modality comparison and processing, it is efficient for the brain to use a *standard quantum* size to represent identical or related feature/information for each of the modalities. In this context, it is highly conceivable that a hexagonal matrix equivalent to approximately a 32 x 32 square matrix may indeed represent the *principal quantum* of feature/information representation (engram) in the brain.

Specific Meaning of Columnar Organization

One of the most significant corollaries of this hypothesis is that the model provides specific meaning to the well recognized columnar organization of the cortex, one of the essential features of the brain (Kandel et al., 1991). The self-organized brain shape based on free convection ensures that flow is always perpendicular to the surface of the brain within the cortical columns. An entropy-vortex wave always travels along with the initial flow and, hence, within the identical column. A *lattice-gas shell* which is self-organized into *Marangoni convection* cells creates the first step of information processing in the form of a two-dimensional hexagonal cellular automata system, and is highly consistent with columnar organization as well. The ELDER system connects the lattice-gas shell to the neuronal network shell, which in turn creates a cascade of multi-step processing in the fashion of stacked two-dimensional hexagonal layer networks. The three-dimensional realization of this stacked multi-step two-dimensional processing is indeed again columnar organization.

A Tangible Definition of Consciousness

Another essential corollary of the hypothesis is a tangible definition of consciousness and a plausible answer to the long time mystery in anesthesia: Why anesthetic effect fades out when the subject is placed in an environment with higher atmospheric pressure. The Vortex Model can readily address these issues.

Steady flow, the basis of an entropy-vortex wave, and the establishment of self-organized *Marangoni hexagonal convection cells* within the lattice-gas shell of the brain require a certain thermodynamic steady state or homeostasis. The thermal gradients required to produce a convective force are dependent on the presence of the appropriate heated solid body core. It is highly conceivable that establishment of the required thermodynamics, and hence, establishment of steady state cortical activities through the ELDER system makes the brain *conscious*.

All major anesthetic agents, including alcohol, are chemical compounds with a low boiling point. By accepting the presented model, it becomes highly plausible that the basic mechanism of anesthetic agents is their effects on the kinetic viscosity (Reynold's number) of the fluid (gas) of the steady flow. The addition of a relatively heavy fluid (gas) such as anesthetic agents produces alteration in the kinetics of the steady flow and, hence, affects self-organization of the lattice-gas shell. At higher atmospheric pressure, the boiling point of these agents is reduced resulting in diminution of their anaesthetic effect. At lower atmospheric pressure, the boiling point of these agents is facilitated resulting in greater effect, explaining the well known phenomenon of increased alcohol effect in airplanes.

Logical Power Supply

Consider logical processing devices such as digital computers. In addition to input and output current, there is always a power supply for each discrete logical circuit. The voltage of this power supply is in principle identical for all the discrete circuits of the system, ensuring correct processing.

Biological systems depend on a biological power supply and have evolved to utilize as supply compounds such as high-energy phosphates (HEP) for cellular survival and function. Therefore, it is intuitively accepted that neuronal networks also utilize HEP as energy source. This is obviously true for certain biological aspects of neurons and other cellular components of the brain. However, it is actually doubtful that the brain utilizes the biological power supply system, needed for information processing as well, as *logical power supply*.

Cortical neurons vary considerably in size and shape. It is very unlikely, therefore, that all neurons possess identical HEP levels all the time. In addition to being a biological organ, the brain is also a highly sophisticated device involved in information processing. It is quite conceivable, therefore, that the brain utilizes two different kinds of power supplies, namely, a *biological* power supply and a *logical* power supply.

The presented model provides for a steady *logical power supply* of identical magnitude to all cortical areas. The steady flow established for the resting conscious state of the cortex produces spontaneous, non-specific, diffuse "steady state" activities of cortical neurons through activation of the ELDER system. Electrically transmitted neuronal inputs are first converted into non-electrical entropy-vortex waves which travel along the cortical columns to reach the lattice-gas shell. There, information is converted back to electrical information through processes within the lattice-gas shell and the ELDER system. The initial cortical output is effectively calibrated by the *logical power supply*, ensuring proper and accurate processing downstream in the neuronal networks.

Endogenous Source of Initiation

A non-electrical activation system of the lattice-gas shell and its conversion into electrical activation through the ELDER system together with spontaneous non-specific diffuse steady state activities of the *conscious* brain allow for activation of neuronal networks without necessity of any sensory input. The system ensures the capability of purely endogenous initiation of electrical information processing from virtually any part of the cortex. It provides another highly plausible explanation for the long-time unanswered question: how the brain initiates endogenous thinking without any sensory input.

Higher Degree of Freedom and Creation

The human brain is capable of creating totally novel abstractive features. It is difficult to conceive that such brain activities can be operated based on a totally deterministic system. It is quite unlikely that brain function such as abstract thinking and/or creativity can be supported by a totally discrete network system of neuronal connectivity regardless of its level of complexity. If the brain requires a high degree of freedom for function, it is likely that it would need to have a system which possesses a high degree of freedom. The Vortex Model provides a much higher degree of freedom in information processing by virtue of cellular automata. Synaptic alteration and, hence, changes in neuronal connectivity (or transmission efficacy) constitute the secondary system for information processing. The former allows for creation, while the latter, execution and memory.

Potentials

Solitons for Binding?

In 1960's, Hama and colleagues introduced a series of publications regarding deformation of a single, strong vortex filament under its self-induced velocity, idealized using an asymptotic analysis (Hama, 1962, 1988). The theory is now generally referred to as localized-induction approximation (LIA). In spite of various shortcomings, LIA solutions have provided useful clues to the large-scale behavior of a concentrated vortex. It is apparent that under well-controlled, small-scale conditions, as in the case of entropy-vortex waves described above, critical concerns for the application of LIA, such as deformation of the vortex core, can actually be treated as negligible. Additionally, the steady flow compartment could be made to meet the boundary conditions required for application of appropriate asymptotic analysis. It is highly plausible, therefore, that one of the behaviors of a vortex in the entropy-vortex waves described in the hypothesis can indeed be represented by LIA.

A significant outcome of LIA type vortex dynamics is the fact that the dynamics of a vortex filament represented by LIA can be transformed into a non-linear Schrödinger equation (NLS) and, hence, supports solitons (Hasimoto, 1972) (see Appendix). The transformation implies that the entropy-vortex waves of the hypothesis can have physical significance in producing certain specific turbulence patterns associated with specific input perturbation, namely, that compatible with solitons, in addition to the perturbation for local input of the *lattice gas shell*.

One of the long-debated subjects in neuroscience is the so-called "binding" problem. Consider a free falling red apple as example. It is well known that each "component" of the entire phenomenon, namely, free falling (motion), red (color), and apple (shape) are processed at different sites in the brain. Nevertheless, one perceives it as a single phenomenon. Therefore, the brain should have some mechanisms to *bind* these apparently independently processed pieces of information. Unfortunately, there has been no definite system identified in the brain that performs this binding process (*binding problem*).

Solitons have the potential to solve the long-range binding problems. The well-known unique characteristics of solitons are: (1) solitons travel long distances without significant dispersion; and (2) solitons continue to travel with virtually identical shape and direction even after they collide. Should entropy-vortex waves introduce solitons in addition to input turbulence to the lattice-gas shell, information processing and long-range binding can be simultaneously taken care of by the identical entropy-vortex waves. The neuronal networks of the brain often utilize collision or coincidental arrival of impulses for learning (synaptic transmission efficacy changes). The cortex may indeed utilize similar strategies for long-range binding, namely, one impulse from the conventional neuronal network and another from solitons travelling in the lattice-gas shell.

Requisiteness of Phylogeny

One of the breath-holding beauties of Mother Nature is phylogeny. With painstaking effort, in time, Mother Nature has created, revised, refined, and perfected structure for the desired function. Development of the frontal lobe and hemispheric specialization represent two unique results of Mother Nature's efforts in creating the ultimate brain. As with any phenomenology in nature, development of the frontal lobe and hemispheric specialization should have a specific essential (or unavoidable) reason.

From the phylogenetic standpoint, reason is not synonymous with purpose or goal. Reason should be a natural occurrence, which by chance produced beneficial effects for the organism. The significant occurrence in the evolution of Homo sapiens compared to the other mammals is our *erect posture*. Any other human feature should represent the result rather than cause of phylogenetic development of the human brain. Can we explain the development of the frontal lobe and hemispheric specialization by the *erect posture*? The answer is yes, providing one accepts the following presented hypothesis.

Given that brain development followed the *self-organization* rule based on free convection, the *erect posture* results in the backward rotation of the axis of the brain[13], generating additional growing space in the frontal area (α in Figure 11). Development of the brain in this functionally "undefined" area[14] may indeed have provided the necessary "degree of freedom" in developing abstractive functions such as working memory and selective attention.

The principal feature of the presented hypothesis is the *vortex*, the traveling wave which possesses vorticity. As seen in the example of a gyroscope, a traveling vortex subjected to a force angled to the surface of vorticity will be affected by torque. Should the direction of vorticity be identical in both hemispheres, torque introduced by the earth's gravitational force will affect the kinetics of the vortex and introduce minute differences in the energy distribution between the right and left hemispheres. Such effects introduce asymmetry within the brain. This provides the essential reason for asymmetry in the brain, a structure which otherwise would have been totally symmetric. Non-human primates are known to have significant anatomical asymmetry of the brain. This observation supports the notion that asymmetry of the brain appears prior to the appearance of hemispheric specialization.

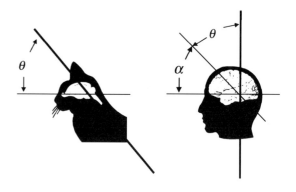

Figure 11

With the development of functional asymmetries in motion (handedness) and audition and vocalization (language) in the human, this pre-existing asymmetry became the basis of the formation of hemispheric specialization. The erect position alters the physical relationship of vortex and gravitational force in certain areas of the brain more than others. The highest effect occurs in the regions located within the axis perpendicular to the earth's surface which passes the center of the rotation and coincids with the areas subserving the primary motor and auditory cortices.

Finale: Mother Nature's Constitution

The hypothesis presented links together virtually all neurobiologic phenomenology, many

---

[13] Clockwise when one views the brain from the left hemisphere.

[14] The sensory and motor areas have been securely defined without the frontal lobe. Therefore, the appearance of additional brain tissue in the frontal lobe can be treated as "luxury", which can be dedicated to other "advanced" functions.

of which thus far unexplained. In this context, the hypothesis represents a grand unification theory of how the brain works. The most important features of the hypothesis are that it is based on the universal concept of complex systems and allows for anatomical realization. There is already substantial evidence to support the hypothesis. The unchallenged part of the hypothesis is fully testable.

Nature represents a highly complex system. The brain is no exception. Accumulating scientific knowledge has slowly disclosed the basic principles to deal with complex, non-linear systems (Lam, 1997). The critical point phenomenon and phase transition provides theoretical as well as practical applications over a surprisingly broad range of science and engineering. Microscopic behavior actually defines macroscopic behavior through the universality and re-organization principles (Davies, 1989). Self-organization phenomena based on the principal rules result in structural realization widely observed in nature in areas as diverse as geology and biology.

Just as Shannon's entropy revolutionized information science (Cover & Thomas, 1991), the concepts originally derived from thermal physics, as illustrated by the Bolzmann machine, now play essential roles in computational neuroscience (Arbib, 1995). Guided by the advancements in modern physics, scientists now clearly realize that all the phenomenology that has been classified into different fields of science actually represent outcomes from identical principal rules of the universe.

Twenty first century science must by necessity be highly multi-disciplinary, an approach that has been shown to be successful in the field of non-linear physics (Lam, 1997). Unfortunately, communication difficulties have by and large prevented effective exchange among scientists across substantially different disciplines. This is especially true between biologists and physicists. I hope (and believe) that, by introducing the current hypothesis, the distance among scientists who want to deal with the brain, regardless of their discipline, has been substantially shortened. We now possess an identical language based on the principal rules of the universe: *Mother Nature's Constitution*.

Acknowledgement

The study was supported by grants from the Ministry of Education (Japan). The manuscript is in part presented at the COE Hearing, National Council of Science (Japan), June 2000.

Suggested Reading and General References

Ablowitz MJ, Segur Harvey: Solitons and the Inverse Scattering Transform, SIAM, Philadelphia, 1981.

Anderson CR, Greengard C (eds.): Vortex Dynamics and Vortex Methods. American Mathematical Society, Providence, 1991.

Arbib MA (ed.): The Handbook of Brain Theory and Neural Networks. The MIT Press, Massachusetts, 1995.

Ames WF: Numerical Methods for Partial Differential Equations. Third Edition. Academic Press, Boston, 1992.

Copard B, Droz M: Cellular Automata Modeling of Pysical System. Cambridge University Press, Cambridge 1998.

Cover TM, Thomas JA: Information Theory. John Wiley & Son, New York, 1991.
Davies P (ed.): The New Physics. Cambridge University Press, Cambridge, 1989.
Doolen GD (ed.): Lattice Gas Methods. The MIT Press, Massachusetts, 1991.
Faber TE: Fluid Dynamics for Physicists. Cambridge University Press, Cambridge 1995.
Kandel ER, Schwartz JH & Jessell TM: Principles of Neural Science. Third Edition. Elsevier, New York, 1991.
Kettenmann H & Ransom BR (eds): Neuroglia. Oxford University Press, Oxford, 1995.
Lam L (ed.): Introduction to Non-linear Physics. Springer-Verlag, New York, 1997.
Landau LD, Lifshitz EM: Fluid Mechanics. Second Edition. Pergamon Press, Oxford, 1987.
Parent A: Carpenter's Human Neuroanatomy. Ninth Edition. Williams & Wilkins, Media 1996.
Remoissenet M: Waves called Solitons. Springer-Verlag, Berlin, 1996.
Saffman PG: Vortex Dynamics. Cambridge University Press, Cambridge, 1992.

Specific References

Cserr HF, Ostrach LH: Bulk flow of interstitial fluid after intracranial injection of blue dextran 2000. Exp Neurol 1974;45:50-60.
Cipra BA; An astrophysical guide to the weather on earth. Science 1989;246:212-213.
Colella P, Woodward PR: The piecewise parabolic method (PPM) for gas-dynamical simulations. J Comput Phys 1984;54:174-201.
Eccles JC, Szentágothai J: The Cerebellum as a Neuronal Machine. Springer-Verlag, New York, 1967.
Lund JS, Yoshioka T, Levitt JB: Comparison of intrinsic connectivity in different areas of macaque monkey cerebral cortex. Cerb Cortex 1993;3:148-162.
Hama FR: Streaklines in perturbed shear flow. Phys Fluids 1962;5:644-650.
Hama FR: Genesis of the LIA. Fluid Dynamics Res 1988;3:149-150.
Hasimoto H: A soliton on a vortex filament. J Fluid Mech 1972;51:477-485.
Hirano A: Astrocyte: Neuropathological viewpoint. In Ikuta F. (ed.) Glial Cells. Kubapro, Tokyo 1999.
Nakada T, Suzuki K: Partial differential equation to define the shape of human brain. Abstract of the 30th Annual Meeting, Society for Neuroscience, New Orleans, 2000.
Rosenberg GA, Kyner WT, Estrada E: Bulk flow of brain interstitial fluid under normal and hyperosmolar conditions. American J Physiol 1980;238:F42-F49.
Rosenfeld A, Kak AC: Digital Picture Processing. Academic Press, New York, 1976.
Sasaki H: Morphology of astrocytes from a phylogenetic viewpoint. In Ikuta F. (ed.) Glial Cells. Kubapro, Tokyo 1999.
Shamma S, Fleshman J, Wiser P, Versnel H: Organization of reponse areas in ferret primary auditory cortex. J Neurophysiol 1993;69:367-383.
Yang B, Brown D, Verkman AS: The mercurial insensitive water channel (AQP-4) forms orthogonal arrays in stably transfected Chinese hamster ovary cells. J Biol Chem 1996;279:4577-4580.
Woodward PR: Piecewise-parabolic methods for astrophysical fluid dynamics. Proceedings of the NATO Advanced Workshop in Astrophysical Radiation Hydrodynamics, Munich, West Germany, August 1982.

Appendix

- Brain and Self-organization -

Three-dimensional simulation of free convection has been performed utilizing a SGI origin 2000 64 CPU system. Governing equations in Eulerian form are given below. Subroutine FORTRAN codes of PPM schema (Woodward, 1982, Colella & Woodward, 1984) was obtained from http://wonka.physics.ncsu.edu/pub/VH-1/. The representative examples of two-dimensional "slice" images of consecutive steps of the simulation were sampled and showed below. Details are reported elsewhere (Nakada & Suzuki, 2000).

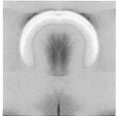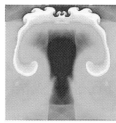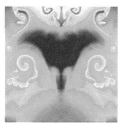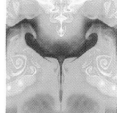

$$\frac{\partial}{\partial t}\rho + \nabla \cdot (\rho \mathbf{v}) = 0$$

$$\frac{\partial}{\partial t}(\rho \mathbf{v}) + \nabla \cdot (\rho \mathbf{v})\mathbf{v} + \nabla p = \mathbf{F}$$

$$\frac{\partial}{\partial t}(\rho \varepsilon) + \nabla \cdot (\rho \varepsilon \mathbf{v}) + \nabla(p\mathbf{v}) = G + \rho \mathbf{v} \cdot \mathbf{F}$$

$$\text{where} \quad \varepsilon = \frac{\mathbf{v} \cdot \mathbf{v}}{2} + \frac{(\gamma - 1)^{-1} p}{\rho}$$

- Bénard convection -

Conventional Bénard convection is described for a fluid layer which lies between two parallel plates. The temperature of the lower plate is higher than that of the higher plate. The fluid undergoes a "peculiar" convection pattern forming multiple blocks (cells).

- Decimation -

One of the simplest examples of coarse-graining is the method of *decimation* (Davies, 1989). Coarse graining by decimation effectively eliminates certain fractions of spins on the lattice producing a new system (coarse-graining) which still contains the long-distance physics of the original system (fine-graining).

The process can be illustrated as follows. The spins of the Ising lattice are divided into two sets: {s} and {s*} as shown. The spin {s} forms a lattice of spacing 2. The appropriate energy function H for s can be defined by performing an average over all the possible arrangements of the s* spins with energy function of $H*$:

$$\exp(-H) = \sum_{\{s*\}} \exp(-H*).$$

Here, a new set of the spin {s} describes the original system by "coarse" lattice.

- Hasimoto Soliton -

Betchov's intrinsic equations are given as:

$$\frac{\partial \kappa}{\partial \theta} + \frac{\partial(\kappa\tau)}{\partial s} = -\frac{\partial \kappa}{\partial s}\tau$$

$$\frac{\partial \tau}{\partial \theta} - \frac{\partial}{\partial s}(\frac{1}{\kappa}\frac{\partial^2 \kappa}{\partial s^2} - \tau^2 + \frac{1}{2}\kappa^2) = 0$$

where κ represents curvature, τ, torsion, θ, time variable, and s, arclength, respectively (Anderson & Greengard, 1991, Saffman, 1992).

As Hasimoto has shown, if a "wave function" is defined as:

$$\psi(s,\theta) = \kappa(s,\theta)\exp\{i\int^s \tau(s,\theta)ds\},$$

the equations can be elegantly transferred into a non-linear Schrödinger (NLS) equation of:

$$i\frac{\partial \psi}{\partial \theta} + \frac{\partial^2 \psi}{\partial s^2} + \frac{1}{2}|\psi|^2\psi,$$

which possesses soliton solutions (Ablowitz & Harvey, 1981, Remoissenet, 1996).

T. Nakada (Ed.)
Integrated Human Brain Science: Theory, Method Application (Music)
© 2000 Elsevier Science B.V. All rights reserved

Chapter I.2

Interactions that Coordinate Cortical Activity

William A. Phillips[a,1] *and Mark E. Pflieger*[b]

[a] *Center for Cognitive and Computational Neuroscience, Department of Psychology, University of Stirling*
[b] *Department of Integrated Neuroscience, Brain Research Institute, University of Niigata*

Overview

Interactions that coordinate activity include dynamic grouping, contextual disambiguation, and selective attention. We first present some general theoretical arguments for their importance, and outline evidence concerning their occurrence in cerebral cortex and the mechanisms by which they are achieved. We then note the potential value of EEG data for studying these issues, and outline five EEG studies as examples of cognitive paradigms and forms of EEG data analysis that can be used for this purpose. These studies were selected because they clearly show ways in which EEG studies can be made relevant to various aspects of coordination, but our review is not intended to be comprehensive. Finally, we note some prospects for EEG studies of coordinating interactions.

Theoretical arguments for coordinating interactions

Functional specialization at the level of local circuits is fundamental to cortical computation. Different cortical regions have different specialized functions, and receptive fields (RFs) within regions adapt so that different cells signal different things. Specialization of function is thus the best established general principle of cortical organization at both regional and cellular levels. Carried to the limit it implies that the goal of cortical computation is to produce outputs that are as independent of each other as possible. Most techniques of multi-variate data analysis, such as Principal Component Analysis, do just that. They simplify masses of data so as to make it more manageable by translating it into independent dimensions of variation, while preserving the information that it contains. This cannot be the goal of cortical processing, however. Motor output is obviously highly coordinated across both space and time. In addition, there are good reasons for supposing that the internal signals transmitted between brain regions must also be highly coordinated,

---

[1] Correspondence: William A. Phillips, Center for Cognitive and Computational Neuroscience, Department of Psychology, University of Stirling, Stirling, FK9 4LA, Scotland, UK, Tel: (44)-1786-467646, Fax: (44)-1786-467641, e-mail: wap1@forth.stir.ac.uk

rather than being composed of sets of independent variables. First, one goal of cortex is to acquire and use knowledge of what predicts what. This requires representations of interdependent variables. Second, forces of disorder, or noise, are ever present and the best way to overcome them is by context-sensitive redundancy. Third, the system as a whole is always choosing between alternative possibilities, e.g. between alternative interpretations or alternative courses of action. Context-sensitivity increases the probability that the local processors involved will make a coherent set of choices. Fourth, only small subsets of the sensory input are similar to previous inputs, and these subsets cannot be prespecified, so dynamic grouping is necessary for action in the present to be based on information from the past, whether of the individual or of the species. Our central working hypothesis is therefore that coordination is just as important as specialization. We assume this to be true of any cognitive system, whether biological or technological, and whether on this planet or elsewhere.

Research outlined in the next section provides evidence for contextual field (CF), or reentrant, input to cortical pyramidal cells that coordinates their activity by modulating the transmission of RF information so as to emphasize signals that are relevant to the context in which they occur, and so as to group them into coherent subsets through synchronization (Phillips and Singer, 1997; Tononi et al., 1992). CF inputs therefore provide a way in which activity can be coordinated, and several computational theories and neural network simulations have explored what they can contribute to the computational capabilities of neuronal systems. They show that coordinating interactions can support various cognitive capabilities, including: figure-ground organization; dynamic perceptual grouping within and between multiple visual feature domains; object recognition; contextual disambiguation in word perception; selective attention; dynamic grouping in working memory, and the binding of events across widely distributed cortical zones. For a review and critical discussion of these theories see Phillips and Singer (1997). Although they differ from each other in detail these various theories all show ways in which contextual coordination at the level of local circuits can produce useful cognitive capabilities.

Coordinating interactions use context to disambiguate and organize activity. They are not necessary for processing to occur, but by making it context-sensitive they enhance its coherence, flexibility, and relevance. One impediment to our understanding of coordination at a cognitive level has been the vagueness of concepts such as 'coherence', 'relevance', and 'context'. Computational theories have now shown how such notions can be given formal precision, and they have done so in a way that explicitly relates the concepts to neurobiology (e.g. Tononi et al., 1992; Phillips et al., 1995; Kay et al., 1997). Computational theories may thus provide a way of bridging the gap between neurobiology and higher mental function. They are important because to understand how the cortex works we must understand the work that it does. This requires us to specify the information processing operations performed by local circuits, and if they are operations of general utility then they must be specifiable in a way that is independent of what the information is about. One way to tackle this problem is by using objective functions that provide general but precise descriptions of the information processing work to be done. One objective, Coherent Infomax, emphasizes the use of contextual information, and it implies that both specialization and coordination are fundamental to cortical computation (Phillips et al, 1995; Kay and Phillips, 1997; Kay, et al 1997).

The conflict between conceptions that emphasize localized specialization of function and conceptions that emphasize integrative synthesis has played a central role in the development of neuroscience. It may seem as though localist conceptions have won their long-standing debate with the more integrative conceptions by relegating the role of coordination to specialized regions of the pre-frontal cortex. This cannot be the whole story, however. Organizing activity by imposing top-down strategic commands upon what would otherwise be anarchy could not by itself be adequate, because any strategic control system must be ignorant of nearly all the details upon which effective local cooperation depends. This suggests that activity must be coordinated through both local and strategic interactions, and evidence outlined in the following section supports this view.

Physiological and psychological evidence for coordinating interactions

Three kinds of coordination that we emphasize are dynamic grouping, contextual disambiguation, and selective attention. They affect the salience and timing of neural signals but without fundamentally changing what those signals transmit information about. There is much relevant evidence for each, but also several unresolved issues. Our goal here is no more than to indicate the essential nature of the coordinating interaction for which there is evidence.

Dynamic grouping refers to processes that organize activity so as to show what goes with what at each moment. Gestalt organization in vision is the best known example, but there are many others. Synchrony of spiking activity has been proposed as a possible code for dynamically formed assemblies of cells whose activities are to be treated as a single coherent group by those regions to which they project, and there is much neurophysiological evidence for this proposal (Singer, 1995). Though first observed in the visual cortex of anesthetized cats, it has since been shown in awake monkeys (Kreiter and Singer, 1997), and within and between parietal and motor cortex (Roelfsma et al., 1997). There is also psychophysical evidence that dynamic grouping in human vision depends on synchrony (Leonards et al., 1996; Usher and Donnelly, 1998).

Local context and selective attention both influence salience. Effects of local context have been directly observed in many studies of vision. Pyramidal cells in primary visual cortex are specialized so as to respond only to stimuli at particular retinal locations, which constitute the receptive field (RF) of the cell. They respond most to bars and edges within the RF that have particular orientations and size. Responses to stimuli within the RF can be amplified or suppressed by stimuli presented well outside of the RF, however. This interaction is highly specific to the geometric relations between the stimuli, with facilitation being strongest when they are collinear. Facilitation is strongest when the RF input is weak, and decreases as the spacing between the stimuli increases. We assume that this highly specific pattern of interaction reflects the relative frequency of co-occurrence of the stimulus elements. This pattern of contextual interaction has been shown by psychophysical (Polat and Sagi, 1993) and electrophysiological studies (Kapadia et al., 1995), between which there is close agreement (Kapadia et al., 1995). Locally specific contextual amplification in primary visual cortex has also been shown to be produced by feedback signals from extrastriate cortex that increase the salience of stimulus elements that are part of the figure rather than of the ground (Zipser et al., 1996). Though in some ways 'top-down', these particular effects were nevertheless still driven by the local stimulus context. Similar effects on salience can also be produced by processes of selective attention (Desimone and Duncan, 1995), and at the level of local circuits they may be produced by mechanisms that are much the same as those that mediate the effects of local stimulus context. It is possible that both attention and stimulus context increase salience by increasing synchronization, and evidence from evoked potentials suggests that attention does indeed increase synchronization as hypothesized (Tiitinen et al., 1993). That evidence will therefore be outlined and discussed in more detail below.

Most of the studies cited above deal with sensory processing, but we assume that the coordinating interactions for which they provide evidence have general relevance. Context-sensitivity and dynamic grouping seem to be of even greater relevance to higher cognitive functions. Indeed, it may be that their importance to such higher functions has inhibited a search for their cellular and molecular bases by encouraging the assumption that they are not implemented at such levels, but rather emerge only at much higher levels of organization. In contrast to such an assumption, we emphasize the possibility that coordinating interactions are of such basic and general importance that it is worthwhile seeking cellular and molecular mechanisms by which they may be implemented. The following section therefore briefly summarizes evidence for such mechanisms.

Mechanisms that coordinate cortical activity

Dynamic grouping and contextual disambiguation seem to occur under such similar conditions that it is possible that both are implemented by the same long-range intracortical projections, if not by the same synapses. Long-range communication within and between cortical regions and also from cortical to noncortical sites is mediated by pyramidal cell axons. Mapping their projections therefore plays a central role in cortical neurobiology. Though complex, they can be made comprehensible by showing how identified regions are interconnected, and by emphasizing patterns of connectivity that are common (Felleman and Van Essen, 1991). Several reviews of intracortical connections and their possible role in contextual integration are available (e.g. Salin and Bullier, 1995; Gilbert, 1992; Singer and Gray, 1995). One pattern of connectivity that seems common involves ascending feedforward RF projections from region to region, which determine what the cell transmits information about, and coordinating connections, which are lateral and descending and which modulate or coordinate RF transmission. Pyramidal cells of necessity receive RF input from only a small subset of cells in the preceding region, and these subsets vary such that there is little overlap between many of them. Neighborhood relations tend to be preserved in the feedforward RF projections from region to region. CF connections could include long-range intraregional connections, as well as descending connections from higher regions. This conception of coordinating connections is similar to that of reentrant connections as developed by Edelman and colleagues (Edelman, 1992; Tononi et al., 1992).

Coordination depends crucially upon interactions between the activity of cells within and between cortical regions. Though in a sense modulatory, such interactions differ from the effects of the cholinergic, adrenergic, dopaminergic, and serotonergic systems because they are locally specific rather than diffuse, and because they must arise from cortical rather than from subcortical sites. Furthermore, coordination must operate rapidly because the cognitive activity that must be coordinated changes rapidly from moment to moment. Coordinating information therefore seems most likely to be carried by pyramidal cells, and by ionotropic rather than by metabotropic synaptic receptor channels. The neurotransmitter used by pyramidal cells is glutamate (GLU), and ionotropic GLU-receptors can be divided into two classes: i) those such as AMPA- and KA-receptors which play the primary role in determining whether the post-synaptic cell will be activated or not because they open whenever GLU binds to them; and ii) NMDA-receptors which are exceptional in being both ligand- and voltage-gated. Many reviews of NMDA-receptors are available, but few focus on their role in ongoing information processing, about which much remains to be determined (Daw et al., 1993). Here we emphasize the possibility that NMDA-receptors play a major role in coordination. They are voltage-dependent because at or below resting levels of post-synaptic membrane potential they are blocked by magnesium ions. Current flow through NMDA-channels thus requires both that GLU binds to the receptor and that the post-synaptic membrane is already partially depolarized. Their main effect on ongoing processing is therefore that of gain-control (Fox et al., 1990). The key point to note here is that because of this voltage-dependency they could contribute to coordination by amplifying activity that is appropriate to the current context, and by suppressing, via GABAergic interneurones, which is inappropriate. Though they open less rapidly than non-NMDA channels, and open for longer, the direct application of NMDA to neural circuits can activate rhythmic bursting (Daw et al., 1993), and they have a rapidly decaying component with a time constant that may be short enough to support fast bursting and synchronization (Jensen et al., 1996). On this hypothesis, therefore, NMDA-channels play a major role in coordinating cortical activity. NMDA-antagonists should therefore not prevent cortical activity but should reduce its coherence. Studies of NMDA-antagonists such as PCP and ketamine provide evidence that this is so (Javitt and Zukin, 1991).

EEG as a window on interactions that coordinate human cognitive functions

Many sources of evidence show that different parts of the cortex have different functions, but coordination of those functions is much less easily observed. Coordination is typically brief and highly dynamic, and changes from moment to moment at the speed of perception, thought, and action. It is also highly context-sensitive, and may adapt rapidly to new circumstances. Furthermore, it involves relations between distinct streams of neuronal processing, and is therefore best observed by studying several distinct streams of processing simultaneously. The requirements for observing coordinating interactions are therefore not easily met, and it is no surprise that evidence for such interactions figures much less prominently in the history of neuroscience than does evidence for functional specialization. One important consequence of the new methodologies that are now being developed may be to correct this imbalance.

Though not without limitations, EEG (and MEG) methods may be better suited than most others to the task of observing coordinating interactions in humans. Both have the temporal resolution required to reflect rapidly changing activities. Furthermore, scalp potentials reflect only the synchronous activity of pools of neurons, and thus may in part reflect the activity of cells that are actively synchronized within the nervous system. This view is encouraged by MEG evidence that coherent and internally generated periodicities at around 40 Hz can be observed in humans during dream states (Llinas and Ribary, 1993) and auditory temporal binding (Joliot et al., 1994), and by patterns of coherence between potentials recorded directly from the cortical surface of patients undergoing brain surgery (Towle et al., 1996).

Two major problems must be solved if EEG data are to be used to monitor coordinating interactions. The first concerns the design of cognitive task paradigms in which well specified coordinating functions are required. The second concerns the development of ways to analyze electrophysiological data recorded from the scalp so as to reveal underlying neuronal interactions that are specifically associated with the coordinating aspects of those tasks. As examples of the progress that is now being made on these issues, each of the following five sections outlines an EEG study of coordinating interactions.

Synchronized Gamma-band activity in auditory selective attention

The physiological evidence for coordinating interactions suggests that input signals perceived as a whole are synchronized, and that more synchronized activities are more salient. This suggests that when a population of sensory signals is attended to as a whole then those signals will be more synchronized than when they are not attended. Tiitinen et al. (1993) used auditory evoked potentials to test this hypothesis. Subjects were presented a sequence of 1,000 Hz tone pips in one ear and a concurrent sequence of 500 Hz tone pips in the other ear. They were instructed to attend to the pips in one ear, or to the pips in the other ear, or to ignore both while reading a book. When listening to the pips their task was to detect occasional tones of slightly deviant frequency in the attended ear. EEG was recorded at five electrodes distributed along the midline, and was Gabor filtered to yield a continuous frequency-specific power distribution over time. This showed clear auditory evoked responses in all three attention conditions, with the response to the pips in the attended ear being larger than that to the unattended ear, and with the response being least when subjects were reading rather than attending to auditory input. The effects of attention were specific to power at a frequency centered at 40 Hz with a standard deviation of 8 Hz. These effects occurred within 25 ms of stimulus onset, and the increased synchrony at around 40 Hz was mostly observed over frontal and central scalp areas.

These results support the hypotheses that attention produces increased synchronization of neural activity and that periodicities within the Gamma-band are of particular relevance to these interactions. They do not provide direct evidence on what is being synchronized with what, but it seems likely to include increased synchronization of the population of signals that transmit information about the auditory stimulus because the EEG analysis was phase-locked to stimulus onset. This does not preclude the possibility that the neural activity directing attention was also being synchronized with the neural activity transmitting information about the auditory input, however, and the frontal involvement in the observed effect suggests that this may be so. The experiment was not designed to test the more specific hypothesis that the increased synchrony depended upon attending to the stimulus as a whole. This could be done, however, by using stimuli that can either be attended to as a single whole or as several separate entities. EEG's could then be recorded to determine whether synchronization within particular frequency bands is greater when the stimulus is attended to as a whole than when it is attended to as several distinct entities or than when only specific aspects of it are attended.

Inter-regional coherence in Gamma-band activity induced by face perception

Rodriguez et al. (1999) used stimuli that subjects can see as a face when presented with the face upright, but only as an array of unrelated blobs when presented with the face upside-down. The subjects' task was to press one button when they saw a face, and another when they did not. EEG was recorded through 30 electrodes, and was analyzed to assess the time course of power at various frequencies, and of phase synchrony between pairs of electrodes. When faces were perceived power in a narrow gamma band and its phase synchrony both first peaked at around 230 msec. They then both reduced to a minimum at around 500 msec, before increasing for at least the next 300 msec. When faces were not perceived gamma power showed a time course similar to that in the face perception condition, but with less power in the first peak at 230 msec. Phase synchrony showed a very different time course, however, with no obvious sign of either the peak around 230 msec or of the minimum around 500 msec. The first peak in phase synchrony was interpreted as reflecting phase interactions between parietal and occipitotemporal regions associated with perceiving the input as a face, and the second increase in phase synchrony was interpreted as reflecting the coordination of motor activities in temporal and central regions. The minimum between these two peaks was interpreted as reflecting a period of desynchronization that serves to separate distinct cognitive assemblies, in this case separating assemblies involved in perceptual organization from those involved in organizing the response.

These interpretations are speculative but the analyses used are important because they can detect synchronization between scalp electrodes that is triggered by the stimulus, though not phase-locked to it. Micro-electrode studies of cross-correlations between the spiking activity produced by cortical neurons provide much evidence for synchronization that is not phase locked to the stimulus (Singer, 1995; Singer and Gray, 1995). This is important because it implies that synchrony as a cue to grouping could be determined by internal criteria and not just by the time at which stimuli are presented. Synchrony that is triggered by but not phase-locked to the stimulus has been called 'induced' to distinguish it from the 'evoked' activity that is phase-locked to the stimulus. Methods to detect induced synchrony in EEG data may thus reflect the dynamic grouping that occurs in a wide range of cognitive tasks. Therefeore, we will now outline and discuss those methods in more detail.

Details of the methods used by Rodriguez et al. (1999) are described in Lachaux et al. (submitted), to whom we are grateful for providing us with an advance copy. Three main objectives of the analysis are: (a) to estimate EEG signal phase (shift relative to stimulus onset) as a function of event-related time in a narrow frequency band of interest; (b) to compute a phase-locking statistic between two EEG channels for a collection of epochs; and (c) to assess the significance of the inter-channel phase-locking statistic.

Objective (a) is achieved in two steps. After the EEG record has been segmented into epochs that are framed with respect to their stimulus events, the data are filtered in a narrow band centered on a frequency of interest f. Next the filtered data are temporally convolved with a complex Gabor wavelet. The real part of the complex Gabor wavelet is obtained by applying a temporal Gaussian window to a cosine function of frequency f. Likewise, the imaginary part is obtained by applying the identical Gaussian window to a sine function of frequency f. The temporal variance parameter that determines the width of the Gaussian window is set using the following rule of thumb: Divide 3.5 seconds by f Hz. Thus, the temporal width of the wavelet narrows as the center frequency of interest increases. These two steps produce a complex number per time point for each epoch and for each channel. At each event-related time point per epoch per channel, the "instantaneous" phase angle is computed as the argument of the complex number, i.e., the arctangent of the imaginary part divided by the real part. This phase angle is relative to the onset of a stimulus event, which defines time zero.

To meet objective (b), an estimate of synchrony between EEG channels A and B is computed as follows. For each epoch and time, the phase angle for channel B is subtracted from the phase angle for channel A. This phase difference (a real number) is converted to a unique complex number of magnitude 1.0 by intersecting the unit circle in the complex plane with the ray of the difference angle. At each latency, all such complex numbers are averaged across epochs; the result must fall within the unit circle. The phase-locking value (PLV), which measures inter-channel synchrony, is defined as the magnitude of this average complex number. The PLV ranges between 0.0 (completely random phase differences) and 1.0 (perfect inter-channel phase synchrony).

How far from 0.0 must a PLV be to achieve statistically significant inter-channel synchrony? Objective (c) is met by employing a nonparametric shuffle test. A Monte Carlo sampling distribution for the null hypothesis of no phase synchrony is generated by shuffling channel A's epochs, computing the PLV for the shuffled data, and repeating this procedure many times. Random shuffling obliterates inter-channel simultaneity within the dataset, so if the fraction of PLVs greater than or equal to the actually observed PLV is smaller than a preset probability of a Type I error, e.g., 0.05, then the observed PLV is considered to be statistically significant.

A closely related method for achieving objectives (a) and (b), complex demodulation and induced coherency, will be discussed in a subsequent section. By contrast to FFT methods, which integrate over time intervals, the methods just described and those to be described below provide "instantaneous" estimates of signal amplitude and phase for a center frequency of interest. Another variation on this theme, i.e., use of a complex Merlet's wavelet transform, is described in a recent review by Tallon-Baudry and Bertrand (1999). The relative time-frequency resolution of these variants, as well as the relationship between the phase-locking value and induced coherency, are technical questions that require further investigation.

In all studies of synchronization at the scalp, special care must be taken to sort out the cortical dynamics of interest from passive volume conduction in the head. For example, a single dipole-like generator with a tangential orientation with respect to the scalp surface, e.g., a synchronized patch of sulcal cortex, can produce the appearance of 180 degree phase-locking between distant electrode sites. A radial source can likewise produce the appearance of 0 degree phase-locking between adjacent electrode sites. At best, volume conduction effects may bias the phase estimates; at worst, they may give the appearance of cortical phase-locking when actually there is none. We therefore note that the shuffling test does not control for volume conduction phase artifacts.

High-density EEG studies of dynamic grouping in vision

Dynamic grouping is essential to cognition because only small subsets of sensory input are similar to past inputs, and those subsets cannot be prespecified when they arise from objects that can stimulate the sensors in any of an indefinitely large number of ways. A central defining feature of dynamic grouping is therefore that it can organize a large body of data into distinct subsets in a flexible, context-sensitive, way such that, although the criteria, or grammar, by which grouping occurs can be prespecified, all possible groupings that can be formed by the operation of those rules are far too numerous to be prespecified. Binding, or combining, a subset of inputs by prespecified feedforward convergence of the data to be grouped is therefore not an example of dynamic grouping, because it is limited to a set of groupings that is small enough to be prespecified by the feedforward architecture.

A paradigm that explores the criteria under which dynamic grouping of simple visual features occurs in human vision was designed by Field et al. (1993), and has since been used by many other investigators. It uses arrays of many small, randomly positioned and oriented, Gabor patches, but with some arrays containing a continuous chain of elements oriented such that they lie along a smooth contour. Gabor patches are used because they are well matched to the RF features to which cells in primary visual cortex are tuned. The activity produced by these stimuli therefore consists of well distinguished streams of processing that must then interact to discover coherent relationships between the distinct stimulus elements. The subjects' task is to detect the arrays containing elements that lie along a continuous contour embedded within the background of randomly positioned and randomly oriented elements. This allows precise determination of the conditions under which the contour is visible, and it turns-out that Gestalt grouping criteria such as proximity, continuity, and closure are all relevant and can be precisely quantified using this paradigm. All possible groupings that could be formed by the operation of these criteria could not possibly be prespecified, however, so this paradigm provides a clear example of a task that requires dynamic grouping.

EEG studies using this paradigm are being performed by Shugo Suwazono working in Tom Nakada's lab in Niigata. These studies are still in progress so we outline them here only to give an example of a psychophysical paradigm in which high-density EEG is being used to explore dynamic grouping. To increase the chance of observing effects of grouping in scalp potentials we have modified the original paradigm of Field et al. (1993) by using stimuli created by George Lovell in the Psychology Department of Stirling University. In these stimuli eight continuous closed contours are embedded within the random background, rather than only one as in the original paradigm. Each array with visible contours is used to produce a matched array differing only in that the contours are made invisible by randomly changing the orientation of the elements along the contours so as to greatly reduce their smoothness and thus their visibility. Pairs of such arrays are presented to the subject for 100 msec with the array containing visible contours occurring randomly either to the left or right of fixation. Responses to the two kinds of array can only differ as a result of dynamic grouping, so major issues being studied are whether, when, where, and how scalp potentials evoked or induced by them differ. These issues are being studied by recording scalp potentials on a high-density array of 256 electrodes, and by analyzing them using a variety of methods, such as those discussed elsewhere in this chapter. As discussed in Pflieger (this Volume), there are grounds for suspecting that relationships between evoked and induced activity may involve higher order spatial frequencies that can more easily be detected using a high-density electrode array. Results obtained from this investigation (to be reported elsewhere) may therefore also provide evidence on whether or not this is so.

VEP studies of contextual disambiguation in vision

Disambiguating effects of context are a major form of coordination, and have been shown to occur in low-level vision by single-unit and psychophysical studies of the effects of various flanking and surrounding contexts on response to small line elements or Gabor patches (e.g. Kapadia et al., 1995; Polat and Sagi, 1993). Detection of low-contrast elements is facilitated by high-contrast elements centered on the axis of the target element and with the same orientation, i.e. collinear with the target, and is suppressed by elements centered on the axis of the target element but with the orthogonal orientation. These effects require the target to be present, but not of high-contrast, i.e. they require evidence for the target to be present but ambiguous. They therefore seem to provide a paradigmatic example of contextual disambiguation.

Polat and Norcia (1996) report visual evoked potential (VEP) studies that provide further evidence on these issues. Low-contrast Gabor patches were presented at fixation, either alone or in the context of other patches, which were centered on the axis of the target element and either collinear or orthogonal to it. These contextual patches were presented at various distances from the target, but were always quite separate from it and did not either overlap or abut the target. If the neural responses produced by these well separated stimuli do not interact then the VEPs produced by displays of target and context combined should be equal to the linear sum of the responses to target and context separately. This was the case when target contrast was either high or very low, but not for intermediate contrasts when evidence for the target was present but ambiguous. Displays with collinear contexts within a certain range of distances produced VEPs that exceeded the linear prediction, and displays with orthogonal contexts produced VEPs that were less than predicted. These results therefore show that evoked potentials can provide a window on the precise conditions under which contextual facilitation and suppression occur in visual cortex.

Polat and Norcia (1996) also calculated the response phase to target plus context, and compared that to the phase predicted by a linear combination of the responses to target and context separately. They found that the phase of response to target plus collinear context led the predicted phase, and that the response to target plus orthogonal context lagged the predicted phase. This study therefore also shows how evoked potentials can be used to study contextual interactions that affect timing. It thus shows how evoked scalp potentials can be used to study the conditions under which there are non-linear interactions between streams of processing that respond to distinct stimuli, including interactions that modify response timing.

EEG combined with PET and MRI in studies of coherence between motor areas

In the context of subcortical and spinal controls, frontal cortex apparently plays an "orchestrating" role in the coordination of motor activity. Since coordinated motor output implies coordinated neural input, and because the neural side of the motor system involves multiple frontal areas, it seems reasonable to suppose that motor areas must coordinate their activities in order to achieve successful preparation, initiation, and completion of a voluntary motor movement. The different phases of motor coordination in frontal cortex might engage both cooperative and competitive dynamics. Can movement-related scalp EEG be used to "image" such large-scale neural network dynamics?

Thatcher and colleagues (1994) attempted to do this based on the prior finding of good correlations between EEG dipole locations and PET activation centers co-registered in anatomical MRI (Toro et al, 1994). Independent of the PET analysis, a spatiotemporal source analysis (Scherg and Berg, 1991) of average movement-related cortical potentials (MRCPs) was performed in the time interval from -212 ms to +200 ms relative to the onset of movement as detected by EMG. In a single subject chosen for further analysis, three equivalent dipoles that accounted for 96% of the MRCP variance were identified as contralateral motor cortex (cMC), ipsilateral motor cortex (iMC), and contralateral supplementary motor area (cSMA). These dipoles co-registered within 3 mm, 10 mm, and 6 mm, respectively, of the centers of significantly active PET regions.

As the first step toward a three-node large-scale network analysis, it was desired to obtain estimates of macroscopic currents at the three cortical locations as a function of time relative to movement onset for the raw unaveraged data. The dipole locations and orientations were used to derive a pseudoinverse spatial filter, which was applied to 134 unaveraged EEG epochs. This procedure reduced 29 scalp channels to 3 "source channels" that were designated cMC, iMC, and cSMA. Thus, the dipole locations and orientations obtained for the average potentials were used to derive 134 epochs of source activity estimates from the raw EEG.

As the second step, the method of complex demodulation was used to convert the relatively broadband (0.1 Hz to 30 Hz) source activities to narrowband time series of complex numbers for a center frequency f of interest, plus or minus some half-bandwidth b, e.g., 7 Hz ± 4 Hz. Complex demodulation computes the real part of the output time series in two steps: The input time series is multiplied by $\cos(2\pi ft)$, which then is lowpass filtered with a highcut of b Hz. Likewise, the imaginary part of the output times series is obtained by multiplying the input time series by $\sin(2\pi ft)$ followed by the same lowpass filter. The resulting complex time series has an "instantaneous" magnitude (square root of the real part squared plus the imaginary part squared) and phase angle (arctangent of the imaginary part divided by the real part) for each time point. Phase delay (seconds) equals the phase angle (radians) divided by $2\pi f$.

The final step taken toward large-scale network analysis was to compute event-related coherency between the three pairs of three sources, i.e., cMC-iMC, cSMA-cMC, and cSMA-iMC. Event-related coherency is the exact complex analog of Pearson's correlation. It is a complex number that has been computed across all epochs at each time point, and normalized to fall within the unit circle of the complex plane. Since the mean is removed in a manner strictly analogous to the ordinary correlation coefficient, and since this event-related mean represents phase-locked activity, it follows that event-related coherency is a measure of induced (non-phase-locked) brain activity. Event-related coherence is the squared magnitude of the event-related coherency; thus, it ranges between zero and one. Event-related relative phase is the phase of the event-related coherency. If two channels have perfect coherence for a frequency f, then they are offset by a fixed relative delay for all epochs. That is, although the absolute phase for both channels varies randomly across epochs, knowledge of the phase at one channel is sufficient to determine the phase at the other channel. On the other hand, if two channels have zero coherence for f, then the relative phase is entirely random across all epochs, i.e., knowledge of the phase at one channel cannot be used to predict the phase at the other channel. Thus, event-related coherence is an alternative measure of phase synchrony between channels.

The strongest event-related coherence effects between estimated source activities were found in the theta frequency band, i.e., 4 to 7 Hz. (Recall that the data had been lowpass filtered below 30 Hz.) Coherence between contralateral and ipsilateral motor cortices was almost zero at -500 ms and increased to about 0.3 at -300 ms, where it remained approximately level (with some oscillations) up to the time of movement onset, after which there was a sharp decline followed by a gradual increase to +600 ms. When cSMA-cMC coherence was compared with cSMA-iMC coherence, Thatcher noted that there was an approximate inverse relationship, both before and after the movement. For example, following the motor movement there was an increase in cSMA-cMC coherence and an almost mirror image decrease in cSMA-iMC coherence. Prior to movement onset, the relative phase for cSMA-iMC was near zero degrees ("in phase" coherence), whereas the relative phase for cSMA-cMC was near 180 degrees ("out of phase" coherence). Based on this inverse relationship, the speed of the coupling dynamics, and the pivotal role that the SMA appeared to play, Thatcher speculated that the SMA may have served in this instance as a "neural network switching element" for governing response competition between contralateral and ipsilateral cortices.

This study illustrates in broad strokes some tantalizing possibilities for the analysis of large-scale brain network dynamics on the basis of EEG source analysis. At the same time, it also raises interesting issues of substance and method. For example, recall that the original source analysis was performed on phase-locked activity, whereas the identical dipoles were subsequently treated as sources of induced activity (event-related coherence and relative phase). In the context of gamma responses, Tallon-Baudry and Bertrand (1999) have recently hypothesized different source models for evoked versus induced cortical activity. Whereas an equivalent current dipole normal to the cortical surface, which represents an open field of pyramidal cells, has often proved to be adequate for characterizing evoked activity, Tallon-Baudry and Bertrand (1999) propose a ring-shaped dipole distribution as a better model for induced activity. "This geometry fits with the idea that interneurons might be involved in a network generating a coherent oscillatory activity." (p. 160). In the context of the Thatcher et al. study, the ring-shaped dipole model for induced activity at once suggests that: (a) an estimator based on a single dipole at the center of a ring may yet pick up considerable induced activity from the surround, although (b) an estimator that suppresses activity in the center while enhancing activity in the immediate surround might be preferable, and could possibly make qualitative differences in the results. One more issue raised by the Thatcher et al. study is the possible effect of misallocation of source activities due to imperfect modeling. For example, if a fourth source is active but not modeled, then its activity will tend to be allocated to the three sources that are modeled, thus introducing an artificial coherence at 0 or 180 degrees. On the other hand, if no attempt at source modeling is made, then scalp coherences with phases of 0 or 180 degrees are already contaminated with common volume conducted activity. Thus, it appears that several challenging theoretical and methodological problems remain to be solved in this arena.

Do induced activities reflect CF processing and evoked activities reflect RF processing?

It has been suggested that induced Gamma activity reflects the formation of object representations and binding more generally (e.g. Tallon-Baudry and Bertrand, 1999; Rodriguez et al., 1999). As dynamic grouping is a major function of the coordinating interactions hypothesized to be implemented via CF connections by Phillips and Singer (1997) this suggests that induced activities reflect CF processing. Evoked activities, being phase-locked to external events, may then predominantly reflect RF processing. If so, then the distinction between RFs and CFs may be relevant to the question of how evoked and induced activity are related, and measures of evoked and induced responses can be employed to test hypotheses about the role of RFs and CFs in cortical computation in the context of specific tasks.

CF activity, by definition, modulates RF activity, so this predicts a relationship between induced EEG and evoked EEG, if these EEG measures reflect underlying activity as hypothesized. Although the EEG component not phase-locked to the stimulus is usually conceived of as a statistically independent "background noise" process from the standpoint of forming average evoked responses, Basar (1983) has demonstrated that spontaneous EEG activity in the prestimulus interval affects the subsequent evoked response. Likewise, it is possible that there are dynamic relationships between evoked and induced responses in the interval following the stimulus. The findings of Tiitinen et al. (1993) provide some support for this, because selective attention, which is an internal coordinating process, was shown to affect evoked EEG activity.

More precisely, if local cortical processors are evolved to approximate the Coherent Infomax objective function, i.e., to maximize transmission of RF information that is predictably related to the CF (Phillips et al., 1995; Phillips and Singer, 1997), then this may be reflected by some suitable higher order coherence measure between evoked responses (localized to different cortical regions) and induced coherences (between these and other regions). The goal of such a measure would be to quantify the relationship between evoked spatial dynamics and concurrent induced spatial dynamics, and to clarify their relation to underlying neuronal activity. No such measure has yet been specified, however, and that is one of the many issues in this area that remain to be resolved.

Prospects for EEG studies of coordination

Though the development of rigorous quantitative theories that relate EEG to underlying brain dynamics raises deep issues that remain to be resolved, substantial progress is being made in the development of such theories (Wright and Liley, 1996). This includes the distinction between activity at various spatial scales, from local to global, and the development of a better understanding of the kinds of two-way interaction across scale that are possible (Nunez, 1999). One possibility that is currently being discussed as part of this development is that coordinating processes of functional integration may be more closely related to global dynamics than are locally specific processes of functional segregation (Nunez, 1999). This then raises the possibility that diffuse modulators of global mental state, such as the cholinergic and adrenergic systems, may be closely intertwined with processes of coordination. Particular states of these systems might therefore be such as to provoke, or allow, thought to proceed, but in an uncoordinated and incoherent way, as in REM sleep. The development of EEG measures of coordinating interactions would provide the means for testing such speculations.

The assumption that coordinating interactions are fundamental to cortical computation is central to the perspective from which this chapter is written. Basic issues that arise from this assumption are critically assessed by Phillips and Singer (1997), together with the associated commentaries and the responses to them. One key issue is whether coordination is indeed achieved in part through NMDA-receptor channels. They are dense in supra-granular layers of neocortex, and sparse in layer IV, which supports the hypothesis, but more evidence is required. Prima facie it may seem as though EEG methods could not possibly contribute to this issue. That may not be so, however. NMDA receptor channels are diverse (Kutsuwada et al., 1992), and different NMDA subunits are expressed at different stages of development, some not being expressed until late post-natal stages (Watanabe et al., 1992). Different subtypes also have very different regional distributions (Watanabe et al., 1992). EEG studies of the developmental and regional distribution of coordinating interactions might therefore provide a rich new source of evidence on these issues. If particular coordinating interactions involve NMDA synaptic-receptor channels of particular sub-types, and if those interactions can be imaged using EEG methods then their developmental and regional distribution should reflect that predicted from the developmental molecular biology of NMDA subunits.

Our concern for coordination emphasizes the need to consider not just the amount of activity in any region, but also the relationships between activity within and between regions. An analytic technique for distinguishing between modulatory and obligatory relationships has been developed using fMRI data (Friston et al., 1995). Their conception of modulatory interactions is based upon the gain-controlling actions of voltage-dependent, i.e. NMDA, synaptic-receptor channels, and the analytic technique developed rests upon this distinction. The basic idea is to see whether the statistical dependence between activity in regions A and B depends upon the activity produced in B by other inputs. If it does not then the connectivity from A to B is described as 'obligatory'. If it does then it is described as 'modulatory'. Such modulatory interactions seem to be much the same as what we think of as contextual disambiguation. They will often occur on a fast time-scale, however, so the development of analogous techniques for distinguishing between modulatory and obligatory relationships in EEG or MEG data might provide a rich new source of evidence on these basic issues.

Low-level perceptual tasks have been prominent in the above discussion, but coordinating interactions such as dynamic grouping, contextual disambiguation, and selective attention are likely to be even more crucial to higher cognitive functions. EEG studies have recently been used to study the role of synchronized activity in the gamma band during sensori-motor coordination, and have shown that this synchronization can be learned through classical conditioning procedures, and removed by extinction trials (Miltner et al., 1999). Such methods therefore hold great promise for studies of both sensorimotor coordination and associative learning. More importantly in relation to higher cognitive functions, they may also have much to contribute to studies of object perception, working memory and language processing, all of which rely heavily upon dynamic grouping and context-sensitivity. An extensive review of studies that relate induced gamma activity to such cognitive functions (Tallon-Baudry and Bertrand, 1999) provides strong support for this view.

Finally, our hypothesis that coordinating interactions are so fundamental to cortical computation that they are implemented at both molecular and neural network levels raises the possibility that genetic, developmental, and drug induced pathologies of these mechanisms may occur, with the consequence of impaired coordination in any of various cognitive domains. EEG methods for studying coordination may therefore provide new insight into such disorders. Tiitinen et al. (1993) note that their techniques might be relevant to schizophrenia in which disorders of selective attention are common. Other psychotic disorders may also involve fragmentation of cortical activity with a consequent reduction in the coherence of various aspects of perception, thought, and action. Indeed, the existence of such disorders provides independent support for the view that there are mechanisms whose primary role is to coordinate activity. Furthermore, recent advances concerning the neuropharmacology of schizophrenia also support the view that these mechanisms include NMDA-channels, because there is evidence that they are under-active in at least some forms of this disorder (Olney and Farber, 1995). EEG techniques for monitoring the coherence of cortical activity may therefore tell us much about many psychotic disorders such as schizophrenia, autism and the drug-induced psychoses. These techniques are also likely to be of relevance to very specific disorders of sensory processing, however. Polat et al. (1997) have shown that abnormal long-range spatial interactions in amblyopia can be revealed using the EEG methods of Polat and Norcia (1996) outlined above. The potential clinical utility of EEG techniques for monitoring processes that coordinate cortical activity may therefore extend from disorders of particular sensory and motor processes to disorders of high-level cognition, such as in psychosis. The development and use of such techniques may therefore be of great value, both to clinicians and to basic cognitive neuroscientists.

Acknowledgments

We thank Tsutomu Nakada for conceiving and facilitating this Volume, and for hosting a fruitful meeting between the authors of this Chapter and the research workers in the Department of Integrated Neuroscience of the Brain Research Institute, University of Niigata. In particular we thank Shugo Suwazono for insightful and helpful comments on a draft of this chapter.

References

Basar E. Synergetics of neuronal populations: A survey on experiments. In E Basar et al. (Eds.) Synergetics of the Brain. Berlin, Springer-Verlag, 1983, pp. 183-200.

Daw NW, Stein PSG, Fox K. The role of NMDA receptors in information processing. Annu Rev Neurosci 1993; 16: 207-222.

Desimone R, Duncan J. Neural mechanisms of selective attention. Annu Rev Neurosci 1995; 18:193-222.

Edelman GM. Bright Air, Brilliant Fire. New York, Penguin Books, 1992.

Felleman DJ, Van Essen DC. Distributed hierachical processing in the primate cerebral cortex. Cerebral Cortex 1991;1: 1-47.

Field DJ, Hayes A, Hess RF. Contour integration by the human visual system: evidence for a local 'association field'. Vis Res 1993: 33: 173-93.

Fox K, Sato H, Daw NW. The effect of varying stimulus intensity on NMDA-receptor activity in cat visual cortex. J Neurophysiol 1990; 64:1413-1428.

Friston KJ, Ungerleider LG, Jezzard P, Turner R. Characterizing modulatory interactions between areas V1 and V2 in human cortex: A new treatment of functional MRI data. Human Brain Mapping 1995; 2: 211-224.

Gilbert CD. Horizontal integration and cortical dynamics. Neuron 1992; 9: 1-13.

Javitt DC, Zukin SR. Recent advances in the phencyclidine model of shizophrenia. Am J Psychiatry 1991; 148: 1301-1308.

Jensen O, Idiart MAP, Lisman JE. Physiologically realistic formation of autoassociative memory in networks with Theta/Gamma oscillations: Role offast NMDA channels. Learning and Memory 1996; 3: 243-256.

Joliot M, Ribary U, Llinas R. Human oscillatory activity near 40 Hz coexists with cognitive temporal binding. Proc Natl Acad Sci USA 1994; 91: 11748-51.

Kapadia MK, Ito M, Gilbert CD, Westheimer G. Improvement in visual sensitivity by changes in local context: Parallel studies in human observers and in V1 of alert monkeys. Neuron 1995; 15: 843-856.

Kay J, Floreano D, Phillips WA. Contextually guided unsupervised learning using local multivariate binary processors. Neural Networks 1998; 11:117-40.

Kay J, Phillips WA. Activation functions, computational goals and learning rules for local processors with contextual guidance. Neural Computation 1996; 9: 895-910.

Kreiter AK, Singer W. Stimulus-dependent synchronization of neuronal responses in the visual cortex of the awake Macaque monkey. J Neurosci 1996; 16: 2381-96.

Kutsuwada T, Kashiwabuchi N, Mori H, Sakimura K, Kushiya E, Araki K, Meguro H, Masaki H, Kumanisis T, Arakawa M, Mishina M. Molecular diversity of theNMDA receptor channel. Nature 1992; 358: 36-41.

Lachaux JP, Rodriguez E, M·ler-Gerking J, Martinerie J, Varela, FJ.Measuring phase-synchrony in brain signals. Submitted.

Leonards U, Singer W, Fahle M. The influence of temporal phase differences on texture segmentation. Vis Res 1996; 36: 2689-97.

Llin R, Ribary U. Coherent 40-Hz oscillation characterizes dream state in humans. Proc Natl Acad Sci USA 1993; 90: 2078-81.

Miltner WHR, Braun C, Arnold A, Witte H, Taub E. Coherence of gamma-band EEG activity as a basis for associative learning. Nature 1999; 397: 434-6.

Nunez PL. Neocortical function and EEG. Behav Brain Sci 1999. In press.

Olney JW, Farber NB. Glutamate receptor dysfunction and schizophrenia. Arch Gen Psychiatry 1995; 52: 998-1007.

Pflieger M. The spatial resolving power of high-density EEG: An assessment of limits. This volume.

Phillips WA, Singer W. In search of common foundations for cortical computation. Behav Brain Sci 1997; 20: 657-722.

Phillips WA, Kay J, Smyth D. The discovery of structure by multi-stream networks of local processors with contextual guidance. Network 1995; 6:225-246.

Polat U, Norcia AM. Neurophysiological evidence for contrast dependent long-range facilitation and suppresion in the human visual cortex. Vis Res 1996; 36: 2099-2109.

Polat U, Sagi D. Lateral interactions between spatial channels: Suppressionand facilitation revealed by lateral masking experiments. Vis Res 1993; 33: 993-999.

Polat U, Sagi D, Norcia AM. Abnormal long-range spatial interactions in amblyopia. Vis Res 1997; 37: 737-44.

Rodriguez E, George N, Lachaux J-P, Martinerie J, Renault B, Varela FJ. Perception's shadow: long-distance synchronization of human brain actiivity. Nature 1999; 397: 430-33.

Roelfsma PR, Engel AK, König P, Singer W. Visuomotor integration is associated with zero time-lag synchronization among cortical areas. Nature 1997; 385: 157-161.

Salin P, Bullier J. Corticocortical connections in the visual system: Sturcture and function. Physiolo Rev 1995; 75: 107-154.

Scherg M, Berg P. Use of prior knowledge in brain electromagnetic source analysis. Brain Topogr 1991; 4: 143-150.

Singer W. Development and plasticity of cortical processing architectures. Science 1995; 270: 758-764.

Singer W, Gray CM. Visual feature integration and the temporal correlation hypothesis. Annu Rev Neurosci 1995; 18: 555-586.

Tallon-Baudry C, Bertrand O. Oscillatory gamma activity in humans and its role in object representation. Trends in the Cognitive Sciences 1999; 3:151-162.

Thatcher RW, Toro C, Pflieger ME, Hallett M. Human neural network dynamics using multimodal registration of EEG, PET, and MRI. In Thatcher RW, Hallett M, Zeffiro T, John ER, Huerta M (Eds.) Functional Neuroimaging. New York, Academic Press, 1994.

Tiitinen H, Sinkkonen J, Reinikainen K, Alho K, Lavikainen J, Ntnen R. Selective attention enhances the auditory 40-Hz transient response in humans. Nature 1993; 364: 59-60.

Tononi G, Sporns O, Edelman GM. Reentry and the problem of integrating multiple cortical areas: Simulation of dynamic integration in the visual system. Cerebral Cortex 1992; 2: 310-335.

Toro C, Wang B, Zeffiro T, Thatcher RW, Hallett M. Movement-related cortical potentials: Source analysis and PET/MRI correlation. In Thatcher RW, Hallett M, Zeffiro T, John ER, Huerta M (Eds.) Functional Neuroimaging. New York, Academic Press, 1994, pp 259-267.

Towle VL, Kinnunen L, Berger C, Plisken N, Erikson R, Milton J. EcoG inter-electrode coherence: Patterns observed during cognitive tasks. J. Clin. Neurophysiol., 1996; 13: 351.

Usher M, Donnelly N. Visual synchrony affects binding and segmentation in perception. Nature 1998; 394: 179-182.

Watanabe M, Inoue Y, Sakimura K, Mishina M. Developmental changes in distribution of NMDA receptor channel subunit mRNAs. NeuroReport 1992; 3:1138-1140.

Wright JJ, Liley DTJ. Dynamics of the brain at global and microscopic scales: Neural networks and the EEG. Behav Brain Sci 1996; 19: 285-320.

Zipser K, Lamme VAF, Schiller PH. Contextual modulation in primary visual cortex. J Neurosci 1996; 16: 7376-7389.

Section II

Advanced Methodology

T. Nakada (Ed.)
Integrated Human Brain Science: Theory, Method Application (Music)
© 2000 Elsevier Science B.V. All rights reserved

Chapter II.1

Myths and Truths in fMRI

Tsutomu Nakada[1]

Department of Integrated Neuroscience, Brain Research Institute, University of Niigata
Center for Tsukuba Advanced Research Alliance, University of Tsukuba
Beckman Institute, California Institute of Technology

This article represents the English version of the review "Principles for fMRI Implementation" published as part of "Lecture Series in Methodology for Physiologists" in the official Journal of Japan Physiology Society (61:155-168, 1999). The article has been rewritten in English by the author to fulfill multiple requests from colleagues outside of Japan.

Introduction

In the few years since its original description, blood oxygenation level dependent (BOLD) contrast functional magnetic resonance imaging (fMRI) has revolutionized *non-invasive* analysis of higher human brain function (Figure 1). With the growing use of fMRI, however, exponentially increasing criticism questioning the validity of fMRI data, especially those data performed on conventional clinical systems, has been raised. The criticism is well founded as many of the studies have been performed without proper validation regarding the effect of experimental paradigms on image statistics. The paucity of rigorous data validation processes indeed is a reflection of the complexity of fMRI. Nevertheless, everybody should bear in mind the constitutional principle of science: *No validation equals no science*.

The majority of validation issues arise during the process of obtaining raw images, which eventually provides raw numerical data for post-processing statistical analysis. Therefore, the principal issues are all related to the unique characteristics of magnetic resonance (MR[2]) and biological complexity, and *not* to paradigm selection or post-processing statistical analysis of

---

[1] Correspondence: Tsutomu Nakada, M.D., Ph.D., Department of Integrated Neuroscience, Brain Research Institute, University of Niigata, Niigata 951-8585, Japan, Tel: (81)-25-227-0677, Fax: (81)-25-227-0821, e-mail tnakada@bri.niigata-u.ac.jp
[2] Following general convention, in this article I refer to the biomedical applications of nuclear magnetic resonance (NMR) as MR.

fMRI time series (Figure 2). As a result, without proper execution of the process of ultra-fast MR image acquisition, "task to functional map" correlation cannot be obtained. Unfortunately, a large number of fMRI studies in the world literature, including those published in the renowned scientific journals Science and Nature lack proper validation assessment.

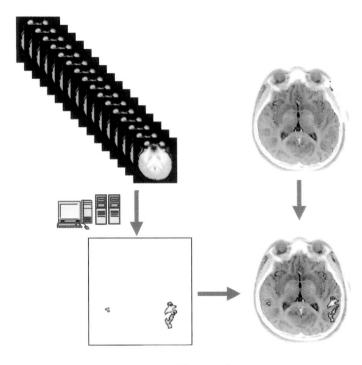

Figure 1

Schematic presentation of the fMRI process using a "language comprehension" task as example. Multiple raw images are obtained by the ultra-fast imaging technique while the subject performs the given task. Scalar values reflecting relative intensities of pixels in arbitrary units determined from raw images are subjected to statistical analysis. Subsequently, those pixels which show statistically significant intensity changes associated with the task are determined. Combining these with the relevant structural image completes an activation map. As shown here, properly performed fMRI is an awesomely powerful technique for neurology and neuroscience.

As illustrated by the fact that the term "susceptibility effect" has long been used virtually synonymous with "the cause of artifact" in the realm of diagnostic imaging, it is evident that T_2^* (pronounced "tee two star") weighted imaging essential for BOLD contrast is inherently artifact prone. In clear contrast to other imaging techniques, MRI images provide relative, *not* absolute, spatial information. Stereotactic neurosurgeons typically utilize computed tomography (CT), *rather than* MRI, for coordinate determination because they have found through experience that MR images lead to misjudgment of pre-surgical coordinates in excess of a few millimeters, a potentially fatal difference in medicine.

This article is organized to provide a distillation of the essential knowledge for the non MR physicist fMRI investigator. It provides salient illustrations of the issues that cannot be modified or ameliorated by any post-processing algorithms. To start, I would like to present three myths often unconditionally accepted as truths.

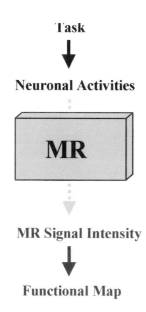

Figure 2

Simplified view of fMRI study. The processes of the same color are considered to be well correlated. To make "Task-Functional Map" correlations, appropriate processes in ultra-fast MR image acquisition are mandatory (green). Unfortunately, many scientists do not pay serious attention to this issue thereby creating "uncertainty" in published fMRI findings.

Three Common Myths

Myth I: $p < 0.01$ is brain activation

This appears obvious. Nevertheless, some investigators blindly believe in the power of "p value." Detection of pixels which show "statistically significant" intensity changes associated with a given set of tasks does not automatically imply that those pixels denote an area of neuronal activation. Rather it simply implies "statistically significant" changes in intensity, a scalar number given by the MR system upon completion of image reconstruction including magnitude correction. Only if and when certain critical conditions (see below for details) are cleared can these "statistically significant intensity changes" be taken as indirect indices of neuronal activation.

This first myth represents one of the pitfalls into which even renowned statisticians and/or positron emission tomography (PET) experts may inadvertently fall. Due to apparent similarity in the given *non-invasive* images, some investigators inadvertently treat the numerical data given by the raw MRI images of fMRI in a way identical to those of other activation studies such as PET, SPECT[3], and Xe-CT. Unfortunately, the fact is not so simple. The basic principles for obtaining MRI images and, hence, values to be utilized for statistical analysis in fMRI are fundamentally different from those for other X-ray or gamma wave based imaging techniques (see "k-space acquisition and quadrature detection" section below for details). One should clearly understand those characteristics unique to MRI prior to statistical consideration or paradigm determination.

---

[3] Single Photon Emission Computed Tomography

Myth II: fMRI spatial resolution = MRI spatial resolution

Technological advancements have quickly lowered the theoretical limit of MRI spatial resolution to the micron level. It is now generally believed that the limit of MRI spatial resolution is approximately 4 micron-cube[4]. Even in the world of clinical MRI, micro-structural analysis of the brain is now possible (Nakada, 1999, Fujii et al., 1999). Unfortunately, this state of the art advancement in MRI technology paradoxically created potentially fatal pitfalls for many fMRI investigators.

Appropriate spatial resolution for fMRI is not determined by the capability of the MRI system. It is limited by various factors, among which those related to the natural physiology of the live human play significant roles. For example, consider physiological brain motion. As stereotactic surgery has unambiguously demonstrated, even under the condition of perfect skull fixation, the brain itself physiologically moves a minimum of 0.3 mm. Should one select a spatial resolution of raw MRI where the intensity changes associated with physiological *non-linear* motion of the brain of 0.3 mm are large enough to affect activation statistics, the resultant activation maps will have no validation (see "pixel misalignment" subsection below).

Myth III: BOLD events = neuronal events

This can be rephrased as "*Myth III: time resolution in MRI imaging = fMRI time resolution.*" This is the pitfall which conventional neurophysiologists are especially prone to fall into.

The BOLD phenomenon is an empirically observed "biological" phenomenon. For MR purposes, BOLD is accepted as the phenomenon primarily introduced by the "weak" susceptibility effects of hemoglobin. Nevertheless, the precise physiological mechanism of the BOLD phenomenon still remains unknown.

The most commonly accepted, simplified view of the BOLD phenomenon associated with "brain activation" is the so-called "oxygen oversupply hypothesis"(Figure 3). This hypothesis states that oxygen supply by activation induced perfusion (flow in Figure 3) exceeds oxygen consumption associated with neuronal activities. As a result, regional deoxy-homoglobin (deoxy-Hb) levels paradoxically increase associated with neuronal activities (brain activation). This alteration in regional deoxy-Hb concentrations can be detected by T2* weighted imaging and represents the basis of fMRI.

As one can immediately understand, the BOLD phenomenon is at best one of the indirect indices of "brain activation" and *not* that of neuronal activities. Furthermore, the BOLD phenomenon is a phenomenon observable in the order of

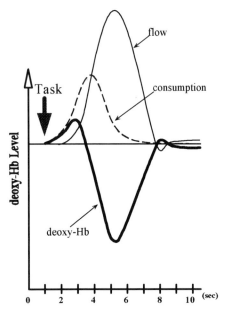

Figure 3

Schematic presentation of the BOLD phenomenon based on the "oxygen oversupply" hypothesis.

---

[4] Based on water molecular diffusivity.

seconds four to ten seconds after cessation of the corresponding neuronal activities. The time course of BOLD phenomena detected by fMRI time series represents, at best, the time course of "brain activation" and *not* the time course of neuronal activities. Neuronal activities associated with a single given task are activities which occur in the order of milliseconds. It is impossible to scientifically accept the time course of the BOLD phenomenon as a reflection of neuronal activities without rigid validation[5].

"Truly Basic" Knowledge of MRI Fundamentals for fMRI Scientists

It is far beyond the scope of this short review to describe all MRI related knowledge necessary for fMRI scientists. Here, only a small portion of the basics, which I believe to be useful for steering scientists away from "misunderstandings" regarding fMRI, is described.

Fundamentals of NMR

There is only one fundamental equation for NMR:

$$v = \frac{\gamma}{2\pi} B \quad \text{(Eqn. 1)}.$$

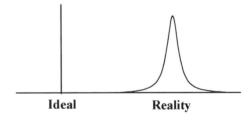

Figure 4

Pure theory is based on the ideal situation where each resonance can be treated as a line. In reality, however, each resonance has a certain shape (in this example: Lorentzian) with width.

Since the gyromagnetic ratio, γ, is a given constant for each nucleus, this equation simply implies that the resonance frequency of the target nucleus is dependent on the magnetic field strength that the target nucleus observes. At this point, the reader should be alerted to the nature of the magnetic field strength, B. Nuclear magnetic resonance (NMR) is a phenomenon of the nucleus and therefore, B has to be considered at the level of the "nucleus", and *not* at the macro level.

In MRI, all three-dimensional spatial information is "encoded" into radio frequency signals. The principle utilized for "encoding" is given by the above equation (Yes, there is a single fundamental for all three dimensional spatial information encoding processes). Image construction processes represent "decoding" based on the principle identical to "encoding." This can be intuitively understood by correlating "axial frequency" to the corresponding axial

---

[5] In this context, it is clear that the term "event-related fMRI (ER-fMRI)" utilized in many fMRI publications is highly misleading to neurophysiologists. Many renowned neurophysiologists who are rather naïve regarding fMRI inadvertently believe that "event-related fMRI" actually reflects "time events" of neuronal activities similar to event-related potentials (ERPs). The methodology described by the term event-related fMRI is actually "event-driven" state-related fMRI. Although I do not elaborate this issue further in this article since other principal issues have a much higher priority, one should clearly understand the nature of so-called ER-fMRI.

spatial position:

$$x \Leftrightarrow v_x$$
$$y \Leftrightarrow v_y$$
$$z \Leftrightarrow v_z.$$

Variation in B (line width)

There is no perfect machine in the world. Therefore, we always have to accept a certain degree of variation. In theory, B can be a scalar number. In practice, however, B has a range:

$$B \Rightarrow B \pm \Delta B.$$

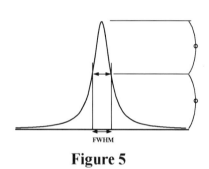

Figure 5

In NMR, this is translated into variation of the resonance frequency (Figure 4):

$$v \Rightarrow v \pm \Delta v.$$

The degree of this variation is quantitatively expressed by full width at half maximum (FWHM), which is also referred to as line width (Figure 5).

As can be easily deduced, in MRI, this frequency variation will appear as variation of the spatial position in images:

$$v_x \Rightarrow v_x \pm \Delta v_x \Leftrightarrow x \pm \Delta x$$
$$v_y \Rightarrow v_y \pm \Delta v_y \Leftrightarrow y \pm \Delta y$$
$$v_z \Rightarrow v_z \pm \Delta v_z \Leftrightarrow z \pm \Delta z.$$

This represents one of the characteristic limitations of MRI and explains why MRI is inherently ambiguous with respect to spatial information. How this ambiguity specifically affects actual images is dependent on the technique applied[6]. The canonical technique of MRI is the two-dimensional Fourier transform (2DFT) method where one spatial dimension is "taken care of" by "slice definition", which, in NMR terms, represents "selective excitation." The remaining two-dimensional information is obtained utilizing the identical method as that for classical two-dimensional NMR. The typical effects of the aforementioned spatial ambiguity in 2DFT MR images will be elaborated later.

---

[6] This is generally referred to as "pulse sequence."

B_0 Field Homogeneity and Magnet Shimming

There are many factors which affect B. Among these, main field, B_0 (pronounced "bee zero"), homogeneity has substantially significant effects. Accordingly, prior to each individual experiment involving any NMR technique, including clinical MRI, B_0 homogeneity is maximized by empirical means. This process is called magnet shimming.

Since the configuration the main magnet cannot be altered, magnet shimming is performed by adding additional small fields at various directions. This is actually accomplished by adjusting small amounts of current run in specially designed hardware called room temperature shim coils. To what extent B_0 homogeneity can be achieved is therefore dependent on (1) the configuration of the shim coils; (2) the conditions of the sample; and (3) the operator's capability to perform magnet shimming.

Shim coils are grouped into "order" according to their significance. High-resolution NMR magnets have an extensive number of shim coils from first through third order. In stark contrast, clinical MRI systems typically lack any specially designed shim coils. Instead, magnet shimming in clinical MRI is performed grossly by using gradient coils (gradient shimming)[7]. Modern MRI systems designed specifically for fMRI[8] has by necessity high order shim coils, at minimum first and second order shim coils[9].

Any sample which contains high levels of a paramagnetic substrate such as oxygen is hard to shim. Any sample in a large volume is hard to shim. Any sample composed of multiple elements is hard to shim. Obviously, therefore, human brains are not great samples for shimming. Additionally, the live human brain undergoes continuous motion (see below). Under the best shimming condition, the line width of small samples studied using high resolution NMR is typically less than 1 Hz. In contrast, however, the line width of live human organs is at best 200-400 Hz[10].

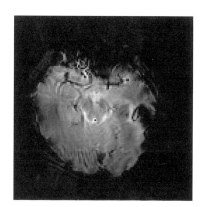

Figure 6

---

[7] This is due to practical reasons. The process of serious magnet shimming requires substantially large amounts of time. In clinical MRI where patient examination time is a critical factor in determining its success as diagnostic tool, techniques to ameliorate field inhomogeneity without serious shimming have been developed. Unfortunately, these techniques cannot be applied to fMRI and magnet shimming is once more the most significant factor in determining the success of MRI experiments.

[8] Examples include 3.0T systems at the University of Niigata (see "Development of 3.0 tesla vertical MRI system for advanced fMRI applications" in this book.

[9] Just considering shim coil configuration and no additional factors, the system for fMRI has substantially different requirements compared to systems designed for clinical MRI. This consideration illustrates the critical importance of having systems especially designed for fMRI for the sucsessful performance of fMRI studies.

[10] Many vendors of clinical systems provide "line width" in echo signals and give deceivingly low numbers, such as a few Hz. This so-called "line width" does not reflect the actual line width for B_0 homogeneity.

Susceptibility Effects and BOLD contrast

The term "susceptibility effect" is a quite ambiguous term. Nevertheless, in MRI, it is generally accepted as the term to describe the effects on images introduced by regional field inhomogeneity due to the presence of substrates with significantly high paramagnetic

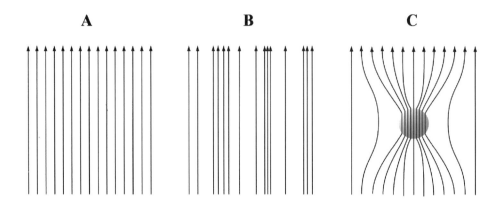

Figure 7

Schematic presentation of the field. A: Ideal homogeneous field, B: Actual B_0 field with minor inhomogeneity, C: Susceptibility effect with regional perturbation. BOLD contrast is based on the susceptibility effects of hemoglobin.

susceptibility or ferromagnetic susceptibility (see Appendix). Prior to fMRI, it was utilized virtually synonymous to a type of artifact, such as that introduced by air sinuses (Figure 6).

For practical purposes, it is useful to simplify the concept in MRI by considering susceptibility effects as *regional* field inhomogeneity in contrast to B_0 inhomogeneity which affects whole field (Figure 7). As discussed earlier, B_0 inhomogeneity can be partially corrected by magnet shimming, while the susceptibility effect cannot. If the susceptibility effect is strong enough to affect a large area of the image matrix, it introduces image deformity similar to that introduced by B_0 inhomogeneity. If the susceptibility effect is reasonably small, it introduces only intensity changes in nearby pixels (see "T2* contrast" subsection below). This is the basis of BOLD contrast.

For conventional structural MRI, image artifacts introduced by either B_0 inhomogeneity or susceptibility effect can be effectively suppressed by applying certain techniques such as fast spin echo. However, for

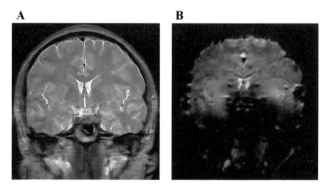

Figure 8

Images obtained under identical conditions but using different sequences. A: The sequence optimized to eliminate unwanted effects by field inhomogeneity (fast-spin echo). B: The sequence necessary to detect "activation" by BOLD contrast in fMRI (gradient echo echo planar). Note significant artifacts due to B_0 inhomogeneity and susceptibility effects underscoring the substantial difficulties in performing proper fMRI.

fMRI where BOLD contrast itself represents "weak" susceptibility effects, any of the techniques used to eliminate susceptibility cannot be applied. As a result, images obtained without special attention to B_0 inhomogeneity and strong susceptibility effects are full of artifacts. fMRI represents a "statistical" method based on intensity changes of each of the pixels within the target structure. As illustrated in Figure 8, without an understanding of the "truly basic" knowledge for MRI, it is virtually impossible to perform proper fMRI.

At this point, the fact regarding B of Eqn.1 (not B_0!) has to be emphasized one more time. This is the field strength observed by the target nucleus and therefore, its homogeneity should be at the nuclear level. All NMR related techniques including fMRI work perfectly as described by theory only if and when "perfect" B is achieved (Figure 2). In practice, however, no NMR experiment can achieve the ideal condition. As a matter of fact, the conditions required for obtaining BOLD contrast for fMRI are far from "ideal." In order to obtain data to which the fundamental theory of fMRI is validly applicable, one has to carefully follow a substantial numbers of conditions. Otherwise, only "look-alike" activation maps (fictitious activation) will be obtained[11].

MRI Contrast Mechanics and T2* Contrast

As with any other digital images, the structures in MRI images are resolved based on contrast resolution (Figure 9). Unlike other imaging techniques, however, the contrast mechanics determining the final intensity of pixels are complex in MRI. While densities of

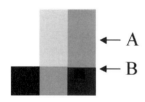

Figure 9

Principles of contrast resolution for digital imaging. The lines between adjacent pixels with identical intensity are not resolved (A). Clarity of the lines between adjacent pixels is dependent on their intensity differences.

pixels in X-ray CT, SPECT and PET are determined by a single factor[12], pixel intensities in MRI are a function of multiple contrast mechanics. In a highly simplified view, this fact can be expressed as:

$$D_{CT} = f(e^-)$$
$$D_{SPECT/PET} = f(\gamma)$$
$$I_{MRI} = f(\rho, T_1, T_2, u)$$

where e^- is electron density, γ, gamma wave strength, ρ, spin density, T_1 and T_2, spin-lattice and

---

[11] They are often published as "fact" even in respected journals.
[12] For CT, the X-ray attenuation coefficient reflecting eletron density; for SPECT or PET, strength of γ waves and hence, quantity of tracer material.

spin-spin relaxation time, respectively, and *u*, motion factor[13].

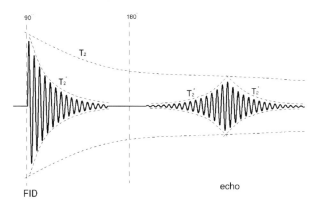

Figure 10

Schematic presentation of T2 and T2*. NMR signals appears to decay and disappear quickly (T_2^*). This apparent disappearance is actually due to dispersion of signals which can be recalled by a 180 degree pulse in the form of an echo. Spin-spin relaxation time (T_2) can then be measured in the envelope connecting the highest intensities of each observable signal. The highest value of echo signals and, in turn, the intensities of pixels are affected by T2 characteristics, implying that T2 is one of the "conventional" contrast mechanics for MRI. On the other hand, T2* does not directly affect the highest value of echo signals. T2* is an index of field inhomogeneity. In ideal situations where the field is perfectly homogeneous, T2* becomes equal to T2. FID: free induction decay.

So, compared to other imaging techniques, MRI contrast mechanics are complex to begin with. Contrast mechanics for fMRI are even more complex.

BOLD contrast is *not* based on "conventional" MRI contrast mechanics. The mechanism is a rather empirical one based on the *weak* susceptibility effects of hemoglobin. In NMR jargon, this is termed apparent T2 or T2* (pronounced "tee two star") and represents dispersion of NMR signals due to field inhomogneity (Figure 10).

T2* is the time constant for the apparent decay of a given signal either in the form of FID or echo (Figure 10, 11). T2* becomes a determinant of pixel intensity because, unlike conventional NMR, MRI utilizes height, instead of area, of frequency domain signals as determinant of pixel intensity (Figure 11). Since the concept of field homogeneity itself is

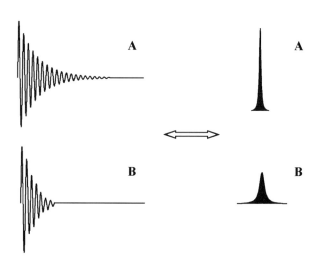

Figure 11

Schematic presentation of T2* contrast. Comparison of long T2* (A) and short T2* (B) signals in the time domain (left) and frequency domain (right). Maximum intensities of FID are represented by area (black-filled), and not by height in a Fourier transformed function. Because of the increase in line width, the short T2* signal in the frequency domain has a lower height than the longer T2* signal in spite of their identical areas. MRI utilizes height for pixel intensity determination irrespective of line width.

---

[13] I will not elaborate on this further. Readers are referred to general MRI textbooks.

qualitative in nature, T2* contrast in MRI cannot be treated quantitatively in the rigid fashion of "conventional" contrast mechanics such as T1 or T2. The latter two are dependent on tissue characteristics and are generally steady. In practice, it is highly prudent for any serious scientist not to treat T2* contrast, and in turn the BOLD phenomenon, too quantitatively. The simplest example is schematically illustrated in Figure 12. The identical amount of deoxy-Hb levels producing the identical amount of susceptibility effects does not result in identical intensity changes in all the pixels. They are strongly affected by target area background field inhomogeneity which is constantly fluctuating. The behavior of *weak* susceptibly effects towards signal intensity is quite difficult, if not impossible, to determine precisely.

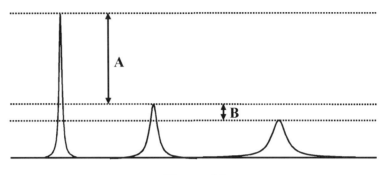

Figure 12

Schematic illustration demonstrating BOLD contrast. Identical amounts of changes in deoxy-Hb levels produce different intensity changes which are determined by curve height (A > B), and not area.

Partial Volume Effect

Another essential basics is actually common to all three representative *non-invasive* imaging techniques, CT, SPECT, and PET: the partial volume phenomenon. Since final images are given in two dimensions, image readers tend to forget about the fact that actual data collection for image construction is done three dimensionally. In the early 1970's when CT became increasingly popular, this caused various misinterpretation and its clinical consequences.

The unit which corresponds to the pixels of the images in three dimensional reality is called voxel: volume-pixel. Pixel intensity actually represents an *average* of the intensities composing the voxel (Figure 13). Brain activation produces intensity changes in the area which undergoes the BOLD phenomenon. If the entire voxel volume is subject to the BOLD phenomenon, this produces intensity changes which reflect

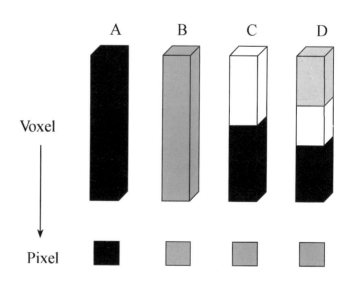

Figure 13

Illustration of the partial volume phenomenon. Pixel intensity actually represents averaged intensities of the volume. Partial filling of the volume with significantly different regional intensities results in deceivingly different pixel intensity.

the actual BOLD phenomenon (A → B in Figure 13). If the volume is partially occupied by high intensity substrates such as cerebrospinal fluid (CSF) in T2* weighted imaging, the pixel will exhibit an artificially high intensity which can be mistaken as "activation" (C in Figure 13). The most likely voxel condition is the mosaic (D in Figure 13), suggesting inherent ambiguity of quantitative correlation between pixel intensity and BOLD contrast.

k-Space Acquisition and Quadrature Detection

One of the characteristics highly unique to MRI is represented by the fact that raw data of MRI images are acquired in the k-space using quadrature detection. A detailed explanation of this characteristic, especially with respect to the resultant images, is above the scope of this brief review[14]. Because of the tremendous significance of a proper understanding of fMRI technology, I take the liberty of listing some salient points related to this MR characteristic here without further explanation. (1) Minor variation in part of the raw data acquisition tends to result in artifacts which affect the entire image, not just part of image. (2) There is no rigid correlation between images on MRI and the physical world[15]. (3) In addition to accuracy of each set of data, accuracy of the "balance" between two sets of data (real and imaginary sets) affects the final image. And so on.

Sample (Human) factors

For any of the state-of-the art technologies, issues related to target sample are unavoidable. In *non-invasive* imaging methods, the samples are live human beings. In practice, human related factors often become significant limiting factors in system performance leading to results which are significantly less optimal than predicted by theoretical specifications[16]. One typical example is fMRI "spatial resolution" which is limited by uncontrollable physiological brain motion rather than by MR system limitations (see "Limits in Spatial Resolution" subsection below). Similar to any other medical technology, the appropriate application of fMRI mandates a minimum amount of clinical experience[17].

Common Problems and Suggested Solutions

Shimming and Slab

Magnet shimming is the heart of NMR experiments including fMRI. For structural MRI,

---

[14] This actually represents one of the most essential issues for a proper understanding of MRI technology and, hence, fMRI methodology. However, the issue is highly technical and difficult to summarize for non-MR scientists in this short review. Accordingly, I decided not to elaborate on this topic further. Any scientist who wishes to perform fMRI competently has to study this unique characteristic of MRI (or at least get expert help).

[15] Simply put, MRI is not a "true" photograph system. MRI may be likened to a system which produces an artist drawn photograph-like picture.

[16] An extensive list of human related factors cannot be given here but are available in various text books.

[17] Humans cannot be treated simply as subject in a study. The author has observed the use of many unnecessarily uncomfortable head restraints for fMRI studies at institutions throughout the world because of inexperience.

various techniques for ameliorating the effects of field inhomogeneity have been developed and implemented (Figure 6). Unfortunately, such techniques cannot be adopted to fMRI (see previous discussion regarding the BOLD phenomenon). In this context, fMRI is similar to conventional high resolution NMR. The success of shimming actually defines the success of the experiments. However, unfortunately, shimming over a large sample volume is quite difficult in any situation. Under *in vivo* conditions, it is virtually impossible to obtain high B_0 homogeneity over the entire brain. The solution is to choose the appropriate slab size for the study[18].

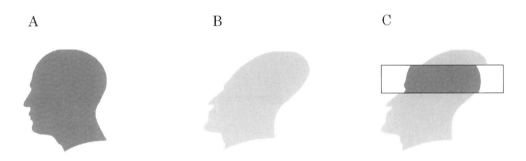

Figure 14

Illustration of slab selection. A: ideal, B: reality, C: shimming over the slab.

The term "slab" literally means thick slice. In MRI, this usually represents the volume over which focused shimming is performed to obtain high homogeneity. The term is well known in magnetic resonance angiograpy (MRA) where, similar to fMRI, high homogeneity of the target volume is required for the success of the study (Figure 14). In order to perform valid fMRI, the appropriate slab size where shimming specific for the MRI system can be comfortably performed[19]. This slab size automatically limits the volume to be studied at once and, hence, the appropriate number of slices obtainable in a single experiment. An example of inappropriate selection of slab size is illustrated in Figure 15. Inappropriately thick slab selection resulted in a minor

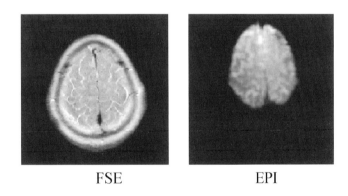

Figure 15

Inappropriately thick slab selection resulted in unavoidable B_0 inhomogeneity. Note significant artificial shift of the slice near the edge of the slice on EPI due to minor perturbation of the field.

---

[18] Multi-slice imaging with a large number of slices to cover the entire brain at once is easily performed by MRI as far as sequences not highly sensitive to shimming conditions are employed. However, it is impossible to perform such studies and obtain valid fMRI results. Unfortunately, many reports in the literatures indicate such fMRI experiments, strongly suggesting that these studies were performed without appropriate regard of MRI basics. Such studies lack validation.

[19] At our institution, slab thickness of approximately 30 mm has been found to yield consistently high field homogeneity appropriate for fMRI experiments. Accordingly, four consecutive 5 mm slices with an interslice gap of 2.5 mm has been chosen as the standard.

perturbation in homogeneity at the edge of the slab. While fast spin echo (FSE), which is highly insensitive to minor field inhomogeneities, provided an unaffected image, echo planar imaging (EPI), which is highly sensitive to minor field inhomogeneities, produced a significant artificial shift of the slice.

Slice Definition

In 2DFT imaging, MRI defines "slice." However, the technique of slice definition in MRI is fundamentally different from other imaging techniques such as CT, SPECT and PET. In MRI, slice is defined by selective excitation based on Eqn. 1. Therefore, as discussed earlier, slice definition becomes subject to variation inherent in the MRI technique because of field inhomogeneity (Figure 16). Realization of these effects is shown using phantom studies in Figure 17.

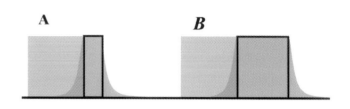

Figure 16

Schematic illustration of MRI slice definition. Slice thickness defined by the black box is always subject to variation (line width) which actually extends into adjacent structures. If the slice is thin (A), signals which come from adjacent structures become significant. Appropriately thick slices are not significantly affected by adjacent structures.

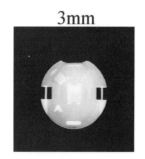
3mm

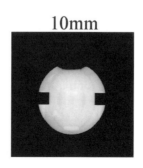
10mm

Figure 17

All images are obtained by FSE which is least sensitive to field inhomogeneity. Proper slice thickness (10 mm) produced an image unaffected by adjacent structures (see also illustration above). In clear contrast, thinner slice definition (3 mm) produced significant partial volume effects from adjacent structures. The example demonstrates the inherent ambiguity of MRI slice definition as detailed before. Generally speaking, slice definition of less than 5 mm is inappropriate for raw fMRI images.

It should be clear to the readers by now that, in order to do appropriate fMRI experiments, selection of appropriate slice thickness is a must[20]. Furthermore, to avoid inadvertent effects from the volume of adjacent slices, the appropriate inter-slice gap should be placed between slices (arrows in Figure 18). Along with the limit of slab thickness for appropriate shimming as discussed earlier, these conditions automatically limit the maximum numbers of slices that can be studied in one experiment (Figure 18)[21].

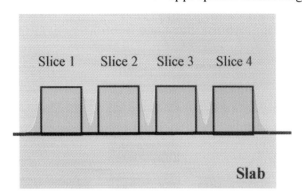

Figure 18

Nyquist Ghost

The unique k-space trajectory of the EPI sequence results in the appearance of a characteristic artifact termed Nyquist ghost[22]. It is characterized by multiple superimposed ghosts on the phase direction and is often described as an artifact due to system imperfection. However, in practice the most common cause of Nyquist ghost is minor field perturbation (Figures 19 and 20).

Nyquist ghost represents one of the most common causes of fictitious activation encountered in fMRI. It is a common pitfall even for the most cautious fMRI scientist. Nyquist ghost may appear only in some raw images during a single experimental session and intensity changes introduced by Nyquist ghost can correlate well with the given paradigm task (see Illustrative Case Study I below) owing to characteristics of the magnetic field.

As Maxwell's equation clearly indicates, a magnetic field always closes. Therefore, contrary to the intuition of many investigators, field perturbation introduced at the site away from the imaging volume can cause field perturbation within the imaging volume (Figure 21).

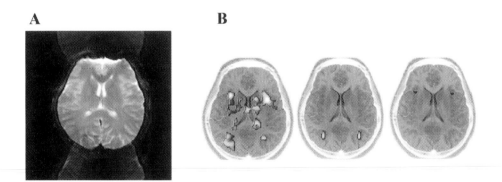

Figure 19

Nyquist ghost (A) and representative fictitious activation (B).

---

[20] In principle, slice thickness 2DFT MRI should not be less than 5 mm. Otherwise, slice definition becomes unacceptably ambiguous.

[21] This further underscores the inappropriateness of multiple thin slice studies covering the entire brain. Unfortunately, the results of such studies have appeared with frequency in the published literature. It should by now be clear to the readers why such studies lack validation.

[22] This is caused by characteristic misalignment of data points on phase direction of the k-space. Therefore, any imaging sequence which produces a similar misalignment may produce a similar ghost.

Therefore, any paradigm which requires motion of a body part should be performed with extreme care. Even hand motion may cause field perturbation within the image volume and produce Nyquist ghost. Classical behavior scientists typically prefer paradigms which require subject response[23]. Direct adaptation of such paradigms to fMRI typically introduces *task-correlated* Nyquist ghost and fictitious activation (see "Illustrative Case Study I" below).

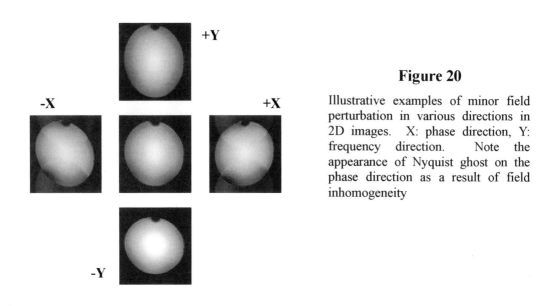

Figure 20

Illustrative examples of minor field perturbation in various directions in 2D images. X: phase direction, Y: frequency direction. Note the appearance of Nyquist ghost on the phase direction as a result of field inhomogeneity

Pixel Misalignment and Limitation of Spatial Resolution

Pixel misalignment represents another common cause of fictitious activation (Figure 22). The primary cause of pixel misalignment is subject motion. Such motion cannot be totally suppressed. Many "motion correction" post-processing algorithms have been developed to ameliorate the problem. Unfortunately, such algorithms are based on two dimensional imaging and not totally applicable to correcting three dimensional misalignments which produce partial volume effects. These algorithms are considered "cosmetic" at best (Nakada et al., 1998).

The theoretical limit of spatial resolution for MRI is approximately 4 micron-cube[3]. Therefore, many believe that microscopic fMRI can be performed. This cannot be farther from the truth. In addition to shimming problems, there are many

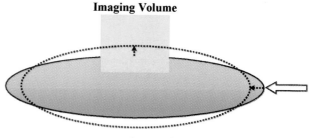

Figure 21

Schematic illustration of field perturbation introduced by subject response. A magnetic field behaves like a balloon. Perturbation introduced at the outside of the imaging volume (open large arrow) still causes perturbation within the imaging volume.

---

[23] Paradigms which require vocal responses represent the worst paradigm for fMRI studies.

subject related conditions which determine the limit of fMRI spatial resolution.

An example of an EPI image of the normal brain is shown in Figure 23. The numbers indicate the relative intensity of cerebral cortex at various sites relative to CSF which is assigned a value of 100 (written in red). The substantial variation in the intensity of cortex is primarily due to the partial volume phenomenon (Figure 13).

Let's do a rough calculation at this point. The maximum increase in intensity due to brain activation utilizing a 1.5T system is 3%[24]. Taking the average cortex intensity as 30 (Figure 23), the intensity of the activated cortex would be approximately 30.9. Cortex is always surrounded by CSF. Therefore, minor motion of the brain results in alteration of the percentage of CSF within the voxel. If 2% of voxel volume is replaced by CSF, the intensity of the corresponding pixel becomes 31.4 and exceeds that of activated cortex. Statistically speaking, the intensity of 30.7 associated with 1 % voxel replacement by CSF is already highly significant, albeit not representing any activation whatsoever.

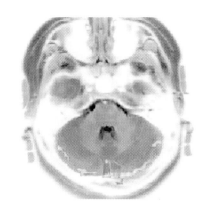

Figure 22

Illustrative example of fictitious activation due to pixel misalignment

As stereotactic surgery has unambiguously demonstrated, the brain physiologically moves a minimum of 0.3 mm even under conditions where the skull is completely fixed by a fixation device. It follows that the voxel volume chosen for fMRI experiments should be sufficiently large such that 0.27 mm^3 represents less than 1% of the total volume. In this schema, the minimum voxel volume should be 27 mm^3 which represents an isotropic voxel size of 3 mm x 3 mm x 3 mm. At this point, the readers should be reminded that this gross estimate is valid only under the ideal condition. In reality, one should determine "reliable" voxel size as part of the calibration studies. In addition, brain motion has to be monitored during every individual session to confirm validation of the study[25]

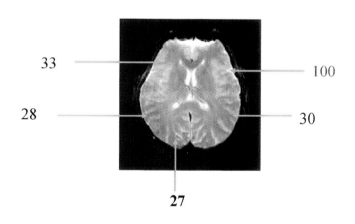

Figure 23

---

[24]Our personal experience indicates that a 3% increase in activation associated intensity at 1.5T is exceptional. It is usually around 1.5 to 2%. Even using a 3.0T system, the maximum increase in activation associated intensity is found to be approximately 4-5%.

[25]At our intitution where two 3.0 T systems optimized for fMRI are operational, physiological brain motion under optimal head fixation conditions is found to range betwen 0.3-0.6 mm. Accordingly, a 3 mm x 3 mm matrix with 5 mm slice thickness (45 mm^3) is chosen as standard. By experience, brain motion monitored pictorially in two dimensions exceeding 0.6 mm produces pixel misalignment artifacts. Data form such studies are discarded.

Limitations of Post-processing Correction Algorithms

It is virtually impossible to discuss all the available post-processing correction algorithms in this brief review. Nevertheless, it must be emphasized that the majority of post-processing algorithms have been developed based on the concept based on two-dimensional images. These algorithms do not adequately address the fundamental characteristics of MRI, especially those related to field inhomogeneity. It is highly prudent for the serious fMRI scientist *not* to rely on post-processing correction algorithms. Each raw image should be utilized for statistical analysis as it is. Post-processing correction algorithms such as "re-slicing, "standardization", or "motion correction" should be avoided. These data manipulations produce further complexity to MRI image data (Figure 24). Instead, investigators should properly perform ultra-fast MRI data acquisition. Extensive calibration of the system is essential. In this sense, collaboration with advanced MR scientists is essential

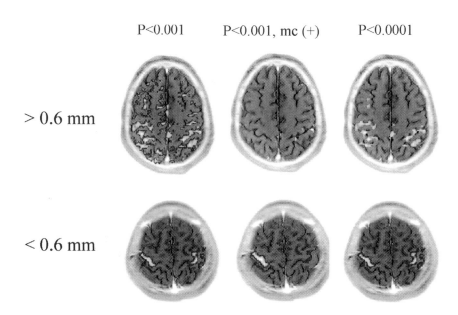

Figure 24

Illustrative example showing the effects of a representative motion correction algorithm of SPM96 (the Wellcome Department of Cognitive Neurology). The activation maps shown were obtained for a bilateral hand motion paradigm using a horizontal 3.0 T system with image voxel resolution of 3 mm x 3 mm x 5 mm. In this setting, acceptable pixel misalignment was determined to be 0.6 mm. Brain motion exceeding 0.6 mm (> 0.6 mm) produced significant pixel misalignment artifact. A motion correction algorithm (mc) wiped out these artifacts as well as actual activation. Change in p-value (P< 0.0001) also eliminated a small cluster of fictitious activation, while some "true" activation remained visible. In contrast, the study performed appropriately with brain motion less than 0.6 mm (< 0.6 mm) provided activation maps of bilateral primary motor cortices. Application of motion correction algorithm (mc) artificially eliminated some "true" activation. Change in p-value did not affect the map.

Some Illustrative Examples

Case Study I

Case History

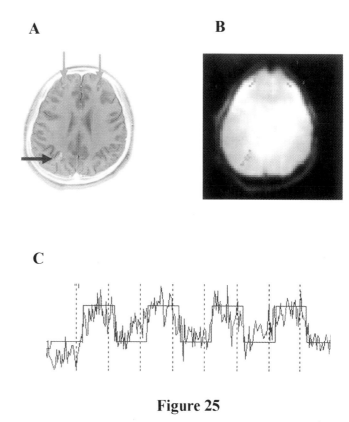

The fMRI study was performed using a three-dimensional motion paradigm. A subtractive approach in boxcar configuration provided an activation map showing activation in the right intra-parietal sulcus (red arrow in Figure 25A). The raw EPI image did not show significant ghost or susceptibility effects suggestive of potential fictitious activation (Figure 25B). Corresponding time series of activated pixels showed intensity changes which were well correlated with the boxcar type of paradigm (Figure 25C). The right intraparietal area is well known to subserve those functions related to three-dimensional motion in primate physiology. As a result, the investigators were convinced that Figure 25A would represent a valid activation map.

Figure 25

A: Activation map on structural (T2R) image. B: Activation map on EPI image. C: Time series of a pixel from activation cluster (Blue: raw data. Red: boxcar model function).

Diagnosis

The senior investigator was consulted and expressed concern regarding the symmetric nautre of the activation in the frontal lobes (yellow arrows in Figure 25A). Further evaluation of the activation map disclosed that the areas thought to reflect activation actually corresponded to Nyquist ghost (Figure 26). The activation map was then determined to be fictitious.

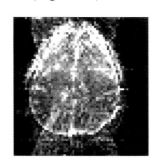

Figure 26

This standard deviation image clearly demonstrates that the area thought to reflect brain activation corresponds to Nyquist ghost.

Comment

This case illustrates the situation where appropriate diagnosis of fictitious activation may in fact be quite difficult for inexperienced fMRI investigators. The clue here was the symmetric natuare of the activation seen in the frontal areas (yellow arrows in Figure 25A), a pattern which is typical for Nyquist artifact, strongly suggesting the presence of Nyquist ghost. The standard deviation image unambiguously confirmed it. Such determinations are in fact rather easy case for senior fMRI investigators who always keep all potential sources of artifacts in mind and are aware that the fMRI process is in fact the process of fighting artifacts.

Case Study II

Case History

The fMRI study was performed using a motor learning paradigm. A subtractive approach in modified boxcar configuration provided an activation map showing activation in the left cerebellar cortex (Figure 27A). The raw EPI image did not show significant ghost or nearby susceptibility effects suggestive of potential fictitious activation (Figure 27B). Corresponding time series of activated pixels showed intensity changes which are well correlated with the modified boxcar type of paradigm (Figure 27C). The cerebellum is well known to be involved in motor learning processes in primate physiology. As a result, the investigators were convinced that Figure 27A was a valid activation map.

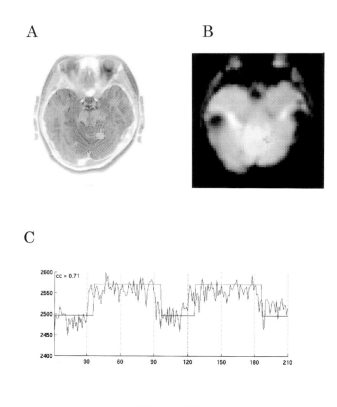

Figure 27

A: Activation map on structural (T2R) image. B: Activation map on EPI image. C: Time series of a pixel from activation cluster (Blue: raw data. Red: boxcar model function).

Diagnosis

The senior investigator was consulted and expressed concern regarding the inclusion of eye movements in the paradigm. Further evaluation of the activation map disclosed that the area thought to be activated actually corresponded to motion ghost (Figure 28). The activation map was then diagnosed as fictitious.

Comment

This case again illustrates the situation where appropriate diagnosis of fictitious activation may in fact be quite difficult for inexperienced fMRI investigators. The clues were: (1) slice selection which included the eyes; and (2) paradigm selection which required eye movements. In addition to Nyquist ghost, EPI images are quite sensitive to motion ghost similar to other MRI imaging techniques. Expanded cross correlation images unambiguously confirmed this. The basic rule is that slices which contain moving parts of the face such as eyes or jaw should never be utilized as analytic slice. Make sure the slice contains no moving parts!

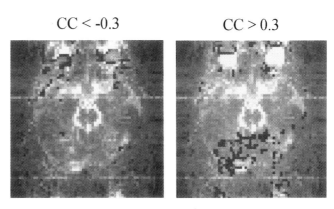

Figure 28

Extended cross correlation activation map on standard deviation image. Note the typical motion ghost runs vertically along with the eyes corresponding closely to the activation map and indicating that the activation map in Figure 27A is actually motion ghost related and fictitious.

Case Study III

Case history and Diagnosis

The final case study represents simple, but common artifacts, namely, artifacts associated with structures with strong susceptibility effects. Such artifacts are in principle not paradigm dependent. Rather they simply represent inappropriate imaging slice selection. Although easily avoidable, This type of artifact is found widespread in fMRI publications and shown at presentations. Some basic examples are shown in Figure 29. The structural image (Figure 29A) was obtained with a FSE sequence which is highly

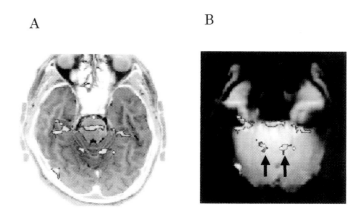

Figure 29

A: Activation map on structural (T2R) image. B: Activation map on EPI image. Fictitious activation occurred adjacent to structures with strong susceptibility (air sinuses and air cells in this case). T2R image based on FSE sequence was not affected by these same structures partially occupying the slice volume. Without examining the actual EPI images, the presence of structures with strong susceptibility may easily be missed. There were true activation sites associated with bilateral hand motion (black arrows).

insensitive to susceptibility effects and provided an undistorted high resolution image (presented as T2 reversed (T2R) image). Identical activation sites were mapped onto the EPI image, the same process with which raw data were obtained (Figure 29B). The structures giving rise to strong susceptibility effects became clearly visible and fictitious activation was easily identifiable. The discrepancy between FSE and EPI images occurs due to the partial volume effect (Figure 13). This kind of artifact may appear not just within its own slice but also in the adjacent slice (interslice partial volume phenomenon). Such examples are shown in Figure 30. For experienced diagnostic imaging physicians, identification of those structures which have a similar configuration as the activation map (red circle in Figure 30A and B) is not difficult. For fMRI scientists with only minor experience in diagnostic imaging, this may be quite difficult to recognize and illustrates the inherently multi-disciplinary nature of fMRI and necessity for knowledge in NMR, MRI, electrical engineering, physics, medical imaging, as well as neuroscience[26].

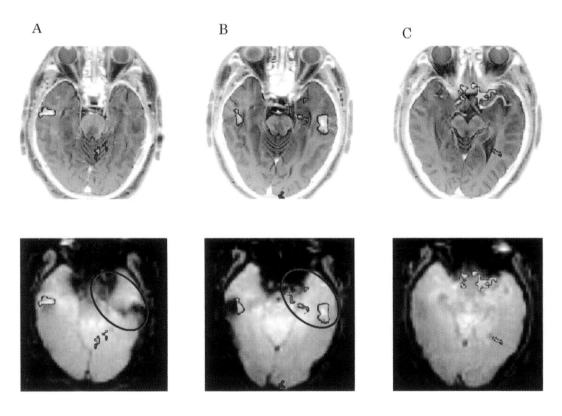

Figure 30

A: Activation map on structural (T2R) image. B: Activation map on EPI image. Fictitious activation occurs adjacent to the structure with strong susceptibility as shown in Figure 28. Again, the T2R image based on a FSE sequence did not show these structures with strong susceptibility which only partially occupies the slice volume. Additionally, there was fictitious activation associated with the structure with strong susceptibility seen adjacent but not in its own slice (red circle). Three consecutive 5 mm thick slices were imaged with an interslice gap of 2.5 mm. Even taking the appropriate precautions in selecting the imaging slice, strong susceptibility effects produced fictitious activation in adjacent slices due to the partial volume phenomenon.

---

[26] If one does not have enough time to be trained in all these disciplines, one needs close collaborations with experts in these disciplines.

Comment

The basic rule for avoiding this kind of artifact is the proper selection of the imaging slice. One should not utilize any slice which shows significant susceptibility effects on EPI images. Air sinuses and air cells are the common structures to be avoided. Ferromagnetic substrates may be unexpectedly found. One should always obtain scout EPI images for determination.

The careful readers have probably realized by now the fact that I did not utilize any coronal slices for functional maps in this article for examples of artifacts. The reason should be obvious by now. Any coronal slice ends up including a large portion of air sinuses or air cells within its image volume. Therefore, coronal cuts are in principle not suitable for fMRI[27]. As shown in this case study, even for axial slices, air sinuses may partially occupy the slice volume and thereby significantly affecting EPI images. It is highly prudent for the serious fMRI scientists to consider that it is virtually impossible to obtain EPI images using coronal slices suitable for fMRI (Figure 31).

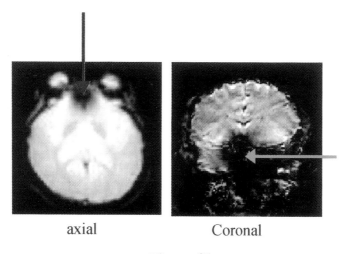

axial   Coronal

Figure 31

Illustration of the effects of strong susceptibility. Even though it only partially comprised the imaging volume (axial slice), the air sinus introduced significant distortion in the EPI image (red arrow) much worsened in a coronal slice where the sinus occupied a larger part of the image volume (blue arrow). In principle, one should never use coronal slices for fMRI.

Closing Remarks

High-field System: Is it an advantage or requirement?

Paramagnetic susceptibility of spin with a gyromagnetic ratio of γ is given by the Brillouin equation as:

---

[27] There are some techniques which can ameliorate this problem. However, the application of such special techniques will result in other limitations. Therefore, I will not elaborate on this issue in this article.

$$\chi = \frac{\hbar\gamma}{2B_0}\tanh(\frac{\hbar\gamma B_0}{2kT}),$$

where k is the Bolzmann constant and T absolute temperature. When one wants to deal with susceptibility effects quantitatively, the quantity of χB_0 is often utilized.

It is clear that susceptibility effect increases exponentially (actually as the hyperbolic tangent) associated with an increase in the main field, B_0, of the system. The higher susceptibility effect affects fMRI positively as well as negatively. BOLD contrast detectable activation (ΔI in Figure 32) will be significantly increased. At the same time, the occurrence of artifact inducing perturbations also increases dramatically. Without a proper understanding of MR, higher field systems may further confuse the fMRI society.

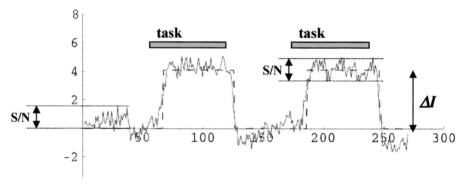

Figure 32

Typical time series of an activated pixel in the primary motor cortex illustrating performance of a horizontal 3.0 T system optimized for fMRI. S/N indicates variation of EPI images, while ΔI, the BOLD phenomenon-induced increase in signal intensity. Red: raw data. Blue: boxcar type model function. Note the delay in the BOLD phenomenon compared to the actual task performance. This time series represents signals from a single voxel volume of 3 mm x 3 mm x 5 mm. The quality of the time series shown here is high enough for detailed time domain analysis without statistical help, demonstrating the power of high-field fMRI (HF-fMRI)

It has become progressively clearer, however, that BOLD contrast obtainable with conventional MR systems, such as the 1.5 T system, may not be sufficient for analysis of complex behavioral tasks. High BOLD contrast may, therefore, be essential. If one exercises proper knowledge of MR and performs fMRI experiments appropriately, the results are indeed extremely rewarding. The time series of a single pixel from a single session can be directly analyzed similar to electrophysiological data without need for statistics (Figures 32). Paradigms aimed at the identical abstract goal, each of which actually requiring performance of the given tasks in differential modalities, can provide almost identical activation maps (Figure 33). These are but a few examples of the power of *properly performed* high-field fMRI (HF-fMRI)[28].

---

[28] It is highly desirable to distinguish fMRI performed on 1.5 T systems from fMRI performed on high-field systems (HF-fMRI), such as a 3.0 T system, as totally different method in neuroscience.

For clinical purposes, it is important to have a technique available at most of the hospitals and, therefore, use of widely distributed systems, such as the 1.5 T system, is important. On the other hand, for advanced neuroscientific investigation, higher field systems are essential. In both situations, the importance of a proper understanding of MR cannot be overemphasized.

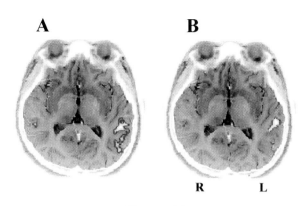

Figure 33

Functional map for comprehension tasks. A: reading. B: hearing. Paradigms requiring tasks of different modalities (visual vs. auditory) provided almost identical activation maps based on the identical abstract concept of "comprehension." This kind of study has become possible with the advancement of high field fMRI which enabled investigators to obtain high quality activation maps in a single subject in a single session.

Science, Medicine, and Morals

Proper performance of fMRI requires proper knowledge in a number of different disciplines including physics, engineering, medicine, and neuroscience. Unfortunately, the number of people in the world who has adequate expertise in all these fields is severely limited. Furthermore, warnings given by the experts are often ignored as noise by the "majority" of fMRI scientists.

I was fortunate to have a background in physics and electrical engineering prior to becoming a neurologist. I was further fortunate in choosing biomedical application of NMR, including MRI, as my main research objective for the last 30 years. However, I am quite unfortunate to find only a few scientists who really want to take validation issues of fMRI seriously. The myth of "productivity" is blowing around the scientific community polluting this final sacred bastion of human morals. Competition without proper philosophy can only be harmful. I sincerely hope that many well meaning scientists, especially the young scientists, will consider the meaning of science one more time. Science, just as medicine with its Hippocratic oath, is supposed to exist for the happiness of mankind.

The key word for 21st century science is "multi-disciplinary" as guided by non-linear physics. fMRI is no exception. The human brain and its function represent the most advanced "complex system." The technology which explores the most sophisticated product of Mother Nature should be utilized properly. By eliminating any of its misuse, we should protect this precious gift given to 21st century neuroscientists and, in turn, all of humanity.

Acknowledgement

The contents of this manuscript represents the essence of lectures given at several international and domestic symposia, including the XVIIIth International Conference on Magnetic Resonance in Biological Systems, the International Rodin Remediation Academy Conference 1999, the International Symposium in Ultra-fast Magnetic Resonance Imaging in Medicine, and the 21st Annual Meeting of Japanese Society for Neuroscience. This work was supported by grants from the Ministry of Education (Japan). The author thanks Drs. Ito, Fujii, and Matsuzawa for preparation of the illustrative examples and Dr. Kwee for her critical reading.

Suggested Readings

Ernst R. R., Bodenhausen G., Wokaun A.: Principles of Nuclear Magnetic Resonance in One and Two Dimensions. Oxford University Press, Oxford, 1987.

Fukushima E., Roeder S. B.: Experimental Pulse NMR: A Nuts and Bolts Approach. Addison-Wesley, London, 1981.

Schmitt F., Stehling M. K., Turner R.: Echo-Planar Imaging. Springer-Verlag, Berlin, 1998.

Stark D. D., Bradley W. G.: Magnetic Resonance Imaging. Third Edition. Mosby, St. Louis, 1999.

Specific References

Friston K. J., Holmes A. P., Worsley K. J., Poline J. P., Frith C. D., Frackowiak R. S. J.: Statistical parametric maps in functional imaging: a general linear approach. Hum Brain Map 1995;2:189-210.

Fujii Y., Nakayama N., and Nakada T.: High-resolution T_2 reversed MRI on a high-field system. J Neurosurg 1998;89: 492-495.

McKeown M. J., Makeig S., Brown G. G, Jung T. P., Kindermann S. S., Bell A. J., Sejnowski T. J.0: Analysis of fMRI data by blind separation into independent spatial components. Hum Brain Map 1995;6:160-188.

Nakada T., Fujii Y., Suzuki K., and Kwee I. L.: High-field (3.0T) functional MRI sequential epoch analysis: An example for motor control analysis. Neurosci Res 1998;32:355-362.

Nakada T.: High-field, high-resolution MR imaging of the human indusium griseum. AJNR 1999;20:524-525.

Appendix

- Magnetic Susceptibility -

In principle, all material is "magnetic" and once placed in a magnetic field, they are magnetized (Figure 34). When magnetization, *M*, and magnetic field, *H*, satisfy the following equation, the coefficient, χ_m, is called magnetic susceptibility:

$$M = \mu_0 \chi_m H,$$

where μ_0 is magnetic permittivity in a vacuum.

When χ_m takes a positive value, such material is termed paramagnetic. When χ_m takes negative value, such material is termed diamagnetic. Diamagnetic susceptibility originats from electrons surrounding the nucleus and, therefore, is a physical property of all material. Magnetic susceptibility which produces significant effects in MRI is in principle paramagnetic[29].

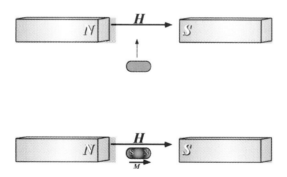

Figure 34

Schematic presentation of magnetization of paramagnetic nature.

---

[29] The behavior of iron metal is *non-linear* showing phase transition. It is specifically called ferromagnetic and can be treated as "gigantic" paramagnetic effects.

Chapter II.2

Development of 3.0 Tesla Vertical MRI System for Advanced fMRI Applications

Tsutomu Nakada[1] and V3T Team

Department of Integrated Neuroscience, Brain Research Institute, University of Niigata

Introduction

For an old time nuclear magnetic resonance (NMR) spectroscopist like myself, NMR magnets equal vertical. We indeed started our first *in vivo* biological investigation in rats in the 1970's using a vertical magnet system. The magnetic resonance imaging (MRI) revolution quickly made the strong impression that magnets should have a horizontal orientation. Physician-scientists soon adopted the idea of horizontal magnet systems as well as the neologism "magnetic resonance" (MR) as the basic standards in this field. This apparent flexibility is reasonable for the purpose of clinical convenience. However, it raises the fundamental questions regarding human brain functional analysis.

Awake humans seldom function in the horizontal position. As a matter of fact, the vertical position is believed to be the main driving force in human evolution. Brain circulation, which is tightly related to brain function, is highly sensitive to body position.

Figure 1
World's First Human Vertical MRI system: *Issorar*

---

[1] Correspondence: Tsutomu Nakada, M.D., Ph.D., Department of Integrated Neuroscience, Brain Research Institute, University of Niigata, Niigata 951-8585, Japan, Tel: (81)-25-227-0677, Fax: (81)-25-227-0821, e-mail tnakada@bri.niigata-u.ac.jp

Without a proper functioning autonomic nervous system, the brain cannot function correctly. Functional analysis of the human brain, however, has thus far been performed in the horizontal position due to system configuration defined by convenience. In addition to perfusion analysis, it is strongly desirable that functional MRI (fMRI) studies, which utilize regional blood flow as index for functional localization, are performed in the natural awake human position, namely, the vertical position. As part of a revolutionary plan for funding academic science launched by the Japanese government called "Center for Excellence (COE)", we proposed to create an MRI system optimized for investigation of human brain function. Here, the first such system, a 3.0 Tesla vertical MRI system, *Issorar*, is presented (Figure 1).

Design Concept

To design a vertical MRI system specifically optimized for fMRI, we emphasized on the following two principles: (1) the most appropriate field strength; and (2) maximization of the degree of freedom for highest versatility in task performance. General Electric (GE) Medical Systems, Milwaukee, MI, offered to integrate the system based on its LX platform, and its Japanese division, GE Yokogawa Medical Systems (GEYMS), Tokyo, Japan, agreed to provide all necessary technological support for system integration. Subsequently, Magnex Scientific Limited, Abingdon, England, was selected as magnet vendor

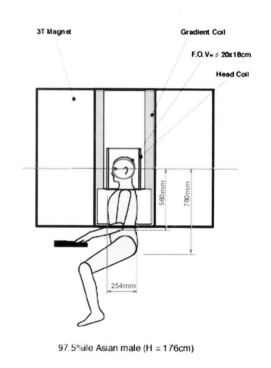

Figure 2

The selected field strength was 3.0 Tesla (T). The selection of this strength was primarily based on the personal experience of the project director in MRI at various fields (Nakada & Kwee, 1986, Nakada et al., 1988, Nakada et al., 1994, Nakada et al., 1999). This field strength appears to be the ideal compromise when one considers the pros and cons of higher magnetic fields. For example, the effects of dielectric resonance becomes much more intense at 4.0 T or higher. Power requirements and, hence, specific absorption rate (SAR) become a significant limiting factor at 4.0 T or higher. The appropriate voxel size of base images is primarily determined by biological rather than technical constraints. A well known example of the former is spontaneous motion of the brain (Nakada et al., 1998b). Under these settings, the signal strength obtainable based on a given spin density provides more than adequate signal to noise (S/N) at 3.0 T. The relationship between the susceptibility effects necessary for blood oxygenation level dependent (BOLD) contrast and shimming difficulties is well balanced at 3.0 T. The implementation of 3.0 T MRI and its suitability for structural as well as fMRI has been well documented (Nakada et al., 1998a, b, Fujii et al., 1998, Nakada, 1999, Nakada et al., 1999, Kwee et al., 1999).

Piano playing was chosen as the target task for working space definition. Subsequently, the configuration for a "head only" system schematically shown in Figure 2 was designed. All other system configurations including clear bore size, radio frequency (RF) coil configuration, and gradient coil strength, are self-determined based on these basic constraints.

Implementation

Magnet

The magnet was manufactured and tested by Magnex Scientific Limited in Abingdon England and was installed in a purpose built facility at the University of Niigata in December 1999.

The system consists of a homogeneous and extremely stable superconducting magnet (3.0 Tesla) housed in a low-loss helium cryostat with an 580 mm room temperature vertical bore. The distance from the base of the magnet to the homogeneous region has been minimised to 525 mm to allow the subject to perform tasks with their hands during an imaging sequence. Field shimming is accomplished using superconducting shim coils and passive shims. Fine-tuning of the homogeneity is made with a set of resistive shims that are mounted on the gradient coil. A cryo-refrigerator is provided to minimize cryogen consumption and eliminate the necessity for a liquid nitrogen reservoir (Figure 3).

The magnet is a five coil design with an additional two reverse wound shielding coils to control the stray field. The field homogeneity is specified to be less than ±6 ppm over a 24cm diameter spherical volume and less than ± 0.5 ppm over a spheroid 20 cm diameter x 18 cm long. Figure 4 shows the homogeneity profile of the magnet plotted along the axis of the magnet and the theoretical field contours viewed from a point transverse to the magnet axis. The 5 gauss contour is approximately 4.1 m axially x 2.5 m radially from magnet center.

The magnet requires a current of 339.2 amps to achieve its maximum operating field of 3.0 Tesla. The total inductance of the windings is 54.7 Henries, the stored energy at maximum field is 3.1 Mjoules. The magnet coils are fully protected from accidental damage due to a quench by a resistor and diode network located within the helium reservoir. In the event of the need to activate an emergency discharge of the magnet a quench heater circuit is incorporated within the windings. Environmental influences and winding tolerances distort the field homogeneity. Corrections for these distortions are made in the first instance by superconducting shim coils located on a former which also

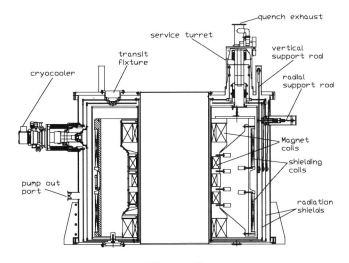

Figure 3

supports the shielding coils, these coils provide correction for the following terms; Z1, Z2, X, Y, ZX, ZY, X2-Y2, XY. Higher order corrections are made with passive shims.

The magnet is wound from wire-in-channel conductor. This consists of multiple filaments of

Niobium Titanium Alloy in a circular copper matrix. The circular wire is laid into a rectangular copper channel that effectively increases the copper cross section of the conductor, thereby improving its ability to withstand the high magnetic forces. The superconducting current carrying properties of this circular wire are determined by the amount of superconducting material (Niobium Titanium), the number of filaments of the superconducting material and the manufacturing process. The properties of the wire are chosen so that the critical current, i.e. the current at which the wire starts to lose its superconducting characteristics for a given field, is well beyond the actual operating current of the magnet. The size of conductor used in different parts of the magnet is chosen to ensure the hoop stresses and axial stresses on the conductor are sufficiently low to prevent wire movement. Two sizes of copper channel were used in the manufacture of the magnet, the largest conductor having insulated dimensions of 2.96 mm x 1.61 mm. The inner former is constructed from glass reinforced polyester wound onto the outer surface of the stainless steel helium vessel bore tube. It is precision machined to tolerance of typically 0.25 mm to ensure that all the coils forming the magnet are concentric and accurately positioned. The main axial stresses on the former come from the end coils on inner former; these coils experience a compressive force of 129 tones. The outer former is constructed from a stand alone glass reinforced polyester tube. The magnet coils were wound from 51 km of conductor and consisted of 17568 turns of wire arranged in 288 layers. The superconducting shims coils consumed a further 26 km of wire.

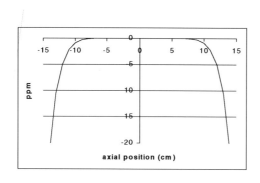 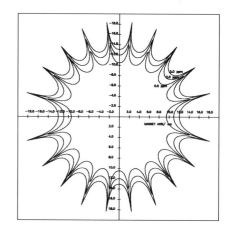

Figure 4
Axial field profile (left) and transverse homogeneity contours (right).

Chair Unit

The chair unit was designed and manufactured by the Mechanical Engineering section of GEYMS. The project represents the first of its kind in the history of MRI. The key words for design concept were "functionality" and "safety."

Overall height of the unit is 998 mm. "Screw and nut" configuration was adopted for the motion mechanics. A 200V three-phase motor was utilized to ensure sufficient torque for a 100 kg subject. The total stroke length is 998 mm. The velocity of the unit motion under an 80 kg load was adjusted to be 17.11 mm/sec and 17.50 mm/sec for upward and downward motion, respectively.

The mobile part contains the chair, floor, and backrest as a single unit. The backrest also contains a specially designed headrest and RF coil holder. The floor can hold upto a 400 kg load, while the chair, 200 kg. The unit maintains structural integrity under 45 N force applied perpendicular to the unit at any point. All the mobile part are safely secluded from the subject. Considering the necessity for passive shimming of the magnet, the entire unit was designed to be easily removable from the magnet.

In principle, low magnetic, low-conductive materials were utilized to construct the entire unit. The materials utilized include A5052 and A5052P aluminum, SUS316, SUS304 stainless steel, C2700 brass, and FRP glass fiber. To further improve functionality, usage of any conductive materials for the unit placed directly under the magnet bore axis was severely minimized.

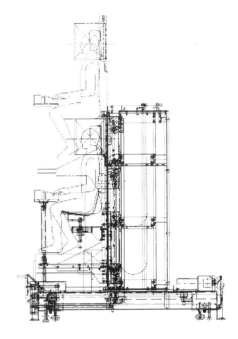

Figure 5

First Study

The first human study was performed on February 9, 2000. The subject was a 52-year-old normal male volunteer (Figure 6). The detailed results of this study will be reported elsewhere[2].

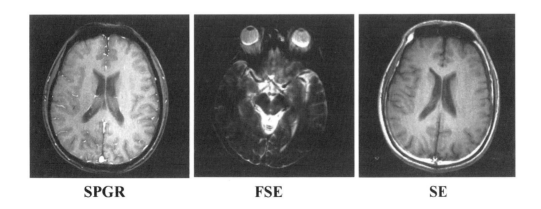

SPGR **FSE** **SE**

Figure 6

---

[2] The results will be in part reported at the 86th Scientific Assembly and Annual Meeting, RSNA 2000.

V3T Team

General Concept and Design
Tsutomu Nakada

Magnet and Gradient Coil (Magnex)

Design
Peter Feenan - Program Management
David Reeves - Engineering Manager
Wayne Lacey - QA and Technical Advice
Chris Cirel - Gradient Project Engineer
Brandon Stoppani - Gradient Mechanical Design
Rory Warner - Magnet Project Engineer
Phil Bellenger - Magnet Mechanical Design
Graham Briggs - Magnet Mechanical Design

Production
Dennis Atkins - Workshop Supervisor
Gerald Bolton - Coil Winding Supervisor
Mike Nichols - Termination Supervisor
Andy Bennet - Cryostat Assembly
Derek Wilkins - Test Supervisor
Alwyn Smith - Installation

Chair Unit (GEYMS)

Design and Production
Akira Imai - Mechanical Engineer
Shigeo Nagano - Electric Engineer

Platform and System Integration (GEYMS)

Advanced Technology Center
Tetsuji Tsukamoto - Program Manager
Nobuyuki Tasaka - Program Integrator
Akira Nabetani - Mechanical Engineer
Hiroshi Sato - Software Engineer
Leon ter Beek - Software Engineer

Engineering Department
Tohru Shimizu - Saftey and Regulatory Representative

Construction Technology
Yoshimi Sakurai - Site engineer

Service Technology
Yuji Suzuki - Service Manager
Tetsuya Hara - Service Representative

Regional (Niigata) Office
Yoshio Isozaki - Management
Masaki Watanabe - Electric Engineer

Clinical Advice

University of Niigata
Yukihiko Fujii
Hitoshi Matsuzawa
Kiyotaka Suzuki

University of California
Ingrid L. Kwee

Acknowledgement

The project was entirely supported by grants from the Ministry of Education (Japan). The manuscript was presented in part at the Eighth Annual Meeting of the International Society for Magnetic Resonance in Medicine, Colorado, April 2000.

References

Fujii Y, Nakayama N, Nakada T: High-resolution T_2 reversed MRI on a high-field system. J Neurosurg 1998;89:492-495.

Kwee IL, Fujii Y, Matsuzawa H, Nakada T: Perceptual processing of stereopsis in human: High-field (3.0T) functional MRI study. Neurology 1999;53:1599-1601.

Nakada T, Kwee IL: Oculopalatal myoclonus. Brain 1986;109:431-441.

Nakada T, Kwee IL, Card PJ, Matwiyoff NA, Griffey BV, Griffey RH: 19-Fluorine NMR imaging of glucose metabolism. Magn Reson Med 1988;6:307-313.

Nakada T, Kwee IL, Igarashi H: Brain maturation and high energy phosphate diffusivity: Alteration in cytosolic microenvironment and effective viscosity. Dev Brain Res 1994;80:121-126.

Nakada T, Fujii Y, Suzuki K, Kwee IL: "Musical Brain" revealed by high-field (3 tesla) functional MRI. NeuroReport 1998a;9:3853-3856.

Nakada T, Fujii Y, Suzuki K, and Kwee IL: High-field (3.0T) functional MRI sequential epoch analysis: An example for motor control analysis. Neurosci Res 1998b;32:355-362.

Nakada T: High-field, high-resolution MR imaging of the human indusium griseum. AJNR 1999;20:524-525.

Nakada T, Nakayama N, Fujii Y, Kwee IL: Clinical application of magnetic resonance axonography. J Neurosurg 1999;90:791-795.

Chapter II.3

Using Adiabatic Rapid Passage to Minimize RF Power Requirements in NMR

Michael Garwood[1]

Center for Magnetic Resonance Research and Department of Radiology, University of Minnesota Medical School

Introduction

Just a decade ago, the concept that *in vivo* NMR would benefit from the use of yet stronger magnetic fields was often debated among experts. Much of the skepticism about high field NMR was based on expectations of multiple technical obstacles. Recent experience with high field NMR has revealed ways to circumvent many of these technical problems so that experimental advantages can now be realized for many different applications. Consequently, the numbers of high field human (≥3 Tesla) and animal (≥7 Tesla) systems in use around the world are rising rapidly.

One potential problem with high field NMR relates to increased magnetic susceptibility (Ψ) and chemical-shift dispersion. While these effects are desirable for fMRI (BOLD contrast) and spectroscopy applications, they can cause errors and artifacts in certain situations. In applications requiring slice selection, for example, frequency shifts such as those induced by Ψ and chemical-shift dispersion have the undesirable effect of displacing the slice position. Such displacement errors can be particularly problematic when performing localized (single-voxel) spectroscopy at high field. Figure 1 illustrates this problem for a spectrum containing two resonances with chemical shifts ω_A and ω_B. As can be seen, the only way to alleviate this problem is to increase the bandwidth (*BW*) of the RF pulses with a concomitant increase of the B_0 gradient strength employed for slice (or voxel) selection, so that the chemical-shift difference ($\Delta\omega_{AB}$) becomes small relative to the spatial frequency-shift. In most cases, the maximum B_0 gradient strength is not the limiting factor (present day gradient hardware for human magnets can achieve upwards from 3 Gauss/cm). On the other hand, with conventional R pulses, *BW* is proportional to peak RF amplitude, B_1^{max}. As such, a

---

[1] Correspondence: Michael Garwood, Center for Magnetic Resonance Research and Department of Radiology, University of Minnesota Medical School, Minneapolis, Minnesota 55455, e-mail: gar@cmrr.umn.edu

linear increase in *BW* requires a *squared increase in peak power*, since peak RF power is proportional to $\left(B_1^{max}\right)^2$.

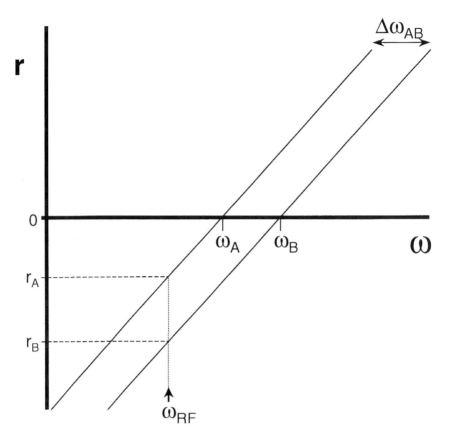

Figure 1
Illustrative plot showing how the precession frequencies of two spins with different chemical shifts (ω_A and ω_B) depend on position along the direction (r) of an applied field gradient. To select a slice for MRI or localized NMR spectroscopy, a pulse is applied at frequency ω_{RF}. With this two spin system, the center of the slices will occur at the spatial locations (r_A and r_B) where the spin precession frequencies are equal to ω_{RF}. The slices generated for spins A and B are thus displaced relative to each other because of the chemical-shift difference ($\Delta\omega_{AB}$). The relative displacement can be reduced by increasing the gradient strength, γG (the slope of the plots), but this will also require a proportionate increase in the pulse bandwidth to maintain the same slice thickness. To increase the bandwidth of conventional pulses, the peak RF power must increase.

Unfortunately, high power RF amplifiers for high frequency operation are very expensive. In addition, RF coils and other electronic components in the RF transmitter channel (e.g., T/R switch) must be designed to withstand higher voltages, often at the expense of compromised performance. Finally, the demand for increased RF power may limit the use of some pulse sequences which already operate near the allowable limits for specific absorption rates (SAR) at lower fields. In particular, pulse sequences that have high duty cycles and include multiple 180° pulses often exceed current SAR guidelines when applied at high field.

Adiabatic full passage (AFP) pulses can overcome peak power limitations in NMR methods that require spin inversion. With an AFP pulse, *BW* is determined by the range of the frequency sweep ($\pm\Delta\omega^{max}$) and the need to satisfy the adiabatic condition (1, 2).

Thus, *BW* is limited only by the time average of $(B_1(t))^2$, not $(B_1^{max})^2$. With this approach, any desired bandwidth can be achieved simply by increasing the pulse length and the frequency sweep range, while B_1^{max} remains the same. This unique property of AFP pulses is not enjoyed by conventional RF pulses, which require *BW* and B_1^{max} to be intrinsically tied to the pulse length T_p. To illustrate this point further, Table 1 lists the numerical constants that link bandwidth and pulse length when using some common pulse shapes and flip angles.

In this work, a vector description of adiabatic rapid passage is reviewed in order to provide insight into how the bandwidth of AFP pulses can increase without increasing B_1^{max}. Results from computer simulations are also used to demonstrate the advantages of AFP pulses over conventional pulses.

Pulse	*Factor* [a,b]
90° square	1.4
180° square	0.8
90° gauss [c]	2.7
180° gauss [c]	1.5
90° sinc (5 lobe)	5.9
180° sinc (5 lobe)	4.5
hyperbolic secant [d] (AFP)	$\Delta\omega^{max} T_p/\pi$

Table 1
Bandwidth factors of different pulse shapes

[a] BW (fwhm) = *Factor*/T_p

[b] BW has units of hertz and is defined as the frequency range in which $0.5 \leq (M_x^2 + M_y^2)^{\frac{1}{2}}/M_0 \leq 1$ for 90° pulses, and $-1 \leq M_z/M_0 \leq 0$ for 180° pulses

[c] Amplitude-modulated function truncated at 1% of maximum

[d] Range of the adiabatic frequency sweep = $\pm\Delta\omega^{max}$ (rad/s)

Principles of Adiabatic Rapid Passage

With most modern NMR spectrometers, adiabatic pulses are usually transmitted using phase modulation instead of frequency modulation. However, to easily visualize the vector motions during an AFP pulse, the pulse is usually described in terms of its frequency-modulated form in a reference frame that rotates with the time-dependent pulse frequency, $\omega_{RF}(t)$. This frame is commonly known as the frequency-modulated (FM) frame, and by our convention, its axes are labeled with primes (x', y', z'). As illustrated in Fig. 2, the effective magnetic field $\mathbf{B}_{eff}(t)$ is the vector sum of the amplitude-modulated (AM) and FM field components in the FM frame,

$$\mathbf{B_1}(t) = B_1(t)\hat{\mathbf{x}}' \qquad [1]$$

$$\mathbf{\Delta B_0}(t) = [(\omega_0 - \omega_{RF}(t))/\gamma]\,\hat{\mathbf{z}}', \qquad [2]$$

where ω_0 is the Larmor frequency and the orientation of $\mathbf{B}_1$ is arbitrarily chosen to coincide with x' (ϕ=0).

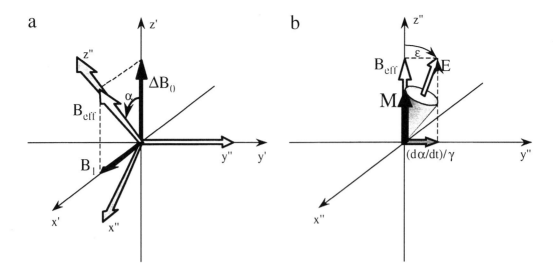

Figure 2
Vector diagrams showing the effective magnetic fields and their components in two different rotating frames of reference. (a) Relationship between the FM frame (x', y', z') and the B_{eff} frame (x", y", z"). (b) Evolution of the magnetization vector **M** in the B_{eff} frame during an adiabatic rapid passage.

If the pulse frequency starts far below resonance ($\omega_{RF} \ll \omega_0$), ΔB_0 is initially very large relative to B_1. Thus, $B_{eff} \approx \Delta B_0$ and the initial orientation of **B**$_{eff}$ is approximately collinear with z'. As $\omega_{RF}(t)$ begins to increase, $\Delta B_0(t)$ decreases and **B**$_{eff}$ rotates toward the transverse plane. **B**$_{eff}$(t) changes its orientation at the instantaneous angular velocity, $d\alpha/dt$, where

$$\alpha(t) = \arctan\left[\frac{B_1(t)}{\Delta B_0(t)}\right]. \quad [3]$$

To accomplish a full passage (inversion), the frequency sweep continues past resonance toward the negative extreme, leading to a final **B**$_{eff}$ orientation along -z'. During the sweep, **M** follows **B**$_{eff}$(t) provided that

$$|\gamma B_{eff}(t)| \gg |d\alpha/dt|, \quad [4]$$

for all time (3). Also known as the "adiabatic condition", this inequality represents the main guiding principle that has been used in the design and optimization of modulation functions for adiabatic pulses (1-11).

The vector analysis in the FM frame is insufficient to explain why **M** follows **B**$_{eff}$(t) during an adiabatic passage. For this purpose, it is necessary to define a second frame of reference which rotates with **B**$_{eff}$(t), called the "B_{eff} frame". Figure 2a depicts the relationship between the FM and B_{eff} frames during an adiabatic passage. The axes of the B_{eff} frame are labeled with double primes, x", y", z". Initially, the two frames are superimposed. As the adiabatic passage proceeds, the B_{eff} frame rotates about y' (=y") with an instantaneous angular velocity $d\alpha/dt$, while **B**$_{eff}$ remains collinear with z". By the end of an AFP pulse, the two frames are related by a 180° transformation around y' (=y").

Figure 2b portrays the relationships between **M** and the magnetic field components in the B_{eff} frame. The rotation of the B_{eff} frame about y' gives rise to a field component along y", which has an instantaneous magnitude equal to $(d\alpha/dt)/\gamma$. Thus, in the B_{eff} frame, the effective magnetic field **E**(t) is the vector sum of $\mathbf{B}_{eff}(t)$ and $[(d\alpha/dt)/\gamma]\,\hat{\mathbf{y}}''$. To simplify the picture, let us choose a pair of modulation functions, $B_1(t)$ and $\Delta B_0(t)$, such that B_{eff} and $(d\alpha/dt)/\gamma$ are constants for the duration of the adiabatic passage. In this case, **M** simply evolves about **E** on the cone of angle ε. As shown in Fig. 2b, **M** never strays beyond an angle 2ε from $\mathbf{B}_{eff}$. When the adiabatic condition (Eq. [4]) is well satisfied, ε is small and the motion of **M** closely traces the trajectory of $\mathbf{B}_{eff}$ in the FM frame.

This analysis has identified the perturbation ($[(d\alpha/dt)/\gamma]\,\hat{\mathbf{y}}''$) that must be minimized in order to perform an ideal rotation with an adiabatic pulse and explains the origin of the adiabatic condition (12). It is also obvious that the B_{eff} frame is an instructive platform to visualize the behavior of adiabatic and other types of pulses. Furthermore, vector analyses in the FM and B_{eff} frames provide an understanding of how AFP pulses are able to induce near perfect inversion using a wide range of B_1^{max} values, which is the basis for their highly touted "B_1 insensitivity".

Computer Simulations

The RF energy needed to implement adiabatic pulses is commonly thought to be excessive in comparison to that required by conventional pulses (e.g., square-shaped or amplitude-modulated pulses). To dispel this misconception, B_1^{max} values and the relative RF energy (E_{rel}) needed to produce 180° flips with square, sinc, and AFP pulses were determined from Bloch equation simulations (neglecting spin relaxation effects). To illustrate how the choice of AM and FM functions can affect pulse performance, two different types of AFP pulses were evaluated and compared with conventional pulses.

The first AFP pulse is the well known hyperbolic secant (HS) pulse having modulation functions (3, 13),

$$B_1(t) = B_1^{max}\text{sech}\!\left[\beta\!\left(2t/T_p - 1\right)\right] \qquad [5]$$

$$\Delta B_0(t) = \left(\frac{\Delta\omega^{max}}{\gamma}\right)\tanh\!\left[\beta\!\left(2t/T_p - 1\right)\right] \qquad [6]$$

where $\Delta\omega^{max}$ is the maximum offset frequency attained in the frequency sweep, and β is a truncation factor, which in the present study was chosen such that sech(β)=0.01. The HS pulse used in the present simulations had an adiabaticity factor R ($\equiv \Delta\omega^{max}T_p/\pi$) of 20. A highly desirable feature of this HS pulse is its insensitivity to changes in B_1^{max} above the minimum value required for adiabaticity. In addition, inversion is achieved within distinct borders and is highly uniform for isochromats Ω inside the frequency sweep range ($-\Delta\omega^{max}<$

$\Omega < \Delta\omega^{max}$), since the adiabaticity is invariant when $\mathbf{B}_{eff}(t)$ crosses the transverse plane (1, 2). This latter property is known as offset independent adiabaticity (OIA).

The second AFP pulse to be studied, called the tanh/tan pulse (14), was not derived from OIA principles, but from a procedure known as NOM which predicts the optimizes the modulation functions from a consideration of the adiabatic condition (Eq. [4]) for all times during the pulse in a specified range of B_1^{max} and Ω values (4, 7). The modulation functions of the tanh/tan AFP pulse are given by

$$B_1(t) = B_1^{max} \tanh\left[\frac{2\xi t}{T_p}\right] \qquad (0 < t < 0.5 T_p) \qquad [7]$$

$$B_1(t) = B_1^{max} \tanh\left[2\xi\left(1 - \frac{t}{T_p}\right)\right] \qquad (0.5 T_p < t < T_p) \qquad [8]$$

$$\Delta B_0(t) = \left(\frac{\Delta\omega^{max}}{\gamma}\right) \frac{\tan\left[\kappa(2t/T_p - 1)\right]}{\tan[\kappa]} \qquad [9]$$

with constants $\xi = 10$, $\kappa = \arctan[18]$, and $R = 40$ ($\equiv \Delta\omega^{max} T_p/\pi$). This pulse was designed for the purpose of inverting spins rapidly in experiments that do not require a sharp transition between inverted and non-inverted spectral regions. With the tanh/tan pulse, the BW is limited by the ability to satisfy the adiabatic condition, not the sweep range ($\pm\Delta\omega^{max}$), which by design is usually much wider than BW. The adiabatic condition can be sufficiently satisfied only for isochromats with Larmor frequencies near the center of the frequency sweep range, since the sweep rate accelerates very rapidly in either direction away from the center of the pulse.

Pulse [a]	T_p (ms)	$\gamma B_1^{max}/2\pi$ (kHz)	E_{rel} [b]
180° square	0.16	2.68	1
Tanh/tan ($R=40$)	0.64	1.41	1
180° sinc	0.94	2.57	0.87
HS ($R=20$)	4.0	1.35	1.20

Table 2
Performance comparison of different pulse shapes used for spin inversion

[a] For all pulses, BW (fwhm) = 5 kHz [b] Relative RF energy, $E_{rel} \propto \int_0^{T_p} B_1^2(t) dt$

In the simulations of the four different pulses (HS, tanh/tan, sinc, and square), pulse parameters were set to produce a 5 kHz bandwidth (fwhm). The B_1^{max} value of each pulse was adjusted to produce 95% inversion ($M_z/M_0 = -0.9$) at the center of its inversion profile.

The resultant values of B_1^{max}, T_p, and the relative RF energy (E_{rel}) used with each pulse are listed in Table 2. It can be seen that the square pulse and the tanh/tan pulse require the same amount of RF energy. However, the square pulse requires B_1^{max} to be almost twofold greater than that required by the tanh/tan pulse, whereas the duration of the tanh/tan pulse is fourfold longer than the square pulse.

The shapes of the AM functions are shown in Fig. 3a. As opposed to modulating the pulse frequency, phase ϕ is usually modulated in practice. Figure 3b shows the $\phi(t)$ function of the sinc pulse, as well as the $\phi(t)$ functions of the HS and tanh/tan pulses, which were obtained from the integrals of Eqs. [6] and [9], respectively, and multiplying by the gyromagnetic ratio, γ. Figure 3c shows the theoretical inversion profiles generated with these pulses. As can be seen, the tanh/tan and square pulses have nearly identical inversion profiles, whereas the frequency selectivity of the HS pulse is superior to all the others.

The tolerance of the pulses to changes in B_1^{max} are shown in Figs. 4 and 5. With a conventional pulse (Figs. 4a and 5a), the shape of the spectral profile is severely affected by changes in B_1^{max}. With the AFP pulses (Figs. 4b and 5b), near perfect inversion continues to be produced despite increases in B_1^{max}. The specific choice of modulation functions used in the AFP determines whether the inversion profile broadens (Fig. 4b) or remains essentially unchanged (Fig. 5b) as B_1^{max} increases.

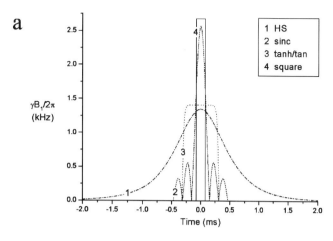

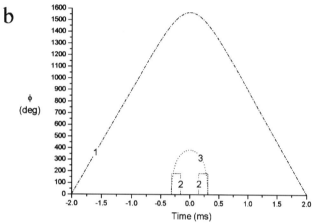

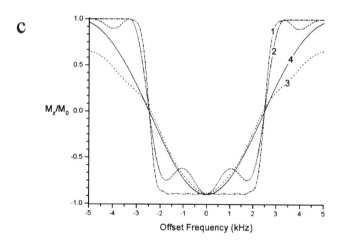

Figure 3

Plots of (a) $B_1(t)$, (b) $\phi(t)$, and (c) M_z/M_0 profiles of four different RF pulses: (1) HS, (2) sinc, (3) tanh/tan, and (4) square pulse. Pulse parameters are given in Table 2 and were set to produce a 5 kHz bandwidth (fwhm) and 95% inversion ($M_z/M_0 = -0.9$) at the center frequency.

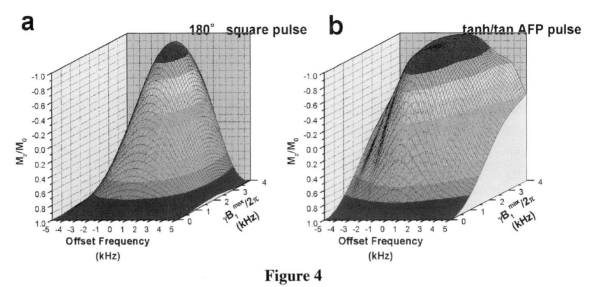

Figure 4

Three-dimensional plots of M_z/M_0 as a function of frequency offset and peak RF amplitude ($\gamma B_1^{max}/2\pi$) produced with the non-selective pulses: (a) square pulse (T_p=0.16 ms) and (b) tanh/tan pulse (T_p=0.64 ms).

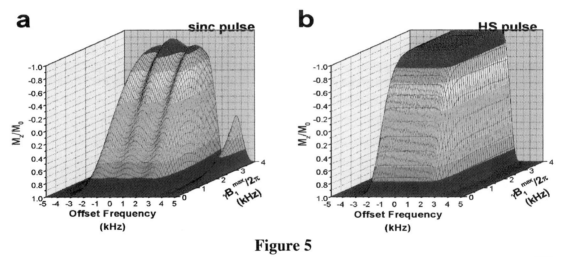

Figure 5

Three-dimensional plots of M_z/M_0 as a function of frequency offset and peak RF amplitude ($\gamma B_1^{max}/2\pi$) produced with the frequency-selective pulses: (a) sinc pulse (T_p=0.94 ms) and (b) HS pulse (T_p=4 ms).

Discussion

In some types of high field NMR experiments, it may be difficult to achieve broad bandwidths with conventional RF pulses due to their bandwidth dependence on peak RF field (B_1^{max}). Inversion pulses that are based on the principles of adiabatic rapid-passage offer a way to circumvent this problem, since their bandwidth is limited by the RF energy contained in the pulse, not the RF amplitude. The ability of an AFP to invert magnetization uniformly across wide bandwidths with arbitrarily low B_1^{max} values has recently been exploited to achieve broadband spin decoupling with minimal sample heating (15-20). Other applications can also benefit in a similar way. For example, we have shown how a single long AFP pulse with a linear frequency sweep can be used to generate multislice inversion in MRI (21). When applying such a pulse in the presence of a gradient, the "time of inversion" is a linear function of position along the gradient direction, which is exactly what is required for multislice inversion-recovery and related imaging sequences (e.g., MDEFT (22)). With this adiabatic approach, the amount of RF energy needed to invert any number of slices is equivalent to that required by a *single* AFP pulse.

Another advantage of adiabatic pulses is their ability to tolerate spatial variation of B_1^{max} (i.e., B_1 inhomogeneity). At the high frequencies used in high field NMR, conditions become favorable for dielectric resonances which can have detrimental effects on B_1 homogeneity (23). With conventional pulses, B_1 inhomogeneity causes flip angle errors and spectral artifacts. Adiabatic pulses can avoid these problems.

In this work, we have shown how the choice of AM and FM functions can profoundly influence the performance of AFP pulses. Certain types of experiments demand the use of RF pulses that are able to rotate spins uniformly across broad spectral bandwidths with sufficiently short duration to avoid significant decay by T_2 relaxation. For fast broadband inversion, we have developed an adiabatic pulse with AM and FM functions based on hyperbolic tangent and tangent functions, respectively (14). These modulation functions were derived using a numerical optimization procedure (4, 7). We have previously demonstrated the utility of the tanh/tan pulse for *in vivo* heteronuclear decoupling (20) and spectral editing (20, 24). At the threshold value of B_1^{max} where this pulse achieves satisfactory inversion ($M_z/M_0 \leq -0.9$), the required RF energy is nearly identical to that needed to produce the same degree of inversion and bandwidth with an ordinary square pulse. For a range of B_1^{max} values above this threshold, the tanh/tan pulse continues to produce complete inversion ($M_z/M_0 \approx -1$), whereas the inversion produced with an ordinary pulse is proportional to $\cos(\gamma B_1^{max} T_p)$. Thus, for many applications that require non-selective spin inversion, the adiabatic pulse should be used instead of a simple square pulse, since the RF energy deposited is the same (when using the minimum B_1^{max} value needed to achieve inversion with the adiabatic pulse) and the adiabatic pulse has the additional feature of continuing to produce inversion in spatial locations that experience a larger B_1^{max} value (e.g., in the regions closest to the RF coil).

Other applications require frequency-selective inversion with sharp borders demarcating the edges of the bandwidth. In certain cases, only one edge of the profile is required to have an abrupt border. For such purposes, the width of this transition region can be narrowed using an asymmetric AFP pulse (11, 25, 26). We recently constructed such a pulse from a combination of tanh/tan and HS pulses and used it to refocus NH groups while avoiding interference from water signal in HSQC experiments (27). This asymmetric pulse may also have advantages in MRI sequences that utilize outer-volume suppression (28-31) or measure blood flow (32).

Acknowledgments

This work was supported by NIH Grant RR08079. The author is grateful for many valuable discussions with Drs. T.-L. Hwang, P.C.M. van Zijl, A. Tannus, and L. DelaBarre.

References

1. Kupce, E., and Freeman, R. Optimized adiabatic pulses for wideband spin inversion. J. Magn. Reson. A, *118:* 229-303, 1996.
2. Tannús, A., and Garwood, M. Improved performance of frequency-swept pulses using offset-independent adiabaticity. J. Magn. Reson. A, *120:* 133-137, 1996.
3. Baum, J., Tycko, R., and Pines, A. Broadband and adiabatic inversion of a two-level system by phase-modulated pulses. Phys. Rev. A, *32:* 3435-3447, 1985.
4. Ugurbil, K., Garwood, M., and Rath, A. Optimization of modulation functions to improve insensitivity of adiabatic pulses to variations in B_1 magnitude. J. Magn. Reson., *80:* 448-469, 1988.
5. Johnson, A.J., Garwood, M., and Ugurbil, K. Slice selection with gradient-modulated adiabatic excitation despite the presence of large B_1 inhomogeneities. J. Magn. Reson., *81:* 653-660, 1989.
6. Town, G., and Rosenfeld, D. Analytic solutions to adiabatic pulse modulation functions optimized for inhomogeneous B_1 fields. J. Magn. Reson., *89:* 170-175, 1990.
7. Garwood, M., and Ugurbil, K., *B_1 insensitive adiabatic RF pulses*, in *NMR Basic Principles and Progress*, J. Seelig, and M. Rudin, Editors. 1991, Springer-Verlag: Berlin. p. 109-147.
8. Skinner, T.E., and Robitaille, P.-M.L. General solutions for tailored modulation profiles in adiabatic excitation. J. Magn. Reson., *98:* 14-23, 1992.
9. Ordidge, R.J., Wylezinska, M., Hugg, J.W., Butterworth, E., and Franconi, F. Frequency offset corrected inversion (FOCI) pulses for use in localized spectroscopy. Magn. Reson. Med., *36:* 562-566, 1996.
10. Tannús, A., and Garwood, M. Adiabatic pulses. NMR Biomed., *10:* 423-434, 1997.
11. Rosenfeld, D., Panfil, S.L., and Zur, Y. Design of selective adiabatic inversion pulses using the adiabatic condition. J. Magn. Reson., *129:* 115-124, 1997.
12. Slichter, C.P. Principles of magnetic resonance. (eds.), Springer Series in Solid-State Sciences, pp. New York: Springer-Verlag, 1990.
13. Silver, M.S., Joseph, R.I., and Hoult, D.I. Highly selective p/2 and p pulse generation. J. Magn. Reson., *59:* 347-351, 1984.
14. Hwang, T.-L., van Zijl, P.C.M., and Garwood, M. Fast broadband inversion by adiabatic pules. J. Magn. Reson., *133:* 200-203, 1998.
15. Luyten, P.R., Bruntink, G., Sloff, F.M., Vermeulen, J.W.A.H., van der Heijden, J.I., den Hollander, J.A., and Heerschap, A. Broadband proton decoupling in human ^{31}P NMR spectroscopy. NMR Biomed., *1:* 177-183, 1989.

16. Bendall, M.R. Broadband and narrowband spin decoupling using adiabatic spin flips. J. Magn. Reson. A, *112:* 126-129, 1995.
17. Kupce, E., and Freeman, R. Adiabatic pulses for wideband inversion and broadband decoupling. J. Magn. Reson. A, *115:* 273-276, 1995.
18. Fu, R., and Bodenhausen, G. Ultra-broadband decoupling. J. Magn. Reson. A, *117:* 324-325, 1995.
19. Hwang, T.-L., Garwood, M., Tannús, A., and van Zijl, P.C.M. Reduction of sideband intensities in adiabatic decoupling using modulation generated through adiabatic R-variation (MGAR). J. Magn. Reson. A, *121:* 221-226, 1996.
20. Pfeuffer, J., Tkac, I., Choi, I.-Y., Merkle, H., Ugurbil, K., Garwood, M., and Gruetter, R. Localized in vivo 1H NMR detection of neurotransmitter labeling in rat brain during infusion of [1-^{13}C]D-glucose. Magn. Reson. Med., *in press:* 1999.
21. Tannús, A., and Garwood, M. (1996) *4th Scientific Meeting of the International Society for Magnetic Resonance in Medicine.* (Abstr.), 362.
22. Ugurbil, K., Garwood, M., Ellermann, J., Hendrich, K., Hinke, R., Hu, X., Kim, S.-G., Menon, R., Merkle, H., Ogawa, S., and Salmi, R. Initial experiences at 4 Tesla. Magn Reson Quarterly, *9:* 259-277, 1993.
23. Barfuss, H., Fischer, H., Hentschel, D., Ladebeck, R., Oppelt, A., and Wittig, R. In vivo magnetic resonance imaging and spectroscopy of humans with a 4T whole-body magnet. NMR Biomed., *3:* 31-45, 1990.
24. Terpstra, M., Gruetter, R., High, W.B., Mescher, M., DelaBarre, L., Merkle, H., and Garwood, M. Lactate turnover in rat glioma measured by in vivo nuclear magnetic resonance spectroscopy. Cancer Res., *58:* 5083-5088, 1998.
25. Rosenfeld, D., Panfil, S.L., and Zur, Y. Analytic solutions of the Bloch equation involving asymmetric amplitude and frequency modulations. Phys. Rev. A, *54:* 2439-2443, 1996.
26. Rosenfeld, D., Panfil, S.L., and Zur, Y. Design of adiabatic pulses for fat-suppression using analytic solutions of the Bloch equation. Magn. Reson. Med., *37:* 793-801, 1997.
27. Hwang, T.-L., van Zijl, P.C.M., and Garwood, M. Asymmetric adiabatic pulses for NH selection. J. Magn. Reson., *in press:* 1999.
28. Singh, S., Rutt, B.K., and Henkelman, R.M. Projection presaturation: a fast and accurate technique for multidimensional spatial localization. J Magn Reson, *87:* 567-583, 1990.
29. Shungu, D.C., and Glickson, J.D. Band-selective spin echoes for in vivo localized ^{1}H NMR spectroscopy. Magn. Reson. Med., *32:* 277-285, 1994.
30. Luo, Y., Tannús, A., and Garwood, M. (1995) *3rd Annual Meeting of the Society of Magnetic Resonance and 12th Annual Meeting of the ESMRMB.* (Abstr.), 1017.
31. Tkac, I., Starcuk, Z., Choi, I.-Y., and Gruetter, R. In vivo ^{1}H NMR spectroscopy of rat brain at 1 msec echo time. Magn. Reson. Med., *in press:* 1999.
32. Edelman, R.R., Siewert, B., Darby, D.G., Thangaraj, V., Nobre, A.C., Mesulam, M.M., and Warach, S. Qualitative mapping of cerebral blood flow and functional localization with echo-planar MR Imaging and signal targeting with alternating radio frequency. Radiology, *192:* 513-520, 1994.

Chapter II.4

An Efficient Method for ICS Analysis of Functional Magnetic Resonance Imaging

Kiyotaka Suzuki[a,b,1], Tohru Kiryu[b], and Tsutomu Nakada[a,c]

[a]*Department of Integrated Neuroscience, Brain Research Institute, University of Niigata*
[b]*Graduate School of Science and Technology, University of Niigata*
[c]*Department of Neurology, University of California, Davis*

Introduction

The analysis of functional magnetic resonance imaging (fMRI) data is one of the most promising applications of independent component analysis (ICA). Some of the spatially independent maps estimated by ICA correspond to independent brain functional units (McKeown et al., 1998). However, the problem of how to extract maps of physiological interest out of a vast number of statistically independent components remains. An fMRI procedure combining sequential epoch analysis (SEA) with ICA proposed to address this has been named *independent component - cross correlation - sequential epoch (ICS) analysis* (Nakada et al., in press). The straightforward way of performing ICS analysis is to calculate the correlation of time series with a reference pattern for all of the independent maps estimated by ICA beforehand. This is a quite inefficient way because only one or two percent of components are significant, yet several days are often required to perform ICA for the complete data set of a single fMRI experiment even with the latest high-performance workstation. In this chapter, we describe an optimized technique for performing ICS analysis that incorporates the optimal initial parameters with the deflation approach of the fixed-point ICA algorithm (Hyvärinen, 1999). Our procedure remarkably improves the computational efficiency of ICS analysis since the physiological components can be selectively and individually calculated.

This chapter is reprinted from *Proceedings of ICA2000, pp. 309-314*, with some modifications.

[1] Correspondence: Kiyotaka Suzuki, Department of Integrated Neuroscience, Brain Research Institute, University of Niigata, Niigata 951-8585, Japan, Tel: (81)-25-227-0681, Fax: (81)-25-227-0822, e-mail: ksuzuki@bri.niigata-u.ac.jp

Theoretical Background

Statistical Parametric Mapping (SPM) as a Contrastive Method

A common approach to analyzing fMRI data is to test the statistical significance of a correlation between an analytical model and the time series of every voxel. An activation map is presented as a contrast between two specified conditions. The contrast is given by a set of weights for the basis functions constituting a specific analytical model. The activation areas are constructed by the voxels that satisfy the predetermined criteria (e.g. $p < 0.01$) for statistics yielded to a given contrast. Statistical parametric mapping (SPM) (Friston et al., 1995) is an analytical tool that utilizes this type of hypothesis-driven approach and is frequently applied to fMRI and PET research. Although this type of method is useful for various fMRI studies, the functional resolution is insufficient for some studies. The power of hypothesis-driven methods is limited mainly by the following: (1) A set of basis functions representing signal components should be determined *a priori* so that it satisfies the completeness to measured data. However, fMRI data include various signal components besides task-related signals such as physiological noise, head movements, and measurement noise, and accurately assuming all of such signal components is usually impracticable. Consequently, analytical errors are inevitable. (2) Similar bases share weights, and are not separated into different physiological components in principle.

Independent Component Analysis (ICA)

Independent component analysis (ICA) is a novel statistical technique that can solve blind source separation (BSS) problems, and it should overcome the limitation of hypothesis-driven methods described above, since it is data-driven. Several algorithms have been proposed for performing ICA (Jutten and Herault, 1991, Burel, 1992, Pham et al., 1992, Cichocki, et al., 1994, Bell and Sejnowski, 1995, Amari et al., 1996, Belouchrani et al., 1997, Cardoso, 1997, Girolami and Fyfe, 1997, Karhunen et al., 1997, Oja, 1997, Hyvärinen and Oja, 1998, Hyvärinen, 1999, Lee et al., 1999). In parallel, the applications can be extended to research and industrial projects such as data communication, speech recognition, and medical sciences.

The discrete model of blind source separation (BSS) problems with a time-invariant linear mixture is represented as the transform

$$\mathbf{x}(n) = \mathbf{A}\mathbf{s}(n), \quad n = 1, \cdots, N \qquad (1)$$

where n is the time index, $\mathbf{s}(n) = \{s_1(n), \cdots, s_K(n)\}^T$ is a set of unknown source signals, $\mathbf{x}(n) = \{x_1(n), \cdots, x_K(n)\}^T$ is a set of observations, and $\mathbf{A} \in \mathbf{R}^{K \times K}$ is an unknown mixing matrix. The goal of ICA is to recover both the mixing matrix and source signals from observations without knowing their properties. The only assumption is that source signals are mutually independent. In practice, ICA seeks for an unmixing matrix $\mathbf{w}$ so that the vector

$$\mathbf{y}(n) = \mathbf{W}\mathbf{x}(n) \qquad (2)$$

is an estimate of source signals $s(n)$, except for a permutation, signs and amplitudes. In the learning process of ICA, the unmixing matrix (or transformation) $\mathbf{W}$ is iteratively updated to minimize dependence among the elements of $\mathbf{y}$. Most of the algorithms estimate the whole of $\mathbf{W}$ at the same time. The fixed-point deflation algorithm (Hyvärinen, 1999) is an exception that extracts independent components one-by-one, and thus is suitable for ICS analysis. In the fixed-point deflation ICA, an approximation of differential entropy is used for the measure of dependence. The contrast function to find one component is given by

$$J_G(\mathbf{w}) = [E\{G(\mathbf{w}^T\mathbf{x})\} - E\{G(v)\}]^2 \qquad (3)$$

where $E\{\cdot\}$ denotes the expectation operator, $\mathbf{w} = (w_1, \cdots, w_K)^T$ is a weight vector under the constraint $E\{(\mathbf{w}^T\mathbf{x})^2\} = 1$, which is a certain row vector of $\mathbf{W}$, G is a non-quadratic function, and v is a normalized Gaussian variable. A single independent component, $y = \mathbf{w}^T\mathbf{x}$, would be found at a maximum of the function $J_G(\mathbf{w})$. If the data has been sphered (decorrelated up to the second-order moment) in advance, the maxima of $J_G(\mathbf{w})$ are obtained at optima of $E\{G(\mathbf{w}^T\mathbf{x})\}$. The data can be sphered by such a classic computational method as principal component analysis (PCA) or singular value decomposition (SVD). The optima of $E\{G(\mathbf{w}^T\mathbf{x})\}$ are solutions of the following equation:

$$\begin{aligned} E\{\mathbf{x}G'(\mathbf{w}^T\mathbf{x})\} - \beta\mathbf{w} &= 0 \\ \beta &= E\{\mathbf{w}_0^T\mathbf{x}G'(\mathbf{w}_0^T\mathbf{x})\} \end{aligned} \qquad (4)$$

where G' is the derivative of G with respect to $y (= \mathbf{w}^T\mathbf{x})$ and $\mathbf{w}_0$ is the value of $\mathbf{w}$ at the optimum. Newton's method can be applied to derive the following fixed-point algorithm:

$$\begin{aligned} \mathbf{w}^+ &= E\{\mathbf{x}G'(\mathbf{w}^T\mathbf{x})\} - E\{G''(\mathbf{w}^T\mathbf{x})\}\mathbf{w} \\ \mathbf{w}^* &= \mathbf{w}^+ / \|\mathbf{w}^+\| \end{aligned} \qquad (5)$$

where $\mathbf{w}^*$ is the updated value of $\mathbf{w}$. Our technique uses (5) as its core.

An ICA Model of fMRI Data

The ICA model (1) assumes that the number of source signals (or channels) K is less than the number of samples in a time domain N. It is impossible to apply the model (1) directly to fMRI data because the size of time series is usually much less than the number of channels, or voxels. To circumvent this problem, McKeown et al. proposed transposing the data matrix in order to allow ICA to estimate *spatially independent* components. Then the data model for fMRI is represented by

$$\mathbf{x}(k) = \mathbf{A}\mathbf{s}(k), \quad k = 1, \cdots, K \qquad (6)$$

where k is the index of pixels, $\mathbf{s}(k) = \{s_1(k), \cdots, s_N(k)\}^T$ is a set of spatially independent images, and $\mathbf{x}(k) = \{x_1(k), \cdots, x_N(k)\}^T$ is a series of acquired images. We call each element of $\mathbf{s}$ *an independent map*. Each column vector of matrix $\mathbf{A}$ then represents a time series of the corresponding map, which can be referred to determine whether or not a map is of interest. Figure 1 is a pictorial expression of (6). It is the value of using ICA that the sets of a map and the associated time series are obtained without any *a priori* information. However, some criteria for selecting significant components should be defined for each fMRI experiment.

Figure 1

fMRI data model

Sequential Epoch Analysis (SEA)

The basic idea of state-related fMRI analyses is subtraction on a specified combination of two states for data of a multiple-state experiment. We describe here the human hand motion paradigm used in [23] as an example to explain the idea in brief. Figure 2 shows the paradigm schematically. The letters r, R, L, and B correspond to rest, right hand motion, left hand motion, and bilateral hand motion, respectively. The duration of each segment (or epoch) is fixed to 30 seconds. The inter-scan interval is one second for each slice. The following three contrasts are considered fundamental for the paradigm: $(R - r)$, $(L - r)$, and $(B - r)$.

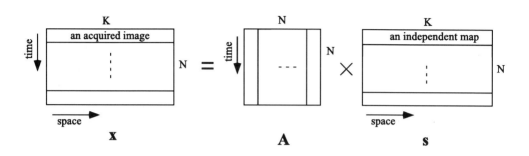

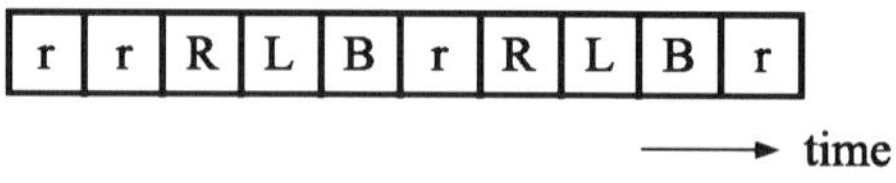

Figure 2

Sequential epoch hand motion paradigm

Sequential epoch analysis (SEA) was proposed to elucidate the specific functional area of interest by combining the statistical images of fundamental contrasts into a single SEA display (Nakada et al., 1998). This analysis has the advantage of identifying brain functions over conventional state-related analyses. The hand motion paradigm has six possible SEA patterns as shown in Fig. 3 (a): (A) right hand M1, (B) left hand M1, (C) non-specific, (D) right hand exclusive, (E) left hand exclusive, and (F) bilateral exclusive. Note that each of these SEA patterns should be interpreted, in its original sense, as the way of combining three resultant maps of fundamental contrasts, not as the contrast weights for the model bases. The concept of SEA is schematically expressed in Fig. 3 (b).

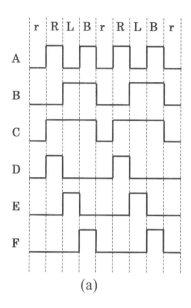 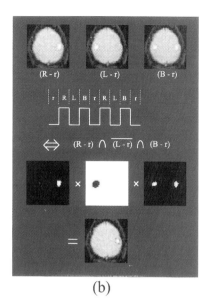

Figure 3

(a) Possible SEA patterns and (b) a schema for the concept of SEA

Concept of ICS Analysis

Since SEA patterns were actually observed in the raw fMRI data of our sequential epoch paradigm with an acceptable level of pixel misalignment using a high-field (3 Tesla) MRI system, Nakada et al. proposed ICS analysis that uses an SEA pattern for the reference time series with intent to extract activation maps out of spatially independent maps obtained by ICA. This technique incorporates a hypothesis-driven approach (correlation with SEA patterns) into the data-driven analysis (ICA). The results of ICS analysis showed that the physiological resolution of fMRI was considerably increased compared with SPM.

The correlation of a time series with an SEA pattern is calculated to identify the activation maps by a given threshold. The time series vector of a map can be denoted by $\mathbf{a}=(a_1,\cdots,a_N)$ and one of the SEA patterns by $\mathbf{b}=(b_1,\cdots,b_N)$. If these vectors are normalized so that $\sum a_i = \sum b_i = 0$ and $\|\mathbf{a}\|^2 = \|\mathbf{b}\|^2 = 1$, the correlation r between these two vectors is simply represented by their inner product:

$$r = \mathbf{a} \cdot \mathbf{b} = \sum_i a_i b_i \ . \tag{7}$$

Only around two percent of the components were selected by the empirical condition, $r \geq 0.7$, for all the SEA patterns of our hand motion paradigm. This led us to exploit an efficient method for performing ICS analysis.

A Direct Way of Component Extraction

We discuss here a direct way of extracting the activation maps. The time series vector $\mathbf{a}$

of a single independent map estimated by the fixed-point ICA (5) is given by a psuedo-inverse of the corresponding unmixing vector $\mathbf{w}$. Since $\mathbf{w}$ is normalized to have unit norm $\|\mathbf{w}\|^2 = \mathbf{w}^T\mathbf{w} = 1$, $\mathbf{a}$ is equal to $\mathbf{w}^T$, and the nearest $\mathbf{w}$ to a given initial vector is obtained by (5). Thus the relationship between the correlation and the distance of two vectors should be examined first. Euclidean distance, or the squared norm of difference of two vectors, is used to measure the distance d:

$$d = \|\mathbf{a} - \mathbf{b}\|^2 \equiv \sum_i (a_i - b_i)^2. \tag{8}$$

If the effects of sphering are not taken into account, the distance (8) is directly related to correlation r in (7) according to the following derivation:

$$\begin{aligned} d &= \sum_i (a_i - b_i)^2 \\ &= \sum_i (a_i^2 + b_i^2) - 2\sum_i a_i b_i \\ &= 2(1 - r) \end{aligned} \tag{9}$$

This linear relationship laid the foundation of our optimized technique. Figure 4 plots the relationships between r and d observed for the right hand M1 pattern (denoted by A in Fig. 3a) in our hand motion study as circles together with the theoretical relationship (9) shown by a solid line.

Sphering should now be considered. The Jacobian matrix of the left side of (4) becomes diagonal if the data are sphered. Therefore, the Euclidean distance can still be a practical measure of closeness. A transfer matrix $\mathbf{B}$ for sphering is estimated by PCA as

$$\mathbf{B} = \mathbf{D}^{-1}\mathbf{E} \tag{10}$$

where $\mathbf{E}$ denotes the (transposed) matrix of eigenvectors of $\overline{\mathbf{xx}^T}$ and $\mathbf{D}$ denotes the diagonal matrix whose elements are the square root of the corresponding eigenvalues. Sphered data χ and the unmixing vector ω for one component in a sphered space are given by

$$\chi = \mathbf{B}\mathbf{x}$$
and
$$\omega = (\mathbf{w}^T \mathbf{B}^{-1})^T, \tag{11}$$

respectively. A time series $\mathbf{a}$ in the original data space relates to ω as

$$\mathbf{a} = (\mathbf{B}^{-1}\omega)^T. \tag{12}$$

As expressed in (11), the relationship (9) is generally broken in the space of ω. Because $\mathbf{E}$ is an orthogonal matrix, distance is preserved by the transformation of $\mathbf{E}$ alone. It is $\mathbf{D}$ that disperses the relationship by multiplying imbalance weights with the elements of ω. However, components having a high correlation should still have a relatively short distance. This was validated by the actual relationships observed in the sphered data of our hand motion study shown in Fig. 5. The observations support the notion that a normalized SEA pattern can be used as an appropriate initial value of $\mathbf{w}$ for the criterion, $r \geq 0.7$.

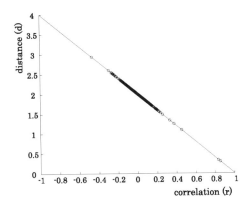
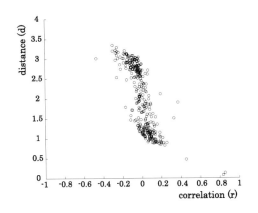

Figure 4
Correlation vs. distance relationship observed in the original ($\mathbf{w}$) space.

Figure 5
Correlation in the original ($\mathbf{w}$) space vs. distance in the sphered ($\boldsymbol{\omega}$) space.

The following procedure is proposed for extracting one every desired component related to a (normalized) target SEA pattern $\mathbf{b}$.

1. Remove the bias of time series from each voxel.

2. Sphere the data.

3. Prepare the initial vector $\boldsymbol{\omega}_{init}$ so that

$$\boldsymbol{\omega}_{init} = \left(\mathbf{Bb}^T\right)^T$$
$$subject\ to\ \|\boldsymbol{\omega}_{init}\| = 1\ .$$

4. Execute the iteration (5) with its initial value $\boldsymbol{\omega}_{init}$ to calculate the single unmixing vector $\boldsymbol{\omega}_0$.

5. Recover the raw time series $\mathbf{a} = (\alpha_1, \cdots, \alpha_N)$ of an estimated component by using (12) and put the normalized vector into $\mathbf{a}$.

6. Calculate the correlation r between $\mathbf{a}$ and $\mathbf{b}$ using formula (7).

7. If $r \geq 0.7$, then the component $y = \boldsymbol{\omega}_0^T \mathbf{Bx}$ is considered an activation map of the target SEA pattern. To search the next component, subtract its time series $(\alpha_1 y, \cdots, \alpha_N y)^T$ from the original data, then repeat from step 2. If $r < 0.7$, then stop searching.

Results

An fMRI experiment was conducted to examine the validity of our method. A right-handed normal volunteer (18-year) participated in our sequential epoch hand motion study. Gradient echo echo-planar images (EPI) were obtained using a General Electric

SIGNA 3.0 Tesla system equipped with an Advanced NMR EPI module. The following parameters were used for data acquisition: FOV 40 × 20 cm; matrix 128 × 64; slice thickness 5 mm; TR 1 sec. Spatial smoothing was applied by convolving with a 5 mm full width at half maximum (FWHM) Gaussian kernel to minimize the effects of pixel misalignment due to brain motion during the experimental session. The optimized ICS technique was tested for the right hand M1 pattern. The third power was selected for the non-linear transfer function G' in (4) through our experience. The first thirty images of the fMRI time series were not used for the analysis to avoid the effects of initial decay. Figure 6 shows all of the extracted maps and the associated time series with their correlation values. The broken line in the time series represents the normalized SEA pattern taken for reference.

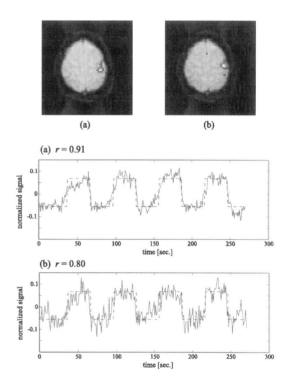

Figure 6

Activation maps and the associated time series extracted by the optimized ICS analysis.

The estimated components were precisely physiological independent units. The two time series are so similar that these cannot be separated into different physiological components by model-based approaches such as SPM. The time efficiency of performing ICS was remarkably improved by our technique. A Sun Microsystems Ultra 60 workstation required only a few minutes to extract all of the significant components.

Concluding remarks

We proposed a validated and efficient technique for ICS analysis that extracts all significant activation maps one-by-one from measured data. The direct search was achieved by incorporating appropriate initial parameters into the iteration rule of fixed-point ICA. Our investigations into the relationship between correlativity in a time domain and distance in the parameter space of ICA ensure the iteration process of convergence upon a desired target. The optimized technique can be utilized generally for fMRI analyses as far as a temporal reference function can be set for physiological units that have an objective task-related function. We emphasize that our technique embodies the concept of ICS analysis itself in its algorithm.

Acknowledgement

The study was supported by grants from the Ministry of Education (Japan).

References

1. Amari, S., Cichocki, A., Yang, H.H., "A new learning algorithm for blind signal separation," *Advances in Neural Information Processing* **8**, 757-763, 1996
2. Bell, A.J., Sejnowski, T.J., "An information- maximization approach to blind separation and blind deconvolution," *Neural Computation* **7**, 1129-1159, 1995
3. Belouchrani, A., Abed-Meraim, K., Cardoso, J.-F., Moulines, E., "A blind source separation technique using second-order statistics," *IEEE Trans. Signal Processing* **45 (2)**, 1997
4. Burel, G., "Blind separation of sources: a nonlinear neural algorithm," *Neural Networks* **5**, 937-947, 1992
5. Cardoso, J.-F., "Infomax and maximum likelihood for blind source separation," *IEEE Signal Processing Letters* **4 (4)**, 1997
6. Cardoso, J.-F., "Blind Signal Separation: Statistical Principles," *Proc. IEEE* **86 (10)**, 2009-2025, 1998
7. Cichocki, A., Unbehauen, R., Rummert, E., "Robust learning algorithm for blind separation of signals," *Electronics letters* **30 (17)**, 1994
8. Comon, P., "Independent component analysis, a new concept?," *Signal Processing* **36**, 287-314, 1994
9. Friston, K.J., Holmes, A.P., Worsley, K.J., Poline, J.P., Frith, C.D., Frackowiak, R.S.J., "Statistical parametric maps in functional imaging: a general linear approach," *Human Brain Mapping* **2**, 189-210, 1995
10. Friston, K.J., Holmes, A.P., Poline, J.P., Grasby, P.J., Williams, S.C.R., Frackowiak, R.S.J., Turner, R., "Analysis of fMRI time series revisited," *Neuroimage* **2**, 45-53, 1995
11. Girolami, M., Fyfe, C., "Stochastic ICA contrast maximisation using Oja's nonlinear PCA algorithm," *International Journal of Neural Systems* **8**, 661-678, 1997
12. Hyvärinen, A., Oja, E., "Independent component analysis by general nonlinear Hebbian-like learning rules," *Signal Processing* **64**, 301-313, 1998
13. Hyvärinen, A., "Fast and robust fixed-point algorithms for independent component analysis," *IEEE Trans. Neural Networks* **10 (3)**, 626-634, 1999
14. Jung, T.-P., Humphries, C., Lee, T.-W., Makeig, S., McKeown, M.J., Iragui, V., Sejnowski, T.J., "Extended ICA removes artifacts from electroencephalographic recordings," *Advances in Neural Information Processing Systems* **10**, 894-900, 1998
15. Jutten, C., Herault, J., "Blind separation of sources, Part I: an adaptive algorithm based on neuro- mimetic architecture," *Signal Processing* **24**, 1-10, 1991
16. Karhunen, J., Oja, E., Wang, L., Vigário, R., Joutsensalo, J., "A class of neural networks for independent component analysis," *IEEE Trans. Neural Networks* **8 (3)**, 486-504, 1997
17. Lee, T.-W., Girolami, M., Sejnowski, T.J., "Independent component analysis using an extended infomax algorithm for mixed subgaussian and supergaussian sources," *Neural Computation* **11 (2)**, 417-441, 1999

18. Makeig, S., Jung, T.-P., Bell, A.J., Ghahremani, D., Sejnowski, T.J., "Blind separation of auditory event-related brain responses into independent components," *Proc. Natl. Acad. Sci. USA* **94**, 10979-10984, 1997
19. McKeown, M.J., Makeig, S., Brown, G.G., Jung, T.-P., Kindermann, S., Bell, A.J., Sejnowski, T.J., "Analysis of fMRI by blind separation into independent spatial components," *Human Brain Mapping* **6**, 160-188, 1998
20. McKeown, M.J., Jung, T.-P., Makeig, S., Brown, G., Kindermann, S.S., Lee, T.-W., Sejnowski, T.J., "Spatially independent activity patterns in functional MRI data during the Stroop color-naming task," *Proc. Natl. Acad. Sci. USA* **95**, 803-810, 1998
21. McKeown, M.J., Sejnowski, T.J., "Independent component analysis of fMRI data: examining the assumptions," *Human Brain Mapping* **6**, 368-372, 1998
22. Nadal, J.-P., Parga, N., "Redundancy reduction and independent component analysis: conditions on cumulants and adaptive approaches," *Neural Computation* **9**, 1421-1456, 1997
23. Nakada, T., Fujii, Y., Suzuki, K., Kwee, I.L., "High-field (3.0 T) functional MRI sequential epoch analysis: an example for motion control analysis," *Neuroscience Research* **32**, 355-362, 1998
24. Nakada, T., Suzuki, K., Fujii, Y., Matsuzawa H., Kwee, I.L., "Independent component – cross correlation – sequential epoch (ICS) analysis of high-field fMRI time series," *Neuroscience Research* (accepted)
25. Oja, E., "The nonlinear PCA learning rule in independent component analysis," *Neurocomputing* **17**, 25-45, 1997
26. Pham, D.T., Garat, P., Jutten, C., "Separation of a mixture of independent sources through a maximum likelihood approach," *Signal Processing VI: Theories and Applications*, 771-774, 1992
27. Pham, D.T., "Blind separation of instantaneous mixture of sources via an independent component analysis," *IEEE Trans. Signal Processing* **44** *(11)*, 2768-2779, 1996

Chapter II.5

Recording the EEG during fMRI: Advantages and Disadvantages

John R. Ives [1]

Department of Neurology, Harvard Medical School

Introduction and History

Functional MRI (fMRI) is just emerging as a new technique for documenting blood flow changes correlated with neuronal activity (Belliveau et al., 1991; Kwong et al., 1992). FMRI consists of special software to generate specific sequences and special gradients to accommodate fast images within the environment of MRI. This scientific area is rapidly changing as new advances and application are being introduced almost daily. EEG on the other hand has been around as a clinical tool since the early 1930's, and is fairly well established. The MRI environment is a harsh environment with a strong static magnetic field, transient RF pulses, and changing magnetic gradients. Yet it is a very sensitive environment as well, where small amounts of magnetic material, or external RF signals can completely alter the images. It is also an environment where loose magnetic materials can be very dangerous. The first impression most investigators have when the question of recording the EEG in the MRI environment is raised, is that it cannot be done nor can it be accomplished without considerable image artifact and/or other substantial problems associated with patient safety due to wires, electrodes coils, heating, burning, etc. The second question sometimes asked is why would you want to do this any way?

During an informal discussion with Dr. Steven Warach in the summer of 1992, concerning the details of the state-of-the-art of fMRI and the fact that blood flow (BOLD) images could be obtained in 40 to 100ms, this author realized that if the EEG of the patient with epilepsy could be monitored while resting in the MRI magnetic, then one could wait until epileptic discharges occurred and then the BOLD images could be obtained. The fMRI BOLD results might correlate with the source of the patient's epileptic focus and then the results could be displayed on a high spatial resolution 3D anatomical image.

Inquires were made to various fMRI centers and to fMRI researchers, but no one had any experience with such a problem and the off-the-cuff response was that it could not be done because of preconceived, significant technology conflicts. Experience with multi-channel EEGs over the years since 1970 (Ives et al., 1995) permitted the simplification and

---

[1] Correspondence: John R. Ives, B.Sc., Associate Professor of Neurology, Harvard Medical School, Director of Neuroelectronics, Beth Israel Deaconess Medical Center, East Campus, Room GZ-522, 330 Brookline Avenue, Boston 02215, MA, USA, Tel: (1)-617-667-3509, Fax: (1)-617-667-7023, e-mail: jives@caregroup.harvard.edu

miniaturization of EEG recording systems so that it was possible to obtain high quality EEGs from almost any environment, and the MRI was just another challenge. I was also aware that there was very little if any magnetic material within the design of modern low powered, battery operated ambulatory EEG data acquisition systems. Even the EEG electrodes were constructed of pure metals such as silver and gold and had been already shown be compatible with images taken in 0.5T systems (Lufkin et al., 1988). With respect to "why" recording the EEG during fMRI studies was important, some groups, as well as ours (Warach et al., 1994; Jackson et al., 1995; Detre et al., 1995) had already demonstrated that in rare cases, where the patient's subtle clinical signs could be used to indicated whether they were involved in a seizure or not, and thus allowed the collection of both "on" and "off" states to permit comparison fMRI images to be taken.

Upon visiting our MRI department we were very pleased to find that a Siemens engineer (Mr. Franz Schmitt) was just finishing the installation of the beta-site Siemens Vision 1.5T machine. Since it was not as yet under hospital management, we were able to do some preliminary test, first on a phantom and then on a EEG recording directly from Dr. Warach's scalp.

The first experiment was to bring a Grass model-8 EEG machine into the MRI shielded room and record directly from extended EEG leads. The results demonstrated that the MRI images were fine and unaffected by the Grass gold electrodes, but the EEG was very bizarre and not readable. We then installed a custom built cable-telemetry 16-channel EEG preamplifier/multiplexor with a matching demultiplexor between very short EEG leads and the Grass model-8 EEG machine. This system provided both very clean MRI images as well as a very readable EEG. We first disclosed the concept of recording the EEG in 1993 (Ives et al., 1993a), where we demonstrated the ability to record a subject's EEG during fMRI echo planar images. See figure 2 (Ives et al., 1993b) as this is the first human EEG recorded from within an MRI system and this occurred on November 10, 1992. At about the same time our Institute decided to patent the technique and shortly thereafter licensed it to Neuro Scan Labs. We then went on to solve the next problem and that was to permit the EEG machine to be relocated from within the MRI shielded room to a location next to the MRI control console that made a more practical experimental set-up. Initially this required that we cross the shielded room with hard-wire cables and this introduced all sorts of problems with image quality, even when we ran cables through the RF wave-guides. This was solved by installing a fiber-optic link (Ives et al., 1994) to permit the analog multiplexed EEG signals to be transmitted via light and not hard-wire.

We were now in a position to submit an IRB and begin to study patients with epileptic discharges. The initial protocol allowed us to set up the desired fMRI BOLD sequence and then wait for two types of patterns in the EEG. The first was the epileptic marker in the EEG of the patient's usual interictal discharge and upon visual identification, the fMRI sequence was activated. The second was to wait for a period in the EEG when no interictal discharges were seen for at least 10 seconds and then to manually trigger the fMRI sequence. Because there is about a 2 second delay (Bandettini et al., 1992) between the BOLD response to a stimuli, manual triggering after "seeing" an event in the on-line EEG proved very effective. The former activity was then classified as positive, while the later sequences were classified as base-line images. Simplistic data manipulation then permitted both an average positive and an average base-line data set to be obtained and then these were simply subtracted from one another to obtain a resulting difference that then could be superimposed on the corresponding anatomical MRI image.

The three images shown in figure 1 illustrate the logistics, the raw data and the results of the original hypothesis and demonstrates that the technique is significant in contributing to the localization of epileptic activity. However, the combination of EEG and MRI is not an easy technique to master. As in our case as well as the situation at other institutions, the "team" approach of involving experts from both the fMRI and the EEG data acquisition field has proven very effective.

Clinical and Research Applications

The ability to record the patient's or subject's EEG while they are within the bore of an MRI unit, basically permits one to trigger functional BOLD sequences in relationship to the state of the EEG. With respect to patients with epilepsy, the BOLD sequence can be triggered whenever an epileptic event is seen in the EEG. On the other hand, one can also collect sequences when it is clear in the EEG that there are no epileptic events and thus permit the saving of comparative sequences that can serve as base line. One can also use this technique to monitor the patient or subject while they are sleeping in the MRI and thus trigger sequences whenever the on-line sleep EEG indicates that they are in any particular sleep stage that is of interest (Ives et al., 1997; Jakob et al., 1998). On the cognitive side, paradigms using both behaviour and EEG criteria or results can now be studied simultaneously. An example here would be to flash a light or any other stimuli and then collect the EP over a 100ms to 500ms period, then begin the BOLD sequence. This could be repeated until enough EP and/or fMRI averages have been obtained to define the desired response.

Artifact

When the EEG is recorded from the bore of a magnet, the most dominant, persistent artifact is the ballistocardiogram. This artifact is caused by micro-movements of the head and thus the electrodes and electrode wires in a strong magnetic field which subsequently causes current to flow and thus a voltage to be seen in the EEG amplifier related to the EKG, but shifted in phase. The source of this artifact is essentially mechanical as 20% of the blood, flows into the head and stops, while the rest makes a 180 degree turn in the aortic arch. Both these expenditures of kinetic energy causes micro-movement of the head and upper body. This ballistocardiogram artifact is very dependent on blood pressure, physical condition of the body, heart rate, etc. It is not possible to predict the amplitude of this artifact before going into the magnet, nor it's distribution. However, a very young, tall, slim, athletic person with a strong, big heart is going to cause the most problem.

Other factors affecting this artifact are the Tesla strength of the magnet, recording method, recording mode, head fixation, etc. The best montage to reduce this artifact is bipolar, while referential recordings are worst. The basic reason for this is in bipolar recordings, all electrodes are relatively equally spaced which contrasts to referential recordings where the electrode distance can vary significantly. Also important with referential recording is the selection of the reference electrode. Probably the best location is near the centroid of the other electrodes, i.e. Cz. Unfortunately, as one moves from 1.5T to 3.0T and 4.0T, the ballistocardiogram artifact goes up linearly. The fixation of the electrodes is also important, as collodion/gauze usually produces better results than simple paste on electrode application.

If the electrodes and the wires are wrapped securely to the head with a confining bandage, then this helps to stabilize the wires from micro-movement impacted on them by the head movements. It also helps to stabilize these electrode wires up to and including the preamplifier. This can be done using simple towels to cover the wires and preamplifier. Theoretically, ridged stabilization of the head coupled with loose stabilization of the body with an air mattress may help in that the ballistocardiogram energy may then dissipate in the body and not the head.

Eliminating as much ballistocardiogram from the original recorded EEG is the first approach, but under certain conditions this artifact my reduce one's ability to properly read the EEG. This can be deal with by using digital processing as out-lined by several groups (Allen et al., 1998). However, in EEG monitoring cases, this artifact can be significantly reduced at the front-end by using bipolar montages instead of referential montages.

The EEG is also sensitive to even the slightest movement and thus talking coughing, swallowing generates significant signals in the fMRI/EEG. A mechanical device sensitive to movement mounted on the forehead works very well to record these mechanical movements and thus permit confirmation of the artifact seen in the EEG as being mechanical movement and not epileptic discharges (Hill et al., 1995).

Recording Electrodes

The essential interface of the EEG is the electrode itself and its fixation to the scalp. With fMRI/EEG recordings, this becomes even more of a critical area. The type of material used has to be completely non ferrous. If the electrode demonstrates any artifact in the MRI, or the fMRI BOLD images, then it simply cannot be used. Thus tin electrodes, or any electrode containing any spring steel are impossible to use in this sensitive environment. We have found that pure silver, or gold plated pure silver electrodes work very well in this environment and produces very little, none or acceptable artifact under most MRI or fMRI BOLD sequences.

We have also had experience with pure silver sphenoidal electrodes as well in the 1.5T systems and they do no cause image problems, but may be more sensitive to ballistocardiogram artifact.

As one moves up in Tesla (3T and 4T) and/or increase the number of channels (32 and 64) and thus electrodes, then there is point where these silver, or gold-silver electrodes become a problems, thus other material and designs have been tried. In our experience (Ives et al., 1998) with recording the EEG during repetitive transcranial magnetic stimulation (rTMS), we have developed conductive-plastic electrodes to operate under rTMS conditions and thus prevent heating. Under test at 3T with 64 channels, the conductive-plastic electrodes produced less artifact than the silver electrodes with sensitive fMRI sequences (Bonmassar et al., 1999, in preparation). It was also subsequently found that these conductive-plastic electrodes seemed to be less sensitive to the ballistocardiogram artifact as well.

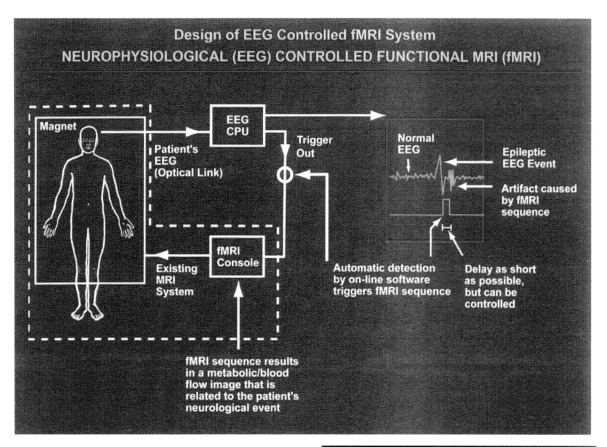

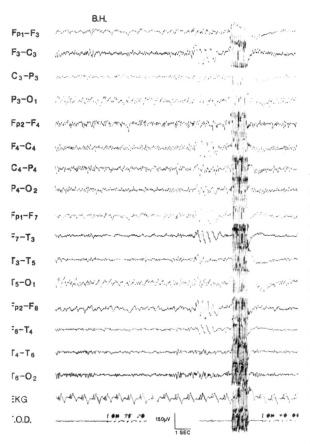

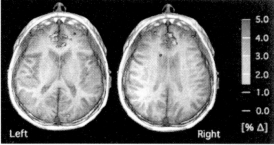

Figure 1

This composite figure summarizes the concept used to generate fMRI/BOLD blood flow correlation with the patient's epileptic discharges. The upper section outlines the block diagram of the process to manually trigger (as automatic triggering has not been introduced) the BOLD sequences when an epileptic event is seen in the patient's on-line EEG. The lower left shows the section of the patient's EEG exhibiting a generalized spike and wave discharge followed shortly there after (about 2 sec) by the artifact generated when the fMRI multi-slice BOLD sequence which was manually triggered. The results shown in the lower right were generated by comparing many base-line (no discharge) BOLD sequences to many active BOLD sequences. In summary, changes in blood flow can be demonstrated that are directly correlated to the patient's epileptic discharges.

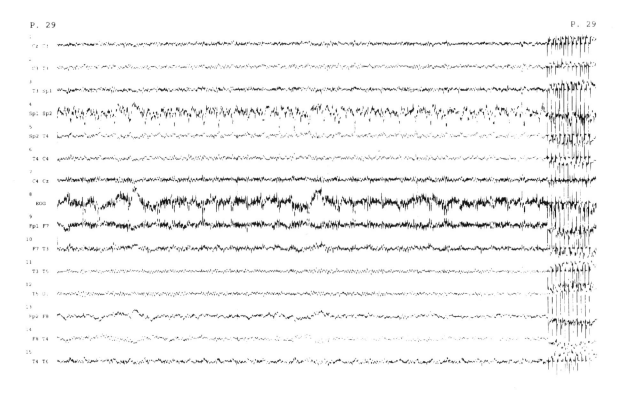

Figure 2

This bipolar, sphenoidal-temporal EEG also taken while the patient rested in the bore of a 1.5T magnet, shows a quiet base-line EEG until just after the 14:51:46 time mark, when an isolated left temporal lobe spike is seen. About 1.5 seconds after the spike, an fMRI BOLD multi-slice sequence was triggered as confirmed by the artifact in the EEG record. In fact, by the number of discrete points in the artifact, one can even determine that this was an eleven slice sequence. Also notable in the EEG is the balistocardiogram artifact that is fairly prominent in the Sp1-Sp2 (channel 4) montage, as this channel has higher impedance and consist of electrodes that are farther apart. The EOG (channel 8) also confirms the presence of the spike as it consist of surface electrodes located at the T1, T2 locations.

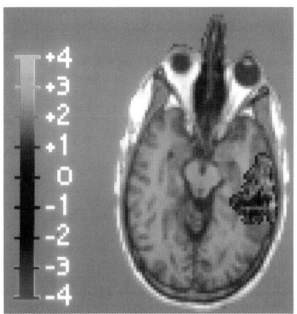

Figure 3

From the analysis of the interictal EEG data collected in figure 2, this figure (left is right, right is left) represents the fMRI/BOLD localization superimposed on an anatomical slice at the same level. However, in this example, there was no averaging done on the positive EEG events, all positive events like the one show were analyzed individually and looked at separately. This particular result was obtained by comparing the data obtained from this one sequence compared to an average of normal base-line sequences. In other words, a single, isolated, epileptic spike generates a blood flow pattern change that is sensitive to fMRI/BOLD analysis.

EEG Montages

As discussed, the better mode of recording is the bipolar rather than referential. The bipolar montage works well if one is just monitoring the EEG and waiting for something interesting to occur. However, the research area usually prefers referential based EEG/EP recordings. There is however, a compromise that works very well in the routine EEG area, but still needs to be shown to be applicable in the EEG/fMRI area. Under certain conditions, one can record the EEG in a specifically designed bipolar montage that subsequently permits off-line remontaging to a referential based recording (Ives et al., 1993c).

The Future of Clinical and Research Applications of EEG/FMRI Techniques

The future is always difficult to predict, as most advances are usually serendipitous like this one was as described above. However, after 7 years of experience, some future applications of EEG/fMRI can be suggested.

As most epileptic centers now routinely rely on 1.5T MRI system for critical anatomical localization data, and most MRI system are now capable of fMRI images, there is a technical base that may support the clinical application of fMRI images taken during the patient's spontaneous epileptic interictal activity. Therefore, the equipment needed to perform the study is more readily available than PET, or MEG localization tests. Its acceptance here also depends on several other factors: 1) the EEG/fMRI set up must be made far more routine. 2) The results must demonstrate that there is significant correlation to the patients epileptic focus and that surgical treatment (or the guidance for placement of invasive electrodes) is significantly aided by this extra effort to document the blood flow changes associated with the patient's interictal morphology. In figure 2, a multi-slice fMRI/BOLD image was taken about 2 seconds after a left temporal lobe spike. The results of analyzing this single event (and not an average of many events) is shown in figure 3. Here one can "see" (radiological presentation, left is right and right is left) the blood flow changes activated by a single transient epileptic discharge. Basically, the EEG/fMRI is very sensitive, although this is not too surprising to have results like this on such a subtle neurological event, but if single light flashes and finger movements show up in fMRI/BOLD studies, why not individual epileptic spikes. However, like PET, and MEG; EEG/fMRI is essentially an interictal localization technique and not an ictal based test as is long term EEG/Video monitoring.

As for research applications in the field of cognitive studies, I think the concept that EEG/EP data recorded simultaneously with fMRI/BOLD is more important or significant than data recorded separately from EEG/EP and later fMRI/BOLD during the application of the same paradigm is still under question. As a research tool, EEG/EP/fMRI studies usually demand high spatial resolution (32, 64, 128-channels), referential based recordings and move up in Tesla from 1.5T, 3T, 4T, and possibly higher. All these factors contribute to the degradation of the EEG and EP, and thus there may be a point were the results are not viable.

References

Allen PJ, Polizzi G, Krakow K, et al. Identification of EEG events in the MRI scanner: the problem of pulse artifact and a method for its subtraction. NeuroImage 1998; 8:29-239.

Bandettini PA, Wong EC, Hinks RS, et al. Time course EPI of human brain functioning during task activation. Magn Reson Med 1992; 25:390-397.

Belliveau JW, Kennedy DN, McKinstry RC, et al. Functional mapping of the human visual cortex by magnetic resonance imaging. Science 1991; 254:716-719.

Bonmassar G, Anami K, Ives J, Belliveau JW. Visual evoked potential (VEP) by simultaneous 64-channel EEG and 3T fMRI. Neuro Report (In preparation).

Detre JA, Sirvan JI, Alsop DC, et al. Localization of subclinical ictal activity by functional MRI: correlation with invasive monitoring. Ann Neurol 1995; 38:618-624.

Hill RA, Chiappa KH, Huang-Hellinger F, Jenkins BG. EEG recording during MR imaging: differentiation of movement artifact from paroxysmal cortical activity. Neurology 1995; 45:1942-1943.

Ives JR, Woods JF. 4-channel 24-hour cassette recorder for long term EEG monitoring of ambulatory patients. Electroenceph Clin Neurophysiol 1975; 39: 88-92.

Ives JR, Warach S, Schmitt F, et al. Monitoring the EEG during an MRI scan., XIII Congress of Electroencephalography and Clinical Neurophysiology, Vancouver, B.C., Aug. 29-Sept. 4, 1993a.

Ives JR, Warach S, Schmitt F, et al. Monitoring the patient's EEG during echo planar MRI. Electroeceph clin Neurophysio. 1993b; 87:417-420.

Ives JR, Mainwaring NR, Gruber LJ, Schomer DL. Remontaging from bipolar recordings: a better way? Muscle & Nerve 16;1993c:1125-1126.

Ives JR, Warach S, Schomer DL. Neurophysiological (EEG) control of functional MRI (fMRI) using a fiber optical cable-telemetry link. Presented at the American Electroencephalographic Society, Sept 16-21, 1994, Chicago, IL.

Ives JR, Thomas R, Jacob PM, et al. Technique and methodology for recording/monitoring the subject's sleep stage during "quiet" functional magnetic resonance imaging (fMRI). Presented at the 11th Annual Meeting of the Associated Professional Sleep Society, June 10-15, 1997, San Francisco, California. p153.

Ives, J.R., Keenan, J.P., Schomer, D.L., Pascual-Leone, A. EEG recording during repetitive transcranial magnetic stimulation (rTMS). Neurology, Supplement 4, 1998;50:A167.

Jackson GD, Connelly A. Cross JH, Gordon I, Gradian DG. Functional magnetic magnetic imaging of focal seizures. Neurology 1994; 44:850-856.

Jakob PM, Schlaug G, Griswold M. et al. Functional burst imaging. Magn. Res. Med. 1998:614-621.

Kwong KK, Belliveau JW, Chester DA, et al. Dynamic magnetic resonance imaging of human brain activityduring primary sensory stimulation. Proc Natl Acad Sci USA 1992; 89:5675-5679.

Lufkin R, Jordan S, Lylyck P, Vinuela F. MR imaging with topographic EEG electrodes in place. Am J Neuroradiol. 1988; 9:953-954.

Seeck M, Lazeyras F, Michel CM, et al. Non-invasive epileptic focus localization. Electroenceph. clin. Neurophysiol., 1998; 106:508-512.

Warach S, Levin JM, Schomer DL, et al. Hyperfusion of ictal seizure focus demonstrated by MR perfusion imaging. Am J Neuroradiol 1994; 15:965-968.

Warach S, Ives JR, Schlaug G, et al. EEG-triggered echo-planar functional MRI in epilepsy. Neurology 1996; 47:89-93.

Chapter II.6

A Framework for the Integration of fMRI, sMRI, EEG, and MEG

Michael Wagner and Manfred Fuchs [1]

Philips Research Laboratories

Introduction

While fMRI (functional Magnetic Resonance Imaging) yields high spatial resolution, brain dynamics are hardly resolved. This is due to two facts: First, there is a trade-off between SNR (Signal-to-Noise Ratio), spatial, and temporal resolution of the imaging process. Secondly, there is an inherent low-pass filter in the event chain that leads from neuronal activity to hemodynamic reactions as they are measured. The combination with EEG (Electroencephalography) and MEG (Magnetoencephalography) source reconstruction techniques promises to add the desired temporal resolution. EEG and MEG sampling times are usually in the order of a millisecond, and the effects of the neuronal activity that are measured (electric potentials and magnetic fields) are of an instantaneous nature. They can be modeled using the quasi-static approximations of Maxwell's equations. However, the localizing power of EEG or MEG alone is limited and typically in the order of one centimeter.

To achieve a combination of these modalities, the results of an fMRI analysis shall be used as constraints in the subsequent EEG/MEG source reconstruction. Doing this, one has to keep in mind that there are differences in the anatomical phenomena and in the necessary experimental setups (a single paradigm vs. statistically significant differences with regard to at least two paradigms). As a result, locations of significant hemodynamic differences (which we shall call "hotspots" subsequently) must not simply be equated with current sources. Instead, several problematic constellations may occur:

---

[1] Correspondence: Michael Wagner, Philips Research Laboratories, Röntgenstr. 24, 22335 Hamburg, Germany, e-mail: M.Wagner@pfh.research.philips.com

- Current sources that do not show up in the fMRI images. This might happen, if neurons are not active long enough to cause increased bloodflow.
- fMRI hotspots without a corresponding current source. This situation could occur in the case of so-called silent sources, i.e. sources or source constellations that produce no measurable fields or potentials.
- A slight displacement between fMRI hotspot and source location. This is to be expected due to the different locations of neurons and involved vessels, as well as due to the fact that an extended cortical patch, if curved, may not even contain the representative source location.

In this paper, we propose two strategies for integrating fMRI with source reconstruction techniques that address these difficulties. One strategy has been developed for the dipole fit methods and another for distributed source models.

Furthermore, one has to deal with the co-registration problem between the imaging modalities involved [Elsen 93]. Depending on the information available, several techniques could be used for finding the best match between sMRI (structural Magnetic Resonance Imaging) and fMRI image coordinates:

- landmark-based registration. Here, corresponding landmarks have to be determined in both modalities. [Fuchs 95] (see Fig.1)
- surface-based registration. Prerequisite is the segmentation of e.g. the skin or the brain surfaces in both modalities.
- volume-based registration. No additional information is required. [Wells 97]
- scanner coordinate-based registration. Here, sMRI and fMRI must have been acquired during the same session, and the images must include position information in scanner coordinates.

For the surface-based registration and the scanner coordinate-based registration, we describe how a reliable match can be achieved.

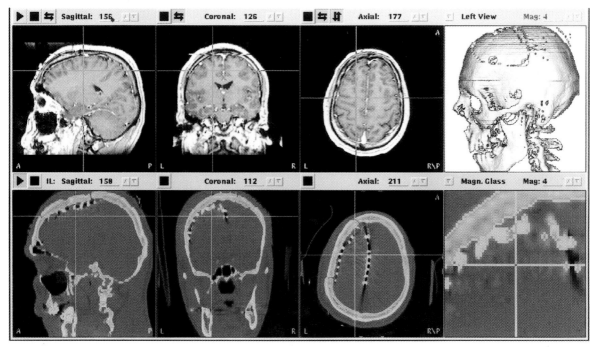

Figure 1
co-registration of MRI and CT data using anatomical landmarks.

Surface-Based Image Modality Co-Registration

The surface registration method is based on a contour fit algorithm [Huppertz 98] that matches a set of N digitized contour points $\mathbf{r}_i$ with a segmented surface such as the skin or the brain. The algorithm minimizes the L_1-norm of the distances of all contour points to the surface. These distances are determined efficiently by using a distance map $D(\mathbf{r})$, which can be pre-computed from the segmented surface (see Fig.2).

$$\min \sum_{i=1}^{N} |D(\mathbf{r}_i)| \qquad (1)$$

The L_1-norm (sum of absolute values) is more robust with respect to outliers than the commonly used L_2-norm (sum of squares). In order to deal with local minima, three translation parameters (shifts) and three rotation parameters (Euler angles) are optimized independently, such that the best-fit translations are found for each set of rotations. This is done by two nested Nelder-Mead simplex algorithms. The translation parameters are initialized with the difference between the center of mass of the segmented surface and the center of a sphere fitted to the set of points. The rotation parameters are initialized by a coarse exhaustive search over the parameter space of the Euler angles with optimized translations.

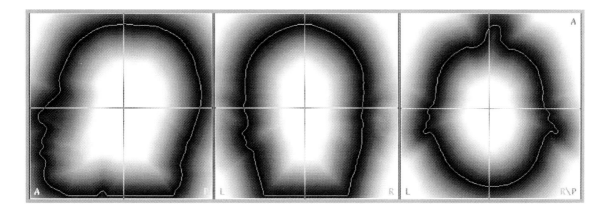

Figure 2
a 3D distance map. Increasing intensity encodes increasing distance from the segmented surface.

For the surface-to-surface registration, one surface is converted into a set of contour points. These contour points must not include areas which are prone to ambiguities or missing correspondences, such as the ears, the neck, or the chin. Thinning is applied in order to decrease numerical complexity, resulting in about 300 contour points. As one surface may be overall larger than the other, e.g. due to the segmentation thresholds used, the average distance of all contour points from the surface is not taken into account in this case, and the problem is now to

$$\min \sum_{i=1}^{N} |D(\mathbf{r}_i) - \overline{D}| \quad \text{with} \quad \overline{D} = \frac{1}{N} \sum_{i=1}^{N} D(\mathbf{r}_i) \qquad (2)$$

Results are shown in Fig.3.

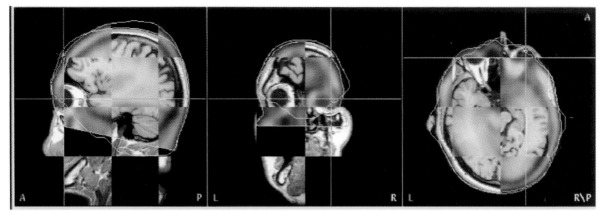

Figure 3
interlaced display of co-registration results obtained for segmented skin surfaces. The first modality is a 3D sMRI T_1 scan with 250 mm Field of View, the second modality is a gross average 3D T_1 data set normalized to Talairach coordinates with 230 mm Field of View. The two surfaces used for registration are shown as white outlines.

Image Modality Co-Registration using Scanner Coordinates

The scanner coordinate-based registration method makes use of the *Image Position* and *Image Orientation* tags which are part of the ACR-NEMA standard and can be found in DICOM images. The *Image Position* is a coordinate vector in [mm]-space that describes the position of the upper left corner of the image with respect to the scanner. The *Image Orientation* are two normal vectors in the same coordinate system. They describe the orientations of the *x* and *y* image axes, respectively. The coordinate system thus defined is tracked down the pipeline of image transformations (shift, zoom, rotation, mirroring, stacking) which are applied when scanner images are read in and can finally serve for co-registration between modalities. The scanner coordinate-based co-registration can be computed on-the-fly when ACR-NEMA or DICOM images are read in. Results are shown in Fig.4. If scanner coordinates are available, their use for co-registration purposes is practical and straightforward.

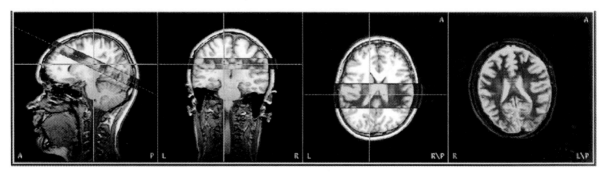

Figure 4
co-registration using scanner coordinates. The first modality is a 3D sMRI T_1 scan with 256 mm Field of View, the second modality are 5 T_2 slices with 250 mm Field of View and a slice-to-slice distance of 4 mm that have been acquired during the same session. fMRI acquisition has been performed for the same slices.

Volume Conductor Modeling

The accuracy of EEG and MEG source localization with respect to source locations has to be in the order of few millimeters, if fMRI priors shall successfully be used. This can – for sources in most parts of the head – only be achieved when using realistic volume conductor models. Two popular modeling techniques are the BEM (boundary element method) and the FEM (finite element method). While the BEM [Fuchs 98a] uses surface triangulations of the borders between compartments of equal isotropic conductivities as a geometric model, most FEM [Buchner 96] implementations work on tetrahedral volume tessellations, where each tetrahedron may be assigned an individual, anisotropic conductivity.

The generation of these geometric models is a nontrivial task, because the segmentation of the compartment borders is not straightforward, and certain minimal mesh element sizes, smoothness constraints, and inter-compartment distances have to be observed in order to avoid instabilities of the algorithms. An automatic procedure can be used to generate BEM or FEM meshes containing the brain, skull, and skin compartment borders (Fig.5) from sMRI. It speeds up model generation and ensures model quality. The procedure performs the following steps:

1. Cortex segmentation. The cortex forms the basis of the model. Segmentation is based upon a 3D region growing algorithm. The segmented white matter, which is smoothed and enlarged, serves as a maximum volume containing the cortical structures. The cortex may also be used for constrained source reconstructions.
2. Inside of the skull. As the inside of the skull does not give reliable contrast in T_1-weighted images, it is derived as the smoothed and slightly enlarged shape of the cortex. This guarantees a minimum distance between cortical sources and innermost compartment border.
3. Outside of the skull. In order to ensure a minimum inter-compartment distance, the smoothed union of the segmented skull-skin transition and the enlarged inside of the skull is used as the outside of the skull.
4. Skin. In a similar way, the skin compartment is defined by the segmented skin and the enlarged outside of the skull. The outside of the skull, strongly enlarged, also serves as a maximum volume for the skin. This avoids the inclusion of the difficult to model and less important basal regions into the skin compartment.
5. Mesh generation (Fig.6). For the BEM, the three compartment borders are triangulated, using individual triangle sizes. For the FEM, a tetrahedral tessellation with regularly shaped tetrahedra is performed.

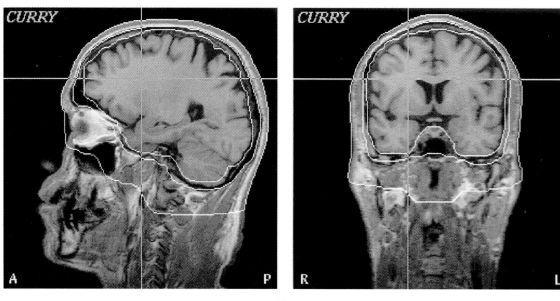

Figure 5
compartment borders (white) overlaid onto MR slices.

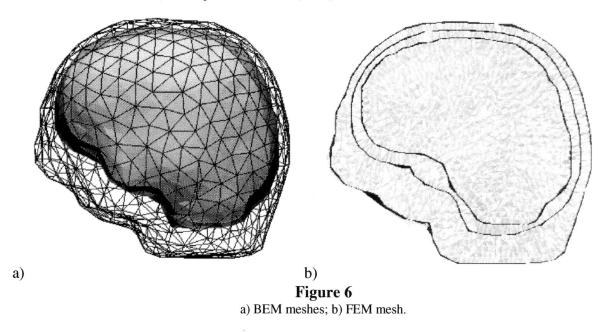

a)   b)

Figure 6
a) BEM meshes; b) FEM mesh.

fMRI-Constrained Dipole Models

Dipole models are parameterizations of the brain activity that use few current dipoles (point sources), whose number, locations, components, and temporal characteristics have to be determined. One dipole is interpreted as an abstraction for an extended patch of cortical tissue that has a predominant orientation. The rationale for this abstraction is, that a single neuron can be modeled as a microscopic current dipole. If location, orientation, and temporal activation of neurons are correlated, they produce a measurable far-field which resembles the one of a current dipole at the same location. A class of neurons for which all of these necessary correlations are certainly given are the pyramidal cells in the cortical gray matter with their aligned orientations perpendicular to the cortical sheet.

Usually, number and temporal characteristics are assumed to be known a priori, while the remaining parameters are fitted. Common temporal characteristics include the moving dipole model, where independent locations and components are determined for each time point, the rotating dipole model, where the location stays constant over a latency range, and the fixed dipole model with constant location and orientation and a time-varying strength only.

This leads to an optimization problem such as

$$\min \Delta(\mathbf{j}) = \|\mathbf{m} - \mathbf{L}\mathbf{j}\| \qquad (3)$$

Here, **m** are the measured data, **j** are the unknown dipole components for given locations **r**, and **L** is the (linear) lead field matrix for given locations **r**. The lead field matrix describes the relation between dipole components and measured data as defined by the conducting properties of the head. **m** and **j** may comprise several time points. For given locations **r**, Δ can be computed non-iteratively using linear algebra, while an iterative solver has to be used for finding the optimal **r**.

For fMRI-constrained dipole models, rotating or fixed dipoles are associated with each fMRI hotspot. To this end, the hotspots are used as seedpoints for the dipole fit, while a maximum distance constraint applies. Dipole locations, components, and timecourses are fitted. In order to account for unexplained data, an additional dipole is fitted, which is not related to any hotspot (Fig.7). A regularized solution is obtained that is an extension to Eq. 3 in the sense that it suppresses source components with small data overlap [Fuchs 98b] and that a penalty term P is added, which assures that the maximum distance criterion is met. The resulting optimization problem is

$$\min \Delta(\mathbf{j}) = \|\mathbf{C}(\mathbf{m} - \mathbf{L}\mathbf{j})\| + \lambda \|\mathbf{W}_{overlap}\mathbf{j}\| + P \qquad (4)$$

Here, **C** is the sensor weighting resp. SNR normalization matrix, and $\mathbf{W}_{overlap}$ is the non-diagonal overlap weighting matrix for given **C**, **m**, and **L**. SNR normalization is an approach that allows for the combination of EEG and MEG, the suppression of noisy channels, and a straightforward assessment of the goodness-of-fit in a χ^2-sense [Pflieger 98]. Due to the noise normalization performed by **C**, the regularization parameter λ can be pre-calibrated with respect to the SNR. The vicinity constraint for N sources and seedpoints $\mathbf{s}_i$, and maximum distances d_i is imposed using the penalty term

$$P = \sum_{i=1}^{N} \begin{cases} 0 & \text{for} \quad |\mathbf{r}_i - \mathbf{s}_i| < d_i \\ c|\mathbf{r}_i - \mathbf{s}_i| & \text{else} \end{cases} \quad \text{with} \quad c \gg \lambda \qquad (5)$$

Again, the optimization problem can be solved analytically for given locations **r**. In the scope of this paper, $d_i = 5$ mm.

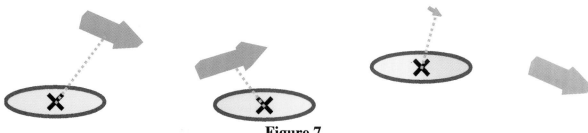

Figure 7

schematic drawing for three hotspots and a total of four dipoles: dipoles are loosely fixed to fMRI hotspots, using the hotspot locations as seeds and constraining the dipoles to stay within a maximum distance from their seedpoints. One additional dipole is fitted, which is not tied to a hotspot.

fMRI-Constrained Distributed Source Models

For fMRI-constrained distributed source models, the cortical sheet segmented from sMRI [Wagner 95] is used as the source space, resulting in some ten thousand fixed locations and normals with distances of 2 to 3 mm. At each of these cortical locations, a dipolar source is assumed that accounts for the neuronal activity of the small surrounding patch. The orientation of the source can be fixed to the pyramidal cell orientation, i.e. perpendicular to the cortical surface.

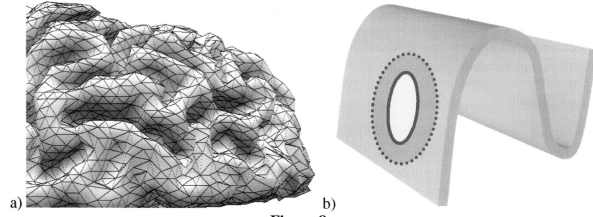

Figure 8

a) a middle layer of the gray matter sheet used as the source space for current density reconstructions (detail). Current dipoles are computed at each vertex location. Their orientations are normal with respect to the triangulated surface; b) spatial relation between cortical sheet, fMRI hotspot, and the somewhat larger area of increased location weights.

A depth-normalized, regularized inverse solution with a minimal weighted L_2-norm of the source strengths is computed:

$$\min \Delta(\mathbf{j}) = \|\mathbf{C}(\mathbf{m} - \mathbf{L}\mathbf{j})\| + \lambda \|\mathbf{W}_{fMRI} \mathbf{W}_{depth} \mathbf{j}\| \qquad (6)$$

The first term is the data term. It measures the discrepancy between measured and forward calculated data. Here, **m** are the measured data, **j** are the unknown currents, **L** is the lead field matrix, and **C** is the sensor weighting resp. SNR normalization matrix.

The second term is the model term. It measures the discrepancy between the computed currents and an implicitly assumed model. Here, $\mathbf{W}_{fMRI}$ is the diagonal fMRI-induced location weighting matrix, and $\mathbf{W}_{depth}$ is the diagonal depth normalization matrix common to weighted least squares methods. The location weights $\mathbf{W}_{fMRI}$ depend on the fMRI-induced significances. Different weighting functions can be used, including a total suppression of non-hotspot activity, a damping of non-hotspot activity, and arbitrary functions of the fMRI correlations (Fig.9). In the scope of this paper, a damping factor of 0.5 is used. Unconstrained solutions are obtained if $\mathbf{W}_{fMRI} = \mathbf{1}$.

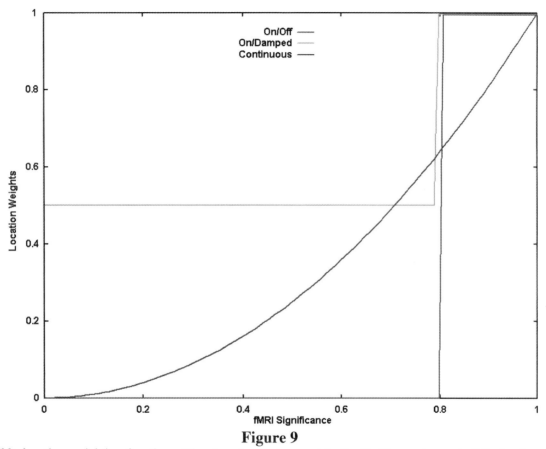

Figure 9
possible location weighting functions. The damping factor used in the On/Damped case is 0.5, the damping factor in the On/Off case is zero.

A Simulation Study

In order to test the constrained source reconstruction methods, forward calculated 61 electrode EEG data were generated for three dipolar sources. Each source showed a depolarization / repolarization sequence as activation pattern. Noise was added to achieve an SNR of 15 for the dipole fits and an SNR of 100 for the current density reconstructions. Functional MRI data were simulated with three hotspots. Two of these hotspots coincided with two of the three source locations, while one hotspot had no corresponding source (Fig.10).

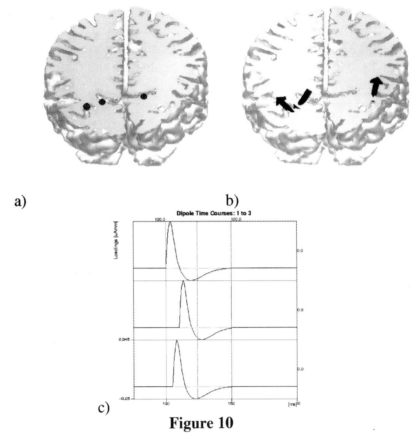

Figure 10
a) three simulated fMRI hotspots (black); b) three simulated sources (black). From left to right: sources 1, 2, and 3; c) source waveforms.

In all 4-dipole fits, three dominating sources at the correct locations were found. Sources 1 and 2 interfere slightly in the rotating dipole fits but not in the fixed dipole fits. Results are shown in Fig.11 and Fig.12.

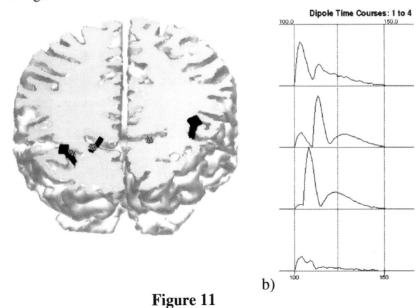

Figure 11
a) fit results of 4 rotating dipole fit (black) and fMRI hotspots (seeds, gray); b) rectified timecourses of the 4 rotating dipoles.

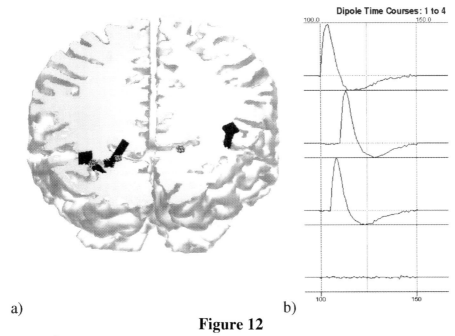

Figure 12
a) results of 4 fixed dipole fit; b) timecourses of the 4 fixed dipoles.

In the fMRI-constrained current density solutions, all source locations were retrieved when using a damping factor of 0.5, even in the case of a source without corresponding hotspot. However, the timecourses of sources 1 and 2 interfere. Results are shown in Fig.13. For comparison, the results of an unconstrained solution with $\mathbf{W}_{fMRI} = \mathbf{1}$ are shown in Fig.14.

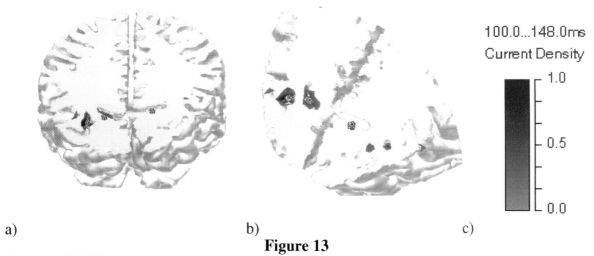

Figure 13
a) results of fMRI-constrained current density reconstruction (maximum over timerange, clipped) and fMRI hotspots (gray); b) detail showing the third source; c) used scale

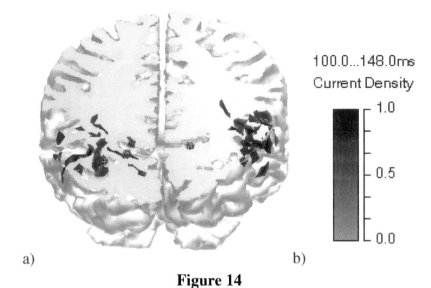

Figure 14
a) results of unconstrained current density reconstruction (maximum over timerange) and fMRI hotspots; b) used scale

Discussion

Depending on the available information, different co-registration methods for merging sMRI and fMRI coordinate systems may be useful. The usage of scanner coordinates seems to be the most straightforward method, if both modalities have been acquired during the same session and the necessary information is part of the image file format. If this is not the case or not possible, other methods have to be applied. Landmark- and surface-based registration both leave the user with the responsibility of choosing matching landmarks resp. surfaces. Volume-based registration [Elsen 93] is independent of such influences, but about an order of magnitude slower than the surface-based approach.

Dipole fits can benefit from fMRI constraints: Meaningful seedpoints for the location fit are obtained, a crucial issue in all multi-dipole fits. A reconstructed dipole in the vicinity of each fMRI hotspot yields the corresponding source timecourse. Spatially unconstrained dipoles are then necessary to account for remaining activity. However, the temporal discrimination of close sources can only be achieved by an additional fixed dipole constraint.

Current density reconstructions react upon fMRI constraints in two ways: Activity in the vicinity of fMRI hotspots is bundled. Remaining activity can only be localized correctly, if the damping factor is not too small, if its field distribution cannot be generated from sources within the hotspots, and if the SNR is very high. In practice, these conditions will hardly be fulfilled. Then, the usability of the method is confined to cases, where all significant sources show up in the fMRI maps. The temporal discrimination of close sources is enhanced with respect to unconstrained reconstructions, if the correct source orientations are dominating within the hotspots.

Outlook

The behavior of fMRI-constrained current density methods other than weighted minimum norm least squares remains to be characterized. Especially, L_1-norm methods with their higher sensitivity, and the LORETA method with its higher localization power – compared to the method used in this paper – are promising candidates for further examinations.

Most important, a systematic application of fMRI-constrained source reconstruction methods to real data has to be performed.

References

van den Elsen PA, Pol EJD, Viergever M: Medical image matching - a review with classification, IEEE Engineering in Medicine and Biology 12: 26-39, 1993

Buchner, H, Knoll, G, Fuchs, M, Rienäcker, A, Beckmann, R, Wagner, M, Silny, J, Pesch, J, Inverse localization of electric dipole current sources in finite element models of the human head, Electroenceph. Clin. Neurophys. 102: 267-278, 1997

Fuchs, M, Drenckhahn, R, Wischmann, HA, Wagner, M, An improved boundary element method for realistic volume-conductor modeling, IEEE Trans. Biomed. Eng. 45: 980-997, 1998

Fuchs M, Wagner M, Wischmann HA, Theißen A: Source reconstructions from combined MEG and EEG data, Electroencephalography and Clinical Neurophysiology 107: 93-111, 1998

Fuchs M, Wischmann HA, Wagner M, Krüger J: Coordinate System Matching for Neuromagnetic and Morphological Reconstruction Overlay, IEEE Transactions on Biomedical Engineering, 42: 416-420, 1995

Huppertz HJ, Otte M, Grimm C, Kristeva-Feige R, Mergner T, Lücking CH: Estimation of the accuracy of a surface matching technique for registration of EEG and MRI data, Electroencephalography and Clinical Neurophysiology 106: 409-415, 1998

Pflieger ME, Simpson GV, Ahlfors SP, Ilmoniemi RJ: Superadditive Information from Simultaneous MEG/EEG Data, Proc. BIOMAG, Santa Fe, in press

Wagner M, Fuchs M, Wischmann HA, Ottenberg K, Dössel O: Cortex segmentation from 3D MR images for MEG reconstructions, Baumgartner C et al, Biomagnetism: fundamental research and clinical applications, Amsterdam, Elsevier/IOS Press 1995, 433-438

Wells WM, Viola P, Atsumi H, Nakajima S, Kikinis R: Multi-modal volume registration by maximization of mutual information, Medical Image Analysis 1: 35-51, 1997

T. Nakada (Ed.)
Integrated Human Brain Science: Theory, Method Application (Music)
© 2000 Elsevier Science B.V. All rights reserved

Chapter II.7

Independent Component Analysis of Simulated ERP Data

Scott Makeig[a,1]*, Tzyy-Ping Jung*[d]*, Dara Ghahremani*[b]*,
and Terrence J. Sejnowski*[b,c,d]

[a]*Naval Health Research Center*
[b]*Howard Hughes Medical Institute, Computational Neurobiology Laboratory, The Salk Institute*
[c]*Department of Biology, University of California San Diego*
[d]*Institute for Neural Computation, University of California San Diego*

Introduction

Event-related potentials (ERPs) are averages of electroencephalographic (EEG) epochs time-locked to a set of similar experimental events. Multichannel electromagnetic recordings from the scalp, including spontaneous EEG or magnetoencephalographic (MEG) records as well as ERP or magnetic event-related field (ERF) averages, have been widely used to study dynamic brain processes involved in perception, memory, selective attention, recognition, and priming. However, the combination of underlying brain processes that produce both spontaneous and event-related potentials and magnetic fields recorded at the scalp is still largely undetermined. Separating ERP sources without a priori knowledge of their number and spatial distribution is called a problem of "blind separation." Since ERPs often sum a complex distribution of activity in overlapping projections from brain and extra-brain generators to the scalp, it is difficult to identify and measure activity arising from each of the contributing sources. Mathematically, the 'inverse problem' of identifying the locations and time courses of activation of brain generators of observed surface potentials is underdetermined. Most existing techniques for attempting ERP source separation employ second-order statistical methods (e.g. covariance, cross-correlation, and principal component analysis) (Chapman & McCrary, 1995), or else assume that sources have a known single- or multiple-dipole architecture (Scherg & Von Cramon, 1986).

---

[1] Correspondence: Scott Makeig, Naval Health Research Center, P.O. Box 85122, San Diego, CA 92186-5122, USA, Tel: (1)-619-553-8416, e-mail: scott@salk.edu *This report was supported in part by the Navy Medical Research and Development Command and the Office of Naval Research, Department of the Navy under work unit ONR.Reimb-6429. The views expressed in this article are those of the authors and do not reflect the official policy or position of the Department of the Navy, Department of Defense, or the U.S. Government. Approved for public release, distribution unlimited.*

Independent Component Analysis (ICA) algorithm we use here (Bell & Sejnowski, 1995) is a blind separation technique based on information-maximization which takes into account higher-order statistical information about the distribution of the input vectors (concurrent field measurements at many spatial locations). Recently, we have shown that the ICA algorithm can also be used to parsimoniously decompose brief ERP data sets into conventional ERP components (Makeig et al., 1997) (e.g., single peaks in the scalp waveforms, eye-movement activity, and steady-state responses (Pantev et al., 1993; Galambos, Makeig & Talmachoff, 1981)), spatially filtering each into a different output channel. Unlike algorithms that seek to both identify and localize ERP sources, the ICA algorithm does not attempt to perform three-dimensional source localization. Instead, it attempts to find the scalp topography of each source and the time course of its activation.

Without prior knowledge of the actual brain source activations that produce ERPs, it is difficult to verify the algorithm's effectiveness. We assume there may be a few strong sources active during given ERP recording epochs, summing with activity generated by a larger number of weaker sources including residual spontaneous EEG sources not time- or phase-locked to the experimental events of interest. To determine whether the ICA algorithm can successfully separate relatively strong ERP components even when mixed with numerous weaker components, we performed several simulation experiments. These complement analyses of much longer (79,000-point) simulated EEG records performed previously (Ghahremani et al., 1996).

The simplest models of ERP generation assume that electrodes placed on the scalp surface record the electromagnetic activity of local or distributed cortical neural networks that can be modeled as effective single- or multiple-dipole sources (Nunez, 1981; Scherg & Von Cramon, 1986; Chapman & McCrary, 1995). Here, we simulate the activities of six simulated EEG components using ERP-length signals projected through a three-shell spherical head model to six model scalp electrodes (Dale & Sereno, 1993) and apply the ICA algorithm to the resulting simulated EEG data. These simulations allow us to investigate changes in ICA algorithm performance with variations in source strength, location, and orientation as well as effects of adding multiple weak EEG sources to the simulated EEG. We use simulated source data drawn from electrocorticographic (ECoG) data collected from the surface of the cortex of a patient undergoing exploratory analysis prior to surgery for epilepsy (Bullock et al., 1995a, 1995b). ECoG data epochs drawn from different channels and time periods in the available data set are used to simulate brief ERP-length EEG components. Further simulations, using artificial component waveforms resembling those produced by ICA decomposition of actual ERPs, clarify differences between ICA decomposition and physical source localization.

Methods

An overview of the process of simulation and ICA decomposition is given in Fig. 1. The ICA algorithm is described in detail elsewhere. Further details and references about the algorithm and its application to EEG data appear in (Bell & Sejnowski, 1995; Ghahremani et al., 1996; Makeig et al., 1997; Jung et al., 1998; Jun et al., 1999; Makeig et al., 1999). Other related approaches and background material are available in (Cover, 1991; Linsker, 1992; Nadal & Parga, 1994; Jutten & Herault, 1994; Amari, Cichocki & Yang, 1996; Cardoso & Laheld, 1996; Karhumen et al., 1997; Lee, 1998).

The Three-shell Spherical Head Model

In our simulations, we use a three-shell spherical head model which projects dipoles at four fixed brain locations onto six scalp electrodes. The projection matrix containing the model parameters is computed using an analytic representation for a three-shell spherical head model (Dale & Sereno, 1993; Kavanaugh et al., 1978) as described in Appendix I.

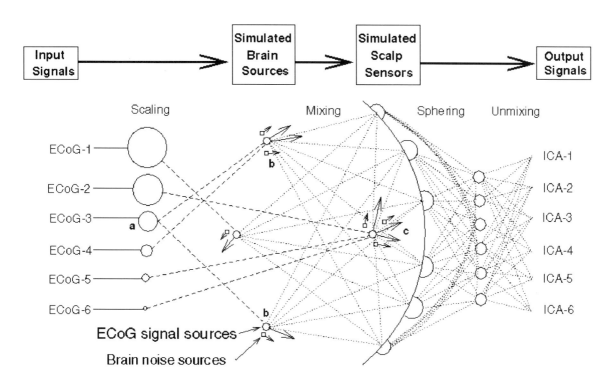

Figure 1
Schematic overview of the simulations. Input signals (ECoG-1 to ECoG-6), minimally correlated ($|r| < 0.085$) 600-point data epochs taken from different times and channels in an available ECoG data set were scaled relative to one another (*Scaling*) and assigned to single- or multiple-dipole brain sources (*longer arrows*). In some conditions, one source signal (a) was projected through bilateral dipole sources (b) approximately simulating a bitemporal source in the auditory cortices. Other signals were assigned to sources modeled as single dipoles with different orientations at the same brain location. Six weak ECoG ('noise') sources (*shorter arrows*) were positioned near the seven signal dipoles. After initial "sphering" of the simulated ERP data, source separation was performed via the "unmixing" matrix produced by the ICA algorithm. Spatial filtering of the sphered simulated ERP data by multiplying with the unmixing matrix produced output component activation waveforms (ICA-1 to ICA-6).

Input Signals

The input signals are six asynchronous 3-sec (600-point) epochs of ECoG data drawn from different channels of a 12-minute, 80-channel ECoG data set (Bullock et al., 1995a) on the basis of being minimally correlated with one another. To simulate sources of varied strengths, in a second set of simulations we scale the input source signal vectors in steps of -3 or -6 dB relative to one another. Simulated ERP-length EEG signals are then derived from the input signals by multiplying by a mixing matrix specifying the projection of each model dipole to each model sensor.

Additional Low-Level Sources

In a second simulation, six additional simulated brain sources are added to the original six ECoG sources to produce simulated ERP-length EEG epochs. In one condition, these sources consist of ECoG data epochs uncorrelated with each other or with the first six ECoG source signals ($|r| <= 0.022$). In other conditions, the low-level source waveforms are synthesized from uniform- or gaussian-distributed white noise. The low-level sources simulate weaker ERP components or residual EEG activity remaining in an ERP after finite averaging. The low-level sources are scaled to -3 or -6 dB below the level of the weakest ECoG signal source, and are then projected through simulated "diffuse" dipoles nearby the stronger source dipoles to the six simulated scalp electrodes. The "diffuse" dipole sources are modeled by adding 1% gaussian white noise to the weights in the mixing matrix specifying the projections of the strong sources to the model electrodes. In a third simulation, the orientations of the dipoles for the strong and low-level sources are independent of each other. In each condition, the mixed projections of the six weak and six stronger source signals are summed to form the simulated EEG data.

ICA Training

Training input consisted of 600 six-channel simulated EEG waveforms. The time order of the input data was reshuffled before each learning step to avoid overlearning. Training block length was 12. The initial ICA learning rate (0.006) was reduced by 15% after each training step when the change in the weight matrix (considered as a 1x36 vector) formed a greater than 90 deg angle to the weight vector change at the previous training step. Training was continued for at least 1024 steps to insure convergence.

Performance Measures

We measured the performance of the ICA algorithm by the correlations between source and output waveforms, and by the difference in maximal signal-to-noise ratio (SNR) of each input signal in the simulated EEG and ICA output data. These SNR measures are described in detail in Appendix II.

Simulations

1. ICA Decomposition is Independent of Source Projections

To demonstrate the ability of ICA to unmix ERP-length data epochs, we first performed ERP simulations using six 600-point (3-sec) ECoG data epochs (Fig. 2a) drawn from different times and channels in a 12-minute ECoG data set (Bullock et al., 1995a, 1995b). The source epochs were selected as minimally correlated with each other (|r| <= 0.085). These source signals were projected through a mixing matrix representing 7 dipolar sources in the three-shell spherical brain model (Fig. 2c). One source projected to two dipoles with simulated bitemporal placement. The resulting simulated EEG scalp waveforms (Fig. 2b) were relatively highly correlated with one another (|r| = 0.24-0.98), and moderately correlated with the input source waveforms (|r| < 0.68). For comparison purpose, the same simulated data were decomposed using Principal Component Analysis (PCA), both with and without Varimax or Promax rotation (see Makeig, 1999 for details).

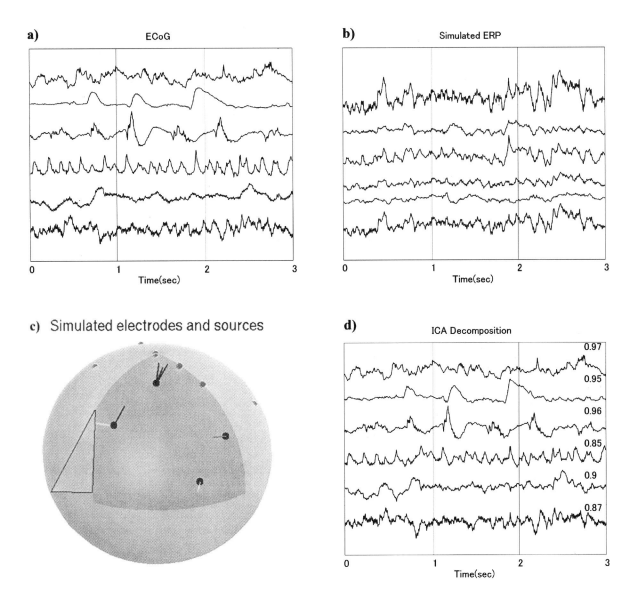

Figure 2

ICA decomposition of simulated brief EEG data epochs derived from six 3-second (600-point) ECoG data epochs drawn from different times and channels in a 12-minute ECoG data set (a) and simulating the activities of six independent brain ERP sources. Projecting this source data through six one- and two-dipole simulated sources in a 3-layer spherical shell head model (c) produces a simulated 600-point EEG data epoch (b). ICA decomposition of this data epoch produces six ICA component waveforms (d). Correlations between each ICA component waveform and its best-correlated input source waveform are shown on the right side of (d).

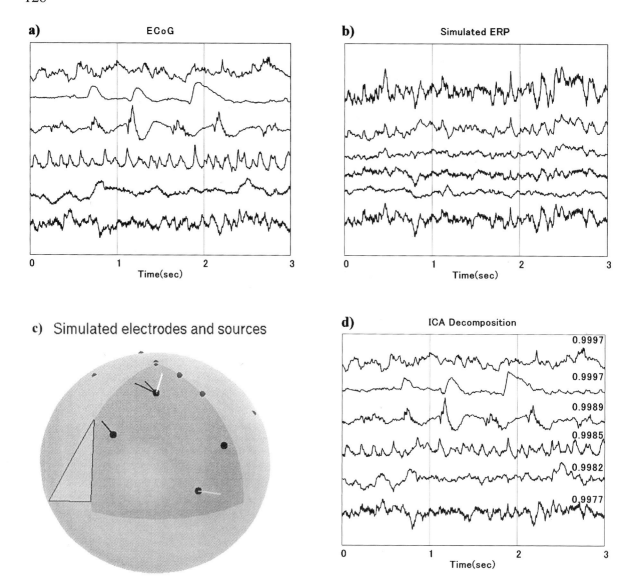

Figure 3

Invariance of the ICA decomposition to changes in source distribution. Decomposition of the six same ECoG input source epochs as in Fig. 2 (a) projected through six one- and two-dipole simulated sources (c) with different locations and orientations than in the first simulation (Fig. 2c), produces a new simulated ERP data epoch (b). ICA decomposition of this new simulated data yields the ICA component waveforms shown in (d). The near-perfect correlations between each component waveform and its matching component waveform in the first simulation (Fig. 2d) are shown on the right side of (d), demonstrating that when the numbers of sources and data channels are equal, ICA decomposition is independent of the spatial distribution of the underlying sources.

Results

The ICA decomposition of the simulated ERP data is shown in Fig. 2d, with the maximal correlations between each ICA output waveform and its best-matching source waveform. These correlations are high in all cases ($|r| > 0.87$), but are less than unity because the input source waveforms are not truly independent. Fig. 3, gives results of a second simulation using the same source waveforms (Fig. 3a) with different dipole assignments and dipole angles (Fig.

3c), producing simulated ERP data (Fig. 3b) only moderately correlated with the simulated ERP data in the first simulation (Fig. 2b). Nevertheless, the ICA decomposition of the second ERP data set (Fig. 3d) is nearly identical to the ICA decomposition of the same data in the first simulation (Fig. 2d, |r| > 0.997).

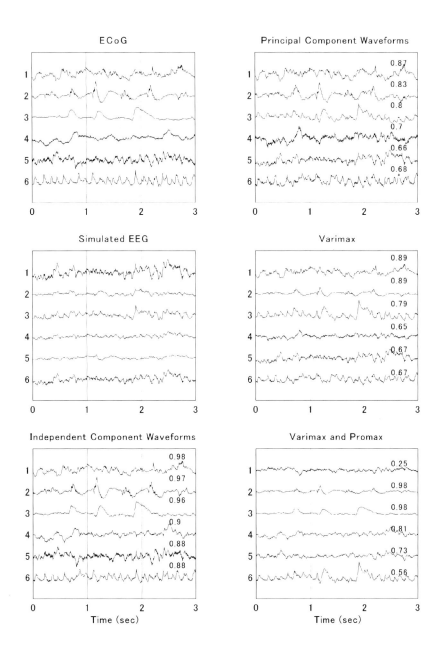

Figure 4
ICA and PCA decompositions of the same ECoG mixtures. ICA (*lower left*) and PCA (*upper right*) decompositions of this data epoch each produce six component activation waveforms. Also shown were the results derived from Varimax (*middle right*) and Promax (*lower right*) rotation methods of the PCA-rotated data. Correlations between each component waveform and its best-correlated input source waveform are shown on the upper right side of each waveform.

Figure 4 shows results of PCA-based decompositions of the same simulated data (*middle left*). PCA itself (*upper right*) rotates the first component so as to account for maximum *variance* in the data. This may sum activity produced by several independent components when these are spatially correlated. Succeeding principal components are both spatially and temporal uncorrelated with preceding components. Therefore, PCA itself is unable to separate sources whose projections to the scalp are spatially correlated. Mean absolute correlation of the 6 PCA component waveforms with the original source waveforms was 0.76. Varimax rotation is an orthogonal rotation method that attempts to minimize the number of temporal or spatial dimensions each component is weighted on. Varimax rotation (*middle right*) also gave a mean absolute correlation of 0.76. Promax is a constrained non-orthogonal rotation method that further concentrates large weights for each component on a few dimensions. Applied to the Varimax-rotated components (*lower right*), Promax produced two components (second and third from top) whose time waveforms were well-correlated (0.98) with two of the source waveforms (*upper left*). However, the mean absolute correlation for all six sources was only 0.72, well short of the 0.93 mean correlation for the ICA decomposition (*lower left*).

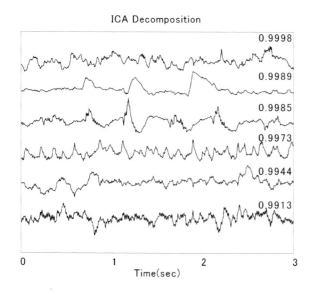

Figure 5

Invariance of the ICA decomposition to spatial source distribution. Direct ICA decomposition of the ECoG input source data, without mixing the source waveforms using any EEG model, produces an ICA decomposition nearly identical to those produced for the previous two mixture models (cf. Figs. 2d and 3d). Absolute values of the correlation coefficients between this decomposition and the corresponding ICA sources in the previous simulation (Fig. 2d), shown on the right, are all above 0.991.

A further demonstration of the invariance of the ICA decomposition from source and sensor placements is the fact that the ICA decomposition of the original source vectors themselves, before mixing (Fig. 5), is nearly identical ($|r| > 0.991$) to the previous decompositions.

Conclusions

First, the ICA algorithm can reliably decompose six-channel, 600-point simulated (noise-free) ERP data sets into independent components. This was a remarkable result, since we assumed that the algorithm required much more input data to converge reliably because of its derivation from information theory. Derivations of the algorithm from maximum likelihood principles support its use on shorter data epochs (Amari, Cichocki & Yang, 1996). Further, our results were obtained using ECoG waveforms recorded directly from the human cortical surface. Presumably, the dynamics and statistical distribution of ECoG data is as near to actual EEG source data as possible.

Second, the results of ICA decomposition do not depend on the mixing matrix, so long as it is non-singular. Singularity cannot be expected to be an issue in applications to real data since the inevitable presence of low-level and recording noise sources make non-singularity highly unlikely.

2. ICA Isolates Strong Signal Sources

When the total number of sources is larger than the number of data channels, changes in source location and orientation can produce large changes in the amplitude of the relative projections of different sources to the scalp, and thus possibly affect the ICA decomposition. To test this, we performed several simulation experiments decomposing simulated ERP-length data synthesized from more sources than the number of model sensors. These tested the ability of the ICA algorithm to accurately decompose simulated ERP-length EEG data even when it summed activity from more low-level sources than the number of simulated data channels.

Six new 3-sec ECoG data epochs, selected to be minimally correlated with each other and with the six epochs used in the first simulation, were used to simulate low-level ERP sources. In actual ERP data, these low-level sources might be either small ERP components or residual spontaneous EEG sources remaining in the data after finite averaging. In these simulations, the six stronger sources were scaled relative to one another before being assigned to model sources (cf. Fig. 2c) and projected via a mixing matrix to the six model scalp electrodes. In the first simulation, the attenuation step size was -6 dB, meaning these sources were scaled -6, -12, -18, -24, and -30 dB below the strongest source. The six additional low-level sources were scaled -36 dB below the strongest source, and were then projected through 'diffuse' dipoles nearby and nearly parallel to the stronger source dipoles (Fig. 1). In another condition, different dipole assignments and orientations were used for the stronger and low-level sources.

Simulated 3-sec EEG epochs summing the stronger and low-level source projections were submitted to ICA decomposition. In each simulation reported below, 50 permutations of input signal attenuation order and model source assignment were used to produce 50 different input data sets. These were then decomposed separately using the ICA algorithm. Mean output SNR for each input source was then plotted as a function of the number of ICA training steps. In each simulation, the ICA algorithm was trained for 16,384 training steps to test for maximum convergence.

Results

Figure 6 shows SNR for the six stronger sources and six low-level sources before, during, and after ICA decomposition. Before training, only the two strongest sources have a positive SNR in any of the simulated scalp channels. After as few as 16 ICA training steps, all six sources are assigned to separate output channels with positive SNR. Output SNR is maximized after about 1000 training steps. Mean dB gain for the six stronger sources is 24 dB. The low-level sources are each largest in the same ICA output as their 'nearby' stronger source, but in each case the SNR difference between the stronger and associated low-level source pairs is larger in the ICA output than in the simulated EEG input data. Effects of permuting the order of attenuation and source assignments are small, as shown by the (<= 3 dB) standard deviations across the 50 permutations (*Fig. 6, right side*).

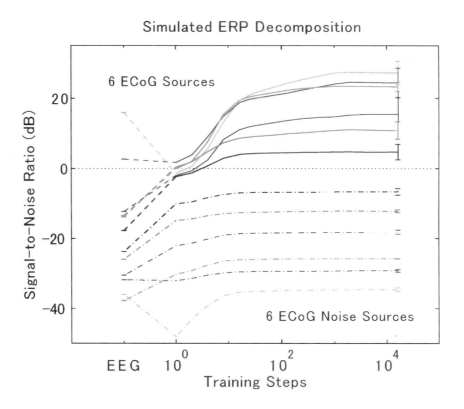

Figure 6

Blind separation performance by infomax ICA. Mean of 50 different permutations of input signal ordering and source assignment. Source attenuation step size was -6 dB. The leftmost column shows the means and standard deviations of maximal signal-to-noise ratio (SNR) in the 6 simulated ERP channels for each source signal. The rightmost column shows the maximal SNR in the six ICA outputs. Each of the six stronger input sources (*solid lines*) are assigned to a different ICA component. ICA component SNR order and range generally reproduce the order and range of source signal attenuation values. SNR gains (from best scalp channel to best output channel) range from 4 dB to 36 dB (mean, 24 dB). Most SNR improvement occurs during the first 16 training steps. The six low-level sources (*dashed lines*), which were scaled -36 dB below the largest source, were strongest in the same output channel as their 'nearby' stronger source. The output SNR difference between 'nearby' stronger and low-level source pairs was 5 dB to 24 dB larger than in the simulated EEG before training (mean, 14 dB).

Figure 7 shows the quite similar results of two more experiments in which (Fig. 7a) uniformly-distributed or (Fig. 7b) gaussian-distributed white noise were used instead of ECoG data to create the time waveforms of the six low-level sources. The similarity of the results in all three experiments implies that the statistical distributions of the low-level source strength values do not have important effect on the performance of the algorithm.

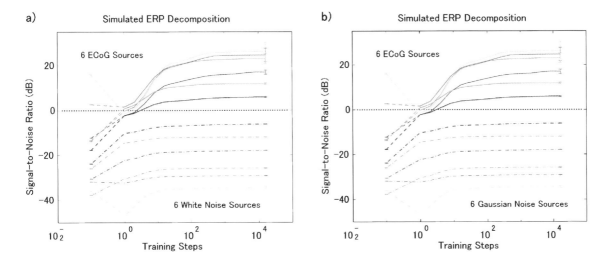

Figure 7

(a) Invariance of the ICA decomposition to changes in low-level source distribution. Blind separation performance by infomax ICA using uniformly-distributed white noise for the six simulated low-level sources instead of ECoG data. Other parameters as in Fig. 6. Results strongly resemble results of simulations using 12 ECoG data sources (Fig. 6). *(b) Blind separation performance by infomax ICA using Gaussian-distributed white noise* for the six simulated low-level sources instead of ECoG data. Other parameters as in Fig. 6. Results again strongly resemble those using 12 ECoG data sources (Fig. 6).

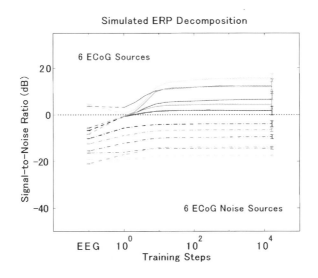

Figure 8

ICA decomposition performance with stronger low-level sources. Blind separation performance by infomax ICA using -3 dB attenuation steps instead of -6 dB steps. Results of 50 simulations using different permutations of input signal order and source assignment. Other parameters as in Fig. 6. Again, the six largest input sources (*solid lines*) are assigned to different output components, and the output SNR order and range generally reproduce the order and range of source signal attenuation values. SNR gains (from best scalp channel to best output component) ranged from 8 dB to 21 dB (mean, 12 dB). The six lower-level sources (*dashed lines*), here scaled to -18 dB below the strongest source, have largest SNR in the same output component as their nearly larger source (*shown by corresponding trace colors*). Output SNR difference between the six stronger and low-level source pairs is 2 dB to 13 dB larger than in the simulated scalp data (*mean, 7 dB*).

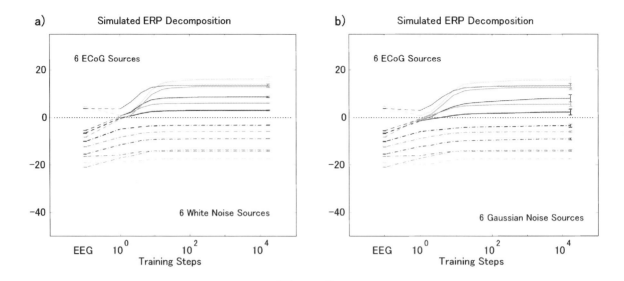

Figure 9

Invariance of the ICA decomposition to changes in low-level source distributions. (a) Blind separation performance by infomax ICA with attenuation step size -3 dB and using *uniformly-distributed white noise* for the six simulated low-level sources. Other parameters as in Fig. 8. Results strongly resemble those in Fig. 8 using 12 ECoG data sources. (b) Blind separation performance by infomax ICA with attenuation step size -3 dB and *using Gaussian-distributed white noise* for the six simulated lower-level sources. Other parameters and results as in Fig. 8.

Figure 8 shows parallel results of another simulation using a -3 dB source attenuation step size and ECoG source data. Again, all six stronger sources are separated into different ICA output channels with positive SNR, with somewhat lower final SNR values than the previous simulations using a -6 dB step size (here, mean SNR gain is 12 dB as compared to 14 dB in the first simulations). Fig. 9 shows the very similar results of simulation experiments using either separate uniformly-distributed (Fig. 9a) or gaussian-distributed (Fig. 9b) white noise epochs to create the low-level source waveforms.

In Figs. 6 through 9, the low-level sources were 'nearby' and near-parallel to the stronger source dipoles. Fig. 10 shows mean results of 50 simulations in which the source dipoles for the stronger and low-level sources were differently oriented. In this simulation, the attenuation step size was -6 dB and the low-level sources were ECoG data epochs. The source configurations of the stronger and low-level sources were those shown in Figs. 2c and 3c. Although in this case only the strongest five sources were separated by the algorithm into different output sources with positive SNR, mean SNR gain for the six stronger sources ranged from 7 dB to 30 dB (mean, 25 dB), quite comparable to earlier results (Fig. 6) using 'nearby' dipole models for the low-level sources. Repeating this simulation using gaussian white noise instead of ECoG data epochs for the low-level sources produced nearly identical results (not shown).

Conclusions

The performance of the ICA algorithm in decomposing linearly-mixed simulated ERP-length EEG data sets degrades gracefully in the presence of additional low-level sources. The degree of separation achieved by the algorithm (measured here in terms of output SNR) depends on the relative amplitudes of the strong and low-level sources, and is weakly affected by their relative placements and orientations.

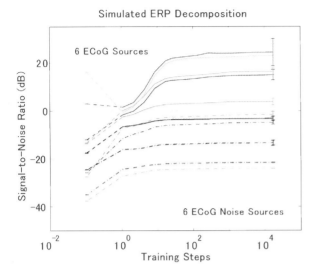

Figure 10

Blind separation performance by infomax ICA with different dipole orientations for the stronger and low-level model source models. Results of 50 simulations using different permutations of input signal order and model source assignment. Signal attenuation step size is -6 dB. The six low-level EcoG sources (*dashed lines*) are assigned to the physical model sources shown in Fig. 3b, while the six stronger sources (*solid lines*) are assigned to model sources shown in Fig. 2b. After 16 training steps, five of the six strong input sources are assigned to different ICA components. SNR gains (from best scalp channel to best output component) ranged from 7 dB to 30 dB (mean, 25 dB), comparable to the SNR gains using in earlier simulations using model sources for the low-level inputs near those of the stronger inputs (Fig. 6).

3. ICA Identifies Spatial Correlations in the Data

For most brain researchers, prototypical brain activity sources are active foci located in single brain structures. It is natural to ask, therefore, how the ICA algorithm treats data generated in several brain structures whose activations are partially correlated in time, since by definition Independent Component Analysis assumes its 'ICA sources' are temporally independent. Brain generators whose activities are partially correlated cannot be independent sources of EEG or ERP data. How, then, does the ICA algorithm decompose their activities?

Figure 11 shows a simulation using six synthetic, partially-correlated simulated brain generator waveforms resembling ICA components produced when the algorithm was applied to actual auditory ERP data. Each simulated generator waveform is plotted as two 310-point response epochs. The period of activation of generator 3 subsumes the activations of generators 1 and 2 (in the *same* sub-epoch, producing overall pairwise correlations of r = -0.460 and r = -0.128, respectively). These correlations were introduced deliberately to test the response of the ICA algorithm to correlated generator input. Other pairs of generator waveforms are nearly uncorrelated ($|r| <= 0.070$). The three waveforms (1-3) roughly simulate an extended activation (3) in one brain structure during which two other brain structures (1 and 2) briefly become active.

Results

Projecting the six simulated generator waveforms through the dipole model shown in Fig. 2c produces the simulated ERP data shown in Fig. 11b. The six simulated ERP channel waveforms contain a wide variety of features that might puzzle a psychophysiologist studying them. The ICA decomposition of this simulated ERP data is shown in Fig. 11c. Note first that the algorithm produces ICA source waveforms very similar to those of generators 1, 2, 4, 5, and 6. However, the waveform of ICA source 3 differs from the waveform of generator 3 in two important respects—during the activation periods of sources 1 and 2, ICA source 3 is silent.

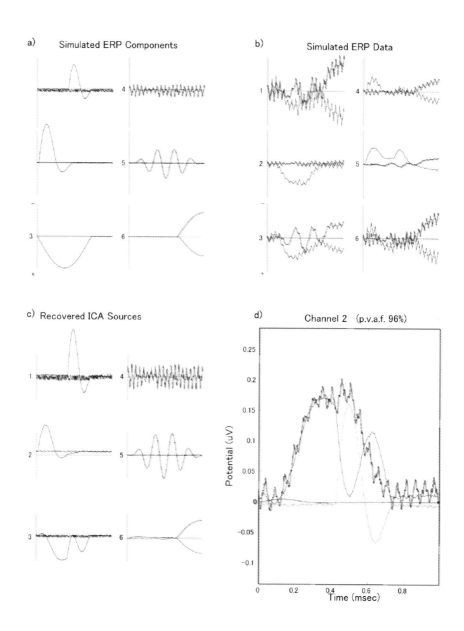

Figure 11

ICA decomposition of simulated partially-correlated source activity (b), synthesized by projecting the six synthetic 620-point brain-generator waveforms (a) through the dipole source model of Fig. 2a. Data are plotted as two 310-point waveforms. Note that time period of activation of source 3 overlaps those of sources 1 and 2. Results of ICA decomposition of the simulated ERP wave forms are shown in (c). The time courses of activation of each source (except source 3) are returned by the algorithm. Plot (d) shows the results of projecting ICA sources 1, 2, and 3 onto simulated ERP channel 2. The simulated ERP waveform for channel 2 is also shown (note polarity change from (b)). The three ICA components account for 96% of the variance in the simulated ERP, and together parse the *correlation structure* of the simulated ERP data into spatially-stable periods of source coactivation. This may or may not be useful for physiological investigation, depending on whether or whether not the coactivations are functional or adventitious (see text).

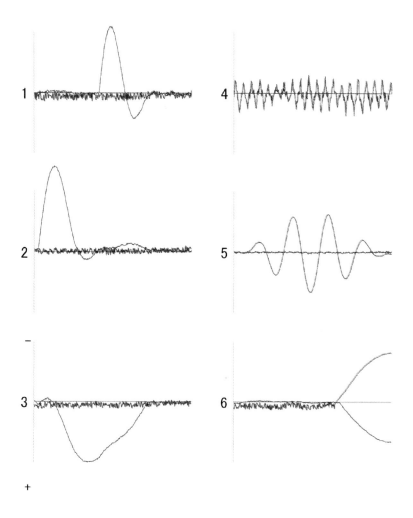

Figure 12

Effects of additional simulated 'no-response' data on the ICA decomposition. ICA decomposition of a simulated ERP data set composed of five consecutive 620-point epochs. The first epoch was identical to that shown in Fig. 11b. The remaining four 'no response' epochs were the result of projecting low-amplitude (0.04 RMS) white noise through the same dipole model of Fig. 2c. The 5 simulated response epochs were decomposed simultaneously by infomax ICA. The figure shows the first 620 points of the resulting ICA component waveforms. Note that the ICA component-3 waveform now follows the source-3 time course (Fig. 11a) more closely (see text).

Essentially, the ICA algorithm parses the data into *periods of stationary spatial correlation*, creating three ICA source components accounting for: (1) the activation of generator 3 alone, (2) the coactivation of generators 1 and 3, and (3) the coactivation of generators 2 and 3. The scalp projections of all three ICA sources, therefore, must also be composed of the scalp projection of generator 3, either alone or summed with the scalp projection of generator 1 or 2.

Figure 11d shows the projections of ICA source components 1 to 3 onto model electrode 2, whose simulated ERP waveform is also shown. Three traces show the projections of ICA sources 1 through 3, respectively. Together, they sum to a waveform accounting for 96% of the variance in the model electrode response. (Remaining variance is explained by the other three ICA source components).

The infomax algorithm parses its input data into the sum of spatially stationary and temporally independent source activations. The decomposition produced by the ICA algorithm reveals, in this case, either: (1) the presence of functional coactivations of generator 3 and generators 1 and 2, (2) coactivations produced in the three generators by some other physiological source, or (3) adventitious correlations with little or no functional significance.

If the correlations in this simulation were indeed functional correlations, the ICA decomposition would actually be more physiologically informative than the physical generator decomposition given in Fig. 11a. The physical generator waveforms only show the time course of activation of the separate brain generators, ignoring their functional relationship, whereas the ICA algorithm parses the data into transient functional units defined by coactivation.

However, it is equally possible to imagine that the event-related coactivations of generators 1, 2, and 3 might be adventitious rather than a result of functional neural cooperativity. In this case, the ICA decomposition (Fig. 11c) would have less physiological significance than the physical generator decomposition (Fig. 11a). The three ICA source components in Fig. 11d each combine the activity of one or two generators (Fig. 11a), with generator 3 common to all the sources. Unfortunately, since the source inversion problem is underdetermined, no algorithm exists that can reliably determine the physical generator waveforms from the simulated scalp data alone.

Note that each ICA source in this simulations is spatially stationary and contains the activity of only one or at most two strong dipoles. This suggests that the ICA decomposition may be a useful preprocessing step prior to applying 'guided' source inversion algorithms that attempt to solve the inverse problem using simplifying assumptions about the number and arrangement of dipolar sources present in the data (Scherg & Von Cramon, 1986; Dale & Sereno, 1993).

Although the ICA algorithm fits a linear model, ICA training is not a linear process. Fig. 12 shows the result of decomposing a data set in which four additional 620-point ('no response') epochs of low-level (0.04 RMS) gaussian white noise were concatenated onto the simulated ERP data epoch of Fig. 11a. The figure shows the first epoch of the ICA source waveforms for the extended data set. The additional degrees of freedom in the appended 'no response' data allows the algorithm to more closely approximate the actual activation time courses of all three generators (Fig. 11a).

Conclusions

Infomax ICA may fail to separate physically separable brain generators when their periods of activation are highly overlapping and input data length is short. In this case, the source decomposition produced by ICA will be physiologically significant only if the correlated activity in fact arises from transient functional correlations between them. *The ICA algorithm parses the correlation structure of its input data into spatially stationary and minimally-correlated pieces, and when possible assigns each piece to a different output channel.*

Anatomic considerations suggest that sensory and cognitive ERPs may be produced largely by sequences of brief activations in different neural networks. Our simulations suggest the ICA algorithm may be used to measure the latency, time course, spatial pattern, and strength of these activations, which may correspond to stages of neural information processing. ICA may be most useful when trained simultaneously on ERP data from several stimulus and/or task conditions, since in this case brief activations with different strengths and/or latencies in different response epochs are more likely to be minimally correlated. This hypothesis is supported by successful results of applying the ICA algorithm to several ERP data sets consisting of 2 to 64 related multi-channel waveforms (unpublished).

General Discussion

The reported effectiveness of the ICA algorithm in separating multiple linearly-mixed simulated audio and EEG sources (Bell & Sejnowski, 1995; Bell & Sejnowski, 1996) has been reproduced in ERP-length EEG simulations using a three-shell head model and six simulated scalp channels, in which stronger 600-point simulated-ERP signals generated from actual ECoG data epochs were successfully and repeatedly separated with SNR gains averaging 18 dB. Results were independent of the orientation or relative positions of the model dipole sources. As we have seen in decompositions of much longer simulated EEG data epochs (Ghahremani et al., 1996), the performance of the ICA algorithm degrades gracefully in the presence of multiple weak independent sources even when the total number of sources is larger than the number of sensors, and is relatively unaffected by source or sensor placements.

That an algorithm based on entropy maximization can be so efficient when given as few as 600 input vectors appears remarkable. Nonetheless, the simulations reported here, as well as recent theoretical work (Lee, 1998), plus published and unpublished results of applying the algorithm to several actual 14- to 128-channel ERP data sets, all imply that the ICA algorithm can correctly decompose ERP data of this complexity into a relatively small number of strong, spatially-stationary sources each accurately separating the correlated activations of one or more physical brain or extra-brain generators.

There is no universal definition of what constitutes an ERP component or source. Most researchers in the field believe that ERPs are composed largely of sequences of brief activations in different brain structures at different times relative to the time locking experimental events (see Makeig et al., 1977). These activations may overlap in time and their projections to the scalp sensors nearly always overlap spatially because of the spatial 'smearing' of EEG signals produced by the high resistivity of the skull. Because waveforms containing single brief activations may be minimally correlated, ICA may be well-suited to ERP decomposition. In fact, our ICA decompositions of several ERP data sets also consist largely of brief sequences of separate source activations lasting 50-300 ms (Makeig et al., 1997; Makeig et al., 1999).

Results of our last simulation (Fig. 12) suggest the possible advantage of applying the ICA algorithm concurrently to ERP data from several stimulus and task conditions. In this case, the overall correlation induced by coincidental coactivations of pairs of sources in some but not all response conditions is lessened. Residual EEG and extra-brain 'noise' in ERP signals may also be used by the algorithm to separate partially correlated but functionally independent sources (as in Fig. 12).

Conclusions

Infomax ICA is a promising tool for the analysis of multichannel EEG or MEG signals. Our results suggest that relatively strong brain source components may be effectively separated from each other and from weaker sources by ICA decomposition with SNR gains above 20 dB. ICA attempts to describe *what* independent sources produce its input data, as defined by their time courses and scalp maps, but not *where in* the brain the sources are located. It is thus compatible with the well-established indeterminacy of the source inverse problem. As the algorithm identifies sources by their spatial correlation structure, neurophysiological interpretation of ICA sources poses both research challenges and opportunities. The neurophysiological importance of transient event-related correlations in the activity of otherwise independent neurons and neural networks is now being studied seriously by theorists and experimentalists. Our simulations suggest the ICA algorithm may be a useful tool for identifying and monitoring spatiotemporal patterns of emergent correlation in brain activity linked to perceptual and cognitive brain processes.

Applications of ICA algorithm to averaged event-related potentials (ERPs) appear quite promising, since response averaging increases the amplitudes of activity time- and phase-locked to experimental events relative to the strength of other spontaneous and ongoing EEG sources. The number of independent strong brain sources contributing to ERP data thus may be smaller than the number of EEG channels typically used to record them (Makeig et al., 1997, 1999). In this case, the ICA algorithm should successfully separate and identify the time courses and scalp projections of the strongest ERP sources. ICA decomposition may be particularly useful for comparing the latencies, time courses, scalp topography, and activation strength of numerous brain generators involved in producing evoked responses to more than one stimulus in multiple experimental and task conditions.

Summary

A recently-derived algorithm for performing Independent Component Analysis (ICA) (Bell & Sejnowski, 1995) based on information maximization is a new information-theoretic approach to the problem of separating multichannel electroencephalographic (EEG) or magnetoencephalographic (MEG) data into temporally independent and spatially stationary sources (Makeig et al., 1996). In a previous report, we have shown that the algorithm can separate simulated EEG source waveforms (independent simulated brain source activities mixed linearly at the scalp sensors), even in the presence of multiple low-level model brain and sensor noise sources (Ghahremani et al., 1996). Here, we demonstrate the ability of the ICA algorithm to decompose brief event-related potential (ERP) data sets into temporally independent components (Makeig et al., 1997) by applying it to simulated ERP-length EEG data synthesized from 3-sec (600-point) electrocorticographic (ECoG) epochs recorded from the cortical surface of a human undergoing pre-surgical evaluation (Bullock et al., 1995a, 1995b).

Six asynchronous single-channel ECoG data epochs were projected through single- and multiple-dipole model sources in a three-shell spherical head model (Dale & Sereno, 1993) to six simulated scalp sensors to create simulated EEG data. In two sets of simulation experiments, we altered relative source strengths, added multiple low-level sources (synthesized from ECoG data and uniform- or Gaussian-distributed noise), and permuted the simulated dipole source locations and orientations. The algorithm reliably separated the activities of the relatively strong sources, regardless of source location, dipole orientation, and low-level source distributions. Recovery of the original component waveforms was much better using ICA than using PCA without or without Varimax or Promax rotation. Thus, the

ICA algorithm should identify relatively strong, temporally independent and spatially overlapping ERP components arising from multiple brain and/or non-brain sources, regardless of their spatial distributions. This shows that the ICA algorithm can decompose ERPs generated by uncorrelated sources.

A third ERP simulation tested how the algorithm treated a simulated ERP epoch constructed using model ERP generators whose activations were partially correlated. In this case, the algorithm parsed the simulated ERP waveforms into a sum of temporally independent and spatially stationary components reflecting the changing topography of correlated source activity in the simulated ERP data. Each of the affected components sums activity from one or more concurrently-active brain generators. This suggests the ICA algorithm may also be useful for identifying event-related changes in the correlation structure of either spontaneous or event-related EEG data. Paradoxically, adding four simulated "no response" epochs to the training data minimized the relative importance of partial correlations in the original data epoch and allowed the algorithm to separate the concurrently active sources. Likewise, submitting ERPs from more than one stimulus or experimental condition to concurrent ICA analysis may allow the algorithm to separate sources from brain generators whose activations are partially correlated in some but not all response conditions.

Acknowledgments

This report was supported in part by grants to S.M., T-P.J. and T.J.S. from the Office of Naval Research, and to T.J.S. from the Howard Hughes Medical Institute. The authors wish to thank Anders Dale for supplying the head model parameters, Dr. Susan Spencer, Dr. Brad Duckrow and Ted Bullock for supplying data used in the simulations, and Tony Bell for assistance and useful discussions.

References

Amari S., Cichocki, A. Yang, H.H. A new learning algorithm for blind signal separation. In *Advances in Neural Information Processing Systems* 8, MIT Press (1996).

Bell, A.J., Sejnowski, T.J. An information-maximization approach to blind separation and blind deconvolution, *Neural Computation* 7, 1129-1159 (1995).

Bell, A.J., Sejnowski, T.J. Learning the higher-order structure of a natural sound. *Network: Computation in Neural Systems* 7, 2 (1996).

Bullock, T. H., McClune, M. C., Achimowicz, J. Z., Iragui-Madoz, V. J., Duckrow, R. B. Spencer, S. S. EEG coherence has structure in the millimeter domain: subdural and hippocampal recordings from epileptic patients. *Electroencephalograph. Clin. Neurophysiol.* 95, 161-77 (1995a).

Bullock, T. H., McClune, M. S., Achimowicz, J. Z., Iragui-Madoz, V. J., Duckrow, R. B. Spencer, S. S. Temporal fluctuations in coherence of brain waves. *Proc. Nat. Acad. Sci. USA* 92, 11,568-72 (1995b).

Cardoso, J-F. Laheld, B. Equivalent adaptive source separation. *IEEE Trans. Signal Proc.* 45:434-444 (1996).

Chapman, R.M. McCrary, J.W. EP component identification and measurement by principal components analysis. *Brain and Language* 27, 288-301 (1995).

Comon, P. Independent component analysis, a new concept? Signal Processing 36, 287-314 (1994).

Cover, T.M. Thomas, J.A. *Elements of Information Theory*, John Wiley (1991).

Dale, A.M. Sereno, M.I. Improved localization of cortical activity by combining EEG and MEG with MRI cortical surface reconstruction - a linear approach. *J. Cogn. Neurosci.* 5, 162-176 (1993

Galambos, R., Makeig, S. Talmachoff P. A 40 Hz auditory potential recorded from the human scalp. *Proc. Natl. Acad. Sci. USA* 78, 2643-2647 (1981).

Ghahremani, D,. Makeig, S., Jung T-P., Bell, A.J., Sejnowski, T. S. *Independent Component Analysis of Simulated EEG Using a Three-Shell Spherical Head Model* Technical Report INC-9601, Institute for Neural Computation, University of California San Diego. La Jolla CA (1996).

Jung, T-P., Colin Humphries, Te-Won Lee, Scott Makeig, Martin J. McKeown, Vicente Iragui, Terrence J. Sejnowski. "Extended ICA Removes Artifacts from Electroencephalographic Recordings" In: D. Touretzky, M. Mozer and M. Hasselmo (Eds*). Advances in Neural Information Processing Systems* 10:894-900 (1998).

Jung, T-P., Makeig, S., Westerfield, M., Townsend, J., Courchesne, E., and Sejnowski, T. J., "Analyzing and Visualizing Single-trial Event-related Potentials," In: *Advances in Neural Information Processing Systems* 11 (1999, in press).

Jutten, C. Herault, J. Blind separation of sources, part I: an adaptive algorithm based on neuromimetic architecture. *Signal Processing* 24, 1-10 (1991).

Karhumen, J., Oja, E., Wang, L., Vigario, R. Joutsenalo, J. A class of neural networks for independent component analysis. *IEEE Trans. Neural Networks* 8:487-504 (1997).

Kavanagh R.N., Darcey T.M., Lehmann D., Fender D.H. Evaluation of methods for three-dimensional localization of electrical sources in the human brain. *IEEE Trans. Biomed. Eng.* 9, 25:421-429 (1978).

Lee, Te-Won, *Independent Component Analysis: Theory and Applications*, Boston: Kluwer Academic Publishers (1998).

Linsker, R. Local synaptic learning rules suffice to maximise mutual information in a linear network. *Neural Computation* 4, 691-702 (1992)

Makeig, S., Anthony J. Bell, Tzyy-Ping Jung and Terrence J. Sejnowski, "Independent component analysis of electroencephalographic data," In: D. Touretzky, M. Mozer and M. Hasselmo (Eds). *Advances in Neural Information Processing Systems* 8:145-151 (1996).

Makeig, S. Tzyy-Ping Jung, Anthony J. Bell, Dara Ghahremani, Terrence J. Sejnowski, "Blind Separation of Auditory Event-related Brain Responses into Independent Components," *Proc. of National Academy of Sciences*, 94:10979-84 (1997).

Scott Makeig, Marissa Westerfield, Tzyy-Ping Jung, James Covington, Jeanne Townsend, Terrence J. Sejnowski and Eric Courchesne, "Functionally Independent Components of the Late Positive Event-Related Potential during Visual Spatial Attention," *J. Neurosci.* 19:2665-2680 (1999).

Nadal, J-P. Parga, N. Non-linear neurons in the low noise limit: a factorial code maximises information transfer. *Network* 5, 565-581 (1994).

Nunez, P.L. Electric Fields of the Brain. New York: Oxford (1981).

Pantev, C., Elbert, T., Makeig, S., Hampson, S., Eulitz, C. Hoke, M. Relationship of transient and steady-state auditory evoked fields. *Electroencephalogr. Clin. Neurophysiol.* 88, 389-396 (1993).

Scherg, M. Von Cramon, D. Evoked dipole source potentials of the human auditory cortex. *Electroencephalogr. Clin. Neurophysiol.*, 65, 344-60 (1986).

Appendix I

The three-shell spherical head model

In our simulations, we used a three-shell spherical head model which projects dipoles at four fixed brain locations onto six scalp electrodes. The projection matrix containing the model parameters was precomputed using an analytic representation for a three-shell spherical head model. Electrode positions were vertices of a triangulated icosahedron located on the model head sphere. At each of the four locations in the head model, we placed one to three dipoles pointing in different directions, giving a total of seven dipoles. We assigned five input signals to single dipoles, and one input signal to two bilateral dipoles (Fig. 2c). As shown in Fig. 2c, two dipoles with different orientations were placed at a single dipole location, and three dipoles with different orientations were placed at another location.

These choices were expressed via a ((4x3)x6) configuration matrix, **C**, which assigned six source signals to the seven dipoles according to the configuration described above. The configuration matrix was then multiplied by the (6x(4x3)) weight matrix, **F**, which projected the seven dipoles (at the four dipole locations) to each of the six simulated electrode sites. The resulting matrix product:

$$\mathbf{M} = \mathbf{FC}$$

was a 6x6 "mixing" matrix specifying the simulated EEG signals as linear combinations of the six input sources. Simulation variables were chosen such that this mixing matrix was non-singular. Note that despite the complexity of the head model, the mixing matrix was a linear 6x6 transformation of the six sources, and therefore satisfied the assumptions of the algorithm.

Source strength adjustment

To simulate sources with varied strengths, the vector of input signals, $s(t)$, were scaled relative to one another in steps of -6 or -3 dB using a 6x6 diagonal attenuation matrix, . Simulated EEG signals, **A**, were derived from the input signals by multiplying by the attenuation and mixing matrices.

$$\mathbf{x}(t) = \mathbf{MAs}(t)$$

Weak brain sources

In some experiments, seven simulated weak brain source signals were added to the simulated EEG. These ("brain noise") sources consisted of uncorrelated random noise with a flat distribution in the [-1,1] interval, scaled to the same level as the weakest input source signal. The seven brain noise sources were assigned to simulated "diffuse" dipoles placed nearby each of the seven brain source dipoles by adding 1% gaussian-distributed noise to the matrix, **M**, before mixing. The mixed brain noise signals were then added to the simulated EEG.

Appendix II

Performance Measures

SNR in the ICA algorithm output

Our measure of the ICA algorithm's performance in these simulations is the signal-to-noise ratio (SNR) of each input signal in the output sources. For each input signal, $s_i(t)$, we defined:

$$\mathbf{s}_i(t) = \begin{pmatrix} 0 \\ \vdots \\ 0 \\ s_i(t) \\ 0 \\ \vdots \\ 0 \end{pmatrix}$$

in which all input signals except for $s_i(t)$ were zeroed out. The output source waveforms for $s_i(t)$ were then defined as:

$$\mathbf{u}_i(t) = \mathbf{WPMA}\mathbf{s}_i(t)$$

The *signal level*, S_{ik}^{ICA}, of the *i*th input signal in the *k*th output source waveform was computed by taking the standard deviation of the *k*th row of $\mathbf{u}_i(t)$. The noise level for each input signal in each output source was computed by letting $\mathbf{s}_i^c(t)$ consist of all input signals except $s_i(t)$:

$$\mathbf{s}_i^c(t) = \begin{pmatrix} s_1(t) \\ \vdots \\ s_{i-1}(t) \\ 0 \\ s_{i+1}(t) \\ \vdots \\ s_n(t) \end{pmatrix}$$

These "complementary" signal vectors were passed through the simulated mixing and unmixing processes with brain noise and sensor noise sources added, giving output source waveforms:

$$\mathbf{u}_i^c(t) = \mathrm{WP}\{\mathbf{MA}[\mathbf{s}_i^c(t)+\mathbf{n}(t)]+\mathbf{r}(t)\}$$

where $\mathbf{n}(t)$ is the weak brain sources and $\mathbf{r}(t)$ is the sensor noise. The *noise level*, $N_{ik}^{ICA}(t)$, was defined as the standard deviation of the kth row of $\mathbf{u}_i^c(t)$. Then, the SNR of the ith signal in the ICA algorithm source waveforms was defined as:

$$SNR_i^{ICA} = \max_{k=1,\ldots,n}(20\log_{10}(S_{ik}^{ICA}/N_{ik}^{ICA}))$$

where n is the number of sources.

SNR in the simulated EEG

The SNR of each input signal in the simulated EEG was computed for comparison with the SNR in the ICA output. The signal level, S_{ij}^{EEG}, for the ith input signal in the simulated EEG signal was defined as the standard deviation of the simulated EEG in the jth recording electrode (i.e. in the jth row of $\mathbf{x}_i(t)$):

$$\mathbf{x}_i(t) = \mathbf{MA}\mathbf{s}_i(t)$$

The noise level, N_{ij}^{EEG}, for the ith input signal was defined as the standard deviation of the jth row of the complementary mixed signal matrix:

$$\mathbf{x}_i^c(t) = \mathbf{MA}[\mathbf{s}_i^c(t)+\mathbf{n}(t)]+\mathbf{r}(t)$$

The SNR of the ith input signal in the simulated EEG was then defined as:

$$SNR_i^{EEG} = \max_{j=1,\ldots,m}(20\log_{10}(S_{ij}^{EEG}/N_{ij}^{EEG}))$$

where m is the number of sensors.

SNR gain from EEG to ICA algorithm outputs

For each input signal, the difference between its SNR^{ICA} and SNR^{EEG} was defined as the SNR gain, G, resulting from ICA algorithm source separation.

$$G_i = SNR_i^{ICA} - SNR_i^{EEG}$$

T. Nakada (Ed.)
Integrated Human Brain Science: Theory, Method Application (Music)
© 2000 Elsevier Science B.V. All rights reserved

Chapter II.8

The Spatial Resolving Power of High-Density EEG: An Assessment of Limits

Mark E. Pflieger[a,b] and Tsutomu Nakada[a,1]

[a]*Department of Integrated Neuroscience, Brain Research Institute, University of Niigata*
[b]*Neuro Scan Labs*

1 Introduction

On the "functional side" of functional neuroimaging, EEG excels on two counts. The first count is temporal resolution: EEG reflects macroscopic neuroelectric processes that can change on the order of milliseconds. The second count is tight linkage with information processing in the brain: Neuroelectric events, some of which are reflected by EEG, implement biocognitive operations that inform cognition and regulate behavior. However, scalp EEG has two notable weaknesses on the "imaging side" of functional neuroimaging. First, spatial resolution is limited by the property that each electrode senses a weighted integral of activity spread throughout the brain volume. Second, by contrast with invasive EEG measurements taken directly in the brain, scalp EEG requires source analysis. 3D images of neuroelectric activity in the brain must be inferred from measurements taken at the sphere-like 2D scalp surface via the application of biophysical and biostatistical models. Because of this strong model dependency, "seeing isn't believing" until we attain confidence in the system of assumptions that are mixed with scalp data to produce a brain image of neuroelectric activity.

So let us regress for a moment to the raw uninterpreted data. A scalp EEG acquisition system records discrete samples of the spatiotemporally continuous electric potential field at the head surface. Potentials with respect to a reference are measured as a function of time at a fixed sampling rate, and as a function of space for a fixed configuration of electrodes having known locations on the scalp. In addition, the acquisition system typically records a stream of events, such as stimuli and behavioral responses, that are co-registered in time with the EEG. This, then, constitutes the entirety of raw data available to the EEG modality for making inferences about "when and where" neuroelectric events occur in the brain while a subject performs a task. Other neuromodalities, primarily anatomical MRI, can be called upon for invaluable assistance in the a priori modeling of processes that are involved in source analysis.

---

[1] Correspondence: Tsutomu Nakada, M.D., Ph.D., Department of Integrated Neuroscience, Brain Research Institute, University of Niigata, Niigata 951-8585, Japan, Tel: (81)-25-227-0683, Fax: (81)-25-227-0822

Nevertheless, all of the a posteriori EEG information available for making 3D images of neuroelectric activity is contained entirely at the scalp. Therefore, the 3D spatial resolution of EEG depends fundamentally on the 2D resolution at the measurement surface. This, in turn, depends on the spatial configuration of electrodes, which encompasses three related parameters: electrode density, area of coverage, and number of channels.

As treated in more detail later, it is generally supposed that the 2D spatial resolution of scalp EEG has a hard theoretical limit, or an asymptote of information content. The basic rationale for this hypothesis is as follows. The head, particularly the skull, acts as a spatial low pass filter between brain and scalp. Thus, although the brain generates signals across a broad band of spatial frequencies, only relatively low spatial frequency signals actually arrive at the scalp surface. From the standpoint of spatial sampling theory, which is analogous to temporal sampling theory, this implies the existence of a density of electrodes that is adequate to sample all information needed to exactly reconstruct the continuous scalp potential field. This theory predicts, in other words, that oversampling beyond a critical density should produce no additional yield of actual information from the data. Our original 256 channel recordings (SF Sands and team, April 1995) were made to conduct an empirical search for this hypothetical limit. What follows is a report on the status of this search.

This chapter has the following plan. We begin by distinguishing six levels of analysis on the larger stage of functional neuroimaging. Three levels in particular are most relevant to the question of EEG spatial resolution. The most basic level is spatiotemporal analysis of 2D scalp data; the second is statistical analysis of relationships between scalp data and task performance; and the third is source estimation or electric source imaging, i.e., 3D reconstruction of neuroelectric activity in the brain by modeling the genesis of the 2D scalp data. Next, we turn attention to the machinery of technical principles and methods for tackling the issue of the limits of EEG spatial resolution. Some of these methods have been applied to our question, and we review several results found in the literature as well as some previously unpublished results. Our final discussion includes an outline for a program of further research to address some outstanding issues and apparent gaps in our knowledge about the spatial resolving power of EEG.

1.1 Six levels of analysis

When we study the human brain at work in order to learn how the human brain works, there usually are three basic ingredients. First, in order to engage a specific mode of brain function, we design a task in the light of more general conceptions about the nature of brain work (e.g., Swets, 1961; Lawrence, 1963; Phillips and Singer, 1997). Second, we record signs of brain activity using one or more "windows on the brain", or neuromodalities, while the subject performs the task. Third, after data acquisition we analyze a relatively sparse set of psychophysical/behavioral data regarding stimulus presentations, overt responses, and the like, together with a massive set of neuromodality data. Our analyses proceed with the object of discovering internal relationships within the data, as well as external relationships of the data to theories about the brain, as mediated by theories about the neuromodality interface. Six progressive levels of analysis are associated with six levels of data relationships that we seek to uncover, viz.:

    (a) Task performance relationships within the psychophysical/behavioral data;
    (b) Spatiotemporal relationships within the neuromodality data;

(c) Correlative relationships between the task performance data and the neuromodality data;
(d) Interpretative relationships between the neuromodality data and structures/processes in the brain (these relationships constitute "neuroimaging");
(e) Correlative relationships between task performance and brain structures/processes; and
(f) Functional relationships between structures/processes in the brain and our conceptions of brain work that is engaged by the task (these relationships constitute "brain theory").

Our ultimate goal—the most difficult to achieve—is to discover functional relationships (f), which illuminate the brain as a solution to a problem. Neuroimaging of structures and processes in the brain (d) is necessary but not sufficient: To understand the brain as a solution, we must show how neural structures and processes solve the brain's problem. Correlative relationships between task performance and brain activity (e) provide a bridge between neuroimaging (d) and brain theory (f). Step (d) is critical: The inferences we make about brain activity from the mere signs of that activity must be valid, otherwise our analysis halts at level (c). Physical and statistical theories of the neuromodality interface enable us to interpret or "image" the spatiotemporal relationships discovered within the measurements taken outside the brain (b) as signs of specific spatiotemporal relationships inside the brain. The raw neuromodality data can embody an astronomical number of spatiotemporal relationships; therefore, step (c) is crucial to keep our analyses focussed on the experiment. That is, spatiotemporal relationships in the neuromodality data (b) are experimentally significant only if they correlate with the task performance data. If no such correlations exist, then there are no experimental effects to explain at higher levels of analysis. Finally, task performance relationships within the pure psychophysical/behavioral data (a) are fundamental: They provide the ultimate objective criteria for successful brain functioning with respect to the given task.

The question of EEG spatial resolution can be addressed at each level beyond (a), especially levels (b), (c), and (d).

1.2 Level (b): Spatiotemporal analysis of scalp EEG

Analysis of neuromodality data begins at level (b), which is conditioned by the task performance analysis of level (a). In the case of EEG, a record of continuous multichannel data is typically segmented into epochs (i.e., intervals of data) that are temporally framed with respect to the occurrence of task events, such as stimulus onsets and behavioral responses. We sort these event-related epochs into ensembles (collections) that are associated with experimentally relevant combinations of task variables, such as the psychophysical parameters of stimuli, or the selection, accuracy, and latency of behavioral responses. Thus, each ensemble corresponds to a task condition, and each data point within the ensemble has three indices: epoch index i, event-related time index t, and channel index j. Epoch indices repeat on the order of seconds (not necessarily at constant intervals) and span a relatively long period of time, typically on the order of minutes or tens of minutes, whereas the event-related time indices are evenly spaced on the order milliseconds, and typically span about one second. Time zero corresponds to the occurrence of a specific type of event as classified by the level (a) analysis. Channel indices are associated with scalp electrodes that typically have been paired with a common reference electrode. Note that the epoch indices and the event-related time indices have a direct relationship to task parameters, whereas the channel indices do not.

In order to extract particular brain signals of interest, event-related statistics typically are computed for each ensemble of EEG epochs by first order and second order summations across the epoch indices. First order summation yields by far the most common statistic, the average event-related potential (ERP), which is computed for each time index and for each channel index. ERP waveforms accentuate activity that is phase-locked to the occurrence of an event, e.g., "evoked" by a stimulus. By contrast, "induced" activity sums to zero; it is not phase-locked to the stimulus, and can be assessed via second order statistics such as variance and covariance. Induced activity is event-related to the extent that it is not stationary within the epoch time frame. Induced covariances can be computed for different time lags (temporal covariance), between pairs of channels (spatial covariance), or both (spatiotemporal covariance). Phase-locked crossproducts can be computed in a similar fashion, but after averaging. That is, whereas induced covariance is based on a sum of products, the phase-locked crossproduct is essentially a product of sums. Oscillatory activity can be studied, whether phase-locked or induced, by restricting the data to one or more frequency bands of interest.

The distinction between phase-locked and induced EEG activity is not an absolute dichotomy: Gradations of phase jitter do occur, and event-related responses can vary from epoch to epoch in other respects as well (Möcks et al., 1984; Pflieger, 1991). Moreover, phase-locked and induced EEG activities occur in parallel, and possibly may share nearly common generator regions in cortex (Tallon-Baudry and Bertrand, 1999). In studies that focus on ERPs, all non-phase-locked activity with an expected mean of zero is usually regarded as "background noise". Conversely, studies of induced activity remove the average phase-locked response. Therefore, these two approaches extract complementary information: They look at exactly the same EEG data from opposite points of view. It is often assumed that phase-locked and induced EEG activities are independently generated in the brain. However, empirical evidence (e.g., Brandt and Jansen, 1991; Brandt, 1997) as well as theoretical considerations (e.g., Phillips and Pflieger, this volume) are reasonably persuasive that these first and second order statistics cannot reflect entirely separable neural processes.

We have briefly discussed alternative ways of extracting spatiotemporal relationships within event-related EEG in order to make the following point: The issue of EEG spatial resolution is necessarily conditioned by our conception of what is regarded as signal and what is not. In other words, our basic question has several varieties such as: "What is the spatial resolution of raw EEG?" and "What is the spatial resolution of phase-locked responses?" and "What is the spatial resolution of induced activity?" These can be qualified further by specifying a frequency band, a time range, and a particular task. In addition, our questions can also take the form of contrasting one type of activity against another, such as: "What is the spatial resolution of phase-locked responses as contrasted against induced activity?" or vice versa. The point, once again, is that the spatial resolution of EEG may vary as a function of the method for extracting "signal" from the data.

Scalp Laplacian. Yet another general way to obtain different views of surface EEG is to apply a linear spatial filter to the data. The scalp Laplacian, the cortical imaging technique (Sidman, 1991; Kearfott et al., 1991; Wang and He, 1998), finite element deblurring (Le and Gevins, 1993), inward continuation via the vector boundary element method (Burik et al., 1998), and like methods are linear operators that have been designed to visually enhance the spatial resolution of surface EEG maps. All approaches except the first are hybrids between level (b) and level (d) analysis because each uses some specific model of the head as a volume conductor. The scalp Laplacian, however, has a special significance for level (b) analysis because it invokes a general principle of volume conduction without using specific volume conductor models. All sources of artifact-free EEG are by definition inside the brain, so the 3D Laplacian of volume conducted EEG must vanish at the scalp (Poisson's equation). Thus, the tangential 2D surface Laplacian at a location on the scalp-air interface must be equal and opposite to the one-sided radial 1D component, which is proportional to scalp current density. In rough conceptual terms, the 2D scalp Laplacian is approximately proportional to the potential at a location with respect to its "average surround potential", as if each electrode had its own local surround as a reference. Thus, it does not depend on the recording reference electrode, i.e., the Laplacian is reference independent; and low spatial frequencies are attenuated to the extent that they do not contribute to the local difference. Conversely, this spatial filter enhances higher spatial frequencies in the data insofar as they contribute to this local difference. Since the true Laplacian is a continuous operator, scalp current density maps from discretely sampled EEG must utilize approximations that rely on various assumptions, some more local and others more global. Hjorth (1975) introduced a local reference based on nearest neighborhoods of electrodes. Most other methods use an interpolating function to obtain a continuous scalar field on the scalp surface, from which the Laplacian is analytically computed, such as spherical harmonic expansion (Pascual *et al.*, 1988), spherical spline interpolation (Perrin *et al.*, 1989), and 3D spline interpolation for ellipsoidal surfaces (Law *et al.*, 1993). Finally, local estimates of the Laplacian for realistic MRI-derived surfaces have been attained (Le *et al.*, 1994; Babiloni *et al.*, 1998). Each Laplacian approximation depends on the density and geometric configuration of electrodes. Indeed, the Laplacian estimate can be sufficiently sensitive to approximation methods and overall electrode configurations that the theoretical advantage of being reference independent is partially lost, especially for relatively low density recordings. These considerations also somewhat complicate the question of the spatial resolution of Laplacian EEG.

Spatial SNR transform. The spatial signal-to-noise ratio (SNR) transform is a linear filter of different design that contrasts an EEG signal of special interest (foreground signal) against another EEG signal (background or control signal) that is treated "as if noise" (Pflieger and Sands, 1996; Pflieger *et al.*, 1996). For example, the foreground signal could be a phase-locked response while induced activity serves as the background signal, or vice versa. A related approach collects a control sample of EEG while the subject is in a "reference state" such as alert eye fixation. This raw EEG sample then serves as a control signal for either phase-locked or induced signals of special interest. Due to the relativity of the foreground signal to the background signal, the spatial SNR transform is nearly reference independent because both signals are affected in the same way by a change of reference. Unlike the scalp Laplacian, which is based on a physical principle and thereby utilizes scalp surface geometry and electrode positions in space, the spatial SNR transform is based on a comparison between two empirically estimated matrices (see equation (6) below) and no physical information beyond the EEG itself is required. Yet the SNR transform, like the scalp Laplacian, spatially sharpens the scalp EEG by selectively attenuating lower spatial frequencies and accentuating higher spatial frequencies in the data. See Figure 1. How is this possible without using auxiliary spatial information? The control EEG signal naturally contains spatial correlations between channels because, like the signal of interest, it also originates in the brain and is volume conducted to the scalp. The spatial SNR transform works by reversing the between-channel correlations that are found in the control signal. This spatially sharpens the signal of interest relative to the control signal, i.e., the foreground signal is "spatially contrasted" against the background signal. As part of this same process, the SNR transform also normalizes the amplitude of the signal of interest relative to the control signal: Physical units of the original EEG measurements (microvolts) are replaced with dimensionless SNR values. Since higher spatial frequencies in the data may be associated with noise, it is important to note that the spatial SNR transform attenuates these unless they contribute more to the signal of interest. This cannot be said of the scalp Laplacian.

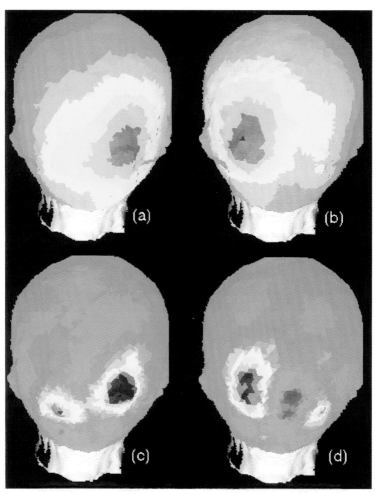

Figure 1
Spatial sharpening via the SNR transform. (a) Visual evoked potential (VEP) map at 62 ms following a checkerboard wedge stimulus presented in the lower left quadrant of the visual field. (b) VEP map at 76 ms following a lower right quadrant stimulus. (c) SNR transform of map (a). (d) SNR transform of map (b). Each map's color scale has been normalized by the standard deviation across all 256 channels, with red most positive, blue most negative, and green at zero. Each SNR transform was computed by estimating the full spatial covariance matrix of the non-phase-locked activity during the same time interval in which the average VEP was estimated. The inverse symmetric square root of this spatial covariance matrix was applied to the phase-locked VEPs. In other words, the phase-locked responses were sharpened via the reverse spatial filter of the natural correlations found in the background activity. SNR maps (c) and (d) selectively enhance higher spatial frequencies, and resolve bilateral responses that are not obvious in (a) and (b).

Suppose that we have selected a particular view of scalp EEG that enhances what we regard as signal of interest and attenuates what we regard as signal of no interest. How can the question of 2D spatial resolution be tackled in this context? Three basic and interrelated approaches are as follows. First we may ask: What is the "information content" of EEG as a function of spatial sampling density? Multichannel information should be a monotonic nondecreasing function of increasing spatial sampling density and/or coverage. How does total information grow as the number of channels increases? Is there an asymptote? A second basic approach asks: What is the spatial power spectrum of EEG at the 2D scalp surface? Higher spatial frequencies require higher electrode densities. Is there a highest effective spatial frequency in the data? Third, we can investigate our ability to predict what is measured at a scalp location as a function of observations at other locations. For example, various interpolation functions predict values at all unsampled locations based on a finite set of sampled locations. How does prediction accuracy improve as a function of spatial sampling density? Is accuracy practically perfect beyond a critical density? Intuitively, we expect these three approaches to be related as follows. If the spatial power spectrum of the signal of interest is negligible beyond some spatial frequency f, then a perfect interpolation function exists for all electrode configurations that are adequate for f. If so, then information content should reach a plateau beyond a critical density, at which point all data required to perfectly reconstruct the continuous scalp function will have been sampled.

1.3 Level (c): Relating scalp EEG to task performance

When scalp EEG is viewed as a covariate of task behavior as it is for level (c) analysis, then we are engaged in cognitive psychophysiology in the sense of Donchin *et al.* (1986). From the perspective of the task, which has been designed to evaluate theories of information processing at the "algorithmic" level, we are primarily interested in finding statistically significant EEG covariates of task performance. Thus, the question of spatial resolution can be translated as: What is the effect of spatial sampling on the statistical power for making EEG detections and discriminations that correspond to conditions and other distinctions in the task? In short, at level (c) the concept of spatial resolving power reduces to the effect of spatial sampling on statistical power. With respect to the statistical power for detecting a task-related EEG response, we ask: How does spatial sampling affect our ability to reject the null hypothesis of no difference between a task condition and background activity (or a control condition)? Similarly, with respect to the statistical power for discriminating between task-related EEG responses, we ask: How does spatial sampling affect our ability to reject the null hypothesis of no difference between two task conditions?

1.4 Level (d): Relating scalp EEG to structures and processes in the brain

Analysis at level (d) implements the critical step of 3D neuroimaging. For the EEG neuromodality, electric source imaging (ESI) is the interface between EEG measurements taken on the outer scalp surface and neuroelectric activity inside the brain. Thus, questions about the 3D spatial resolution of EEG must be addressed within the theoretical framework of ESI.

ESI techniques for estimating the generators of neuroelectric activity in the brain are based on four basic elements: (i) scalp data, (ii) a biophysical model of the head as a volume conductor, (iii) a neurophysiological model of the brain as a source space, and (iv) a statistical estimation procedure. Figure 2 depicts how these four elements fit together within a general framework for solving inverse problems (see legend).

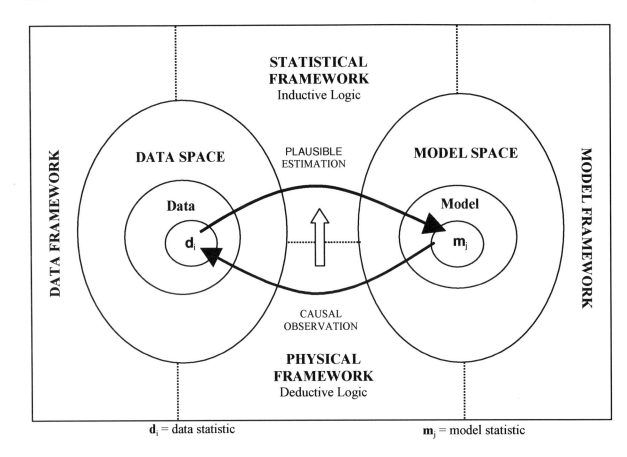

Figure 2

A four-fold framework for ESI theory. Electric source imaging (ESI) is a body of theories and techniques for solving the EEG inverse problem, which has the abstract structure shown in this diagram. There are four parts to ESI theory: a data framework for thinking about the signal/noise composition of EEG measurements at the scalp; a model framework for thinking about the statics and dynamics of neuroelectric generators in the brain; a physical framework, utilizing deductive logic, for predicting observations (EEG measurements) that are caused by assumed source configurations; and a statistical framework, utilizing inductive logic, for estimating plausible source models given the data. The three rings of data represent: an a priori data space, the a posteriori raw data, and a computed data statistic that enhances a signal of interest (SOI). Similarly, the three rings of models represent: an a priori model space, a specific model or distribution of models that generate the data, and a model statistic for the SOI generator. The arrow from source models to scalp data represents a solution to the forward problem via a biophysical model of volume conduction in the head. Note that a "head model" of volume conduction is constructed in the physical framework and is separate from "brain models" of neuroelectric generators. The arrow from scalp data to source models represents a solution to the inverse problem via an estimation procedure that has been designed to optimize a statistical objective function. The vertical arrow in the center of the diagram indicates that inverse estimation depends on a solution to the forward problem, but not vice versa.

Two general points are amplified in Figure 2. From the perspective of physical causation: The spatial complexity of the observed scalp data is driven by the complexity of the neuroelectric generation process. And from the perspective of statistical estimation: Source model accuracy and resolution depend fundamentally on the quality and resolution of the scalp data. These "forward" and "inverse" viewpoints give us two basic approaches for studying the 3D spatial resolution of EEG. Considering first the one-way forward approach, we can simulate different source configurations in a brain model, project to the scalp surface via a head model, and study the resultant patterns of simulated EEG. Two 3D source configurations are empirically distinguishable if, and only if, the difference between their resultant 2D scalp patterns cannot be attributed to measurement error. If the 2D spatial patterns are indistinguishable for a given electrode configuration, then there is no hope of making a 3D distinction, and the one-way forward approach will have found an ultimate limit of 3D resolution. If they are resolvable at the scalp, then we can continue with the inverse phase of a two-way approach to study 3D accuracy and resolution in the source space. That is, we apply an inverse estimator to the simulated scalp data. 3D accuracy and resolution are closely related: Accuracy is the ability of an inverse procedure to make a model that comes close to the original simulated source configuration, and 3D resolution is (in a particular sense) the ability to distinguish two source estimates when the original simulations were different. Stability is another crucial property that can complicate the full two-way approach. If a small change is made to the original source simulation that produces a detectable difference at the scalp, an inverse estimation procedure possibly could produce source models that are far apart, or even qualitatively different. In summary, then, the one-way approach is a general method for studying the ultimate limits of 3D spatial resolution, whereas the two-way approach must be used to assess the actual stability, accuracy, and resolution of particular inverse estimation procedures.

Figure 3 (see legend) depicts the major components of ESI more concretely than does Figure 2. Although differently labeled and arranged, the same basic elements are present.

To oversimplify a bit, there are two general categories of EEG inverse methods: overdetermined and underdetermined. This distinction is based primarily on the number of estimated brain sources relative to the number of measured scalp sensors. Overdetermined methods attempt to find fewer sources than sensors, whereas underdetermined methods attempt to solve for more sources than sensors. Both locations and activity moments are unknown for overdetermined estimators; thus nonlinear optimization is required. By contrast, underdetermined estimators can utilize linear operators because all possible source locations and orientations are specified a priori; thus the problem is to solve for an activity distribution in the source space. When discrete sources are assumed in the overdetermined case, a unique minimum exists. However, in practice it is generally difficult to find this global minimum for more than a handful of sources, even without noise. When distributed sources are assumed in the underdetermined case, a linear solution can easily be calculated. However, numerous so-called "unique" solutions are possible, depending on extra assumptions and requirements that must be added to make the problem determinate. A third notable category of inverse methods is region-of-interest (ROI) estimation, or 3D spatial filtering: The goal is to estimate source activity independently for each ROI, such that contributions or "cross-talk" from the remainder of the source volume are minimized. See, for example, section 2.3.2 below on the Backus-Gilbert theory.

Each inverse method requires a model of the head as a volume conductor, and thus 3D spatial resolution fundamentally depends on head model accuracy. Head model variations include models with simple geometries that have exact analytic or series solutions (spherical and spheroidal, with 1-4 shells), and flexible (realistic) geometry models that have approximate numerical solutions (finite difference method, boundary element method, finite element method). There are also hybrid approaches (e.g., multiple spheres, where a separate sphere is fitted for each electrode). The boundary element method (BEM) is presently the most widely used realistic geometry head model (van den Broek, 1997; Fuchs et al., 1998).

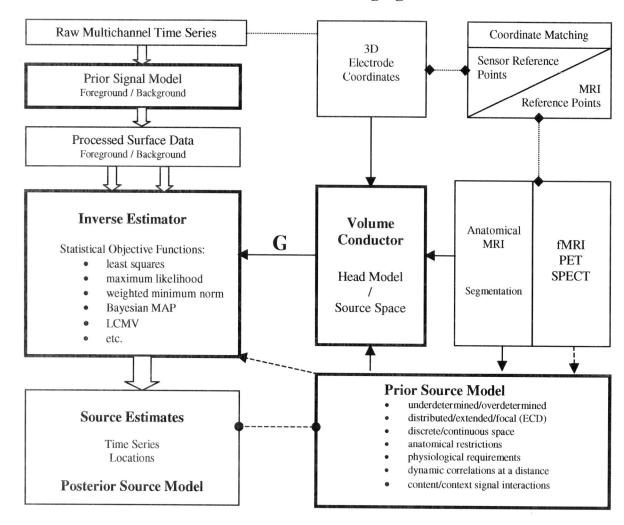

Figure 3

Major components of ESI. The main flow of inverse estimation is depicted on the left side of this diagram, from raw multichannel EEG time series (top) to estimates of source time series and locations (bottom). Prior to inverse estimation, which optimizes some statistical objective function, the raw EEG is processed in a way that enhances a prior conception of the signal of interest (SOI). Thus, the SOI is amplified as a foreground signal, whereas background signals are attenuated. The only contact EEG has with physical space comes from measurements of the electrode 3D coordinates. Anatomical MRI provides a realistic geometric context that is otherwise missing from EEG, although a 3D digitizer by itself can produce a realistic representation of the scalp surface. Co-registration procedures are required to match electrode coordinates with MRI coordinates. The prior source model, which can be informed by anatomical and functional images, embodies all conceptions about the brain as a generator of neuroelectric signals. The volume conductor occupies a key position: It receives input from electrode coordinates, MRI segmentation of major head compartments (scalp, skull, CSF, brain), and the prior source model. All of this information is ultimately reduced to a gain matrix **G** that is utilized by inverse estimation procedures to derive specific posterior source models. See Simpson *et al.* (1995).

Volume conduction is passive, and all volume conductor models presently in use are static models. On the other hand, EEG genesis in cortex (and other parts of the brain) is active and dynamic. Although there is no appreciable frequency dependence of transmission from cortex to scalp, i.e., capacitance can be neglected in the volume conductor, there are tendencies in the cortex itself for high temporal frequencies to be correlated with high spatial frequencies (Pfurtscheller and Cooper, 1975; Nunez, 1981; Nunez, 1995). Thus, since higher spatial frequencies are relatively more attenuated by the skull, higher temporal frequencies tend to be more attenuated at the scalp relative to cortex. The natural coupling between spatial and temporal scales can sometimes break down, however, as in the observation of high-amplitude (>200 µV) gamma activity in scalp EEG (Baldeweg et al., 1998). Most a priori source models that are currently in use for solving the inverse problem do not attempt to incorporate cortical dynamics, aside from the assumption in some distributed source models that nominal correlations of activity in the brain tend to increase with decreasing spatial separation. Equivalent current dipole (ECD) models assume strongly synchronous activity on a local scale, approximately on the order of millimeters to one or a few centimeters.

Specification of the 3D resolution of EEG is difficult in general because of the variety of inverse approaches, and ultimately because of the ill-posed nature of the EEG inverse problem itself. For each scalp pattern, there is an equivalence class of source configurations that could have produced that pattern. The size of an equivalence class in the source space depends not only on measurement errors at the scalp but, more fundamentally, on the nonuniqueness property: In general, two qualitatively different source distributions can generate equivalent scalp patterns. On the other hand, each particular inverse estimation procedure is restricted to a particular class of source models, and therefore it gives solutions belonging to that class only. There may be a unique solution in a restricted class of source models. In this regard, Amir (1994) proved that if source models are restricted to point sources such as dipoles, then they are uniquely determined by their continuous scalp maps. Two distinct source distributions can produce the same scalp map, but only if they intersect, in which case there is a unique minimum volume generator of the map, which represents the equivalence class of all generators that produce the same map. Amir emphasized that these proofs require continuous, i.e., infinite density, scalp maps rather than discrete spatial samples. Thus, there may be no hard theoretical upper limit on the spatial sampling density.

2 Methodological Principles

This section reviews technical principles that can be applied to study the spatial resolving power of EEG.

2.1 2D spatial resolution at the scalp surface

Three interrelated methods for assessing 2D spatial resolution at the scalp surface were introduced at the end of section 1.2: information content measurement, spatial spectrum analysis, and prediction/interpolation. We now consider these in some further technical detail.

2.1.1 Information content measurement and the spatial SNR transform

The method for measuring information content presented in this section is closely linked to the signal-to-noise ratio (SNR) of the data (Pflieger and Sands, 1995; Pflieger and Sands, 1996; Pflieger et al., 1996; Pflieger and Halgren, 1999). Other information theoretic approaches can be based on the estimation of probability densities (Scott, 1992), but are not considered here. The basic concept is considerably simplified for a single channel. If SNR is obtained as the square root of an estimated signal energy (µV units) divided by an estimated noise standard deviation (also µV units), then the estimated information content in dimensionless units of "bits above noise" equals the base 2 logarithm of the SNR. Thus, if the SNR is 1 (2, 4, 8, 16, etc.), then the information content is 0 (1, 2, 3, 4, etc.) bits above noise. Note that if the SNR is 1, then unless other information is available that has not been incorporated into the SNR computation, signal cannot be discriminated from noise in the data; consequently, the information content is 0. Also note that information such as differences in the (temporal) frequency spectra between signal and noise can be incorporated by computing the SNR in the frequency domain (Möcks, 1988; Pflieger, 1991). Increasing SNR reflects our improving ability to discriminate signal from noise. Thus, the bits-above-noise measure of information is fundamentally designed to reflect our capacity to empirically distinguish signal from noise. Implementation of this measure depends on our choice of a theoretical distinction between signals of interest and signals of no interest, as discussed in section 1.2.

To estimate multichannel information content, we utilize the spatial SNR transform that was introduced in section 1.2. Of the various ways of selecting a signal of interest and contrasting it against a background (or control) signal of no interest, the definition of SNR transform given below uses the standard ERP model of signal and noise, i.e., phase-locked foreground signal versus non-phase-locked background signal. However, the spatial SNR transform can be adapted as well to accommodate other signal/noise models, e.g., induced activity following a stimulus contrasted against ongoing activity in the prestimulus baseline, etc.

The standard ERP concept assumes an additive data model:

$$\mathbf{d}_i(t) = \mathbf{f}(t) + \mathbf{b}_i(t) \qquad (1)$$

where $\mathbf{d}_i(t)$ is an M-channel vector for raw data epoch i at time t; $\mathbf{f}(t)$ is the channel vector for the foreground (phase-locked) brain signal at time t, which is assumed to be homogeneous across epochs; and $\mathbf{b}_i(t)$ is a channel vector for a background (non-phase-locked) brain signal for epoch i at time t, which is assumed to be generated by a zero-mean, stationary, multivariate Gaussian process. To the extent that the foreground signal is not homogeneous (Möcks et al., 1984; Pflieger, 1991), signal/noise model (1) leads to misallocation of trial-to-trial variability of the foreground signal into the background variance. On the other hand, although the assumption of Gaussian background activity may or may not be justified on single trials (Elul, 1969), the Central Limit Theorem pushes the residuals after averaging to approach Gaussian behavior as the number of epochs increases.

The crossproduct of sums and the sum of crossproducts matrices are defined as

$$\mathbf{CP}_{\text{sum}} \equiv \sum_{t=1}^{T} \left(\sum_{i=1}^{I} \mathbf{d}_i(t) \right) \left(\sum_{i=1}^{I} \mathbf{d}_i(t) \right)^t \qquad (2)$$

$$\mathbf{SUM}_{\text{cp}} \equiv \sum_{i=1}^{I} \sum_{t=1}^{T} \mathbf{d}_i(t) \mathbf{d}_i(t)^t \qquad (3)$$

where I is the number of epochs included in the condition and T is the number of points in the time interval of interest. Following Pflieger and Halgren (1999), an unbiased estimate of the foreground (phase-locked) signal crossproducts matrix is

$$\mathbf{CP}_f \cong (\mathbf{CP}_{sum} - \mathbf{SUM}_{cp}) / TI(I-1) \qquad (4)$$

with an associated estimate of the background (non-phase-locked) covariance matrix

$$\mathbf{COV}_b \cong (\mathbf{SUM}_{cp} - \mathbf{CP}_{sum} / I) / TI(I-1) \qquad (5)$$

which is scaled as a residual after averaging over the I epochs. Note that equation (4) is "almost" the crossproduct matrix for an average ERP in the time interval of interest, except for the subtraction of the sum of crossproducts matrix, which introduces a correction for the contribution of the estimated residual background activity.

The spatial SNR transform is performed by applying the inverse symmetric square root of the background covariance matrix $\mathbf{COV}_b^{-1/2}$ to the data, which is a spatial whitening, or sphering, operation. The inverse symmetric square root of a matrix is formed here by taking reciprocal square roots of its eigenvalues and reconstructing via its eigenvectors.

In order to estimate information content, an SNR matrix is computed as

$$\mathbf{SNR} \equiv \mathbf{COV}_b^{-1/2} \mathbf{CP}_f \mathbf{COV}_b^{-1/2} \qquad (6)$$

and a principal component analysis is performed:

$$\mathbf{SNR} = \mathbf{UWU}^t \qquad (7)$$

where the eigenvector matrix $\mathbf{U}$ is orthogonal and the eigenvalue matrix $\mathbf{W}$ is diagonal. Note that the root eigenvalues of the SNR matrix are dimensionless signal-to-noise ratios for their associated scalp map eigenvectors, which are linearly independent. For our purposes, it is important to observe that the SNR matrix is invariant with respect to an arbitrary reversible linear reformatting of the data. In particular, a re-referencing transformation is linear, and it can be reversed even though it contains one singularity, because the data itself has one singularity to match at the reference electrode. Thus the SNR matrix is reference invariant, and its eigenvalues are also uniquely determined.

The total information contained in the event-related data, in units of bits above noise, is computed as

$$H = \frac{1}{2} \sum_{w_j > 1.0} \log_2 w_j \qquad (8)$$

where w_j is the jth SNR eigenvalue. In the ERP formulation specified by (2) – (6), the eigenvalues have the special property of decreasing to a plateau at SNR=1.0, as illustrated in Figure 4. This information measure inherits from the SNR matrix the property of invariance with respect to "information preserving" linear transformations of the data, such as a change of reference.

Analogous to the square root law for SNR improvement when averaging noisy ERP epochs, a nominal amount of SNR improvement for multiple channels can be attributed to the reduction of uncertainty due to repetitive sampling of the same underlying signals. Although this source of information can be useful, such as for improving single trial SNR, we shall discount it here by dividing the total SNR by a factor of $\sqrt{M}$. The adjusted information measure we shall use is therefore

$$H' = H - \frac{1}{4}\log_2 M. \qquad (9)$$

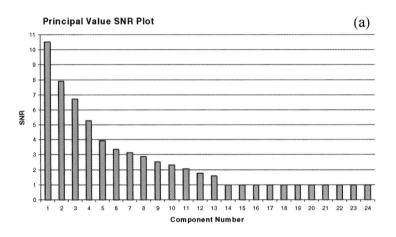

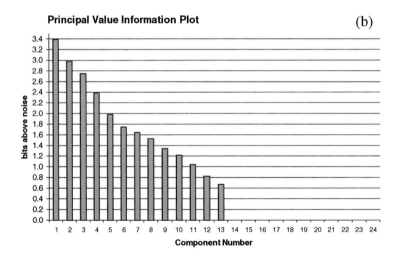

Figure 4
Linearly independent principal components of the SNR matrix. (a) SNR eigenvalues. Note that a plateau at SNR=1.0 is reached after 13 components in this case. (b) Corresponding information plot. The information content drops to zero beyond the thirteenth component. These illustrative plots are from a combined EEG + MEG dataset (Pflieger and Halgren, 1999).

2.1.2 Spatial spectrum analysis

The simplest form of spatial spectrum analysis utilizes a one-dimensional chain of closely and equally spaced electrodes, which mimics the sampling of time series data (Spitzer *et al.*,

1989; Gevins, 1990). This straight chain sampling technique permits the use of standard methods of spectrum analysis, such as the fast Fourier transform (FFT). Two limitations of this technique are: (a) the resulting spatial spectrum depends on the orientation of the linear chain, which is selected a priori; and (b) the actual highest spatial frequencies are to be found along curved paths that are normal to the isopotential contours. Therefore, the one-dimensional method can miss the highest spatial frequencies that are available at the scalp.

Two dimensional spatial spectrum analysis can be performed in rectangular coordinates that utilize a flat projection of the scalp surface, or in spherical coordinates that utilize a more natural spherical projection. Nunez (1981, pp. 249-268) used rectangular coordinates and also included the time dimension in his discussion of frequency-wavenumber spectral analysis, which utilized sinusoids as basis functions in three dimensions (temporal plus two spatial). Wavenumber equals 2π divided by the spatial wavelength, and thus it measures spatial frequency. Spectral power as a function of three coordinates is estimated based on the complex covariance matrix, which is computed from all channel pairwise cross-spectral densities via the FFT. Nunez noted (p. 255) that the power estimation formulas are not invalidated by head curvature so long as actual inter-electrode distances on the scalp surface are plugged in. However, the sphere-like scalp curvature does affect our interpretation of spatial frequency itself, considering that rectilinear coordinates do not preserve global geometric relationships on the scalp. Paranjape *et al.* (1990) interpolated to a regular rectangular grid on a flat map projection in order to apply an algorithm for 2D maximum entropy power spectrum estimation. This method differs in several respects from Nunez (1981), but let us note in particular that the Nunez approach does not require an interpolation function, and that every interpolation function makes implicit assumptions about the spatial spectral content of the data. It would appear, then, that use of an interpolating function for spatial spectrum analysis introduces a factor that begs the question of EEG spatial resolution.

Spherical harmonics. Spherical harmonics are the natural basis functions for spatial spectrum analysis when the scalp surface is projected to spherical coordinates (Shaw, 1989; Srinivasan *et al.*, 1996; Pflieger, 1996). They are defined for complex-valued functions on the sphere (Press *et al.*, 1992, section 6.8), but we are dealing strictly with real-valued functions, so our definition is modified accordingly. The (real-valued) spherical harmonic of integer degree d ($d \geq 0$) and order r ($-d \leq r \leq d$) is defined as

$$Y_{dr}(\theta,\phi) \equiv \sqrt{\frac{(2d+1)(d-r)!}{4\pi(d+r)!}} P_d^{|r|}(\cos\theta) \begin{cases} \cos r\phi, r \geq 0 \\ \sin r\phi, r < 0 \end{cases} \quad (10)$$

where θ and ϕ are the spherical coordinate system "altitude" and "azimuth", respectively, and $P_l^m(x)$ is the associated Legendre polynomial of degree l and order m, with $-1 \leq x \leq 1$. Degree d is related to the spatial period λ by the spherical circumference c via the relation $\lambda = c/d$ so that, for a head of circumference 56 cm, for example, degree 2 corresponds to a spatial period of 28 cm, degree 4 corresponds to 14 cm, degree 8 corresponds to 7 cm, degree 16 corresponds to 3.5 cm, etc. Thus, spherical harmonic degree is proportional to spatial frequency (cm^{-1} units), which is the reciprocal of spatial period. With this understanding, we shall sometimes use the terms "spatial frequency" and "spherical harmonic degree" interchangeably. Figure 5 depicts the 36 spherical harmonics up to degree 5 as maps projected to the back of a head.

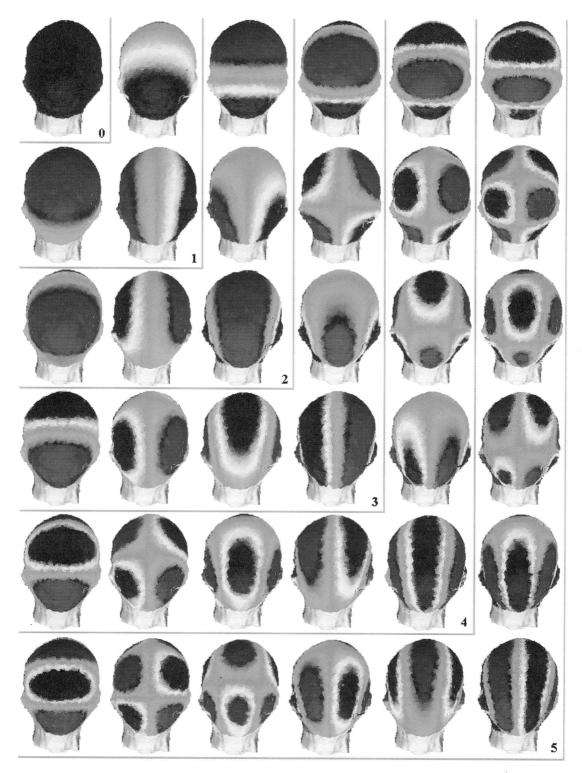

Figure 5

Spherical harmonics up to degree 5 projected to the back of a head. Spatial frequency can be identified with spherical harmonic degree. There are 3 harmonics of degree 1, 5 harmonics of degree 2, 7 harmonics of degree 3, etc. A spherical coordinate system was utilized here with the poles at the ears.

The $\sqrt{M}$ constraint. There is one spherical harmonic of degree 0—a constant function. There are 3 spherical harmonics of degree 1, 5 harmonics of degree 2, and in general $2d+1$ harmonics of degree d. The total number of harmonics up to (and including) degree d is $(d+1)^2$. Therefore, since the spherical harmonics are linearly independent functions, a bare minimum of $(d+1)^2$ EEG channels are required to independently represent all harmonics up to degree d. Put the other way around, M channels are "saturated" by the complete set of harmonics up to degree $d = \sqrt{M} - 1$. For example, 16 channels cannot yield information about spatial frequencies above degree 3 ($\lambda > 18.7$ cm for a head circumference of 56 cm) that is independent of the accumulated information about spatial frequencies below degree 4. In this sense, all 16-channel configurations are strictly limited to the spherical harmonics up to degree 3. Likewise, 36 channels are restricted to harmonics up to degree 5 ($\lambda > 11.2$ cm); 64 channels up to degree 7 ($\lambda > 8$ cm); 121 channels up to degree 10 ($\lambda > 5.6$ cm); 256 channels up to degree 15 ($\lambda > 3.7$ cm); etc. This is a hard constraint: For M channels, power estimates for spherical harmonic degrees greater than or equal to $\sqrt{M}$ cannot be independent of the collected power estimates for spherical harmonic degrees less then degree $\sqrt{M}$.

Nyquist criterion for spherical data. In general, recalling that $\lambda = c/d$, the minimum spatial period supported by M electrodes is

$$\lambda = \frac{c}{\sqrt{M}-1} = \frac{2\pi r}{\sqrt{M}-1} \qquad (11)$$

where r is the spherical radius (roughly 9 cm) and, as before, c is the head circumference (roughly 56 cm). The Nyquist criterion for one-dimensional data, such as time series, requires that the sampling frequency should be twice or more the highest frequency in the data. If this criterion carries over to spherical data in a straightforward way, then the average nearest-neighbor distance for M evenly distributed points on a sphere should approach $\lambda/2$, because λ is the least spatial period that can be supported by M points. Indeed, we can demonstrate this to be the case via a simple algorithm that evenly distributes M points on a sphere (Bourke, 1996): For M about 20 and higher, the computed average nearest-neighbor distance closely approximates $\lambda/2$. Therefore, the following criterion is substantially equivalent to the Nyquist criterion for spherical data:

$$M \geq \left(\frac{c}{\lambda}+1\right)^2 = \left(\frac{2\pi r}{\lambda}+1\right)^2 \qquad (12)$$

where M is the minimum number of electrodes, and λ is the smallest spatial period that contributes to the data.

Relaxation of the $\sqrt{M}$ constraint for partial coverage. One might think that M can be reduced in a straightforward way by covering a smaller area such as a hemisphere, which is closer to actual EEG practice. Intuitively, perhaps M can be reduced by the fraction of total spherical coverage. Although it is locally true that the density in a decreasing area of coverage can remain constant while the number of electrodes decreases, note that the global requirement for M channels remains in force so long as all spherical harmonics up to degree c/λ are retained. At this juncture, there appears to be a breakdown in the analogy with one-dimensional sinusoidal expansion of time series. If the total time epoch of analysis is cut in half and an adequate sampling rate is maintained, then the number of sampled points is also cut in half. The highest frequencies in the time series are still represented, but this is paid for by a loss of frequency resolution, i.e., the spacing between adjacent frequency bins doubles. In time series analysis, the spacing between frequencies equals the fundamental frequency, which equals the reciprocal of the epoch interval, which can be chosen at will. But in spherical harmonic analysis, the fundamental spatial frequency is set by the radius of a sphere fitted to the head, not by the area of coverage. However, an alternative is to eliminate some of the harmonic functions. For hemispheric coverage, Vaidyanathan and Buckley (1997) argue that the negative order harmonics of equation (10), i.e., those with sine terms, can be dispensed with if an even periodic extension is assumed. In this case, the total number of harmonics up to degree d equals $(d+1)(d+2)/2$ rather than $(d+1)^2$, and so the minimum number of required channels is cut almost in half. Thus, our original intuition that M can be reduced by the fraction of total spherical coverage appears to be approximately correct after all.

The electrode-harmonic design matrix. Various electrode configurations are possible. The $\sqrt{M}$ constraint (or its relaxed version for partial coverage) sets up an absolute barrier for all M-electrode configurations, but it does not quantify the actual capacity of a particular configuration to independently cover all spatial frequencies up to some degree $d < \sqrt{M}$. Let (θ_i, ϕ_i) be the spherical coordinates of electrode i, and let $Y_{(h)}$ be the spherical harmonic function of degree d_h and order r_h for harmonic indices h up to $(d+1)^2$. That is, we are considering all harmonics up to degree d. If we define the M-by-$(d+1)^2$ cumulative electrode-harmonic design matrix **D** such that $D_{ih} = Y_{(h)}(\theta_i, \phi_i)$, then the $\sqrt{M}$ constraint says that **D** must be singular if it has more columns (harmonics) than rows (electrodes). If **D** is a square matrix, or if it has more rows (electrodes) than columns (harmonics), then it might still be effectively singular for numerical computations, i.e., ill-conditioned. If a singular value decomposition (SVD) is performed on the design matrix **D**, then its condition number is defined as the ratio of the largest singular value to the smallest (Press *et al.*, 1992, p. 61). **D** becomes ill-conditioned as the reciprocal condition number approaches a computer's floating point precision, e.g., about 10^{-12} for double precision computations. Thus, we can use the reciprocal condition number of **D** as an index to quantify how well a particular electrode configuration is able to independently cover all spatial frequencies up to degree d: The closer to 1, the better the independent coverage.

Spatial power spectrum estimation. Spherical harmonics (10) are continuous functions that are mutually orthogonal when their products are integrated over the sphere. Therefore, if an EEG potential function could be known continuously over the sphere, then its spatial power could be estimated independently for each harmonic via numerical integration. Srinivasan *et al.* (1996) take this approach: The discrete EEG data is converted to a continuous function via 3D spline interpolation, followed by Monte Carlo integration over the portion of the sphere covered by the electrodes (see their equation 32). However, as already noted above, and as the authors themselves show in their figure 5 and Appendix A, every interpolating function makes assumptions about the spatial frequency content of the data to be interpolated, which is the very thing we are trying to estimate. Moreover, in addition to continuity, mutual orthogonality of the spherical harmonics requires integration over the complete sphere, which cannot be performed due to the partial coverage of EEG electrodes. In other words, inability to extrapolate over the complete sphere prevents the property of mutual orthogonality from being realized, which may have been a reason for interpolating in the first place. Thus, it appears desirable to estimate spatial spectra in a way that uses only the actually sampled scalp locations without interpolation. One way of doing this (Pflieger, 1996) is to compute the percentage of the total spatial power in the data that can be accounted for using only the spherical harmonics of degree d. An M-by-$(2d+1)$ electrode-harmonic design matrix is set up separately for each degree d, which is a submatrix of the cumulative design matrix defined in the last paragraph, and all power in this $(2d+1)$-dimensional column subspace of the M-dimensional measurement space is extracted from the data.

Spectral leakage and spatial aliasing. Since the sampled data are not continuous over the entire sphere, the design matrices are not linearly independent for two different spatial degrees d_1 and d_2: There will inevitably be some overlap in the power estimate for degree d_1 and the power estimate for degree d_2. Consequently, when these power estimates are expressed as percentages of total spatial power, the sum over a range of spatial frequencies will generally be greater than 100%. In other words, each estimate of power at a spatial frequency will be more or less inflated. The amount of this inflation can be used to index the spectral leakage due to imperfect orthogonality of the discretely sampled harmonics. In addition, if electrode configuration A is a subset of electrode configuration B, then inflation of the configuration A estimate relative to the configuration B estimate shows that there is some misallocation of spatial variance (i.e, aliasing) in configuration A. To find out if there is also spatial aliasing in configuration B, we would need a configuration C that is a superset of B.

2.1.3 Prediction and interpolation

Since interpolation predicts what should be observed at a location based on observations at other locations, the question of interpolation error arises. Perrin *et al.* (1987) and Perrin *et al.* (1989) calculated RMS interpolation errors (using surface and spherical splines) for data simulated by a dipole situated at various locations in a head model. Koles and Paranjape (1988) used actual recordings to estimate the accuracy and precision of interpolation for ongoing EEG. Extra test locations were added to a standard electrode montage, and the scalp data were digitally bandpass filtered into several bands of interest. For each band, the magnitude and magnitude squared were each interpolated (using bilinear and bicubic splines) at every instant to the test locations. Accuracy was defined as the average difference between the interpolated and measured quantities in a time interval, and precision was defined as Pearson's correlation coefficient in the time interval.

A limitation of these methods for our purposes is that they depend on specific interpolating functions that are not necessarily based on the theory of EEG volume conduction. From the 3D viewpoint, the only perfect interpolation function is one derived from the forward projection of known sources, which is linear from sources to the scalp per equations (13)-(15) below. Except for simulated data and special circumstances, we do not know the sources; but if we had a best linear estimator of the sources from the scalp data, we could in turn project the estimated sources back to any scalp location. We do not generally know the best linear estimator either, but whatever it is, it determines a best linear prediction from the M measured scalp locations to any test location. That is, there is an M-dimensional vector $\vec{h}$ such that the best prediction at the test location given scalp data $\vec{v}$ is $\vec{h}^t \vec{v}$. For any test location, $\vec{h}$ can be estimated via linear regression on the scalp data in a time interval, and the prediction errors (i.e., goodness-of-fit) can be studied. In theory, no interpolating function should ever outperform this best retrospective linear predictor. Note that since linear regression is based on the data covariance matrix, there is a close relationship between the best retrospective linear predictor and the linear information content measure discussed in section 2.1.1.

There is also a tight relationship between interpolation and spherical harmonic spectrum estimation (section 2.1.2): Representation of the data as a spherical harmonic expansion produces a natural interpolating function (Pascual *et al.*, 1988).

2.2 Statistical power

One might expect that increasing the number of EEG channels should automatically increase the statistical power for comparing scalp maps. However, Bellman's "curse of dimensionality" (1961) has to be contended with, which presents itself in a number of guises (e.g., see Scott, 1992, chapter 7). For example, as Karniski *et al.* (1994) put it (p. 203): "Ironically, the increased spatial resolution which is realized by increasing the number of measured points leaves investigators in a paradoxical position. As spatial resolution increases with the addition of more measured points in space, traditional statistical tests for analyzing the large amount of data have become less powerful and more compromised (Blair and Karniski 1994)." The reference to traditional tests includes omnibus tests of global differences such as Hotelling's multivariate T^2 (Faux and McCarley, 1990) and multiple comparison procedures for localizing differences such as Bonferroni correction of multiple univariate t-tests. These problems are caused by several factors. As the number of channels M increases, the following also increase: (i) sensitivity to violations of assumed multivariate normality; (ii) the number of observations required to obtain an estimate of the channel covariance matrix (M at bare minimum, >4M recommended); and (iii) conservatism of corrections due to inadequate treatment of increasing correlations in the data as inter-electrode spacing decreases.

Statistical power is defined as the probability of correctly rejecting the null hypothesis. In a detection problem, the null hypothesis states that there is no signal, i.e., only noise. For example, we might ask if an event-synchronized average differs from the residual background EEG. If signal is actually present, then the statistical power of a test equals the probability that it correctly detects the signal when it is actually present. Note that power is a function of the signal to be detected, noise characteristics, the detection test employed, and the decision criterion (i.e., the controlled probability of a false positive or Type I error). In a discrimination problem, the null hypothesis states that there is no difference between signals in datasets obtained under two (or more) conditions. For example, we may want to know if there are differences in ERPs that were obtained using two stimuli, or differences in event-related coherence, etc. If two signals actually are different, then the statistical power of a test is the probability that it correctly discriminates them. In this case, power depends on the two signals, the noise characteristics of each dataset, the discrimination test, and the criterion.

The omnibus tests just mentioned utilize all spatiotemporal data to detect or to discriminate signals, but not to answer local "when" or "where" questions, e.g., "Which electrodes and/or latencies show significant differences between conditions?" To address such local questions, multiple univariate tests, one per time-space element, are constructed to form a dynamic statistic image, and then a procedure is employed to assess which elements have statistics that are significant at a level that controls for the global probability of a Type I error (cf., statistical parametric or nonparametric mapping per Frackowiak *et al.*, 1997, chapters 4 and 5). For example, Babiloni *et al.* (1994) have used multiple univariate tests as a technique for spatial data reduction, i.e., channel selection, prior to omnibus multivariate statistical analysis via Hotelling's T^2. The curse of dimensionality dictates this sequence, i.e., multiple univariate prior to multivariate, because Hotelling's T^2 statistic cannot be computed unless the number of observations (e.g., experimental subjects) exceeds the number of channels (i.e., data vector dimension). Ideally, however, one might wish to answer the more general questions (Can we detect a signal? Is there a difference?) before proceeding to ask the more specific questions of location (Where is the signal? Where are the differences?). Even better, one might wish for an integrated test that handles the global/local questions in one step.

Randomization tests, particularly random permutation tests (Edgington, 1987), can overcome some of the curses of dimensionality (Blair and Karniski, 1994; Karniski *et al.*, 1994). They are distribution free; the number of observations can be less than the number of channels; and correlations in the data channels are automatically taken into account. In addition, permutation tests do not assume that the observations are independent and identically distributed (i.i.d.), which suggests that they can be applied to sets of observations taken from a single subject, e.g., EEG epochs from a single recording session (Edgington, 1987). Finally, permutation tests are flexible: They can be applied to just about any data statistic that can be imagined, from ordinary event-related averages to more complicated statistics such as event-related coherences.

A random permutation test of the difference between conditions works by blindly remapping observations among the available conditions. Suppose we have a collection of observations, such as multichannel EEG epochs, and each observation is assigned an experimental label, such as a stimulus type. Consider all possible ways that the observations can be relabeled, and randomly select a relatively large number of these, for example 1000. Reassignments can be implemented by randomly permuting the labels while keeping the observations in fixed order, or vice versa. A test statistic, which could be almost any single number computed from the datasets, e.g., the maximum absolute univariate Student's *t*-statistic across all channels, is computed as a function of the observations and their assignments. First we compute the chosen statistic for the actual experimental assignment. Then we compute the statistic 999 more times for the random reassignments, and count the number of times that a value is observed to be greater than or equal to the obtained statistic. This number divided by 1000 is the estimated *p*-value. The null hypothesis in this case is that there is nothing special about the actual experimental assignment. In other words, if the *p*-value is not small, it is likely that the EEG statistic we have computed does not reflect meaningful differences between the experimental conditions.

Statistical power can be estimated via Monte Carlo simulation techniques that generate many datasets for each condition of interest. More will be said about data simulation methods in section 2.4 below, so suffice it to say here that the main task is to simulate plausible brain signals of interest, and to embed them in quasi-realistic background activity, correlations and all. A statistical test with a controlled false positive rate is applied to all simulated datasets, and the estimated probability of signal detection, i.e., the power of the test, can be plotted as a function of signal intensity or SNR. In addition to power, the related measure of statistical sensitivity can be estimated from the operating characteristic, which is plotted by varying the Type I error criterion (Swets, 1961). Similar considerations apply for the case of discriminating two signals.

2.3 3D spatial resolution in the brain volume

In this section, we review of some basic principles of ESI, apply the Backus-Gilbert theory to the question of 3D spatial resolution, discuss how 3D errors are quantified, and briefly speculate on the duality of overdetermined and underdetermined estimators.

2.3.1 Basic principles of ESI

Relationship between scalp potentials and source currents. The biophysical foundation of ESI methods is the instantaneous linear causation from active brain currents to surface EEG potentials (quasi-static approximation; Plonsey and Heppner, 1967). We start with equation (13) below, which is based on Poisson's equation and ultimately derives from Maxwell's equations (Geselowitz 1967; Barnard *et al.* 1967; Plonsey 1969; de Munck 1989; van den Broek 1997). If $\vec{v}_R(t)$ represents an *M*-channel surface ERP/EEG vector at time *t*, and **j** represents a time-varying vector field of active current sources in the brain volume *V*—that is, $\mathbf{j}(\mathbf{p},t) = [j_x(\mathbf{p},t), j_y(\mathbf{p},t), j_z(\mathbf{p},t)]^t$ is the active current source vector at time *t* and point **p** in *V*—then

$$\vec{v}_R(t) = \mathbf{R} \int_{\mathbf{p} \in V} \vec{g}(\mathbf{p}) \cdot \mathbf{j}(\mathbf{p},t) d\mathbf{p} + \vec{n}_R(t) \tag{13}$$

where **R** is a referencing matrix that maps *M'* electrodes to *M* channels (*M'*>*M*); $\vec{n}_R(t)$ is an *M*-dimensional residual noise vector with an expected value of zero; and $\vec{g} = [\vec{g}_x, \vec{g}_y, \vec{g}_z]$ is the gain operator, where $\vec{g}_x(\mathbf{p})$ is the vector of *M'* electric potentials (with respect to infinity) produced by a unit source current at point **p** in the *x* direction, and similarly for $g_y(\mathbf{p})$ and $\vec{g}_z(\mathbf{p})$. (The notation *d***p** here is shorthand for $dp_x dp_y dp_z$.) When the gain operator $\vec{g}$ is viewed from the standpoint of each electrode, it is equivalently known as the lead field operator. The lead fields thus connect what is known (i.e., a finite dimensional vector of macroscopic surface measures) with the unknown generators (i.e., a continuous field of microscopic source currents). By dividing the brain volume *V* into suitably small volume elements that collectively make a source space, we can derive a discrete approximation for equation (13) of the form

$$\vec{v}_R(t) \cong \mathbf{R}\mathbf{G}\vec{j}(t) + \vec{n}_R(t) \tag{14}$$

where the gain matrix **G** maps unit currents from a large number *N* of "microdipoles" in a discretized source space (brain volume grid, or discretized cortical surface) to surface potentials at *M'* scalp electrodes. If the source space is discretized as a 3D grid within the head, then three orthogonal dipoles are placed at each grid location. For simplicity of notation, the explicit referencing matrix **R** can be suppressed, with the understanding that it is absorbed into the gain matrix **G**. The time index can also be suppressed, so that (14) is simplified as follows:

$$\vec{v} \cong \mathbf{G}\vec{j} + \vec{n} . \tag{15}$$

Electrodes are insensitive to solenoidal currents. For the electric case, Grave de Peralta Menendez and Gonzalez Andino (1998a) have argued that the source current distribution **j**, a vector field, can be reduced to the gradient of a scalar field ϕ, i.e.,

$$\mathbf{j} = \nabla \phi \tag{16}$$

because (a) any **j** that vanishes at infinity can be written as the sum of an irrotational part and a solenoidal part (Helmholz theorem), (b) the solenoidal part is undetectable in the electric case, and (c) the irrotational part can be reduced to the gradient of a scalar potential field. Thus, without loss of generality, we can solve for ϕ instead of **j**, which reduces the solution space three-fold.

Half-sensitivity volume. One might suppose that a single EEG channel with a pair of electrodes has no spatial resolving ability, but there is a sense in which it does. The lead field of a pair of scalp electrodes is the vector electric current field produced inside the head when unit current is passed between the leads. By the Helmholz reciprocity theorem, this is equivalent to the sensitivity of the electrode pair when currents are generated in the head (Rush and Driscoll, 1969). Thus, the gain operator of equation (13) is also called the lead field operator; likewise, the gain matrix of equation (14) is also called the lead field matrix. Malmivuo *et al.* (1997) defined the half-sensitivity volume (HSV) as "the volume of the source region in which the magnitude of the detector's sensitivity is more than one half of its maximum value in the source region" (p. 197). In other words, the HSV is the volume inside the isosensitivity surface with sensitivity equal to one half of the global maximum sensitivity. Thus, the HSV measures the ability of a pair of electrodes to concentrate its measurement sensitivity in space. For a pair of closely spaced electrodes, the HSV is composed of a single region; but for sufficiently large separation, the HSV splits into two regions, each beneath its electrode. The two-electrode configuration is most sensitive to currents along the line between them, which tends to be tangential to the scalp surface. A three-electrode linear configuration with the middle electrode as one terminal and linked lateral electrodes as the other has a half-sensitivity volume that is radially oriented beneath the middle electrode. By extension, one can imagine a pseudo-Laplacian configuration that has a central electrode as one terminal and an annular ring electrode as the other. Generalizing further, one can digitally construct half-sensitivity volumes for linear combinations of channels.

2.3.2 Backus-Gilbert theory: Resolution and stability

Whereas the half-sensitivity volume concept looks at the entire source volume from the standpoint of a single measurement channel, we now look at all measurement channels from the standpoint of a single voxel in the source volume.

Backus and Gilbert formulated an inverse theory in the context of geophysical applications (Backus and Gilbert, 1968; Backus and Gilbert, 1970) that has been applied to the MEG/EEG inverse problem by Robinson and Rose (1992) and Grave de Peralta Menendez *et al.* (1997). In this subsection, we consider the Backus-Gilbert theory as an approach for treating the question: How does 3D spatial resolution vary as a function of 2D sensor density? In the Backus-Gilbert formulation, one solves for a linear spatial filter, i.e., a vector of scalp channel weights, that optimizes a tradeoff between spatial resolving power and solution stability at a given 3D location in the brain. Unlike underdetermined linear inverse methods that solve a single simultaneous system of equations for all points in the source space volume, e.g., variants of the minimum norm approach, the Backus-Gilbert method independently solves for an optimal spatial filter at each location. Other notable spatial filter designs are linearly constrained minimum variance (LCMV) beamformers (Spencer *et al.*, 1992; Van Veen *et al.*, 1997) and synthetic aperture magnetometry (Robinson and Vrba, 1998). The following treatment of the Backus-Gilbert theory owes a debt to the clear presentation by Press *et al.*, 1992 (pp. 815-818).

Our first task is to define a reasonable quantitative measure of "spatial resolution" in a 3D sense. We aim to estimate brain electrical activity at any given point $\mathbf{p}$ in the brain volume on the basis of an M-dimensional electric potential measurement vector $\vec{v}$ taken at the scalp. All of our candidate estimators are linear spatial filters, each of which is an M-dimensional nonzero vector $\vec{h}(\mathbf{p})$ such that the estimated activity at location $\mathbf{p}$ is $\vec{h}^t(\mathbf{p})\vec{v}$, i.e., the dot product of the spatial filter vector with the measurement vector. Now consider the following thought experiment. Imagine that a single source is active at a point $\mathbf{p'}$ located at a distance $|\mathbf{p'}-\mathbf{p}|$ from $\mathbf{p}$, with no other sources active in the brain volume. If we have a perfect volume conductor model of the head and if we can measure scalp potentials without error or noise, then we can exactly predict the resulting measurements, $\vec{v}(\mathbf{p'})$. Unless $\mathbf{p'}=\mathbf{p}$, there is no activity at location $\mathbf{p}$. Nevertheless, our linear spatial filter estimates that the activity at $\mathbf{p}$ is $\vec{h}^t(\mathbf{p})\vec{v}(\mathbf{p'})$, which cannot be zero unless $\vec{h}(\mathbf{p})$ is orthogonal to $\vec{v}(\mathbf{p'})$. Thus, there is generally an unavoidable spread from actual activity at point $\mathbf{p'}$ to estimated activity at point $\mathbf{p}$. The mean square spread over the entire source volume V, weighted by the square of the estimated activity at target location $\mathbf{p}$, is

$$\sigma_s^2(\mathbf{p}) = \frac{\int_V \left|\vec{h}^t(\mathbf{p})\vec{v}(\mathbf{p'})\right|^2 |\mathbf{p'}-\mathbf{p}|^2 d\mathbf{p'}}{\int_V \left|\vec{h}^t(\mathbf{p})\vec{v}(\mathbf{p'})\right|^2 d\mathbf{p'}} . \qquad (17)$$

The square root of this quantity, $\sigma_s(\mathbf{p})$, the root mean square spread, measures the average 3D spatial spread of the linear estimator $\vec{h}(\mathbf{p})$ in units of distance, such as millimeters. The reciprocal of the spatial spread is the spatial resolution, which has units of spatial frequency.

Equation (16) treats activity at each point in the source volume as a scalar rather than as a vector. If the source solution space is viewed as a vector field of currents throughout the source volume as in (12), then spatial orientations must be treated in addition to locations. In this case, we must somewhat arbitrarily decide how to penalize the spread from one orientation to a different orientation, e.g., from the y orientation at location $\mathbf{p'}$ to the z orientation at location $\mathbf{p}$, etc. Grave de Peralta Menendez *et al.* (1997) suggest alternatives for doing this, but instead we will utilize equation (16) to work directly with scalar fields. See also the following reference, which was discovered during final editing: Grave de Peralta Menendez and Gonzalez Andino (1999).

If we write equation (13) from the standpoint of a particular channel i and substitute equation (15), the result is:

$$v_i = \int_{\mathbf{p}\in V} \mathbf{g}_i(\mathbf{p}) \cdot \nabla \phi(\mathbf{p}) d\mathbf{p} + n_i \qquad (18)$$

where explicit referencing and the time index have been dropped. If we define

$$\gamma_i \equiv \mathbf{g}_i \cdot \nabla \qquad (19)$$

then equation (18) takes the simpler form

$$v_i = \int_{\mathbf{p}\in V} \gamma_i(\mathbf{p})\phi(\mathbf{p}) d\mathbf{p} + n_i . \qquad (20)$$

Note that $v_i(\mathbf{p'})$ from equation (17), i.e., the ith channel of $\vec{v}(\mathbf{p'})$, equals $\gamma_i(\mathbf{p'})$, which is the potential observed at channel i when the only active source is located at $\mathbf{p'}$. The potential vector observed across all channels is $\vec{\gamma}(\mathbf{p'})$, so the estimated activity at location $\mathbf{p}$ due to a point source at $\mathbf{p'}$ is

$$\kappa(\mathbf{p},\mathbf{p'}) \equiv \vec{h}^t(\mathbf{p})\vec{\gamma}(\mathbf{p'}) \qquad (21)$$

where we have applied the linear spatial filter $\vec{h}(\mathbf{p})$ that has been designed to estimate activity at $\mathbf{p}$. The function $\kappa(\mathbf{p},\mathbf{p'})$ is called the resolution kernel (or averaging kernel), which contains a point spread function (or impulse response) for each source location $\mathbf{p'}$. In the ideal noiseless case, we want the resolution kernel to be as close as possible to the identity function by making $\sigma_s^2(\mathbf{p})$ as small as possible, per equation (17).

In the case of noisy data with an additive multivariate Gaussian background process characterized by an M-by-M covariance matrix $\mathbf{C}$, we would also like to maximize the stability of the estimate at $\mathbf{p}$. That is, we want to minimize the variance of the estimated activity due to noise, which is given by

$$\sigma_n^2(\mathbf{p}) = \vec{h}^t(\mathbf{p})\mathbf{C}\vec{h}(\mathbf{p}) \ . \qquad (22)$$

The standard deviation of the noise variance, $\sigma_n(\mathbf{p})$, is thus reciprocally related to stability.

The most desirable estimation filter $\vec{h}(\mathbf{p})$ should simultaneously minimize the spatial spread variance $\sigma_s^2(\mathbf{p})$ and the noise variance $\sigma_n^2(\mathbf{p})$; but there is a tradeoff between these two. Therefore, we introduce a tradeoff parameter λ that ranges between 0 and 1 such that

$$(1-\lambda)\sigma_s^2(\mathbf{p}) + \lambda\sigma_n^2(\mathbf{p}) \qquad (23)$$

is minimized, subject to the constraint that the denominator of equation (17) equals 1. The optimal solution is given by (Press *et al.*, 1992):

$$\vec{h}(\mathbf{p}) = \frac{[(1-\lambda)\mathbf{W}(\mathbf{p}) + \lambda\mathbf{C}]^{-1}\vec{r}}{\vec{r}^t[(1-\lambda)\mathbf{W}(\mathbf{p}) + \lambda\mathbf{C}]^{-1}\vec{r}} \qquad (24)$$

where the spread matrix $\mathbf{W}(\mathbf{p})$ is defined by

$$W_{ij}(\mathbf{p}) \equiv \int_V |\mathbf{p'}-\mathbf{p}|^2 \gamma_i(\mathbf{p'})\gamma_j(\mathbf{p'})d\mathbf{p'} \qquad (25)$$

and $\mathbf{r}$ is defined by

$$r_i \equiv \int_V \gamma_i(\mathbf{p'})d\mathbf{p'} \ . \qquad (26)$$

In practice, the integrals of equations (25) and (26) must be approximated over a discrete source volume, e.g., they could be replaced by summations over a set of N voxels. Given a discrete gain matrix $\mathbf{G}$ of dimension M-by-$3N$ per equation (15), a discrete gradient matrix $\mathbf{D}_\nabla$ of dimension $3N$-by-N can be formed based on local differences in x, y, and z directions, such that discrete approximations to the integrals equations (25) and (26) can be based on summations across the columns of the M-by-N matrix

$$\Gamma \equiv \mathbf{GD}_\nabla \ . \tag{27}$$

How, then, can this theory be applied to study the dependency of 3D spatial resolution on 2D spatial sampling? For each location **p** there are two independent variables, i.e., sampling density and tradeoff parameter, λ, and there are two dependent variables, i.e., resolution, $\sigma_s^{-1}(\mathbf{p})$, and stability, $\sigma_n^{-1}(\mathbf{p})$. To study the effect of sampling density on resolution and stability, we first compute an optimal spatial filter per equation (24), and then calculate resolution via equation (17) and stability via equation (22). These computations require a gain matrix **G** computed from a theoretical head model, and a background signal covariance matrix **C** that should be estimated from actual EEG data. Intuitively, the 3D spatial resolution should improve as the 2D sampling density increases, but at what rate? Is there an effective asymptote of improvement? We might also expect stability to improve with increasing M since the correlation structure of the background activity becomes more interconnected.

2.3.3 Quantification of 3D error

In the last section, we considered the design of a linear spatial filter to estimate activity at a voxel of interest in the brain. From that perspective, we were not interested in source localization per se, but rather source activity time series estimation. Even so, the ability to accurately estimate neuroelectric activity in a prespecified region of interest is closely linked with the spatial resolving ability of EEG. When we have the luxury of knowing the true source activity function, such as for simulated data, the estimation error for a single time point is the difference between the true and the estimated activity. Estimation bias is the mean of the signed errors in a time interval (mean bias). Estimation inaccuracy can be quantified as the average absolute error after removing the bias. Estimation instability (standard error) is the standard deviation of errors (which implicitly removes the bias), and estimation imprecision is reflected by the standard deviation of the absolute errors (which implicitly removes the inaccuracy). Bias reflects systematic error in a particular direction (positive or negative), inaccuracy reflects mean error magnitude (nonnegative), instability reflects the variability of the signed errors, and imprecision reflects inconstancy of error magnitudes.

For data generated by a single equivalent current dipole, the localization performance of an algorithm can also be quantified by its bias, accuracy, stability, and precision. These are defined parallel to the above, but note in this context that bias is a 3-vector (location offset) and stability is a 3-by-3 matrix (representing an error ellipsoid in space), whereas accuracy and precision are always scalars. Orientation, the other spatial attribute of dipole analysis, is determined by three orthogonal dipole moments. Thus, orientation estimation is equivalent to estimation of three directed source activities at a given location, which leads back to the case of time series estimation of the previous paragraph.

Detection probability curves. Vrba and Robinson (1998) introduced the concept of a detection probability curve (DPC), which is "the probability with which a single equivalent current dipole can be localized within a specified set of parameters" such as the radius of a sphere centered on the dipole. That is, given a dipole, a configuration of sensors, a noise level, and a localization algorithm, the DPC is the probability of localizing the dipole within a sphere as a function of the radius. The authors related the DPC analytically to the error function (erf) of the ratio of radius to the standard deviation of reconstructed dipole position.

Quantification of localization performance for two or more dipoles can be severely complicated by the problem of assignment: Given simulated dipoles A and B, and localized dipoles X and Y, should the location of X be compared with the location of A, or rather with the location of B? To solve this problem, we may adopt the convention of making the most charitable assignment, i.e., the one that minimizes total localization error magnitude. In this case, however, the probability of coming close to the true source configuration just by chance increases with the number of dipoles; consequently, a control for chance performance should be included in the analysis (Pflieger, 1998).

Cramer-Rao lower bound for error variance. In an insightful paper, Mosher *et al.* (1993) applied the Cramer-Rao Inequality Theorem to study the fundamental spatial errors inherent in the spatiotemporal approach to multiple dipole source modeling. The theorem states that the inverse Fisher Information Matrix (FIM) sets a lower bound for the error covariance matrix for unbiased estimates of deterministic parameters. In particular, the authors studied lower bounds for the error variance (instability) when estimating the following parameters from the data: dipole locations, dipole moments as a function of time (source activity), and noise variance. The FIM is based on the partial derivatives, with respect to the source parameters, of the log-likelihood of observing the data given the parameters. Out of all possible estimators, the error variance of the parameters cannot be less than the Cramer-Rao lower bound. The authors theoretically demonstrated that the noise variance at the sensors can be estimated independently of the source parameters, but that there is cross-coupling between the location and moment parameters that always increases their lower error bounds. In other words, location errors increase moment errors, and vice versa. Consequently, estimation of source temporal activity (via the moments) is inextricably bound with localization. An algorithm that improves estimates of source activity should also contribute to improved source localization, and vice versa. We note here that estimation of moments is linear, whereas estimation of locations is nonlinear. Therefore, the authors have shown that there is an interaction between the linear/temporal and the nonlinear/spatial parts of the overdetermined problem.

Distributed solutions. For distributed underdetermined source models, global performance can be assessed by integrating (in practice, summing) some function of the estimation errors over the source volume, e.g., distributed analogs of bias, inaccuracy, instability, and imprecision. For a review of various figures of merit for distributed solutions, see Grave de Peralta Menendez and Gonzalez Andino (1998a, section 3.1; 1998b).

2.3.4 *Duality of overdetermined and underdetermined estimators*

The prototype overdetermined inverse is the least-squares estimator, and the prototype underdetermined inverse is the minimum norm. Considering that these prototype inverses are mathematically dual (Rao and Mitra, 1971, theorem 3.2.4), it may be prove beneficial to speculate about further dual relationships between the overdetermined and the underdetermined problems. In particular, note that the spatial part of the overdetermined problem is nonlinear, whereas the spatial part of the underdetermined problem is linear (at least in its prototype form). In the overdetermined problem, strong spatial constraints are applied, i.e., that the sources are (typically) a few point dipoles, whereas these spatial constraints are relaxed in the underdetermined problem. In the overdetermined case, temporal estimation is linear, and it is not necessary to make special assumptions about temporal dynamics in order to estimate source activities. Therefore, let us entertain a conjecture "by duality" that the temporal part of the underdetermined problem should be treated as highly constrained nonlinear. Will the addition of strong nonlinear constraints on the temporal dynamics of distributed source activities improve the spatial resolution of the linear aspect of underdetermined estimation?

2.4 Simulation of quasi-realistic datasets

Simulated datasets live in the gap between pure theoretical analysis and actual experimental measurements. We may use simulated data to assess the 2D spatial resolution of EEG (per section 2.1), to estimate the power of a statistical test as a function of number of channels (per section 2.2), or to study the accuracy of 3D inverse estimators (per section 2.3). For all three purposes, we seek an optimal cross between the realism of actual data and the ideal world of theory. The resulting hybrid is "quasi-realistic" simulated data.

Suppose, for example, that we wish to study the 3D accuracy of inverse estimators. Data used for the purpose of validation needs to meet two simultaneous demands: (a) it must reflect a realistic situation in the brain, and (b) the composition of the data must be completely known, so that estimated sources can be compared against the actual sources. Although data acquired using invasive *in vivo* methods can be obtained under specified task conditions in order to satisfy (a), such data is incomplete from the standpoint of explaining scalp potentials, which are generated by the brain volume integral of equation (13). In other words, invasive probes can sample a few points in the source space, but scalp potentials are generated by summation over the entire brain volume. By contrast, synthetic data constructed by simulating sources, signals, and noise satisfies requirement (b); but the usual methods for simulating data fall far short of meeting requirement (a). The simulated sources (a few point dipoles), signals (could be a sine wave or any other mathematical function), and noise (usually independent and identically distributed Gaussian white noise) typically have little relation to actual brain functioning under specified task conditions.

Can we construct simulations that are more realistic? Equation (13) and its approximation (15) state that scalp potentials are produced as the sum of signals that project to the scalp, plus a residual "background noise" process observed at the sensors. This suggests that we can simulate background activity at the scalp from actual data without assuming a theoretical model of the noise generation process in the brain (e.g., random dipoles; de Munck, 1989). One way to accomplish this in the context of ERPs is to use a plus/minus resampling plan (Pflieger *et al.*, 1995), which is a randomized extension of Schimmel's (1967) plus/minus reference.

On the other hand, the scalp signals of interest cannot be simulated without modeling the brain signals of interest; but what actual data could be used to model these brain signals? Exclusive use of ERP/EEG data must rely on inverse computation methods to model the brain sources. However, this introduces circularity into the logic of validating these very methods. Alternatively, fMRI raw statistic images from the same-subject/same-task could be used to simulate the number, locations, extensions, and shapes of synthetic neuroelectric sources. Although hemodynamic-neuroelectric coupling relationships remain to be elucidated, it is nevertheless plausible to suppose that the correlation is sufficient for our limited objective of simulating quasi-realistic source configurations. By contrast with the usual method of semi-arbitrarily picking a few point dipoles to simulate ERP generators, auxiliary fMRI could be used to non-subjectively approximate a number of physiologically plausible generators that are well located in functionally active parts of the brain and that are extended in space with realistically shaped contours. A threshold parameter applied to each fMRI raw statistic image can be used to generate a family of simulated brain source configurations ranging from "diffuse" (low threshold) to "focal" (high threshold). For each threshold value, the number of sources equals the number of "islands" of activation, and the extent of each source follows the "coastline" of each island. All microdipoles in the anatomical source space model that fall outside the islands are inactive, whereas microdipoles within the same island could be considered to be synchronously active (except when simulating contiguous asynchronous sources). Same-task/same-subject ERP/EEG data can then be used to model brain signal waveforms using linear overdetermined estimation, and these can in turn be forward projected to simulate the scalp signals. Finally, these deterministic signals can be added to stochastic plus/minus background activity to generate a large set of synthetic ERP/EEG epochs. In summary, this approach for simulating quasi-realistic ERP/EEG data using auxiliary fMRI could be used to drive Monte Carlo studies of EEG spatial resolution (e.g., Supek and Aine, 1993).

3 *Review of Results*

Some of the methods outlined in section 2 have been used to study the issues of EEG spatial resolution; others have been partially implemented; and a few have yet to be applied. This section reviews several results that have been obtained to date.

3.1 The quest for adequate spatial sampling

2D spatial resolution studies have been driven primarily by the search for an electrode density that is adequate to capture nearly all of the information available at the scalp surface, and which simultaneously guarantees that relatively high spatial frequencies are not aliased, i.e., misallocated to lower spatial frequencies, when making interpolated topographic maps. We could determine this density if we knew the highest effective spatial frequency in the data, if we found a plateau in information growth as a function of number of channels, or if we found a density that is adequate for near-perfect reconstruction via interpolation.

The Nyquist spatial sampling frequency, which is just adequate to capture all information and to avoid aliasing, is twice the highest effective spatial frequency in the data. Spitzer *et al.* (1989) estimated the spatial Nyquist frequency for somatosensory evoked potential (SEP) data using two one-dimensional chains of electrodes, one coronal and the other parasagittal, with 16 electrodes per chain and an inter-electrode spacing of 0.9 cm. Spatial spectra for each chain were obtained at different latencies between 16 and 41 ms. The Nyquist inter-electrode distances for the coronal chain ranged from 2.9 cm to 3.6 cm, which implies that the highest spatial frequencies were approximately 1/6 cycles per cm. The parasagittal chain contained higher spatial frequencies, and the authors' conservative estimates for Nyquist inter-electrode distances ranged from 1.5 cm to 1.8 cm. Thus, the highest spatial frequencies in the parasagittal chain were approximately 1/3 cycles per cm—about two times the coronal maximum. In both cases, the higher spatial frequencies had a declining contribution to the total spatial power. The authors did not offer a quantitative criterion for determining the cutoff point. Gevins (1990) performed a similar analysis of four-quadrant pattern reversal visual evoked potential (VEP) data, and reported that the 20 dB point in the spatial spectrum occurred at 1/4 cycles per cm, which corresponds to a spatial period of 4 cm and a Nyquist inter-electrode distance of 2 cm.

The information content analysis approach of section 2.1 has been applied to ongoing EEG (Pflieger and Sands, 1995) and four-quadrant retinotopic VEP data (Pflieger and Sands, 1996), as well as to simultaneously recorded MEG/EEG datasets (Pflieger *et al.*, in press; Pflieger and Halgren, 1999). Starting with a relatively large number of channels, the method of spatial subsampling was used in all studies for selection of data subsets with smaller numbers of channels. In the 128-channel ongoing EEG study, the signal of interest included everything of brain origin; all else was regarded as noise. Stringent criteria were applied to reject data that contained ocular and EMG artifacts, and the remaining noise was assumed to have environmental and electronic sources that could be assessed via a recording of "subject-shorted noise". EEG covariance matrices were estimated from a large number of time samples (25,000) for single points (i.e., instantaneous spatial covariance), and multiple (2 and 4) consecutive points (i.e., spatiotemporal covariance). Information (bits above noise) was plotted as a function of number of channels, and we subsequently observed that information growth increased approximately with $\sqrt{M}$ (M = number of channels) and $\sqrt{T}$ (T = number of consecutive time points) with some scaling factor. No asymptote of information growth was evident in these data, which prompted us to record 256 channels. After observing that approximate $\beta M^{0.5}$ information growth continued up to 256 channels for ongoing EEG (the fitted exponent was 0.506), we turned our attention to the case of ERP data, for which the signal of interest is phase-locked and the background signal is also generated in the brain, but is not phase-locked. When the off-diagonal elements of the background activity covariance matrix were set to zero, we observed an ERP information growth curve fitted by $2.31 M^{0.463}$, which has somewhat poorer than square root growth performance. (See Figure 6 and legend.) This procedure normalizes the ERP at each channel by its own background activity variance, but it ignores correlations between channels, even though this "noise" has a spatial structure caused by active generation in the brain and passive volume conduction through the skull. Utilization of the full background covariance matrix is equivalent to applying the spatial SNR transform to the ERP. In this case, we observed an SNR-ERP information growth curve fitted by $2.31 M^{0.629}$, which has substantially faster growth compared with a nominal exponent of 0.5. We attribute this to an improved estimate of the total background activity: The assumption of no correlations between channels inflates the estimate of total noise in the multichannel system because summing variances per channel produces the highest possible total.

We turn next to studies of simulated scalp data that use spatial models of source activity placed inside a volume conductor model of the head. Nunez *et al.* (1991) and Law *et al.* (1993) simulated source activity using 4,200 radial dipoles (local potential differences) on a hemispheric cortical layer with about 3 mm spacing to emulate the spatial scale of macrocolumns. Source magnitudes were distributed between ± 200 μV across the macrocolumns to mimic spontaneous ECoG. Clumping patterns of source activity with different levels of spatial complexity were used to generate EEG potentials and surface Laplacian EEG (analytically computed) at 648 scalp locations via a three-concentric spheres head model. The results were first mapped without interpolation: Here the authors noted that the spatial patterns of the Laplacian maps mirrored the major features of source clumping patterns, whereas the potential maps did not. In the 1993 paper, the resulting scalp data were resampled with 48, 64, 88, and 118 electrodes of diameter 1 cm, and a 3D spline interpolation method was used to reconstruct the surface Laplacian EEG at the original 648 scalp locations. A percent RMS error measure was used to compare the analytic (true) versus the reconstructed Laplacian magnitudes, i.e., the raw RMS errors were expressed as a percentage of the total energy of the true magnitudes. For a source configuration with relatively low spatial complexity, the percent RMS errors were rather high, as follows: 94% for 48 electrodes, 65% for 64 electrodes, and 32% for 118 electrodes. As expected, the percentages were higher for another source configuration with relatively higher spatial complexity: 75% for 64 electrodes ($r=0.778$), 60% for 88 electrodes, and 37% for 118 electrodes ($r=0.936$). (The correlation coefficients r were reported in Nunez, 1995.) It bears emphasizing that these are errors for reconstructing Laplacian information that is available at the scalp surface itself; i.e., they are not errors of estimated source activity at the cortical layer. Although relative errors were not reported for the potential maps, the authors noted in their 1991 paper that: "The improvement in spatial resolution which parallels increased electrode density is much more evident for the Laplacian maps than for the potential maps" (p. 158). This is to be expected on the grounds that the Laplacian puts more emphasis on higher spatial frequencies, which are most susceptible to inadequate spatial sampling. Consequently, we can infer that the relative errors should be less for potential maps than for Laplacian maps.

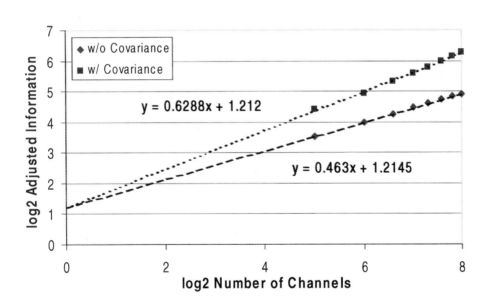

Figure 6

Log_2-log_2 ERP information growth: Covariance vs. no covariance. ERP/EEG information growth curves obey a power law $H' = \beta M^\alpha$, where M is the number of channels, and information is computed as bits above noise, with an adjustment to remove nominal growth due to repetitive sampling. The log-log plot is linear: $\log_2 H' = \alpha \log_2 M + \log_2 \beta$. The lower plot ("w/o Covariance") was obtained for subsampled 256-channel VEP data where each channel was normalized by the variance of its own background activity. The upper plot ("w/ Covariance") was obtained via the spatial SNR transform of the data, which utilized the full covariance matrix for the background activity. β is the average amount of information for single channels considered separately; consequently, both lines have a common intercept, because the 1-by-1 covariance matrix equals the variance for a single channel. The exponent α determines the rate of information growth with increasing number of channels. Incorporation of background activity covariances increases the rate of ERP information growth considerably, well above the nominal exponent of 0.5 that appears to hold for "typical" ongoing EEG data.

Srinivasan *et al.* (1996) used a four-shell head model to analytically compute the spatial transfer functions from an inner spherical surface of radial dipole sources to the outer scalp surface. A transfer coefficient obtained for each spherical harmonic degree reflects the relative magnitude of transfer from the inner surface to the outer surface. Transfer functions were obtained for scalp potentials (their Figure 2) and for scalp Laplacians (their Figure 10) using several choices of head model parameters. The transfer functions for scalp potential show a steep decline of relative magnitude with increasing spatial frequency, which is primarily caused by the high ratio of brain conductivity to skull conductivity. The Laplacian transfer functions increase to a maximum at spherical harmonic degree 3 or 4, and decline thereafter at a gradual rate relative to the potential transfer functions. The Laplacian transfer functions are considerably more sensitive to the choice of head model parameters compared with the potential transfer functions.

Using the method described at the end of section 2.1.2, we have estimated spatial spectra for 256-channel VEPs in a visual spatial attention task. Two 128-channel subsets were formed by selecting the odd and even numbered channels, as shown in Figure 7. Figure 8a plots the spectra for scalp potentials, whereas Figure 8b plots the spectra for the same spatial SNR transformed data. The differences between the two spectra are striking. For potentials, the 128-channel partition tracks nicely with the 256-channel partition until the tail end, where misallocation occurs. The spectrum peaks immediately at the first spatial frequency, and has a rapid falloff that is reminiscent of the scalp potential transfer function. For the SNR spectra, by contrast, spatial misallocation already occurs at relatively low spatial frequencies as evidenced by the overestimation of relative power for the 128-channel partitions. Relative power rises to a peak at spatial frequency number 7, after which it falls off slightly, but not appreciably.

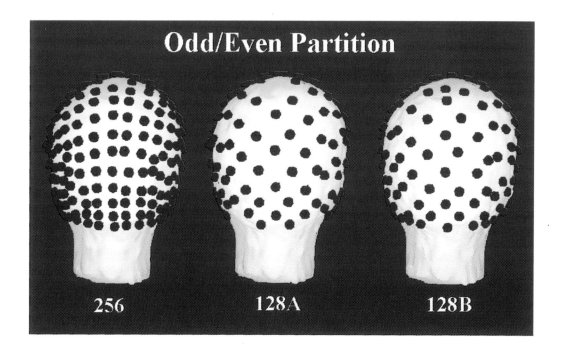

Figure 7
Odd and even partitions of 256 electrodes, back views.

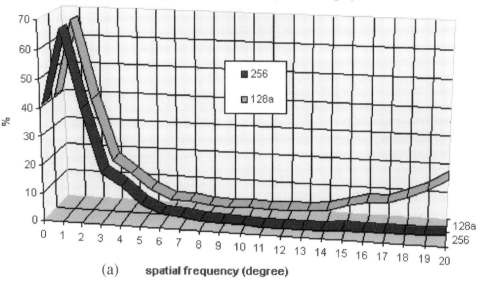

Figure 8

Spatial spectra of potential and SNR maps. Event-related data from a visual spatial attention condition were analyzed for the data partitions shown in Figure 7. (a) Spatial spectra of potential data for 256 channels and 128-channel partition A, which was substantially the same as 128-channel partition B. (b) Spatial spectra of SNR transformed data for all three partitions. Overestimation for the 128-channel partitions is caused by misallocation of spatial variance.

3.2 Power of statistical tests

In a Monte Carlo simulation study of multiple correlated variables, which made no attempt to model the actual characteristics of EEG, Blair and Karniski (1994) showed that the power of several permutation tests remained stable as the number of variables increased, even up to five times the number of available observations. By contrast, the power of Hotelling's T^2 decreased up to the number of observations, beyond which point the statistic could not be computed due to singularity of the covariance matrix. In the same paper, the authors demonstrated increasing power of the t^2 permutation test for discriminating two dipoles in a homogeneous spherical head model, as a function of dipole separation and number of subjects for 21 channels. The t^2 statistic here is the sum across channels of squared univariate t statistics. Additional random components were added to simulate between and within subject variability. For 20 subjects, this test could discriminate maps generated by dipoles with 0.67 cm separation more than 90% of the time. Although the authors did not report the effect of varying the number of channels in this simulation, the implicit suggestion seems to be that power would remain relatively constant, as in the aforementioned simulation.

3.3 Accuracy of source localization

Wang and He (1998) studied the effect of number of channels (16, 40, 128, and 301 electrodes evenly spaced over a hemisphere) on the cortical imaging technique (CIT; Sidman, 1991; Kearfott *et al.*, 1991). As we have previously noted, CIT and like methods fall somewhere between 2D scalp surface methods and full 3D methods: The solutions are 2D at the cortical envelope inside the skull, and a biophysical head model is utilized. The simulated sources were two radial dipoles, from which potentials at the simulated cortical and scalp surfaces were computed using a 3-shell spherical head model. Two levels of noise (5% and 10%) were added to the simulated surface data. For each number of channels and each noise level, relative error was computed for the CIT solution with respect to the true cortical potential at a given eccentricity. At 70% eccentricity, which may approximate the cortical surface, and with 10% noise, the relative errors (from Figure 8 of Wang and He, 1998) were approximately: 55% for 16 channels, 51% for 40 channels, 44% for 128 channels, and 35% for 310 channels.

Mosher *et al.* (1993) estimated Cramer-Rao lower bounds for the standard deviation of dipole location errors in a four-shell spherical head model for four different electrode configurations: 127 electrodes distributed over the upper hemisphere (broad coverage array), 127 electrodes distributed over roughly half of the upper hemisphere (dense array, <1 cm spacing), 37 electrodes distributed over the same area as the dense 127 array, and a clinical 10-20 montage. Results were restricted to tangential dipoles (due to simultaneous comparison with MEG coil arrays that are insensitive to radial dipoles in a sphere); however, the authors gave one example to show that the results are substantially the same when all orientations, including radial, are averaged. Assuming a dipole strength of 10 nAmp-m and a noise standard deviation of .4 µV, Cramer-Rao lower bound isocontours were plotted inside the head for location error standard deviations (in cm). The error isocontours were approximately concentric for the broad coverage 127 array, with error increasing with depth. The error bound for the most superficial (i.e., accurate) sources was 0.4 cm; but note that this value is inversely proportional to the signal-to-noise ratio, i.e., dipole strength divided by noise standard deviation. For the dense 127 array, the concentric circular appearance of the isocontours was lost due to limited coverage. The error bound for the most superficial and central sources was 0.2 cm, although errors increased rapidly for deep central sources due to the limited range of coverage. The isocontour lower bounds for the 37 channel array were similar in appearance to those for the dense 127 array, except that the errors themselves were approximately twice as large. The clinical 10-20 array, which provides wide coverage with poor sampling, had distinctively different error contours: A source directly underneath an electrode had no better accuracy than radially deeper sources. The authors explained: "Although a shallow source generates a significantly stronger signal at the surface, the spatial undersampling is such that only one nearby sensor receives a significant signal. One sensor cannot adequately locate the source, regardless of source intensity. The deeper sources generate a signal across enough surface sensors to compensate for their relatively weak surface signal. The overall effect of this sparse array of sensors is a relatively flat and larger lower bound error surface compared with that of the other studies."

The authors proceeded to demonstrate how two simultaneous dipoles considerably degrade localization performance. One dipole was located directly under the center of the electrode array, while the other was moved throughout the source volume. All orientations for both dipoles were permitted, and the best case and worst case orientations were plotted. For the broad coverage 127 array in the best case, the fixed location dipole distorted the isocontour error bounds up to several centimeters distant. In the worst case, the distortions were especially severe, with a region of especially high lower error bounds (about 4 cm) about 2 cm distant from the fixed source. The dense 127 array permitted the two dipoles to come substantially closer together before hitting the high error region, but this held only directly underneath the limited area of coverage.

Applying the framework of nonlinear regression to source estimation, Huizenga and Molenaar (1993) conducted simulations to study the effect of number of electrodes (19 versus 41) on errors (bias and standard error) for estimating dipole location and orientation. In addition, they studied the validity of statistical tests for assessing model-to-data goodness-of-fit at the scalp, and for determining confidence regions in the source volume. Datasets were simulated using a single dipole (either radial or tangential) in a homogeneous spherical head model, and adding i.i.d. noise (three levels) to the measurements. Depending on the noise condition, the mean localization bias (inaccuracy) ranged from 0.62 cm to 1.46 cm for 19 channels, which improved to 0.23 – 0.54 cm for 41 channels. The 41-channel parameter estimates were also more stable: The standard error of the estimates for 19 channels was a factor of about 2.9 times the standard error for 41 channels. Orientation errors were higher than localization errors. The authors compared empirical distributions for goodness-of-fit of the model to the data (at the scalp) with the F-distribution, and found that the 41-lead data was in substantially better agreement with the theoretical distribution. They also compared actual confidence regions with those estimated by the asymptotic Wald test (which assumes an infinite number of leads) and found that the estimated confidence volume was underestimated for 19 channels, but not for 41 channels.

By contrast with the results of Blair and Karniski (1994) as described in section 3.2, Huizenga and Molenaar (1992) found that statistical power for discriminating two nonsimultaneous dipoles of fixed separation was greater for 41 channels than for 19 channels. Two dipoles could be statistically discriminated with the Wald test using 19 channels if separated by 2.0 – 8.0 cm (depending on SNR), whereas they could be discriminated with 41 channels if separated by 1.0 – 2.0 cm. Note that Blair and Karniski performed tests on the scalp measurements directly, without modeling the data; whereas Huizenga and Molenaar effectively reduced the dimensionality of the measurements by estimating source parameters first. Therefore, the results are related, but not strictly comparable.

4 Discussion

4.1 What are the limits of 2D EEG spatial resolution?

This question can be approached from theoretical, empirical, technical, and practical points of view, which we desire to combine into an integrated perspective.

The practical viewpoint reminds us that EEG is always a means to another end, such as clinical diagnosis or basic research. With a specific end in view, the question becomes: What spatial sampling density and coverage is necessary and sufficient to attain the end? For example, whereas the broad/sparse 10-20 montage may be quite adequate for routine evaluation of normal or abnormal patterns, a dense grid with local coverage may be advantageous for presurgical localization of the central sulcus. In research applications, augmentation of behavioral indices in cognitive psychophysiological experiments may require just a few well-placed leads, whereas electric source imaging and integration with anatomical and functional MRI will certainly benefit from broad coverage and dense sampling. The practical point of view also considers a cost-benefit analysis. The extent to which more channels provide extra beneficial information is our main topic for further discussion, but the added costs are obvious: More channels are more expensive; preparation time takes longer to achieve a criterion level of impedance; the probability of a single channel going bad increases with the number of channels; more data storage is required; analysis time takes longer; and the "curse of dimensionality" has to be confronted. Thus, the practical limit must be set at an affordable point that is just adequate for the specific purposes at hand.

By contrast, from a purely theoretical perspective, we can state that there is no absolute maximum spatial frequency produced at the scalp by single or multiple current sources in the brain. The analytic solution for a single dipole in a multi-shell spherical head model can be expressed as an infinite series of spherical harmonics, where the coefficients for the higher degree terms approach zero, but never exactly vanish. Since spatial frequency can be identified with spherical harmonic degree, it follows that each spatial frequency contributes something to the exact solution. In practice, of course, good approximations are achieved by truncating the series; but the theoretical fact remains that there is no spatial frequency beyond which all terms vanish exactly. In order to achieve a given degree of accuracy, more harmonic terms are required as a single dipole is moved from the sphere center to the eccentric edges of the inner shell. When two dipoles are active simultaneously, the relative contribution of the harmonics may shift in the direction of higher spatial frequencies, so that more terms may be required to adequately distinguish them. As Amir (1994) showed, any two distinct sources with zero volume, such as dipoles, can theoretically be distinguished at the scalp, no matter how closely spaced, so long as the entire continuous function is preserved, i.e., so long as potentially all harmonics in the expansion can be retained. As more dipoles are added, the small coefficients of the higher degree harmonics become relatively more significant. Thus, from the theoretical point of view, there is no smallest coefficient that might not become relatively significant for a sufficiently complex set of sources.

From a technical perspective, though, there is always a smallest quantity that can actually be measured with a given instrument. Digitization quantizes the data and, moreover, electronic and environmental noise are inevitable. And of course there are technical limits to the spacing of electrodes on the scalp without bridging. However, it can also be said that these technical limits have not yet been reached. Average inter-electrode spacing for a broad coverage 256 configuration is about 1.9 cm with an electrode radius of 0.25 cm (0.45 including nonconductive casing). A 512 configuration would have an estimated average inter-electrode spacing of about 1.3 cm, and so it falls within (but near) the limits of feasibility using existing technology. With sufficient medical justification and use of a topical anesthetic, ultra-high density grids of needle electrodes with limited coverage are within the realm of feasibility, e.g., a 7.5-by-7.5 cm grid with 0.5 cm spacing (256 channels). If better spatial coverage is required, a patchwork of IC needle electrode grids can be envisioned.

These considerations lead directly to the empirical perspective: There is no point in attempting higher densities if we have already reached the effective limits of EEG spatial resolution with existing wide area configurations of 256 channels and less.

The prevailing ideas that have shaped our thinking about high-density scalp EEG recordings are well summarized by Srinivasan *et al.* (1998). It is supposed that the skull acts as a natural anti-aliasing filter that effectively cuts out all spatial frequencies above some maximum. Thus, there exists a spatial Nyquist frequency, which is twice the effective maximum frequency, and which permits adequate spatial sampling of all EEG information that is available at the scalp. The consequence of sampling below the spatial Nyquist is aliasing: High spatial frequencies will be misconstrued as lower spatial frequencies when making interpolated topographic maps. Thus, for any application that involves topographic mapping, the prevailing theory implies that the spatial Nyquist criterion must be observed. In other words, this theory implies that there is no distinction between the minimal practical limit, which is just adequate for the purposes at hand, and the ultimate theoretical limit: They are one and the same.

Some points in favor of this prevailing view are as follows. For 256-channel ERP data, there is a rapid decline of spatial power that reaches something like a plateau after about degree 7 (Figure 8a). The two 128-channel subsampled partitions have nearly identical spectra, which track closely with the 256-channel spectrum. The theoretical transfer function from radial sources in a simulated cortex layer to scalp potentials also shows a steep decline that is similar to Figure 8a. Finally, the very concept of diminishing interpolation error seems to imply an asymptote on a priori grounds, since zero error equals perfect reconstruction.

The information growth results present a different picture: There is no hint of a limit. Instead of treating the absolute measurements, this analysis first normalizes the data by estimates of noise or background EEG activity. Thus, the information measure is relative: Small signals with associated small noise are put on an equal footing with large signals with associated large noise, so long as the SNRs are equal. As Figure 6 shows, total information fits a power law curve, which has no asymptote. The apparent contradiction between this result and the concept of diminishing interpolation error can be resolved by plotting the first derivative of total information, which shows the declining rate of information per additional channel. See Figure 9. The rate of information growth per channel has an asymptote at zero, as does interpolation error. However, the area under the information growth rate curve does not converge to a finite total information content as the number of channels increases. Total information has no asymptote.

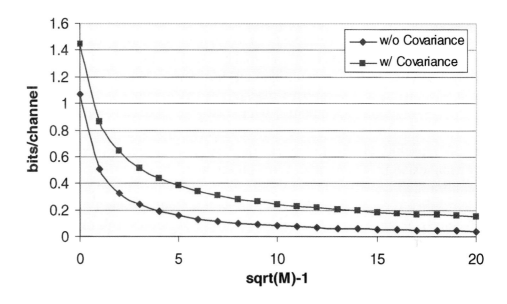

Figure 9

Declining rate of information growth per channel. Shown are first derivatives of the information growth curves derived from Figure 6. Bits gained per additional channel are plotted as a function of $\sqrt{M}-1$, where M is the number of channels. The abscissa has a general correspondence with the maximum spatial frequency (spherical harmonic degree) that can be represented with M channels, per section 2.1.2. These curves have an asymptote at zero. However, the area under such curves as M approaches infinity does not converge to a finite total information content.

The spatial spectrum of SNR transformed data shown in Figure 8b shows misallocation (aliasing) for the 128-channel subsampled data, and apparently no ultimate decline of spatial spectral content. However, for 256 channels, the spectral estimates of spatial frequencies (degrees) at 16 and greater cannot be independent of the accumulated spectral estimates up to degree 15 (per section 2.1.2). We can only say that 256 channels have apparently not captured the highest spatial frequency for SNR transformed data. The SNR transform has spatial filtering properties similar to the Laplacian; however it is determined from the EEG data itself, whereas the Laplacian is determined by electrode spatial coordinates and scalp geometry. The SNR transform depends on an estimate of the covariance of the background activity or another reference sample (e.g., ongoing EEG during a control condition), which requires a nominal number of observations at least four times the number of channels. We assumed that the ERP background covariance was stationary in order to combine many time points and epochs for covariance estimation. However, the important question of the effect of covariance estimation errors on the SNR transform has yet to be studied. This question appears to be closely related to the issue of the statistical power of SNR transformed data, which has also not been studied.

In conclusion, an alternative to the prevailing hypothesis is proposed: The growth of total information with additional channels is theoretically and (by extrapolation) empirically unlimited. The rate of information yield per additional channel does not decay rapidly enough to limit the total information content. The highest information growth rates come from spatial contrasts (e.g., SNR transforms), which maintain significant content from higher spatial frequencies. Laplacians are similar in this regard, although they do not control for signal-to-noise ratios that vary as a function of spatial frequency. On the other hand, an average ERP discards spatial content that could otherwise be provided by non-phase-locked activity; as a consequence, most ERP spatial content is concentrated in low spatial frequencies. Thus, for ERP mapping purposes, an "effective" spatial Nyquist could reasonably be set that assures sufficiently small interpolation errors. Nevertheless, there is no particular spatial Nyquist frequency that guarantees adequate sampling of all scalp information content, especially for SNR transformed data. That is, there is no "magic number" of channels that fits all needs. Therefore, we return to the practical point of view: The choice of an electrode configuration ultimately depends on the purpose for recording EEG.

4.2 The bigger picture

Raw scalp information content must ultimately be utilized for statistical discrimination of task-related brain states and for source analysis. The complexity of task-related brain states can be considerable, whereas we have so far been able to quantify the 3D spatial resolution of ESI for relatively simple source models. Mosher *et al.* (1993) reported an elegant study of single dipole lower error bounds, but also gave a glimpse of the problems that confront us when treating just two dipoles simultaneously. Thus, there appears to be a considerable deficiency at present in our ability to quantify spatial errors for source models that are adequately complex to accommodate task-related brain states of interest to cognitive neuroscientists. Returning to the scheme of six levels of analysis presented in section 1.1, we must bridge the gap between levels (c) and (d), i.e., between scalp-level discrimination of task-related brain states and source analysis, in order to advance to level (e), which relates task performance to brain structures and processes. Level (e), in turn, is a stepping stone to our ultimate objective at level (f), which is to understand how general-purpose cortical functions recruit and coordinate specialized cortical functions to accomplish the assigned task.

General theory at level (f) provides us with a crucial guide to design the task in the first instance, and to direct our efforts towards the integration of data and theory across all six levels of analysis. For example, the concept of large-scale neurocognitive networks (e.g., Mesulam, 1990) serves to bridge the analysis of structure/process at level (d) with the analysis of function (i.e., "design specification" for brain work) at level (f). Specific functionality to perform a task, which engages specialized regions in a wide-area cortical network, somehow emerges dynamically via generalized coordinating functionality that is common across all regions of cortex (e.g., Phillips and Singer, 1997). At least, this is one possible way of thinking about what may actually be taking place in the brain. The point is, a general conception such as this serves to guide the direction of research toward consummation at level (f).

For example, Phillips and Pflieger (this volume) entertain the hypothesis that (i) evoked EEG signals that are phase-locked reflect receptive field (RF) communication of specific information contents between local cortical processors, and (ii) induced EEG signals that are non-phase-locked reflect general-purpose contextual field (CF) coordination between local cortical processors. With a few notable exceptions (e.g., Brandt and Jansen, 1991; Brandt 1997), phase-locked responses and non-phase-locked activity are analyzed in mutual isolation, even though they are parallel activities in the brain. A theory of cortical computation (Phillips and Singer, 1997) proposes functional relationships between RF and CF inputs and outputs. Assuming this theory, the hypothesis predicts that echoes of RF-CF relationships should be detectable in relationships between simultaneous evoked responses and induced activity in a suitable task (e.g., dynamic grouping). Tallon-Baudry and Bertrand (1999) propose complementary source models for evoked responses and induced activity: For evoked responses, the usual equivalent current dipole oriented normally to the cortical surface (generated by pyramidal cells), and for induced gamma activity, a novel ring-shaped distribution of tangentially oriented dipoles (generated by the synaptic activity of horizontal interneurons). Since the CF inputs (induced activities) are thought to modulate RF input-output transformations (evoked responses), the dual source models suggest that at least two things are happening simultaneously in a local region of cortex. If separate source analyses are first performed on the evoked response and the induced activity statistics, then study of their interactions could subsequently proceed by returning to the raw EEG epochs, which contain both evoked and induced responses. Using a combined source model, we can attempt to tease apart two different source activities at essentially one location. If this can be achieved at all, then it seems likely that a high-density EEG recording may be required.

4.3 Problems for further study

Intuitively, the least possible interpolation error should be linearly related to the declining rate of information growth per additional channel. Can a quantitative relationship be derived?

The contrast signal covariance matrix utilized in the SNR transform is imperfectly estimated. How do covariance estimation errors affect the information content measure?

The spatial SNR transform discussed here has not utilized information about the (temporal) frequency spectra of the contrast signal. How much additional information can be extracted by taking temporal correlations into account?

The Monte Carlo study of Blair and Karniski (1994) can be extended to study statistical power as a function of number of channels when the background activity has the spatial properties of ongoing EEG. Will the spatial SNR transform increase statistical power? Is it possible to devise a test statistic that has increasing power with the addition of channels?

The Backus-Gilbert approach to 3D spatial resolution as a function of number of channels is ripe for implementation. A noise covariance matrix derived from actual EEG should be utilized.

A set of quasi-realistic datasets, derived from same-subject/same-task ERP and fMRI data, can be constructed for public domain access via the Internet for benchmark validation studies.

If the spatial SNR transform is applied to an ultra-high density scalp grid prior to ECoG, will the scalp coherence patterns approximate the cortex coherence patterns reported by Towle *et al.* (1998), which can be used to identify the sensory/motor area?

Acknowledgements

Shugo Suwazono's facilitation and comments have been greatly appreciated. It was the vision and know-how of Steve Sands and Ron Stubbers that made 256-channel EEG datasets possible. Informative discussions with Greg Simpson and Herb Vaughan, Jr. have propagated to this paper, and several helpful suggestions by Robert Aasal have been incorporated in the final draft. Finally, Bill Phillips' special emphasis on the mechanisms of brain coordination has provided a fascinating target for focusing our ever-advancing methodologies of dynamic neuroimaging.

References

Amir A. Uniqueness of the generators of brain evoked potential maps. IEEE Trans Biomed Eng 1994; 41: 1-11.

Babiloni F, Babiloni C, Cecchi L, Onorati P, Salinari S, Urbano A. Statistical analysis of topographic maps of short-latency somatosensory evoked potentials in normal and Parkinsonian subjects. IEEE Trans Biomed Eng 1994; 41: 617-624.

Babiloni F, Carducci F, Babiloni C, Urbano A. Improved realistic Laplacian estimate of highly-sampled EEG potentials by regularization techniques. Electroencephalogr Clin Neurophysiol 1998; 106: 336-343.

Backus GE, Gilbert, F. The resolving power of gross earth data. Geophys J Royal Astron Soc 1968; 16: 169-205.

Backus GE, Gilbert, F. Uniqueness in the inversion of gross earth data. Phil Trans Royal Soc London A 1970; 266: 123-192.

Baldeweg T, Spence S, Hirsch SR, Gruzelier J. γ-band electroencephalographic oscillations in a patient with somatic hallucinations. The Lancet 1998; 352: 620-621.

Barnard ACL, Duck IM, Lynn MS. The application of electromagnetic theory to electrocardiology I. Derivation of the integral equations. Biophys J 1967; 7: 443-462.

Bellman R. Adaptive Control Processes. Princeton University Press, 1961.

Blair RC, Karniski W. Distribution-free statistical analyses of surface and volumetric maps. In: Thatcher RW, Hallett M, Zeffiro T, John ER, Huerta M, editors. Functional Neuroimaging: Technical Foundations. San Diego: Academic Press, 1994: 19-28.

Bourke P. Distributing points on a sphere. Internet 1996: http://www.mhri.edu.au/~pdb/geometry/spherepoints/ .

Brandt ME, Jansen BH. The relationship between prestimulus alpha amplitude and visual evoked potential amplitude. Int J Neurosci, 1991; 61: 261-268.

Brandt ME. Visual and auditory evoked phase resetting of the alpha EEG. Int J Psychophysiol, 1997; 26: 285-298.

Burik MJv, Multer MC, Stinstra JG, Peters MJ. Inward continuation of the scalp potential distribution by means of the vector BEM. Proc IEEE/EMBS 1998; 20: 2135.

de Munck JC. A mathematical and physical interpretation of the electromagnetic field of the brain. PhD thesis, University of Amsterdam, 1989.

Donchin E, Karis D, Bashore TR, Coles MGH, Gratton G. Cognitive psychophysiology and human information processing. In: Coles MGH, Donchin E, Porges SW, editors. Psychophysiology: Systems, Processes, and Applications. New York: The Guilford Press, 1986.

Edgington ES. Randomization Tests (Second Edition). New York: Marcel Dekker, 1987.

Elul R. Gaussian behavior of the EEG: Changes during mental task performance. Science 1969; 164: 328-330.

Faux SF, McCarley RW. Analysis of scalp voltage asymmetries using Hotelling's T^2 methodology. Brain Topography 1990; 2: 237-245.

Frackowiak RSJ, Friston KJ, Frith CD, Dolan RJ, Mazziotta JC. Human Brain Function. San Diego: Academic Press, 1997.

Fuchs M, Drenckhahn R, Wischmann H-A, Wagner M. Bioelectric and biomagnetic reconstructions with an improved boundary element method for realistic volume conductor modeling. IEEE Trans Biomed Eng 1998; 45: 980-997.

Geselowitz DB. On bioelectric potentials in an inhomogeneous volume conductor. Biophys J 1967; 7: 1-11.

Gevins AS. Dynamic patterns in multiple lead data. In: Rohrbaugh J, Johnson R, Parasuraman R, editors. Event-Related Potentials of the Brain. New York: Oxford University Press, 1990: 44-56.

Grave de Peralta Menendez R, Hauk O, Gonzalez Andino SL, Vogt H, Michel C. Linear inverse solutions with optimal resolution kernels applied to the electromagnetic tomography. Human Brain Mapping 1997; 5(6)

Grave de Peralta Menendez R, Gonzalez Andino SL. Distributed source models: Standard solutions and new developments. In: Uhl C, editor. Analysis of Neurophysiological Brain Functioning. Berlin: Springer-Verlag, 1998a.

Grave de Peralta Menendez R, Gonzalez Andino SL. A critical analysis of linear inverse solutions to the neuroelectromagnetic inverse problem. IEEE Trans Biomed Eng 1998b; 45: 440-448.

Grave de Peralta Menendez R, Gonzalez Andino SL. Backus and Gilbert method for vector fields. Human Brain Mapping 1999; 7: 161-165.

Hjorth B. An on-line transformation of EEG scalp potentials into orthogonal source derivations. Electroencephalogr Clin Neurophysiol 1975; 39: 526-530.

Huizenga H, Molenaar P. The effect of number of leads on equivalent dipole estimation and statistical testing. [Abstract.] Brain Topography 1993; 5: 457.

Karniski W, Blair RC, Snider AD. An exact statistical method for comparing topographic maps, with any number of subjects and electrodes. Brain Topography 1994; 6: 203-210.

Kearfott RB, Sidman RD, Major DJ, Hill CD. Numerical tests of a method for simulating electrical potentials on the cortical surface. IEEE Trans Biomed Eng 1991; 38: 294-299.

Koles ZJ, Paranjape RB. Topographic mapping of the EEG: An examination of accuracy and precision. Brain Topography 1988; 1: 87-95.

Law SK, Nunez PL, Wijesinghe RS. High-resolution EEG using spline generated surface Laplacians on spherical and ellipsoidal surfaces. IEEE Trans Biomed Eng 1993; 40: 145-153.

Lawrence DH. The nature of a stimulus: Some relationships between learning and perception. In: Koch S, editor. Psychology: A Study of a Science (Vol. 5). New York: McGraw-Hill, 1963: 179-212.

Le J, Gevins AS. Method to reduce blur distortion from EEG's using a realistic head model. IEEE Trans Biomed Eng 1993; 40: 517-528.

Le J, Menon V, Gevins A. Local estimate of surface Laplacian derivation on a realistically shaped scalp surface and its performance on noisy data. Electroencephalogr Clin Neurophysiol 1994; 92: 433-441.

Leahy RM, Mosher JC, and Ermer JJ. Source estimation for MEG studies involving test and control conditions. Biomag98 Abstracts 1998: 97.

Malmivuo J, Suihko V, Eskola H. Sensitivity distributions of EEG and MEG measurements. IEEE Trans Biomed Eng 1997; 44: 196-208. Correction, 44: 430.

Mesulam MM. Large scale cognitive networks and distributed processing for attention, language and memory. Annals of Neurology 1990; 28: 597-613.

Möcks J, Pham Dinh Tuan, Gasser T. Testing for homogeneity of noisy signals evoked by repeated stimuli. The Annals of Statistics 1984; 12: 193-209.

Möcks J, Gasser T, Köhler W. Basic statistical parameters of event-related potentials. J Psychophysiol 1988; 2: 61-70.

Mosher JC, Spencer ME, Leahy RM, Lewis PS. Error bounds for EEG and MEG dipole source localization. Electroencephalogr Clin Neurophysiol 1993; 86: 303-321.

Nunez PL. Electric Fields of the Brain. Oxford University Press, 1981.

Nunez PL, Pilgreen KL, Westdorp AF, Law SK, Nelson AV. A visual study of surface potentials and Laplacians due to distributed neocortical sources: Computer simulations and evoked potentials. Brain Topography 1991; 4: 151-168.

Nunez PL. Neocortical Dynamics and Human EEG Rhythms. Oxford University Press, 1995.

Paranjape RB, Koles ZJ, Lind J. A spatial power spectrum analysis of the electroencephalogram. Brain Topography, 1990; 3: 329-336.

Pascual R, Gonzalez S, Valdes P, Biscay R. Current source density estimation and interpolation based on the spherical harmonic Fourier expansion. Int J Neuroscience, 1988; 43:237-247.

Perrin F, Pernier J, Bertrand O, Giard MH, Echallier JF. Mapping of scalp potentials by surface spline interpolation. Electroencephalogr Clin Neurophysiol 1987; 66: 75-81.

Perrin F, Pernier J, Betrand O, Echallier JF. Spherical splines for scalp potential and current density mapping. Electroencephalogr Clin Neurophysiol 1989; 72:184-187; Corrigenda 1990; 76: 565-566.

Pflieger ME. A theory of optimal event-related brain signal processors applied to omitted stimulus data. PhD thesis, The Ohio State University, Columbus, 1991. Dissertation Abstracts International 52: 9201737.

Pflieger ME, Sands SF. Information growth in multichannel EEG. [Abstract.] Human Brain Mapping 1995; Sup 1: 99.

Pflieger ME, Simpson GV, Vaughan HG Jr. Improved estimation of ERP source activities in the presence of realistic background EEG. [Abstract.] Human Brain Mapping 1995; Sup 1:101.

Pflieger ME, Sands SF. 256-channel ERP information growth. [Abstract.] NeuroImage 1996; 3: S10.

Pflieger ME. Spatial subsampling studies with 256 channel ERP/EEG data. Workshop on Multimodal Neuroimaging, St. Anne's College, Oxford, September 27-28, 1996.

Pflieger ME, Simpson GV, Ahlfors SP, Ilmoniemi RJ. Superadditive information from simultaneous MEG/EEG data. In: Aine CJ, Flynn ER, Okada Y, Stroink G, Swithenby SJ, Wood CC, editors. Advances in Biomagnetism Research: Biomag96, New York: Springer-Verlag, in press.

Pflieger ME, Halgren E. Complementary MEG/EEG information increases withthe number of combined experimental conditions. In: Yoshimoto T, Kotani M, Kuriki S, Karibe H, Nakasato N, editors. Recent Advances in Biomagnetism, Sendai: Tohoku University Press, 1999: 294-297.

Pflieger ME. Local maxima in EEG tomographies and their relation to multiple point sources. Proc IEEE/EMBS 1998: 20: 2139-2142.

Pfurtscheller G, Cooper R. Frequency dependence of the transmission of the EEG from cortex to scalp. Electroencephalogr Clin Neurophysiol 1975; 38: 93-96.

Phillips WA, Singer W. In search of common foundations for cortical computation. Behav Brain Sci 1997; 20: 657-722.

Phillips WA, Pflieger ME. EEG studies of interactions that coordinate cortical activity. This volume.

Plonsey R, Heppner DB. Considerations of quasi-stationarity in electrophysiological systems. Bull Math Biophys 1967; 29: 657-664.

Plonsey R. Bioelectric Phenomena. New York: McGraw-Hill, 1969.

Press WH, Teukolsky SA, Vetterling WT, Flannery BP. Numerical Recipes in C: The Art of Scientific Computing (2nd Edition). Cambridge University Press, 1992.

Rao CR, Mitra SK. Generalized Inverse of Matrices and its Applications. New York: John Wiley, 1971.

Robinson SE, Rose D. Current source image estimation by spatially filtered MEG. In: Hoke M, Erne S, Okada Y, Romani G, editors. Biomagnetism: Clinical Aspects. Amsterdam: Excerpta Medica, 1992: 761-765.

Robinson SE, Vrba J. Functional neuroimaging by Synthetic Aperture Magnetometry (SAM). In: Yoshimoto T, Kotani M, Kuriki S, Karibe H, Nakasato N, editors. Recent Advances in Biomagnetism, Sendai: Tohoku University Press, 1999: 302-305.

ush S, Driscoll DA. EEG electrode sensitivity – An application of reciprocity. IEEE Trans Biomed Eng 1969; 16: 15-22.

Schimmel H. The ($\pm$) reference: Accuracy of estimated mean components in average response studies. Science 1967; 157: 92-94.

Scott DW. Multivariate Density Estimation. New York: John Wiley, 1992.

Shaw, GR. Spherical harmonic analysis of the electroencephalogram. PhD thesis, University of Alberta, Edmonton, 1989.

Sidman RD. A method for simulating intracerebral potential fields: The cortical imaging technique. J Clin Neurophysiol 1991; 8: 432-441.

Simpson GV, Pflieger ME, Foxe JJ, Ahlfors SP, Vaughan HG Jr, Hrabe J, Ilmoniemi RG, Lantos G. Dynamic neuroimaging of brain function. J Clin Neurophysiol 1995; 12: 432-449.

Spencer ME, Leahy RM, Mosher JC, Lewis PS. Adaptive filters for monitoring localized brain activity from surface potential time series. Proc 26th Asilomar Conference on Signals, Systems & Computers, IEEE, Pacific Grove, CA, Oct 1992: 156-161.

Spitzer AR, Cohen LG, Fabrikant J, Hallett M. A method for determining optimal inter-electrode spacing for cerebral topographic mapping. Electroencephalogr Clin Neurophysiol 1989; 72: 355-361.

Srinivasan R, Nunez PL, Tucker DM, Silberstein RB, Cadusch PJ. Spatial sampling and filtering of EEG with spline Laplacians to estimate cortical potentials. Brain Topography 1996; 8: 355-366.

Srinivasan R, Tucker DM, Murias M. Estimating the spatial Nyquist of the human EEG. Behavioral Research Methods, Instruments, and Computers 1998; 30: 8-19.

Supek S, Aine CJ. Simulation studies of multiple dipole neuromagnetic source localization: Model order and limits of source resolution. IEEE Trans Biomed Eng 1993; 40: 529-540.

Swets JA. Is there a sensory threshold? Science 1961; 134: 168-177.

Tallon-Baudry C, Bertrand O. Oscillatory gamma activity in humans and its role in object representation. Trends in the Cognitive Sciences 1999; 3: 151-162.

Towle VL, Syed I, Berger C, Grzeszczuk R, Milton J, Erickson RK, Cogen P, Berkson E, Spire J-P. Identification of the sensory/motor area and pathologic regions using ECoG coherence. Electroencephalogr Clin Neurophysiol 1998; 106: 30-39.

Vaidyanathan C, Buckley KM. A sampling theorem for EEG electrode configuration. IEEE Trans Biomed Eng 1997; 44: 94-97.

van den Broek SP. Volume conduction effects in EEG and MEG. PhD thesis, University of Twente, 1997.

Van Veen BD, van Drongelen W, Yuchtman M, Suzuki A. Localization of brain electrical activity via linearly constrained minimum variance spatial filtering. IEEE Trans Biomed Eng 1997; 44: 867-880.

Vrba J, Robinson SE. Detection probability curves for evaluating localization algorithms and comparing sensor array types. In: Yoshimoto T, Kotani M, Kuriki S, Karibe H, Nakasato N, editors. Recent Advances in Biomagnetism, Sendai: Tohoku University Press, 1999: 97-100.

Wang Y, He B. A computer simulation study of cortical imaging from scalp potentials. IEEE Trans Biomed Eng 1998; 45: 724-735.

T. Nakada (Ed.)
Integrated Human Brain Science: Theory, Method Application (Music)
© 2000 Elsevier Science B.V. All rights reserved

Chapter II.9

All Phrase Event-Related Potentials during Listening to Music

Shugo Suwazono[a,1], Mark E. Pflieger[a,b], Russell E. Jacobs[a,c], and Tsutomu Nakada[a]

[a] Department of Integrated Neuroscience, Brain Research Institute, University of Niigata
[b] Neuro Scan Labs
[c] Biological Imaging Center, Beckman Institute, California Institute of Technology

Introduction

Space-time visualization of higher brain functions is one of the ultimate goals of neuroscience research. In order to engage the attentive interest of human subjects in such studies, it is desirable to employ complex stimuli of relatively long duration, such as spoken sentences or musical phrases. Such long-duration stimuli often have motifs that convey information to be understood by an audience. Processing of motifs by the brain can vary in terms of when and how they are interpreted across subjects and, moreover, can change dynamically from trial to trial within individual recipients. For example, information conveyed by a certain phrase of spoken words could be understood differently, and at different moments in time, by various listeners in an audience. Even within an individual listener, the timing and results of motif processing might be expected to vary across stimulus repetitions.

When using such complex stimuli of long duration, we are immediately faced with a difficult problem of data analysis. The brain necessarily performs feature analysis for each stimulus unit (e.g., phoneme). Because these feature analyses are probably handled by the brain as a series of overlapping parallel processes, any physiological time series data will inevitably contain multiple overlapping brain responses. To minimize this overlap problem, most physiological studies based on the electroencephalogram (EEG) and/or event-related potentials (ERP) averaging technique have utilized phasic stimuli of short duration. Brief durations also insure that evoked brain events are sufficiently time locked to the stimuli to warrant the use of averaging. Several exceptional studies have used musical phrases to

---
[1] Correspondence: Shugo Suwazono, M.D., Ph.D, Department of Integrated Neuroscience, Brain Research Institute, University of Niigata, Niigata 951-8585, Japan, Tel: (81)-25-227-0679, Fax: (81)-25-227-0682, e-mail: shugo@bri.niigata-u.ac.jp

investigate the context-sensitive processing of short-duration endings that violate subjects' expectations (Besson et al. 1994 and 1997, Nittono et al. 2000, Verleger 1990).

Despite the problem of overlap, there are two reasons why EEG analysis (including the ERP averaging technique) may yet be suitable for investigating the brain processes that occur while human beings listen to music: (1) it is possible that music 'resets' the EEG responses at the onset of each note; and (2) this technique achieves time resolution on the order of milliseconds. The second characteristic is actually the most important feature for the analysis of music, which is intrinsically a phenomenon of high time resolution. Accordingly, this study aimed to explore the usefulness of EEG measurements and event-related potential analysis for tracking the brain processes that occur during music listening in some spatio-temporal detail.

Methods

Subjects and Procedures

Two subjects (males, 35 and 18 years old) gave informed consent to participate in the experiment. They had no special musical experiences or training other than regular classes at school. They sat in a comfortable chair in an air conditioned, electrically shielded, and sound attenuated room that was dimly lit during the recording sessions. An Electro-Cap (Electro-Cap International Inc.) with 256 channels (7 of which were placed around the eyes to monitor electrooculogram) was applied using the standard ECI conductive gel. All channels were recorded against a non-cephalic balanced sternovertebral electrode (Stephanson and Gibbs 1951). Using a system of eight 32-channel SynAmps (Neuroscan Labs), 16-bit electroencephalogram (EEG) data were acquired at a gain of 500, sampled at a rate of 1 kHz, and digitally filtered with a bandpass of 0.05-100 Hz. Total experiment time from start (cap application) to finish (cap removal) was four hours.

Stimuli

The stimulus sequences consisted of musical phrase series and prolonged initial sound series. The musical phrase series included: silence (3 s), white noise (4 s), the musical phrase (9 s, see below), silence (3 s), and a simple tone (1000 Hz, 200 ms duration, 10 ms rise/fall time windowed by the Blackman function) (Fig. 1). The white noise had an envelope of linearly increasing amplitude starting at zero and reaching a plateau at 1 second after its onset. The musical phrase was taken from the "energy flow" composed by Mr. Ryuichi Sakamoto, and was familiar to both subjects. It was digitally recomposed by a music editor on a personal computer using a timbre close to that of a piano. The initial sound series included: silence (3 s), white noise (4 s), the prolonged initial sound of the musical phrase (1 s), silence (3 s), and a simple tone (1000 Hz, 200 ms duration). The prolonged initial sound had the same timbre and onset note as the musical phrase, and had a long duration of 1 second.

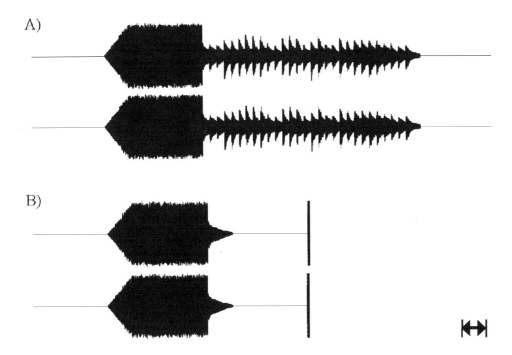

Figure 1
The time course of the stimuli used in the current study. The upper trace shows the left channel, and the lower shows the right. The scale shows one second. A) The sequence including the musical phrase. B) The sequence including the prolonged initial sound of the same musical phrase.

One hundred and fifty sequences for each of the above two series were randomly presented through earphones binaurally. The stimulus sequences were blocked into 25 sessions, making each last for about three minutes. The subjects took a short rest between sessions. The instructions to the subjects were to relax, minimize blinks, and listen carefully to the sounds and musical phrases.

Analysis

All analysis was done offline after the recording. EEG was epoched for each type of stimulus. The analysis period included a 200 ms baseline prior to the stimulus onset. Epochs with artifacts (i.e., containing any point with amplitude 150 µV or greater) were excluded from further analysis. Averages time locked to stimulus onset were made for each type of stimulus, and the results were low-pass filtered at 30 Hz.

In order to remove the slow activity and focus on the phasic activity, additional averages were made by high-pass filtering at 1Hz for each response to the musical phrase. A movie consisting of thirty frames per second was made based on the selected isopotential topographical maps (one per 33.33 milliseconds) from the responses made by the above filtering procedure.

Results

Responses to simple tone, white noise and the prolonged initial sound

A fronto-centrally distributed negative-positive activity was observed in the averaged responses to the simple tone (Figs. 2 and 3). The latency at channel 127 (close to Cz of the conventional 10-20 method) of the negativity was 104 ms and its amplitude was 13.0 μV in subject 1. This activity is comparable to the auditory N1 component. Such components were also found in the responses to the prolonged initial sounds of the musical phrase and white noise. There were no remarkable differences among responses to these three stimuli, except that the latency of the N1 component to white noise tended to be longer compared to the others.

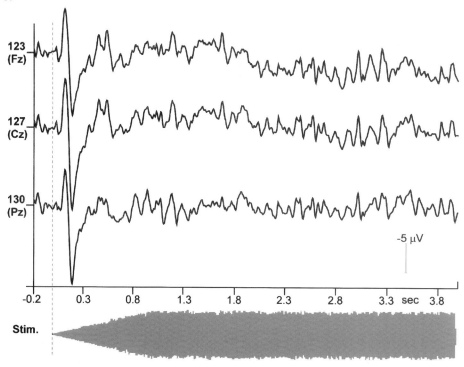

Figure 2
Averaged waveforms at the selected channels in response to white noise in the subject 1. The waveform at the bottom indicates the time course of the stimuli.

Responses to the Musical Phrase

The N1 component was also observed in the responses to the musical phrase. A widely distributed slow negativity followed, which was superimposed by repetition of negative and positive phasic activities. The amplitude of the musical phrase N1 component was larger than those to other stimuli, especially in the parietal region (Figs. 4a, 4b, and 3). The slow negativity was distributed widely and symmetrically (Fig. 4a). This activity reached a plateau in amplitude (about 5 μV) at 900 ms after the stimulus onset, and decayed for about 5

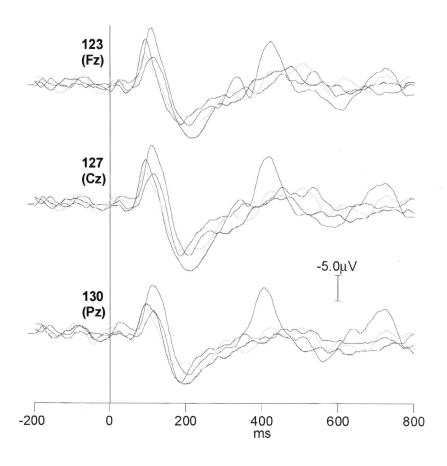

Figure 3
Averaged waveforms at the selected channels to the simple tone (black), the white noise (blue), the prolonged initial sound of the musical phrase (green) and the musical phrase (red) in the subject 1.

seconds before returning to baseline.

The repetitive phasic activity was not found in the responses to the prolonged initial sound of the musical phrase, white noise, or simple tone (Fig. 3). The first phasic negative-positive response, which is comparable to the N1 component of the auditory evoked potential, also appeared in the responses to the musical phrase, with a symmetric centro-frontal distribution (Figs. 4a and 3). The repetitive negative and positive peaks followed, decreasing in amplitude within 1 second after the onset of the stimuli, and later reached a state with smaller amplitudes and shorter durations. These repetitive activities had comparable latencies about 100 ms after the occurrence of the attack of the note (see the movie). They were scarcely observed in the responses to the prolonged initial sound of the musical phrase, white noise, or simple tone (Fig. 4). The scalp topographies of these second and third phasic activities had a parietal distribution, which is apparently different from that of the auditory N1 component (see Fig. 3 and the movie). At a later phase of the second phasic activity, the maximal point of the scalp distribution changed in time from a parietal area to a frontal area.

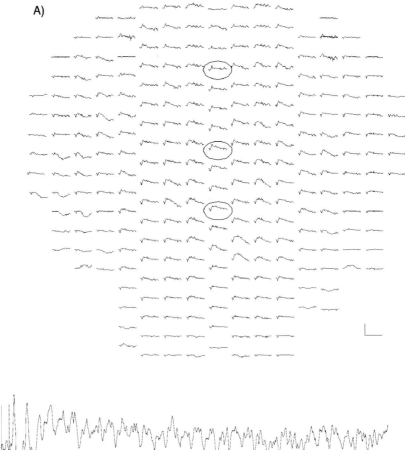

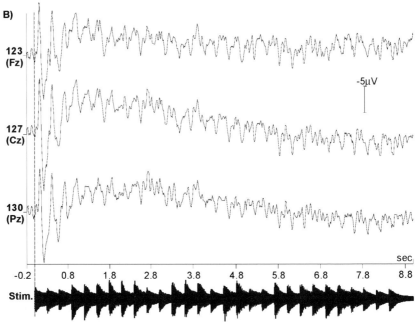

Figure 4

Averaged waveforms at 256 channels from 200 ms prior to the musical stimuli to 9 seconds in the subject 1. The Neurosurgeon's view is taken for A), which means the right side of the figure is the left side of the subject. The scale indicates 50 μV (negative up) for the ordinate and 9 seconds for the abscissa. Circled channels on A) are enlarged on B). The waveform at the bottom of B) indicates the time course of the stimuli (the musical phrase).

Discussion

In addition to the time dimension, the space-time visualization can take two or three dimensions from space. Although electroencephalogram (EEG) originally has three dimensions, the isopotential topographical maps are projected onto a plane of two dimensions. Therefore, the data as presented had three dimensions, consisting of two from space and one from time. The actual visualizations of the data were prepared by making movies, of which the video track consisted of successive isopotential topographical maps of event-related potentials at their latencies from the onset of stimuli. By contrast with most prior physiological studies, the musical phrase had a relatively long duration (about 9 seconds).

The main results of this study included slow negativity lasting for several seconds after the onset of the musical phrase, and the repetitive phasic activities time-locked to the note onsets (Figs. 4a, 4b and 3). The latency of the N1 component to white noise tended to be longer compared to others (Fig. 3). This difference may have been caused by the 1 second amplitude ramp on the white noise, so that the synchronization of neural activity differed from the other stimuli (Fig. 1). The N1 component to the musical phrase showed its largest amplitude especially at the parietal region (see Fig. 3 and the movie). This scalp distribution is different from that of the ordinary auditory N1 component. Therefore this component could be associated with some brain events related to music processing. The slow negativity in the responses to the musical phrase were distributed widely and symmetrically at a centro-frontal region (Fig. 4a). Recent studies have revealed that the scalp topography of slow negativity, especially for some memory tasks, is closely related to the cortical areas subserving the task (Rösler et al. 1997). In the current study, however, it was difficult to localize the area of activation, since the slow negativity was widely distributed. Repetitive positive and negative peaks were observed in the responses to the musical phrase, but not for the prolonged initial sound of the musical phrase, white noise, and the simple tone (Figs. 4b and 3). Therefore, this repetitive activity could be associated with the attack of each note, at least partially. Interestingly, the negative peak of the second phasic activity had a parietal maximum scalp distribution which differed from that of the N1 component.

Main limitations of our current methods include: (1) long recording time, with many iterations of the same musical phrase; (2) slow frame rate of the movie; and (3) poor evidence of correlation between the stimuli and the responses. As for the first limitation, it is essential when using the averaging technique to have a sufficient number of samples to achieve an adequate signal-to-noise ratio. Both subjects reported that, compared with other kinds of stimulation, it was rather easier to maintain attention to the musical stimuli, even for such a long session; i.e., musical stimuli were inherently more interesting. It might be possible to reduce recording time by utilizing additional technical advancements such as single trial analysis (Suwazono et al. 1994). The second limitation was purely technical. Because currently available software permitted us only up to 30 frames per second to make a movie, it was needed to skip many of the topographical maps in order to fit with the speed of the sound track (from 1 millisecond sampling to 33 millisecond sampling). Finally, regarding the third limitation, it was not yet examined the cross-correlation between the stimuli and the music-related EEG responses, primarily because the sampling rate for EEG (1 kHz) is far slower than that for the musical phrase (44.1 kHz). This will be a very interesting project in the future, probably using fewer EEG channels to achieve a higher digitization rate.

References

Besson M, Faita F, Requin J. Brain waves associated with musical incongruities differ for musicians and non-musicians. Neurosci Lett. 1994;168:101-5.

Besson M, Faita F, Czternasty C, Kutas M. What's in a pause: event-related potential analysis of temporal disruptions in written and spoken sentences. Biol Psychol. 1997; 46:3-23.

Nittono H, Bito T, Hayashi M, Sakata S, Hori T. Event-related potentials elicited by wrong terminal notes: effects of temporal disruption. Biol Psychol. 2000; 52:1-16.

Rösler F, Heil H, Röder B. Slow negative brain potentials as reflections of specific modular resources of cognition. Biol Psychol. 1997; 45:109-141.

Stephenson WA, Gibbs FA. A balanced non-cephalic reference electrode. Electroenceph Clin Neurophysiol. 1951; 3:237-240.

Suwazono S, Shibasaki H, Nishida S, Nakamura M, Honda M, Nagamine T, Ikeda A, Ito J, Kimura J. Automatic detection of P300 in single sweep records of auditory event-related potential. J Clin Neurophysiol. 1994; 11:448-60.

Verleger R. P3-evoking wrong notes: unexpected, awaited, or arousing? Int J Neurisci. 1990; 55:171-179.

Chapter II.10

A New Technique in Magnetoencephalography and Its Application to Visual Neuroscience

Keisuke Toyama [a,1], Kenji Yoshikawa [a], Sadamu Tomita [b], and Shigeki Kajihara [b]

[a] *Departments of Physiology and Neurology, Kyoto Prefectural University of Medicine*
[b] *Technology Research Laboratory, Shimadzu Corporation*

Introduction

Until recently, psychophysics has been the only feasible method in visual neuroscience for studying human visual functions. However, the recent development of noninvasive recording techniques such as positron emission tomography (PET), functional magnetic resonance imaging (fMRI) and magnetoencephalography (MEG) has made available alternate methods for mapping neuronal activities during visual perception. A more powerful approach would be to combine the two methods and investigate the correlation between brain activities observed by these noninvasive methods and the visual percepts determined by visual psychophysics. Ideally, the noninvasive method should have both good temporal as well as spatial resolution. While MEG, which measures magnetic signals produced by electric current generated as a consequence of neuronal activities, has an inherently high temporal resolution, PET and fMRI, which measure changes in blood flow linked to brain activities secondarily, or even as a higher order of consequence, have low temporal resolution.

The main limitation of conventional MEG is its low spatial resolution, resulting from the fact that a single current source dipole yields magnetic fields spreading out broadly from the source dipole. It is therefore difficult to uniquely solve the inverse problem of resolving MEG signals detected by an array of magnetometers placed above the cranium into component electrical activities arising in various areas of the brain associated with visual perception (Tripp, 1983; Hamalainen et al., 1984; Sarvas, 1987). A common approach to resolving the inverse problem has been the use of current source analysis to estimate major current dipoles

---

[1] Correspondence: Keisuke Toyama, Department of Physiology, Kyoto Prefectural University of Medicine, 465 Kawaramachi-Hirokoji, Kamigyo, Kyoto 602-8566, Japan, Tel: (81)-75-251-5313, Fax: (81)-75-241-1499, e-mail: toyama@basic2.kpu-m.ac.jp Abbreviations 2D: two-dimensional, 3D: three-dimensional, GF: goodness of fit, HM: homogenous motion, MEG: magnetoencephalogram or magnetoencephalography, MRI: magnetic resonance imaging, ms: millisecond, SF: spatial filter, SFP: SF plane, SM: segmented motion, S/N: signal-to-noise ratio

accounting for the MEG signals. This approach requires implementation of assumptions concerning the structure of the current sources, such as single (Cuffin et al., 1977; Tripp, 1983) or multiple dipole models (Tripp, 1983; Nunez et al., 1986;Mosher et al., 1992; Supek et al., 1993; Lutkenhoener, 1995), minimum-norm estimates (Crowley et al., 1992), or two-dimensional current source models (Roth et al., 1989; Alvarez, 1990).

A disadvantage of these methods is that they only estimate major current sources contributing to the MEG at particular moments. Therefore, these methods provide no effective way for recording neural activity in the brain with high spatio-temporal resolution, which is an ultimate goal of noninvasive recording techniques. In addition, the estimates of current sources stringently depend on the assumptions of the models, which have frequently been rather unrealistic. Also such estimates have been feasible only in cases in which the signal-to-noise ratios (S/N) of the MEG signals were high (Robinson et al., 1992). These disadvantages have limited the application of MEG to the study of human brain function, demanding selective recording of activities in individual cortical areas. A linear approach using spatial filters (SFs) to resolve the MEG into local brain currents by linear summation has opened a new possibility for high spatial resolution MEG (Robinson et al., 1992; Dale et al., 1993; Peralta-Menendez et al., 1998; Toyama et al., in press). In the present study a high spatial resolution MEG system was constructed by combination of three-dimensional (3D) magnetometers measuring the direction as well as the intensity of the magnetic fields with SFs, and the power of the high spatial resolution MEG to estimate the brain currents with high temporal as well as spatial resolution was proved by simulation and phantom experiments. Application of this system resolved MEG responses to visual stimuli into electrical responses in individual human visual cortical areas including V1-V3 and V5. A part of the present results have been reported previously (Toyama et al., in press).

Results

1. Construction of a high spatial resolution MEG system

The high spatial resolution MEG system was constructed of 3D magnetometers (Shimadzu SBI-100) and SFs (Toyama et al., in press). The magnetometers consisted of 43 sets of a 3D sensor coil complex (DC1-DC3 in Fig. 1A, DC1-DC129 in Figs. 1B and C), each of which included three elementary coils oriented 60° each measuring the intensity of the magnetic field passing through the coils. The 3D magnetometers detected magnetic field vectors roughly across a quarter (subtended angle, 87°) of the brain surface (Fig. 1B). Signals from the 3D magnetometers were fed to a computer through an AD converter (16 bits at 1K Hz) and were stored on a hard disc. The signals were analyzed off-line by a single SF or a 21x 21 SF array (SFn in Fig. 1C and SF_1 - S_{441} in Fig. 1D), which recorded local brain currents at a single point or at 21 x 21 nodal points (internodal intervals, 9° or 4.5°) on a SF plane (SFP) that covered the entire or a part of hemisphere at a variable distance below the scalp. The SFs were linear SFs resolving the MEGs into local brain currents by linear summation.

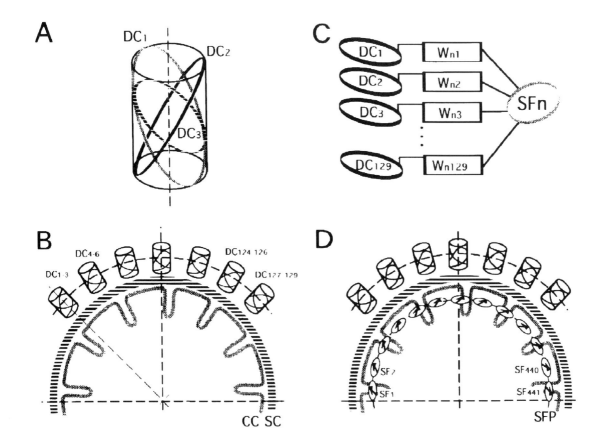

Figure 1

Schematic diagrams for high spatial resolution magnetoencephalography (MEG). A, 3D-magnetometer sensor element consisting of 3 sensor coils (DC_1-DC_3) which are tilted 90° each other and 54.7° to the axis of the sensor element (interrupted line). Diameter of coil bobbin, 20 mm. B, 3D-magnetometer sensor array consisting of 3 x 43 sensor coils (DC_1-DC_{129}) which covered roughly a quarter (subtended angle, 87°) area of the brain. SC, scalp; CC, cortical surface. C, Spatial filter (SF_n) taking a linear sum of magnetometer outputs with variable weights (W_{n1-129}). D, 3D-magnetometer sensor and SF array consisting of 21 x 21 elements (SF_{1-441}) arranged on a spatial filter plane (SFP), covering a half hemisphere (subtended angle, 180°) of the brain (Toyama et al., in press).

2. Simulation study

Optimization of SF

A computer simulation was conducted using a spherical model brain (radius, 90 mm). Single test current source dipole was assumed to be located at a point 10 mm (the origin) below the surface of the model brain, oriented along the equator (x axis) of the model brain, and the magnetometer array was assumed to be distanced 40 mm from the test source dipole. It was further assumed for optimization of the SF weights (Wn1 ·· n129 in Fig. 1C) that the

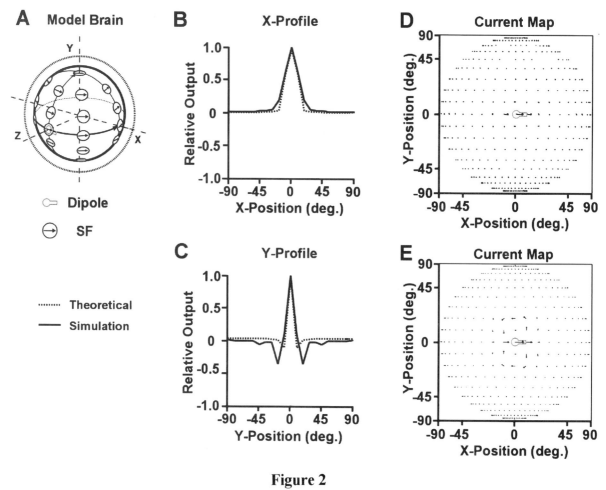

Figure 2

Simulation study of responses of a high spatial resolution MEG system to a single source dipole. A, 3D diagram illustrating positions of a test source dipole and an SF array on the x, y and z axes (along the equator, meridian and radial line of the model brain, respectively). Dotted and solid circles represent scalp and SFP, respectively. B and C, x- and y-output profiles (solid line) of a single SFx as a function of the distance of the SFx from the position of the test source dipole (origin), compared to the theoretical x- and y-current profiles determined for the test source dipole according to Wilson and Bayley's equation (Wilson et al., 1950) (dotted line). The x and y positions are expressed as angles to the y- and x- axes (internodal intervals, 9°), respectively. Ordinate, relative strength of SF outputs. D and E, theoretical 3D current maps and those estimated by the 21 x 21 SFx and SFy arrays (internodal intervals, 9°). Equation (12) in the Appendix indicates that the SFx and SFy exhibit sinusoidal tuning to the orientation difference between the FS and test source dipole, and therefore that the local currents can be estimated as vector sums of the SFx and SFy outputs. Arrows in the maps represent intensity and direction of the current vectors (Toyama et al., in press).

model brain contains no conducting medium, and therefore generates no current flow around the dipole (Equations 1-3 in Appendix). Simulation was conducted to determine the magnetic signals detected by the 3D magnetometers placed above the model brain and the outputs of each SF focused to a current dipole of a particular orientation and position on the SFP of the model brain (Equation 4).

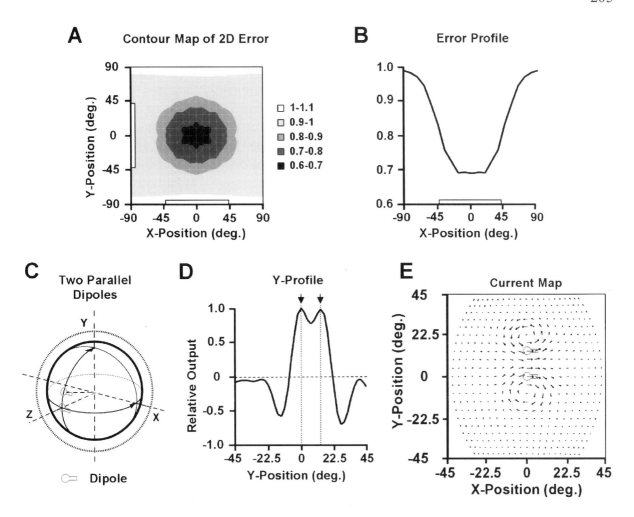

Figure 3

2D errors for the SF array and responses of the high spatial resolution MEG system to two parallel source dipoles. A and B, a contour map and x-profiles of the estimate errors integrated over SFP (Equation 5 in Appendix) for the 21 x 21 SFx array. Internodal intervals, 9°. Blank bars on the x and y axes in A and on the x-axis in B indicate the area covered by the magnetometer array. C, a diagram illustrating two source dipoles placed in the model brain. D and E, y-current profiles estimated by the SFx and a current map estimated by the SFx and SFy arrays, respectively. Internodal intervals, 4.5°. Red circles and rods in C and E indicate the two x-oriented source dipoles distanced 13.5° along the y-axis. Downward arrows in D also indicate locations of the source dipoles.

The SF weights were optimized so that the error function representing an integrated sum of square errors of the current estimates, that is deviations of the SF outputs from the real current source dipole, is minimized over the space of the model brain (Equations 4-12). Ideally, the optimization should be conducted over the 3D space. However, this was difficult due to nonlinear sensitivity of magnetometers along the depth axis. Instead, the optimization over 2D space along the SFP yielded reasonably good results. The 2D error was practically minimized over the entire area covered by the magnetometer sensors (Figs. 3A and B), and rapidly increased outside that area, indicating that the performance of the high spatial resolution MEG is only maintained inside the area covered by the magnetometer sensors.

Performance of SF

Simulation was further conducted for the case where the model brain is filled with a conducting medium (Equations 13-18). Equation 12 in Appendix indicates that output of a SF is a function of the difference between the position and orientation of the source current dipole and those to which the SF is focused. The position selectivity of the SF was studied by moving the focal positions along the x and y axes (along the meridian) on the SFP and (Fig. 2A). Figures 2B and C illustrate, respectively, the x- and y-output profiles of a single SFx whose orientation selectivity is focused on the x direction, representing how the outputs of the SFx change as the focal position of the SFx deviates from the origin along the x and y axes, respectively. The outputs of SFx (solid line in Fig. 2B) peaked at the origin and sharply declined with deviations along the x-axis (half width, 9° and 12.6 mm). The y-output profiles of the SFx was even sharper than the x-output profiles (half width, 5.4° and 7.5 mm) and associated with deep side flanks (solid line in Fig. 2C). The x- and y-output profiles of the SFx determined by the computer simulation agreed the current distributions that were theoretically predicted for the test source dipole in the model brain (dotted lines in Figs. 2B and C) (Wilson et al., 1950), except that the x-output profiles were only slightly broader (the ratio of the half width of the estimated x- and y-profiles to the theoretical ones, 1.1 and 1.2), and the side-flanks of the y-output profiles were much more pronounced.

Figure 2E shows the results of the simulation study on SF arrays, consisting of SFx and SFy arrays whose elements were focused on the x- and y- oriented current dipoles at 21 x 21 nodal points on the SFP (internodal intervals, 9°), respectively. The currents at each nodal point were estimated as a vector sum of the corresponding SFx and SFy. The current map estimated by the SF arrays also agreed very well with the theoretical map predicted for the test source dipole (Wilson et al., 1950) shown in Fig. 2D. These results of the simulation study indicate that the high spatial resolution MEG is capable of mapping currents arising from single current sources in the model brain with a spatial resolution only slightly less than that theoretically possible.

A further simulation study was conducted for the case in which two source dipoles of the same intensity and orientation were placed in parallel at variable distances along the meridian of the SFP (Fig. 3C). The y-current profiles and current map (Figs. 3D and E) for the case where the two dipoles were separated 13.5° (18.8 mm) clearly revealed two current peaks corresponding to the locations of the two source dipoles (downward arrows and dotted lines in Fig. 3D). Simulation studies also confirmed that the high spatial resolution MEG is capable of discriminating two source dipoles separated with the same distance as those for Figs. 3A and B but oriented opposite each other (Fig. 4A). The y-current profile and the current map (Figs. 4B and C) showed positive and negative current peaks at the location of the positive and negative source dipoles, respectively. Likewise, the high spatial resolution MEG was also capable of discriminating two source dipoles oriented 90° each other as is shown in the y-profiles of the strength of the current vectors and the current map (Figs. 4D-F).

It was generally found that the current profiles for multiple dipole cases are generally predicted as the sum of those for the individual dipoles if they are located on the same SFP, implying that high spatial resolution MEG is capable of correctly resolving multiple source dipoles if they were separated by more than a minimum distance of 12.6° (17 mm). Further simulation studies indicated that the SF lacks depth tuning due to the 2D optimization, the sensitivity decreasing in reverse proportion to the square of the distance between the source dipoles and the center of the magnetometer array. Thus this system maps not only the current dipoles exactly located on the SFP on which the system is focused, but also those off-set above or below from the SFP. However, it can only correctly estimate the intensity of the currents on the SFP, and over- or underestimates that off-set above or below, respectively.

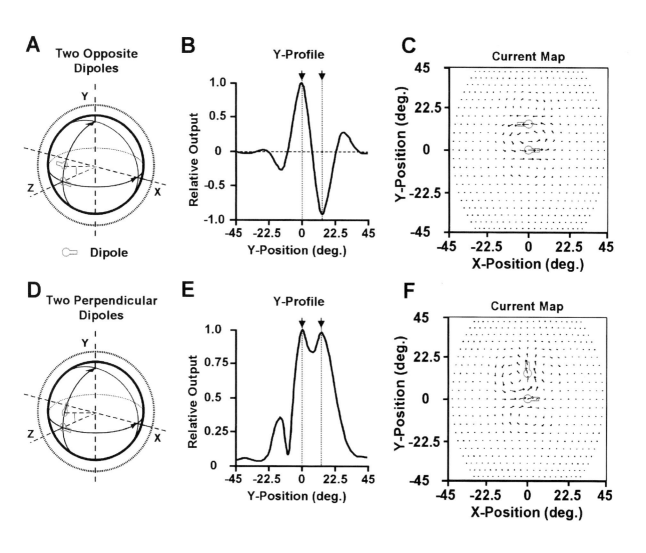

Figure 4

Responses of the high spatial resolution MEG system to two opposite and orthogonal source dipoles. A-C and D-F illustrate responses of the SF array to two opposite and orthogonal source dipoles in the same way as for Figs. 3C-E. The y-profiles in E plot the strength of the vector sum of the SFx and SFy outputs, in order to show two current peaks streaming along the x and y axes corresponding to the two orthogonal dipoles (cf. Fig. 4F), while those in B plot SFx outputs as in Fig. 2C and Fig. 3D. The two dipoles were distanced 13.5° along the y-axis in both cases.

It is an interesting question to what extent current estimates are improved by 3D rather than the conventional radial MEG. This question was answered by conducting simulation studies for radial and 3D magnetometers based on exactly the same current source model. It was found that both major peak and spurious side flanks in the x- and y-current profiles were considerably broader for the radial magnetometer system (half width for x- and y-profiles, 11.5° and 16 mm, and 7.2° and 10 mm) than those for the 3D system (cf. solid and dotted lines in Figs. 5A and B).

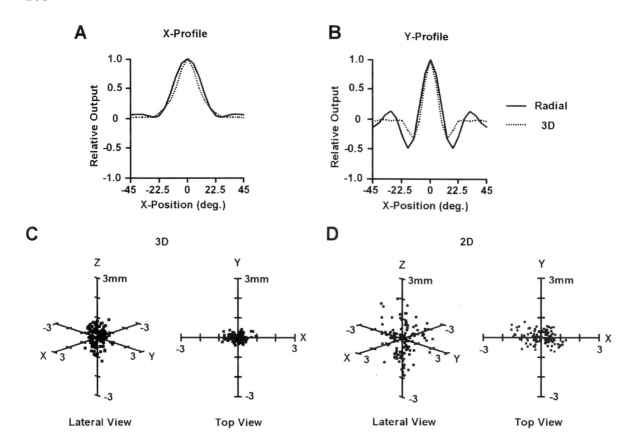

Figure 5

Comparison between 3D and conventional radial MEG. A and B, x- and y-current profiles of SFx outputs for a conventional radial magnetometer SF system (solid line) to the same single source dipole as that for Fig. 2. Those for 3D magnetometer SF system in Fig. 2B and C are shown for comparison (dotted line). C and D, scatter diagrams representing the estimated locations of the current source for one hundred trials of conventional current source analysis conducted for 3D and radial MEGs containing the same intensity of Gausian noise (S/N = 10). Simulation for the radial magnetometer SF system was conducted for the same physical conditions for the source dipole (dipole position and strength), and the MEG system (sensor coil diameter, number, density, and the distance between magnetometer array and the surface of model brain). The origin of the x, y and z-axes represents the location of true source dipole.

A comparable simulation study for the conventional current source analysis using the same dipole model as that for the SF system but implementing Gaussian noise (S/N = 10) in the magnetometer signals also revealed the advantage of the 3D over radial MEG. Figures 5 C and D compare the estimated dipole locations for one hundred simulation trials of the current source analysis by the 3D and radial MEG. The estimates scattered around the true location (the origin) of the source dipole for both cases, and the scatters were generally greatest along the z-axis, modest along the x-axis and smallest along the y-axis for both cases. However, there were generally greater scatters for the radial than for the 3D MEG. The mean estimate error (deviations of the estimates from the origin) was roughly half for the 3D MEG (0.52 ±

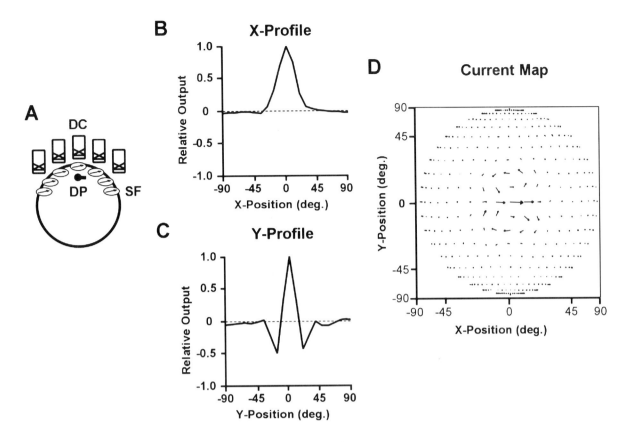

Figure 6

Phantom study of SF array. A, schematic diagram illustrating a phantom experiment. B and C, x- and y-output profiles of the SFx similar to those shown in Figs. 2B and C. D, 3D current map similar to that in Fig. 2E, estimated as vector sums of outputs in the SFx and SFy arrays. A test source dipole (interpolar distance, 10 mm, AC current, 14 μA at 10 Hz) was placed at a point (the origin) of the SFP 10 mm below a model brain surface, to which the central elements of the SFx and SFy arrays were focused. Signals of the magnetic fields produced by the test source dipole were averaged for 1000 trials of recording by the magnetometer array.

The noise tolerance of the SF was enhanced by regularizing and modifying the inverse matrix Γ^{-1} as

$$\Gamma^{-1} = (GG^T + \gamma I)^{-1} \qquad (1)$$

where I represents unit matrix and γ is a value sufficiently smaller (roughly 1/100) than the maximum value of GG^T, and G is expressed by (2) and (3)

$$G = \begin{pmatrix} G_1 \\ G_2 \\ \\ G_M \end{pmatrix} \qquad (2)$$

$$\vec{G}_i(r_{21(k-1)+l}) = \vec{L}_i(\vec{r}_{21(k-1)+l}) \sqrt{\cos \vartheta_l} \qquad (3)$$

(Toyama et al., in press).

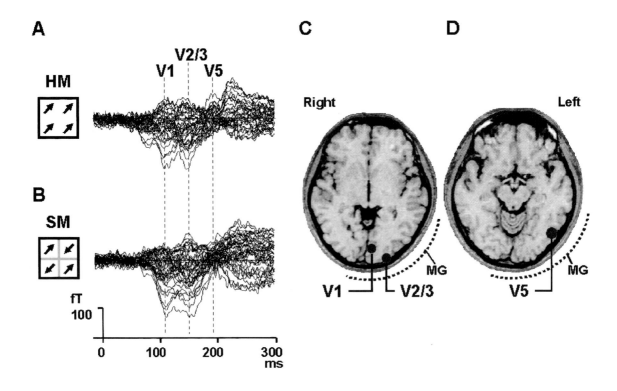

Figure 7

MEGs evoked in human visual cortex by visual stimulation. A and B, Radial MEG components observed in subject NG for homogenous (HM) and segmented (SM) random dot motion stimuli, respectively. The random dot stimuli (grain size, 0.08°; density, 10%; contrast, 2.6 log units) were constructed by a computer, displayed on a screen by a liquid-crystal projector and presented to the right central visual field (horizontal and vertical subtended angles, 36° and 36°) of an experimental subject through a surface mirror and goggles. For the HM stimuli all dots were moved transiently (15 ms) in 4 oblique directions (±45°, ±135°; velocity, 15.6°/s). For the SM stimuli the dots were moved in a similar way to that for HM stimuli, but they were separated into 6 x 6 segments so that the dots in neighboring segments moved in opposite directions. The SM stimuli yielded vivid perception of subjective contours separating the groups of dots moving in opposite directions, although no such contours were physically present. HM and SM stimuli were presented in a quasi-random sequence of quasi-random intervals (5 – 8 s). Stationary random dots were constantly presented during the intermission period of the motion stimuli. MEGs were recorded with high frequency cut-off at 100Hz by the 3D magnetometer array (sampling rate, 1 K Hz) were averaged for 200 trials of HM or SM stimulation in four directions. C and D, locations of the current sources (red circles) determined by conventional current source analysis of the MEG shown on two axial slices of MRI images of the brain of subject NG through V1, V2/3 and V5 (solid lines V1, V2/3 and V5). The current source analysis was conducted for the three peaks of the MEG (dotted lines in Figs. 5A and B) by the method of the least-squares error fit between the observed MEGs and those predicted by the single dipole source model. Eight trials of the least-squares error fit was conducted for each peak, starting from randomly chosen initial conditions for the dipole model, and the one giving the best goodness of fit (GF) determined as $GF = 1 - \sqrt{\sum_{i=1}^{129}(S_i - D_i)^2 / \sum_{i=1}^{129} S_i^2}$, where S and D are the observed and predicted MEGs for 3 x 43 magnetometers, respectively, was chosen as the estimate of the current dipole. The GF was 94, 95 and 85 % for the 1st, 2nd and 3rd peak, respectively. Dotted lines MG in C and D indicate the sensor plane of the 3D magnetometer (Toyama et al., in press).

0.43 mm) of that for the radial MEG (0.98 ± 0.80 mm). These results are consistent with the findings that the SF lacks the depth selectivity and that the current selectivity of the SF is broader for the x axis than for the y axis, and indicate that 3D MEG significantly improves the performance of the conventional current source analysis as well as the SF system.

Altogether the simulation studies indicate that the high spatial resolution MEG yields reliable estimates of brain activities, if it is combined with the conventional current source analysis.

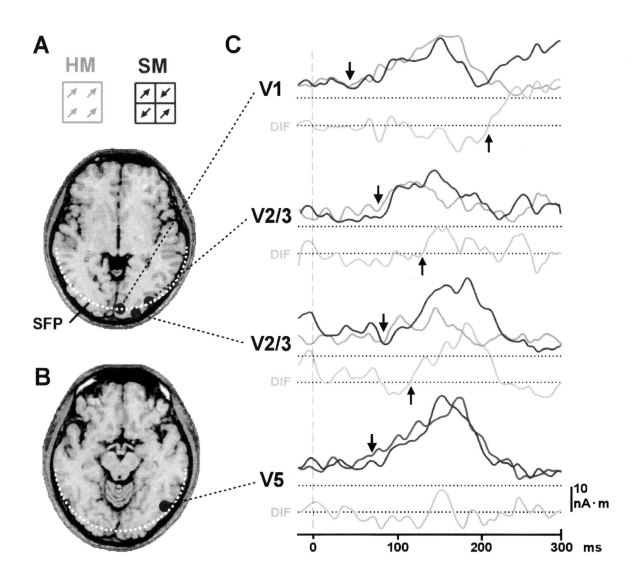

Figure 8

Individual responses in visual cortical areas resolved from MEGs by the SFs. A and B illustrate the focal positions of SFs whose outputs are shown in C. White interrupted line indicates the SFP for determination of the current maps shown Fig. 9. C, responses to HM and SM stimuli (blue and red traces, respectively) in V1, V2/3 and V5 of subject NG. The amplitude of responses represents that of the summed vectors between the outputs of the SFx and SFy focused to each cortical area. DIF, differential responses (shown in green) between HM and SM stimuli. Moving average was conducted for all traces with a 5ms window. Downward and upward arrows indicate onsets of responses and response enhancement, respectively (Toyama et al., in press).

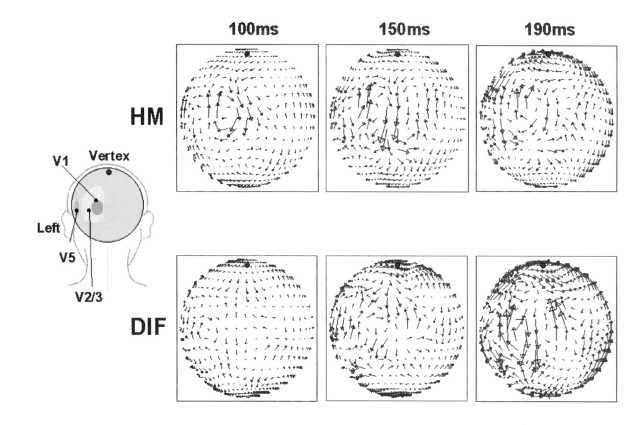

Figure 9
Time-lapse response maps of human visual cortex to visual stimulation. HM and DIF, current maps for the responses illustrated in Fig. 8 C to the HM stimulus and the differential responses between HM and SM stimuli 100, 150 and 190 ms after the HM and SM stimuli. The response vectors were estimated as vector sums of 21 x 21 SFx and SFy arrays whose central elements were focused to the point of major current source in V1, 27 mm below the scalp. The SFP (white interrupted line in Figs. 8A and B) roughly included all focus points for responses in Fig. 8C. Red circles represent the points corresponding to the vertex of the cranium. Inset diagram indicates the areas corresponding to V1, V2/3 and V5 on a rear view of the cranium (Toyama et al., in press).

3. Phantom experiments

The results of the simulation study were confirmed in phantom experiments in which a phantom brain was constructed of a plastic spherical ball filled with saline solution (radius, 84.8 mm). Test bipolar electrodes constructed of twisted platinum wires coated with polyethylene except at the tips (interpolar distance, 10 mm) were placed at a point on the SFP 10 mm below the surface of the phantom brain, oriented along the x axis (Fig. 6A). The test source dipole was generated by passing AC current ($14 \mu A$, 10 Hz) through the bipolar electrodes. The magnetic fields produced by the test current were detected by the 3D magnetometers whose center element was aligned to the radial axis through the test source dipole. The signals from the magnetometers were fed to the 21 x 21 SFx and SFy array whose center elements were focused to the test source dipole and other elements to those at the other nodal points on the SFP.

The x- and y-output profiles of the SFx were in close agreement with those for the simulation study (cf. solid line in Figs. 2B and C with Figs. 6B and C). The y-output profiles were sharper than x-output profiles and bounded with side-flanks in the phantom as well as in the simulation study. It was also found that the current map, determined as vector sums of the SFx and SFy outputs in the phantom experiments, exactly agreed with those for the simulation study (cf. Fig. 2E with Fig. 6D). These results indicate that the actual 3D magnetometer-SF system is in fact capable of mapping real world electric currents, produced by a single pair of bipolar electrodes placed in the conducting medium of the phantom brain at the SFP, exactly as predicted mathematically by the simulation study.

4. Recording of neural activities from visual cortical areas

MEGs evoked by visual stimulation were recorded in 3 right-handed male subjects (NG, KY and SW) aged 29-32, by focusing the 3D magnetometers in the left occipital cortex (dotted line MG in Figs. 7C and D). Experiments were conducted according to the experimental regulations of the Human Experiment Committee in Kyoto Prefectural University of Medicine, including taking informed consents from experimental subjects who understood the aim and experimental procedures of the study (HRC-5). Three flat positioning coils (diameter 11.4 mm, 10 turns) each of which contained a plastic disc filled with 0.1 % NiCls in the center were attached to subjects, one on the mid-frontal skin and two on the left and right temporal skin. The NiCl discs served as markers for calibration of the coil positions by magnetic resonance imaging (MRI: Shimadzu, Magnex-alpha). Calibration was conducted before and after MEG measurement by passing currents (0.2 - 5 mA, 10 Hz) through the coils. The coil positions were determined in the MRI images as well as for the MEGs by conventional current source analysis and were used for the 3D coordinates of SF current profiles for superimposition on the MRI images.

The visual stimulation exhibited homogenous motion (HM in Fig. 7 insets) in which all random dots were transiently moved in a uniform direction, or segmented motion (SM) in which random dots were segmented into 6 x 6 fractions, and moved in opposite directions between neighboring segments (Lamme et al., 1993). These stimuli were presented to the right visual field (subtended angle, 36 x 36°) of the subject in a quasi-random sequence and at quasi-random intervals using a liquid-crystal projector (Sanyo LP-9200). The HM stimuli yielded perception of uniform motion of the random dots, while the SM stimuli yielded vivid perception of subjective contours separating random dots moving in the opposite directions, although they were not physically present. Figures 7A and B illustrate averaged MEGs observed in one subject (NG) for HM and SM stimuli, respectively. Both responses contained three peaks, and conventional current source analysis (Toyama et al., in press) showed the major current sources for the early peak (latency, 102 ± 7 ms; n = 3) to be in the middle part of the upper bank of the calcarine sulcus, with a reasonable high reliability (goodness of fit > 92%), and middle (131 ± 14 ms) and late peaks (170 ± 11 ms) in the posterior part of the medial occipital gyrus and lower segment of the posterior bank of the anterior occipital sulcus in the left hemisphere with slightly lower reliability (> 85 %). The areas roughly corresponded to the areas identified as V1, V2/3 and V5 (red circles in Figs. 7C and D), respectively in a previous study (Sereno et al., 1995). Likewise MEGs in other two subjects (KY and SW) consisted of similar three peaks, and current source analysis revealed the major current sources at V1, V2/3 and V5 (goodness of fit, 95-85%).

The SFs focused on the three sites of the major current sources for the MEGs in V1, V2/3 and V5 and to an additional site in V2/3 (red circles in Figs. 8A and B) resolved the MEGs into the individual responses in each cortical area. The responses to the HM stimulus (blue traces in Fig. 8C) occurred earliest in V1 (latency, 56 ± 8 ms; downward arrow) and later in V2/3 (75 ± 13 ms) and V5 (81 ± 12 ms). The V1 responses to the SM stimulus (red traces) were similar to those to the HM stimulus in the initial half of the sampling periods but significantly became larger in the later half (cf. blue and red traces, and also see green traces DIF) than the HM responses. The response enhancement as the difference of the SM from HM responses was strongest in V2/3 (106 ± 22 % of peak HM responses), and variable among subjects in V1 (15 ± 88 %) and in V5 (-58 ± 30 %). The onset of response enhancement was earlier in V2/V3 (latency, 125 ± 8 ms; upward arrows in green traces) than in V1 (latency, 217 ± 4 ms), and was significantly delayed from that for both HM and SM stimuli in V5.

The time course of response enhancement of the SM response compared to the HM responses corresponded to that reported for the monkey V1 (Lamme, 1995) and the spatial distribution resembled that reported for the human brain using functional MRI (Reppas et al., 1997). The enhanced responses probably represented signals of contextual modulation underlying the perception of subjective contours separating the random dots in the segmented stimulus (Lamme et al., 1993; Lamme, 1995; Reppas et al., 1997).

The spatio-temporal patterns of responses to the HM and SM stimuli were studied by constructing time-lapse current maps using the SF array, with the SFP roughly including the sites of the major current sources indicated by a conventional current source analysis (27 mm below the scalp, dashed white lines in Figs. 8A and B). The current maps for the responses to the HM stimulus (HM maps in Fig. 9) revealed that the major current sources were initially (at 100 ms) localized in V1 and a part of V2/3, and later spread to all areas of V2/V3 and V5 (150 and 190 ms), but any significant activity was neither detectable in the contralateral V1 nor in the ipsilateral parietal cortex neighboring to V2/3 throughout the entire observation period. These observations support the view that the high spatial resolution MEG is capable of estimating current sources in the brain with a reasonably high spatial resolution to resolve activities in individual cortical areas.

The response enhancement (DIF maps in Fig. 9) only appeared in later periods (150 and 190 ms) rather broadly in the area of V2/3 after activation of V5. The spatio-temporal patterns of the responses to the HM stimulus and response enhancement for the HM stimulus are consistent with a view that the signals for contextual modulation are back-propagated from V5 to V2/3 and further to V1. Responses to the HM and HM stimuli appeared in V1, V2/3 and V5 of two other subjects (KY and SW) with similar spatio-temporal patterns.

Altogether, both simulation and phantom studies indicated that the high spatial resolution MEG is capable of mapping multiple current sources distributed in the brain with spatial resolution almost comparable to any other functional brain imaging techniques, but with far greater temporal resolution. Application of this technique to the MEGs evoked by visual stimulation in human subjects demonstrated neural responses in V1, V2/3 and V5, as well as the signals of contextual modulations in V1 and V2/3, with a temporal resolution in milliseconds. All of these results indicate that the high spatial resolution MEG is capable of recording neural activities in the human brain with a combination of a high spatial as well as temporal resolution. This method opens a new way of studying the dynamics of human brain function.

Summary

A high spatial resolution MEG system was developed to study human brain activity with high spatio-temporal resolution. The system supplement the high temporal resolution of magnetoencephalography with high spatial resolution achieved by a combination of three-dimensional magnetometers capable of measuring the direction and intensity of magnetic fields arising in the brain and spatial filters estimating electric currents in the local brain from the magnetic signals of the magnetometers. Simulation and phantom studies indicate that the system is capable of mapping multiple current sources in the brain with spatial resolution comparable to that of any other brain functional imaging techniques while maintaining millisecond temporal resolution. Application of the high spatial resolution MEG to the human brain resolved MEG responses evoked by motion stimuli on a millisecond scale into responses occurring in visual cortical areas V1, V2/3 and V5, and also revealed signals related to contextual modulation in V1 and V2/3. The high spatial resolution MEG makes a new way of studying the dynamics of human brain function.

References

Alvarez RE. Biomagnetic Fourier imaging. IEEE Trans Biomed Eng 1990; 9: 299-304.
Crowley CW, Greenblatt RE, Khalil IS. Minimum norm estimation of current distributions in realistic geometries. In: Williamson SJ editor. Advances in Biomagnetism, New York: Plenum Press, 1992: 603-05.
Cuffin BN, Cohen D. Magnetic fields of a dipole in special volume conductor shapes. IEEE Trans Biomed Eng 1977; 24: 372-81.
Dale AM, Sereno MI. Improved localization of cortical activity by combing EEG with MEG cortical surface reconstruction: A linear approach. J Cogn Neurosci 1993; 5: 162-176.
Greiner W, Classical Electrodynamics. New York:: Springer-Verlag, 1998: 186
Hamalainen MS, Ilmoneimi RJ. Interpreting measured magnetic fields of the brain: estimates of current distributions. Helsinki Univ Tech Report 1984; TKK-F-A559.
Lamme VA.F, van Dijk BW, Spekreijse H. Contour from motion processing occurs in primary visual cortex. Nature 1993; 363: 541-43.
Lamme VAF. The neurophysiology of figure-ground segregation in primary visual cortex. J Neurosci 1995; 15: 1605-15.
Lutkenhoener B. Dipole and multidipole source analysis of magnetic fields: Possibilities and limitations. In: Baumgartner C, Deecke L, Stroink G, Williamson S, editors. Biomagnetism: Fundamental Research and Clinical Applications. Amsterdam: Elsevier, 1995: 376-80.
Mosher JC, Lewis PS, Leahy RM. Multiple dipole modeling and localization from spatio-temporal MEG data. IEEE Trans Biomed Eng 1992; 39: 541-57.
Nunez PL. The brain's magnetic field: some effects of multiple sources on localization methods. Electroencheph Clin Neurophysiol 1986; 63: 75-82.
Peralta-Menendez RG, Gonzalez-Andino SL. A critical analysis of linear Inverse solutions to neuroelectromagnetic inverse problem. IEEE Trans Biomed Eng 1998; 45: 440-48
Reppas JB, Niyogi S, Dale AM, Sereno MI, Tootell RBH. Representation of motion boundaries in retinotopic human visual cortical areas. Nature 1997; 388: 175-79.
Robinson SE, Rose DE. Current source image estimation by spatially filtered MEG. In: Hoke

M, editor. Biomagnetism: Clinical aspects. New York: Elsivier Science Publishers, 1992: 761-765,

Romani GL, Williamson SJ, Kaufman L. Biomagnetic instrumentation. Rev Sci Instrum 1982; 53: 1815 - 45.

Roth BJ, Sepulveda NG, Wikswo JP Jr. Using a magnetometer to image a two-dimensional current distribution. J Appl Phys 1989; 65: 361-72.

Sarvas J. Basic mathematical and electromagnetic concepts of the biomagnetic inverse problem. Phys Med Biol 1987; 32: 11 -22.

Sereno MI, Dale AM, Reppas JB, Kwong KK, Belliveau JW, Brady TJ, Rosen BR, Tootell RBH. Borders of multiple visual system areas in humans revealed by functional magnetic resonance imaging. Science 1995; 268: 889-93.

Supek S, Aine CJ Simulation studies of multiple dipole neuromagnetic source localization: Model order and limits of source resolution. IEEE Trans Biomed Eng 1993; 40: 529-40.

Tesche CD, Uusitalo MA, Ilmoniemi RH, Huotilainen,M, Kajola M, Salonen O. Single-space projection of MEG data characterize both distributed and well-localized neuronal sources. Electroencephalogr Clin Neurophysiol 1995; 95: 189-200.

Toyama K, Yoshikawa K, Yoshida Y, Kondo Y, Tomita S, Takanashi Y, Ejima Y, Yoshizawa S. A new method for magnetoencephalography (MEG): Three dimensional magnetometer-spatial filter system. Neuroscience in press.

Tripp JH. Physical concepts and mathematical models. In: Williamson SJ, Romani GL, Kaufman L, Modena I, editors. Biomagnetism: An interdisciplinary Approach, New York: Plenum Press, 1983: 101-39.

Wilson FN, Bayley RH. The electric field of an eccentric dipole in a homogenous spherical conducting medium. Circulation 1950; 1: 84-92.

Appendix

Optimization of the spatial filters (SFs) was conducted assuming that the model brain contains no conducting medium. The magnetic fields produced by a source current dipole in free space is given by Biot-Savart equation (Greiner 1998)

$$\vec{b}(r) = \frac{\mu_0}{4\pi} \frac{\vec{Q}_d \times (\vec{r} - \vec{r}_d)}{|\vec{r} - \vec{r}_d|^3}, \qquad (1)$$

where $\mu_0, r, \vec{r}_d$ and Q_d represent the magnetic permeability in free space, the position of magnetometers, the position and moment of a source current dipole, respectively.

The lead field $\vec{L}_i$ of the ith magnetometer is given by

$$\vec{L}_i(r) = \frac{\mu_0}{4\pi} \frac{(\vec{r}_i - \vec{r}) \times \vec{n}_i}{|\vec{r}_i - \vec{r}|^3}, \qquad (2)$$

where $\vec{r}_i, \vec{n}_i$ are position and orientation of the ith magnetometer.

Then magnetic signals $\vec{S}(S_1,\cdots,S_M)$ is determined as

$$S_i = \vec{b}(\vec{r}_i)\cdot\vec{n}_i, \tag{3}$$

where M represents the number of magnetometers (43 x 3).

According to Robinson and Rose (1992), the dipole in the model brain is estimated as the output of the spatial filter SF

$$SF = \vec{S}\cdot\vec{w}, \tag{4}$$

where $\vec{w}$ is the optimized weights of the SF by minimizing the error function f given in by (5).

$$f = \int_\Omega \left[\sum_{i=1}^M \vec{L}_i(\vec{r})w_i - \vec{\delta}(\vec{r})\right]^2 dr^3, \tag{5}$$

where Ω and δ represent the integration space and a delta function, respectively. $\vec{w}$ is transformed as

$$\vec{w} = \Gamma^{-1}\vec{B}, \tag{6}$$

where $\vec{w}$, Γ^{-1} and $\vec{B}$ are given by (7)-(9).

$$\vec{w} = \begin{pmatrix} w_1 \\ w_2 \\ \vdots \\ w_M \end{pmatrix}, \quad \vec{B} = \begin{pmatrix} B_1 \\ B_2 \\ \vdots \\ B_M \end{pmatrix} \tag{7}$$

$$\Gamma_{ij} = \int_\Omega \vec{L}_i(\vec{r})\cdot\vec{L}_j(\vec{r})\,dr^3 \tag{8}$$

$$B_i = \int_\Omega \vec{L}_i(\vec{r})\cdot\vec{\delta}\,dr^3 \tag{9}$$

Theoretically, f should be minimized over the 3D space as indicated by the equation (5). However, this yielded no good estimates of the test current dipole. Therefore, f was minimized over the SFP and the equations (8) and (9) were approximated as (10) and (11).

$$\Gamma_{ij} = \sum_{k=1}^{21}\sum_{l=1}^{21} \vec{L}_i\!\left(\vec{r}_{21(k-1)+l}\right)\cdot\vec{L}_j\!\left(\vec{r}_{21(k-1)+l}\right)\cos\theta_l \tag{10}$$

where

$$\theta_l = \frac{\pi}{2} - \frac{\pi}{20}(l-1) \tag{11}$$

The equation (9) was converted to (12) more explicitly expressing the focus point and orientation of the SF.

$$B_i = \left|\vec{L}_i(\vec{r}_p)\right| \left|\vec{e}_u\right| \cos\varphi \tag{12}$$

where φ, $\vec{r}_p$ and $\vec{e}_u$ represent the angle and position of the dipole to which the SF was focused and unit vectors, respectively.

In case the model brain contains homogenous volume conductor, magnetic fields produced by a source current dipole is given by Sarvas equation (Sarvas, 1987).

$$\vec{b}(\vec{r}) = \frac{\mu_0}{4\pi F^2}\left(F\vec{Q}_d \times \vec{r}_d - \vec{Q}_d \times \vec{r}_d \cdot \vec{r}\, \vec{\nabla}F\right) \tag{13}$$

where the parameters in (13) are given by (14) – (17).

$$F = a\left(r\,a + r^2 - \vec{r}_d \cdot \vec{r}\right) \tag{14}$$
$$\vec{a} = \vec{r} - \vec{r}_d \tag{15}$$
$$a = |\vec{a}| \tag{16}$$
$$r = |\vec{r}| \tag{17}$$
$$\vec{\nabla}F = \left(\frac{a^2}{r} + \frac{\vec{a}\cdot\vec{r}}{a} + 2a + 2r\right)\vec{r} - \left(a + 2r + \frac{\vec{a}\cdot\vec{r}}{a}\right)\vec{r}_d \tag{18}$$

Magnetic signals for the model brain containing volume conductor are determined by replacing $\vec{b}(\vec{r})$ in (3) with that in (13).

Computation of (1) - (18) was conducted for a spherical model brain (radius, 90 mm) with a single test current dipole at a point 10mm (the origin) below the surface of the model brain and with a single spatial filter focused to a dipole at variable positions (x and y) and orientations (φ) to the x axis on the SFP (Fig. 2A) and at variable depths (d) from the origin along z axis (Fig. 2A).

Chapter II.11

Lambda Chart Analysis and Eigenvalue Imaging

Hitoshi Matsuzawa and Tsutomu Nakada[1]

Department of Integrated Neuroscience, Brain Research Institute, University of Niigata

Introduction

There are three kinds of molecular motions, namely, translation, rotation, and vibration. If one consider the target molecule as "solid", then vibration can be effectively omitted. Magnetic resonance (MR)[2] represents one of the rare technologies capable of *non-invasively* measuring molecular motion *in vivo*. In principle, MR deals with rotation in relationship with relaxation, whereas translation with diffusion. Relaxation is described by scalar time constant. In contrast, diffusion, which is a phenomenon in three spatial dimensions and described by second order tensor, can be the subject of tensor analysis. The attractive part of diffusion tensor analysis in the context of brain science is the fact that anisotropism of the water molecule observable in the human brain arises from axons and, in turn, anisotropy analysis can provide various information regarding axons *in vivo*. The *non-invasive* nature of MR technology further underscores the significance of MR based diffusion tensor analysis of the brain.

Molecular Diffusion

Fick's law for particle flux density is given by:

$$\mathbf{J} = D\nabla n$$

where D is the diffusion coefficient and n is the concentration of the particle. The equation of

---

[1] Correspondence: Tsutomu Nakada, M.D., Ph.D., Department of Integrated Neuroscience, Brain Research Institute, University of Niigata, Niigata 951-8585, Japan, Tel: (81)-25-227-0677, Fax: (81)-25-227-0821, e-mail tnakada@bri.niigata-u.ac.jp

[2] Following the current general convention, in this article, we utilize the term "magnetic resonance" to describe biomedical application of nuclear magnetic resonance (NMR).

continuity,

$$\frac{\partial n}{\partial t} + \nabla \cdot \mathbf{J} = 0$$

assures that the number of particles is conserved. The diffusion equation is then given by,

$$\frac{\partial n}{\partial t} = D\nabla^2 n$$

where ∇^2 represents Laplacian.

It is apparent that, in three-dimensional diffusion, the diffusion coefficient D has to be given in a second order tensor $\mathbf{D}^\xi$, which can be expressed in the form of,

$$\begin{pmatrix} \lambda_1 & 0 & 0 \\ 0 & \lambda_2 & 0 \\ 0 & 0 & \lambda_3 \end{pmatrix}.$$

with three eigenvalues of $\lambda_1 > \lambda_2 > \lambda_3$ (Borisenko & Tarapov, 1968, Arfken, 1985).

Physical Realism of Diffusion Variables

The canonical solution to the diffusion equation is a normalized Gaussian function, which in one dimension (x-component) can be expressed as,

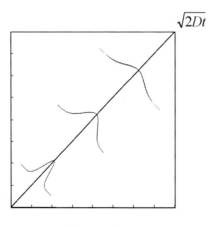

Figure 1

$$\theta(x,t) = \frac{1}{\sqrt{4\pi D_x t}} \exp(-\frac{x^2}{4D_x t}).$$

By evaluating the Gaussian integral, it becomes apparent that the mean square value of x can be expressed simply by,

$$\langle x^2_{(t)} \rangle = 2D_x t,$$

and in turn, the root mean square value is by,

$$x_{rms} = \sqrt{\overline{x_{(t)}}^2} = \sqrt{2D_x t}.$$

In physical realism, this value can be treated as the statistical average distance traveled by the particle in one dimension. The relationship between the Gaussian function and x_{rms} is graphically presented in Figure 1. Here, dispersion occurs with time along with a mean distance of $\sqrt{2Dt}$.

In three-dimensions, ellipsoid expression can further provide "intuitive" impression of physical realism of molecular diffusion (Figure 2). Here, ellipsoid can be seen as three-dimensional distribution of particles at time *t*, all of which are initially located at the origin. To define ellipsoid, the direction of the principal axes and magnitude at each principal axis need to be given. The former corresponds to eigenvector, $|\lambda\rangle$, whereas the latter to eigenvalue, λ (Borisenko & Tarapov, 1968, Arfken, 1985).

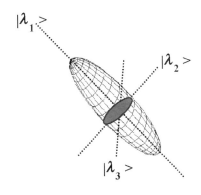

Figure 2

Axiomatic Condition for Biomedical Applications

Biomedical applications of diffusion tensor analysis is based on the widely accepted observation that diffusion anisotropism of the brain arises almost exclusively from neuronal fibers, especially axons. This biological constraint immediately provides an axiomatic condition: *Anisotropism should be observed only in one direction*: $\lambda_1 \geq \lambda_2 = \lambda_3$ (Nakada & Matsuzawa, 1995, Matsuzawa et al., 1995). The concept is easy to understand if one considers the cylinder-like shape of neuronal fibers (Figure 3). Potential sources of anisotropism, including structural restriction or axoplasmic flow, all possess unidirectional characteristics. It is apparent that, regardless of the precise underlying mechanisms responsible for observed anisotropism, anisotropism has to be unidirectional.

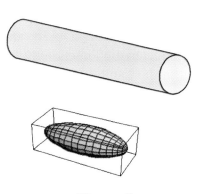

Figure 3

One immediate corollary of this axiomatic condition is the fact that anisotropism can be quantitatively expressed as anisotproic angle, θ, illustrated in Figure 4. Combining θ with the principal quantitative variable of the tensor, trace (*Tr*), properties of the target tissue with respect to its diffusion characteristics can be fully defined by a single function of $\Psi(Tr, \theta)$, which we refer to as *Diffusion Characteristic Function*.

Lambda Chart

In order to effectively analyze diffusion characteristics, we have developed a specific scatter chart method, which we term *Lambda Chart Analysis*. Based on the axiomatic condition described above, the diffusion system in question can be characterized by only two values, λ_α and λ_β, which is defined as,

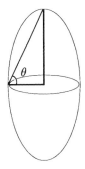

Figure 4

$$\lambda_\alpha = \lambda_1$$
$$\lambda_\beta = \frac{\lambda_2 + \lambda_3}{2}.$$

λ_β was so defined to take into account the minor variation and potential estimation errors of two small eigenvalues, λ_2 and λ_3, which need to be determined experimentally. Subsequently, λ_α of target pixels can be plotted against λ_β in two-dimensional chart, *Lambda Chart* (Figure 5).

Lambda Chart should be examined in polar coordinates. Anisotropic angle, θ, is directly read from the polar angle. The diagonal line at $\theta = \pi/4$, therefore, represents the isotropic line (blue line in Figure 5). Trace can be found using the following formula,

$$Tr = r\,(Sin\theta + 2Cos\theta)$$

where r is radius in polar coordinates. All the isotrace lines (red lines in Figure 5) appear as lines which cross the isotropic line with approximately 71.57° angle. Diffusion Characteristic Function $\Psi(Tr, \theta)$ is expressed in the Lambda Chart with the polar coordinates as,

$$\Psi(r(Sin\theta + 2Cos\theta), \theta).$$

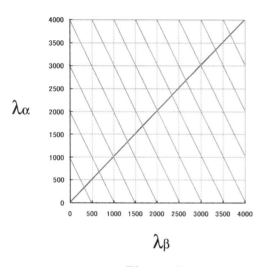

Figure 5

Materials and Methods

Eigenvalue Determination

The methodological description of eigenvalue determination in MR tensor analysis has previously been described by multiple authors. The concept can be easily understood by examining the scalar product $\langle \mathbf{G}|\mathbf{D}^\xi|\mathbf{G}\rangle$ where $\mathbf{G}$ is gradient vector and $\mathbf{D}^\xi$ is diffusion tensor. Utilizing closure relationship, $\sum_i |u_i\rangle\langle u_i|$, scalar product, $\langle \mathbf{G}|\mathbf{D}^\xi|\mathbf{G}\rangle$ can be expressed:

$$\langle \mathbf{G}|\mathbf{D}^\xi|\mathbf{G}\rangle = \sum_{ij}\langle \mathbf{G}|u_i\rangle\langle u_i|\mathbf{D}^\xi|u_j\rangle\langle u_j|\mathbf{G}\rangle$$
$$= \sum_{i,} G_i^* D_{ij} G_j$$

where asterisk signifies the conjugate. It foolws that by applying the appropriate combinations of gradients, all the matrix components can be effectively extracted into a set of linear equations and, in turn, all the matrix components can be determined. Subsequently, the principal eigenvector direction of each tensor matrix can be obtained.

Data Acquisition

Five normal volunteers and a patient with Wallerian degeneration of the pyramidal tract participated in the study. A General Electric (Waukesha, Wisconsin, USA) Echo Speed 1.5 T system was utilized. Diffusion weighted images (DWIs) were obtained using spin echo echo-planner Tetrahedral Stejskal-Tanner sequences and the following parameter settings: FOV 23 cm x 23 cm; matrix 128 x 128; slice thickness 5 mm; TR 5 sec: TE (minimum) 101 m sec; NEX 4. b-value was 500 for each axis. The parameter utilized for MPG gradients were: amplitude 1.518 g/cm; ramp time 1200 μ sec; δ 31.0 m sec; Δ 39.088 m sec.

Lambda Chart Analysis and Eigenvalue Images of the Normal Brain

Representative ellipsoid presentation of the diffusion tensor (full tensor analysis) of the entire slice is shown in Figure 6A (full tensor image). For clear visualization, a small portion of the full tensor image (delimited by square in Figure 6A) was extracted (Figure 6B). A conventional T2 weighted image of the identical slice is shown for comparison (Figure 6C).

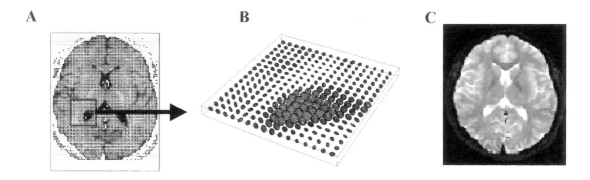

Figure 6

Lambda chart analysis of the identical full tensor image is shown in Figure 7. There are two distinctive groupings observed: one along the isotropic line and one along the isotrace line. Pixels mapped along the isotropic line reflect gray matter, whereas those mapped along the isotrace line, white matter.

In order to further illustrate physiological significance of the lambda chart analysis, four groups of pixels on the lambda chart were re-plotted back onto the original two-dimensional image matrix: *Eigenvalue Images* (Figure 8). These images clearly demonstrate that lambda chart can effectively segregate pixels into several groups of different physiological characteristics. Two

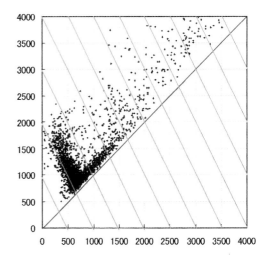

Figure 7

anisotropic groups (A and B on Figure 8) appear to represent white matter with long and short tracts, respectively. Two isotropic groups (C and D on Figure 8) appear to represent gray matter consisting of relatively small and large neurons, respectively. In this context, eigenvalue images can be considered "function-weighted" images where lambda chart analysis plays the role of contrast mechanism.

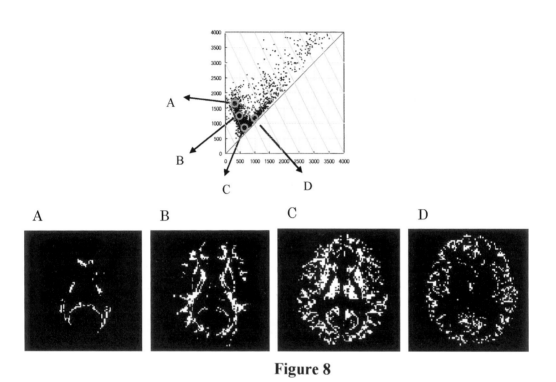

Figure 8

Lambda Chart Analysis of Localized Processes

Lambda chart analysis and eigenvalue images based on all pixels of a given imaging slice as demonstrated above is highly effective under various physiological and pathological conditions. However, for physiological analysis of certain localized processes, lambda chart analysis of the selected structure can be more effective. Figure 9 demonstrates such an example in analysis of the pyramidal tract. Data from the cerebral peduncle were mapped on the lambda chart. While "+" indicates data from normal volunteers, circles (open from the normal side, solid from the pathologic side) are data from a patient. Not only lambda chart analysis did detect abnormalities in pyramidal tract physiology it also provided information as to the characteristics of the pathology. Shift

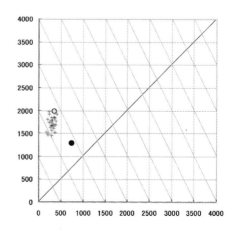

Figure 9

from the normal group toward the isotropic line along with the isotrace line indicated that the normal structures were likely to have been replaced by small isotropic structures, presumably representing gliosis.

Conclusion

A new powerful technique for analyzing diffusion tensor of the human brain is presented. The technique, termed Lambda Chart Analysis, can *non-invasively* provide hitherto unobtainable information of the brain segregated based on regional cell composition. Pictorial presentation (Eigenvalue Images) provides a novel type of functional imaging, the potential of which ranges neuronal density measurement in Alzheimer's disease to quantitative analysis of brain maturation.

Acknowledgement

The study was supported by grants from the Ministry of Education (Japan). The authors thank Drs. Hitoshi Saito and Naoki Nakayama at Azabu Neurosurgical Hospital for their help in data acquisition and Drs. Yuji Suzuki and Kenshi Terajima for data processing.

References

Arfken G: Mathematical Methods for Physicists. 3rd Edition. Academic Press, San Diego, 1985.
Borisenko AI, Tarapov IE: Vector and Tensor Analysis. Dover Publications, New York, 1968.
Matsuzawa H, Kwee IL, Nakada T: Magnetic resonance axonography of the rat spinal cord: Postmortem effects. J Neurosurg 1995;83:1023-1028.
Nakada T, Matsuzawa H: Three dimensional anisotropy magnetic resonance imaging of the rat nervous system. Neurosci Res 1995;22:389-398.

Chapter II.12

Single Ellipsoid Diffusion Tensor Analysis

Naoki Nakayama[a], Yukihiko Fujii[b], and Tsutomu Nakada[b,1]

[a]*Department of Neurosurgery, Hokkaido University*
[b]*Department of Integrated Neuroscience, Brain Research Institute, University of Niigata*

Introduction

Diffusion tensor analysis of the human brain represents an attractive new technique for evaluating structural as well as functional aspect of axons. Our first application of diffusion tensor analysis was directed towards fiber tract orientation. For this purpose, only directional information of the principal eigen*vector* is needed. Nevertheless, the mathematical formalism for extracting directional information out of the diffusion tensor involves the determination of all three eigen*values*. Such processes require complex data acquisition which is highly demanding on gradient hardware. Furthermore, as illustrated by T1 and T2 value images during the early stages of magnetic resonance imaging (MRI) development, the quality of the reconstructed images based on estimated parameter values is inherently disappointing compared to other directly obtained images. Accordingly, realistic clinical applications of principal eigen*vector* imaging ($PE_{vec}I$) has been primarily performed using the "optical method", namely, three dimensional anisotropy contrast magnetic resonance axonography (3DAC MRX) (Nakada et al., 1994, 1999).

Primarily to meet the needs for functional MRI (fMRI), recent advancements in gradient technology and ultra fast imaging techniques have made implementation of echo planar imaging in the clinical setting highly realistic. As a result, diffusion tensor analysis is now ready to be included as routine diagnostic examination tool. Given that diffusion anisotropy in the brain derives primarily from axons, it follows that eigen*vector* provides information on axonal direction (connectivity) while eigen*value*, axonal function. Since the deterministic method (full tensor analysis (FTA)) for obtaining eigen*vector* is always accomplished by the determination of eigen*values*, not only anatomic but also functional information of axonal network should be available. Nevertheless, the principle problems associated with FTA,

---

[1] Correspondence: Tsutomu Nakada, M.D., Ph.D., Department of Integrated Neuroscience, Brain Research Institute, University of Niigata, Niigata 951-8585, Japan, Tel: (81)-25-227-0677, Fax: (81)-25-227-0821, e-mail tnakada@bri.niigata-u.ac.jp

namely, strenuous hardware demand and serious variance in its numerical data, remain to be overcome (Pierpaoli et al, 1996).

For pictorial analysis, high spatial resolution plays a significant role. On the other hand, for quantitative analysis, numerical reliability represents a much more important factor in affecting clinical judgment. Therefore, the technique should be focused on simplicity in reliable quantification. One of the techniques often employed in functional imaging for the purpose of decreasing variance is "voxel averaging" by "volume of interest (VOI)" method. In FTA, VOI can be targeted to anatomically meaningful single structures, namely, tracts and nuclei. We evaluated the clinical usage of such a method, single ellipsoid diffusion tensor analysis.

Materials and Methods

Determination of Diffusion Ellipsoid

Taking into account the condition where translational micromobility of water molecules possesses orientational dependency (*an*isotropic motion), the signal intensity, *I*, of a given pixel in the presence of gradient pulses obtained using a spin echo diffusion weighted imaging (DWI) sequence (Figure 1) can be expressed as (Stejskal & Tanner, 1965):

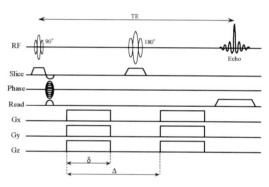

Figure 1

$$I = I_{(\infty,0,0)} \cdot (1 - 2\exp[-\frac{(T_r - T_e/2)}{T_1}] + \exp[-\frac{T_r}{T_1}]) \cdot \exp[-\frac{T_e}{T_2}] \cdot \exp[-\gamma^2 G^T D^\xi G \delta^2 (\Delta - \frac{\delta}{3})],$$

where $I_{(\infty,0,0)}$ is the intensity value for $Tr = \infty$, $Te = 0$, $|G|^2 = 0$; and Tr represents repetition time; Te, echo time; T_1, longitudinal relaxation time; T_2, spin-spin relaxation time; γ, nuclear gyromagnetic ratio; D^ξ, diffusion tensor; G and G^T, gradient vector matrix and its transpose; δ, pulse duration; and Δ, interval between the gradient pulses. In *in vivo* imaging studies, D^ξ is replaced by the apparent diffusion tensor, D_{app}^ξ, which, in addition to pure diffusion, includes the effects of other factors which cause intravoxel phase dispersion, such as axoplasmic flow.

D_{app}^ξ is a second order tensor which takes the general matrix form of:

$$\begin{pmatrix} Dx & Da & Db \\ Da & Dy & Dc \\ Db & Dc & Dz \end{pmatrix}.$$

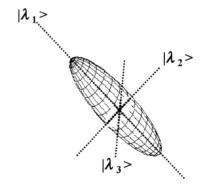

Figure 2

Since $D_{app}{}^\xi$ is real and symmetric, $D_{app}{}^\xi$ has three real eigenvalues, $\lambda_1, \lambda_2, \lambda_3$, and three corresponding eigenvectors, $|\lambda_1\rangle, |\lambda_2\rangle, |\lambda_3\rangle$. While the eigenvectors of $D_{app}{}^\xi$ define the principal axes of $D_{app}{}^\xi$, eigenvalues of $D_{app}{}^\xi$ represent D_{app} along the corresponding principal axes (Borisenko & Tarapov, 1968). This relationship can be readily expressed in pictorial from by diffusion ellipsoid (Figure 2).

Utilizing closure relationship, $\sum_i |u_i\rangle\langle u_i|$, the scalar product, $\langle G|D_{app}{}^\xi|G\rangle$ can be expressed as

$$\langle G|D_{app}{}^\xi|G\rangle = \sum_{ij}\langle G|u_i\rangle\langle u_i|D_{app}^\xi|u_j\rangle\langle u_j|G\rangle$$
$$= \sum_{ij} G_i^* D_{ij} G_j$$

where asterisk signifies the conjugate (Arfken, 1985). This equation signifies that by applying the appropriate combination of gradients, all the matrix components can be effectively extracted into a set of linear equations and, in turn, all the matrix components can be determined. Subsequently, eigenvalues/vectors can be obtained. Since the diffusion tensor contains six independent matrix elements, including zero gradients, a minimum of seven images have to be obtained for their determination. In practice, in order to minimize technical errors, averaging of multiple images for each gradient combination is necessary.

Data Acquisition

A General Electric (Waukesha, Wisconsin, USA) Signa-3.0 T system equipped with an Advanced NMR (ANMR) EPI module was used. Diffusion weighted images (DWIs) were obtained using spin echo echo-planner sequences using the following parameter settings: FOV 23 cm x 23 cm; matrix 128 x 128; slice thickness 5 mm; TR (cardiac gated: QRS x 2); NEX 8; b-value 200 for each axis.

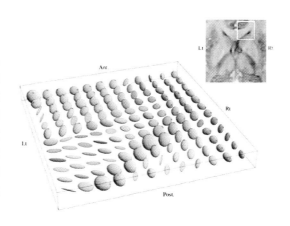

Figure 3

Results and Discussion

A representative "conventional" ellipsoid image is shown in Figure 3 (Pierpaoli et al, 1996). In order to better appreciate each ellipsoid, only a part of the image (12 x 12 pixel out of 128 x 128) is shown. The area presented is indicated by the white square in the 3DAC image shown on the right corner. It is apparent that, paradoxically, this "original" presentation of FTA provides less information for clinical judgment.

A "single ellipsoid" image of the identical object is shown in Figure 4. Each ellipsoid corresponds to the square within the 3DAC image by its number. The characteristics of anisotropic diffusion of anatomically discrete structures are clearly demonstrated. The technique is especially useful for smaller structures as illustrated in Figure 5. The red nucleus, which is primarily gray matter, shows an essentially isotropic (spherical) ellipsoid, whereas the cerebral peduncle shows strong anisotropism. The substantia nigra has fiber directionality of the medial leminiscus.

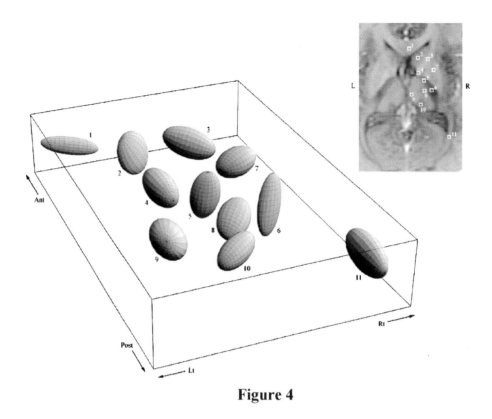

Figure 4

An example illustrating pathology is shown in Figure 6. Wallerian degeneration of the pyramidal tract in the left internal capsule shows an isotropic ellipsoid (a) which is clearly distinguishable from the normal side (b). In contrast to its significantly altered shape, the size

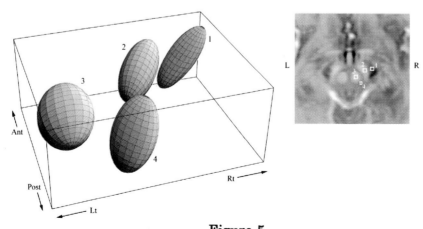

Figure 5

1: cerebral peduncle, 2: substantia nigra, 3: red nucleus, 4: medial leminiscus of the ellipsoid on the pathological site is not significantly changed. This implies that although anisotropism is lost, diffusional property of this structure is not significantly different from that of normal tissue. This finding is highly consistent with gliosis (not liquefaction).

Conclusion

Single ellipsoid tensor analysis is presented. The technique effectively ameliorates various shortcomings of full tensor analysis (FTA), especially its inherently high variance. With further development of ultra-fast imaging techniques, FTA will gradually become a part of the menu for routine MR examination in the clinical setting. The single ellipsoid method can provide an effective means of FTA suitable for pathological evaluation.

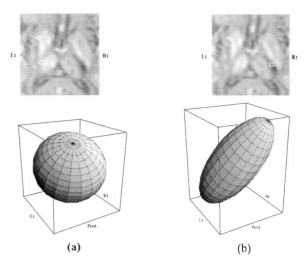

Figure 6

Acknowledgement

The study was supported by grants from the Ministry of Education (Japan).

References

Arfken, G. Mathematical Methods for Physicists, 3rd ed, San Diego: Academic Press, 1985.
Borisenko AI, Tarapov, IE. Vector and Tensor Analysis, New York: Dover Publications, 1968.
Nakada T, Matsuzawa H, Kwee IL. Magnetic resonance axonography of the rat spinal cord. NeuroReport 1994;5: 2053-2056.
Nakada T, Nakayama N, Fujii Y, Kwee IL. Clinical application of magnetic resonance axonography. J Neurosurg 1999;90:791-795.
Pierpaoli C, Jezzard P, Basser PJ, Barnett A, Di Chiro G. Diffusion tensor MR imaging of the human brain. Radiology 1996;201:637-648.
Stejskal EO, Tanner, JE. Spin diffusion measurements: Spin echoes in the presence of a time-dependent field gradient. J Chem Phys 1965;42:288-292.

Chapter II.13

Principal Eigenvector Imaging (PE$_{vec}$I): Comparison of Deterministic vs. Optical Methods

Hitoshi Matsuzawa[a], Yuji Suzuki[a], Hisatoshi Saito[b], and Tsutomu Nakada[a,1]

[a]*Department of Integrated Neuroscience, Brain Research Institute, University of Niigata*
[b]*Azabu Neurosurgical Hospital*

Introduction

Diffusion tensor analysis of the human brain represents an attractive new technique for evaluating structural as well as functional aspects of neuronal fiber networks. The technique is based on the empirical observation that directional dependency of apparent diffusion (anisotropic diffusion) is highly correlated to fiber orientation, especially that of axons (Douek et al., 1991, Hajnal et al., 1991, Nakada et al., 1994, Pierpaoli et al., 1996). Diffusion tensor analysis yields two sets of discrete information, namely, eigenvalues and the corresponding eigenvectors. Given that diffusion anisotropy in the brain derives primarily from axons, it follows that eigenvector provides information on axonal direction (connectivity) while eigenvalue,

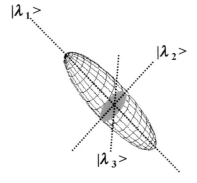

Figure 1

---

[1] Correspondence: Tsutomu Nakada, M.D., Ph.D., Department of Integrated Neuroscience, Brain Research Institute, University of Niigata, Niigata 951-8585, Japan, Tel: (81)-25-227-0677, Fax: (81)-25-227-0821, e-mail tnakada@bri.niigata-u.ac.jp

axonal function. Considering the cylinder-like structure of axons, the most important information is the largest eigenvalue (principal eigenvalue) and its corresponding eigenvector (principal eigenvector) (Figure 1). In contrast to engenvalue, eigenvector is inherently qualitative in nature and, therefore, it should be treated in three dimensional (3D) "physical space." In the context of clinical medicine, this directly implies pictorial display, namely, images. In this article, we review the basic methods for providing principal eigenvector imaging ($PE_{vec}I$).

Materials and Methods

Apparent Diffusion Tensor

Apparent diffusion tensor, D_{app}^{ξ}, is a second order tensor which takes the general matrix form of:

$$\begin{pmatrix} Dx & Da & Db \\ Da & Dy & Dc \\ Db & Dc & Dz \end{pmatrix}.$$

Since D_{app}^{ξ} is real and symmetric, D_{app}^{ξ} has three real eigenvalues, $\lambda_1, \lambda_2, \lambda_3$, and three corresponding eigenvectors, $|\lambda_1>, |\lambda_2>, |\lambda_3>$. While eigenvectors of D_{app}^{ξ} define the principal axes of D_{app}^{ξ}, eigenvalues of D_{app}^{ξ} represent apparent diffusion coefficients along the corresponding principal axes (Borisenko & Tarapov, 1996). This relationship can readily be expressed in pictorial from by diffusion ellipsoid (Figure 1).

Deterministic Method: Full Tensor Analysis (FTA)

Utilizing closure relationship, $\sum_i |u_i\rangle\langle u_i|$, scalar product, $\langle G|D_{app}^{\xi}|G\rangle$ can be

$$\langle G|D_{app}^{\xi}|G\rangle = \sum_{ij} \langle G|u_i\rangle\langle u_i|D_{app}^{\xi}|u_j\rangle\langle u_j|G\rangle$$
$$= \sum_{ij} G_i^* D_{ij} G_j$$

where asterisk signifies the conjugate (Arfken, 1985). This implies that by applying appropriate combinations of gradients, all the matrix components can be effectively extracted into a set of linear equations and, in turn, all the matrix components can be determined. Subsequently, principal eigenvector direction of each tensor matrix can be obtained. Since diffusion tensor contains six independent matrix elements, including zero gradients, a minimum of seven images have to be obtained for their determination.

Optical Method: 3DAC MRX

Optical method is represented by three dimensional anisotropy contrast magnetic resonance axonography (3DAC MRX). The method is based on two postulates, namely: (1) The nervous system is composted of the linear sum of cellular elements which possess either isotropic or anisotropic apparent water diffusion: $\{D_{app}^{\xi}\}_{cns} = \{D_{app}^{\xi}\}_i + \{D_{app}^{\xi}\}_j$; and (2) Anisotropic elements possess anisotropy in only one principal axis direction of D_{app}^{ξ}: $\lambda_1 > \lambda_2 = \lambda_3$ (Nakada et al., 1994, 1999, Nakada & Matsuzawa, 1995).

Based on postulate 2, D_{app}^{ξ} of anisotropic components can be expressed by the linear sum of two terms, namely, isotropic and anisotropic terms on principal axes coordinates. Accordingly, an entire system of D_{app}^{ξ} for the nervous system, $\{D_{app}^{\xi}\}_{cns}$, can be expressed as:

$$\{D_{app}^{\xi}\}_{cns} = \{\lambda_i^{iso} I\}_i + \left\{ R_j \left(\begin{array}{ccc} \lambda_j^{\alpha} & 0 & 0 \\ 0 & 0 & 0 \\ 0 & 0 & 0 \end{array}\right) + \lambda_j^{\beta} I \right) R_j^T \right\}_j$$

$$= \{\lambda_i I\}_i + \left\{ R_j \left(\begin{array}{ccc} \lambda_j^{\alpha} & 0 & 0 \\ 0 & 0 & 0 \\ 0 & 0 & 0 \end{array}\right) R_j^T \right\}_j + \{\lambda_j^{\beta} I\}_j$$

where R and R^T are a unitary rotation matrix and its transpose, respectively, necessary to ensure a single observation axes for all components of $\{D_{app}^{\xi}\}_{cns}$. Under the condition where the isotropic terms of anisotropic components vanish ($\{\lambda_k I\}_k \rightarrow 0$), D_{app}^{ξ} of anisotropic components can be represented by a three dimensional anisotropic term vector, D_{app}^A, the elements of which along the principal axes are represented by the diagonal elements of anisotropic terms. For 3DAC, the process of isotropic elimination can be accomplished optically (Nakada & Matsuzawa, 1995, Matsuzawa et al., 1995).

$$\{D_{app}^{\xi}\}_{cns} \Rightarrow \left\{ R_j \left(\begin{array}{ccc} \lambda_j^{\alpha} & 0 & 0 \\ 0 & 0 & 0 \\ 0 & 0 & 0 \end{array}\right) R_j^T \right\}_j \Rightarrow \{D_{app}^A\}_j$$

Data Acquisition

A General Electric (Waukesha, Wisconsin, USA) Echo Speed 1.5 T system was utilized. Diffusion weighted images (DWIs) were obtained using spin echo echo-planner sequences using the following parameter settings: FOV 23 cm x 23 cm; matrix 128 x 128; slice thickness 5 mm; TR 5 sec: TE (minimum) 101 m sec; NEX 4. b-value was 500 for each axis. Parameter for MPG gradients utilized were: amplitude 1.518 g/cm; ramp time 1200 μ sec; δ 31.0 m sec; Δ 39.088 m sec.

Results

PE$_{vec}$I obtained by the two representative methods described above, namely FTA and 3DAC MRX based on the identical raw DWIs are presented below. Figure 2A is a 3D vector presentation of PE$_{vec}$I which shows directional information of the principal eigenvector by arrows, the length of which is proportional to the corresponding eigenvalue (A). Figure 2B is the transitional image towards "color mapping" (Douek, 1991), where pixels of Figure 2A were color coded according to direction. Red corresponds to axons running in the x-axial direction

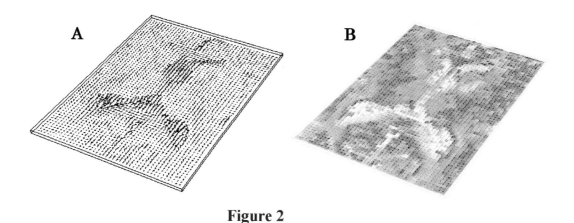

Figure 2

(right-left on axial plane), green to axons running in the y-axial direction (up-down on axial plane), and blue to axons running in the z-axial direction (perpendicular to the axial slice). The direction of axons coursing otherwise is represented by a hue, which is the weighted mixture of the three primary colors described above.

FTA based color mapping and 3DAC MRX based on identical raw DWIs are shown in Figure 3 for comparison. 3DAC MRX has a definite advantage as clinical imaging tool over FTA based color mapping.

 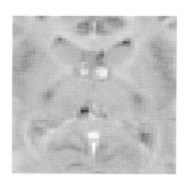

Color Mapping   **3DAC MRX**

Figure 3

Discussion

It is apparent that, as structural imaging method, 3DAC MRX is significantly superior to FTA based color mapping in spite of the fact that the necessary raw DWI images for 3DAC MRX is only 3/7 that for FTA in number. This results primarily from the processing algorithm. While 3DAC MRX performs extraction of principal eigenvector information optically (pictorial mathematics), FTA relies on numerical estimation processes. Considering the various constraints imposed in the clinical setting such as limited examination time, limited hardware configuration, and limited CPU power, 3DAC MRX should be the study of choice for $PE_{vec}I$ examination (Nakada et al., 1999).

Owing to recent dramatic advancements in gradient technology, serious implementation of echo planar imaging in clinical setting has now become realistic. $PE_{vec}I$ is ready to be included as a routine diagnostic examination tool. Such studies will have serious impact on basic neuroscience as well. While advanced applications of non-invasive functional imaging is slowly replacing non-human primate neurophysiology, evaluation of the pathological brain in humans becomes an essential equivalent to lesion studies of non-human primates. The role of non-invasive methods capable of providing information regarding axonal connectivity under physiological and pathological conditions will take the role of histologic examination. The potentials of $PE_{vec}I$ are likely to exceed beyond the current imagination.

Acknowledgement

The study was supported by grants from the Ministry of Education (Japan).

References

Arfken, G. Mathematical Methods for Physicists, 3rd ed, San Diego: Academic Press, 1985.
Borisenko AI, Tarapov IE. Vector and Tensor Analysis, New York: Dover Publications, 1968.
Douek P, Turner R, Pekar J Patronas N, Le Bihan D. MR color mapping of myelin fiber orientations. J Comp Assist Tomogr 1991;15:923-929.
Hajnal JV. Doran M. Hall AS. Collins AG. Oatridge A. Pennock JM. Young IR. Bydder GM. MR imaging of anisotropically restricted diffusion of water in the nervous system: technical, anatomic, and pathologic considerations. J Comput Assist Tomogr 1991;15:1-18.
Matsuzawa H., Kwee IL, Nakada T. Magnetic resonance axonography of the rat spinal cord: Postmortem effects. J. Neurosurg 83:1023-1028, 1995.
Nakada T, Matsuzawa H, Kwee IL. Magnetic resonance axonography of the rat spinal cord. NeuroReport 1994;5: 2053-2056.
Nakada T, Matsuzawa H. Three dimensional anisotropy magnetic resonance imaging of the rat nervous system. Neurosci Res 1995;22:389-398.
Nakada T, Nakayama N, Fujii Y, Kwee IL. Clinical application of magnetic resonance axonography. J Neurosurg 1999;90:791-795.
Pierpaoli C, Jezzard P, Basser PJ, Barnett A, Di Chiro G. Diffusion tensor MR imaging of the human brain. Radiology 1996;201:637-648.

T. Nakada (Ed.)
Integrated Human Brain Science: Theory, Method Application (Music)
© 2000 Elsevier Science B.V. All rights reserved

Chapter II.14

NIRS Imaging

Yoko Hoshi [1]

Biophysics Group, Research Institute for Electronic Science, Hokkaido University

The coupling between neuronal activity, hemodynamics and metabolism (Roy and Sherrington, 1890; Sokoloff, 1981) allows indirect evaluation of changes in neuronal activity by measuring the activity-dependence of these changes. Several neuroimaging techniques such as positron emission tomography (PET) and functional magnetic resonance imaging (fMRI) are based on this idea. Near-infrared spectroscopy (NIRS) is a new neuroimaging technique that was originally developed as a tool for non-invasive clinical monitoring of tissue oxygenation. There are three main measurement types used in NIRS. Each type of measurement is associated with a different type of instrumentation: continuous intensity, time-resolved, and intensity-modulated. NIRS imaging with continuous-intensity instruments, which include most commercially available NIR spectrometers, will be presented here (functional NIRS). With this type of instrument, we can see activity-related concentration changes in oxygenated ([oxy-Hb]), deoxygenated ([deoxy-Hb]) and total hemoglobin ([t-Hb]) in the brain like we see changes in neuronal electrical activity on an electroencephalogram (EEG). Optical topography and tomography will be briefly discussed.

I. Principles of NIRS

The principles of NIRS depend on two properties of NIR light. One is the relative transparency of tissue to light in the NIR region. The other is that only a few chromophores such as hemoglobin (Hb), myoglobin (Mb), and cytochrome oxidase (cyt. ox.) in mitochondria absorb the NIR light in the range between 700 and 900 nm and the absorption spectra vary with the oxygenation and oxidation states of these chromophores.

---

[1] Correspondence: Yoko Hoshi, M.D., Ph.D., Biophysics Group, Research Institute for Electronic Science, Hokkaido University, Sapporo 060-0812, Japan, Tel: (81)-11-706-3371, Fax: (81)-11-706-4964, e-mail: yhoshi@imd.es.hokudai.ac.jp

The absorbance (A) of light (at a given wavelength) as it passes through a non-scattering homogeneous medium is expressed as

$$A = ECL \quad \text{(the Beer-Lambert law)} \tag{1}$$

where E is the wavelength-dependent molecular absorption coefficient of the chromophore, C is the concentration of the chromophore, and L is the distance that light travels through the medium. Since the linear dependence of the changes of absorbance at a certain wavelength on [oxy-Hb] or [deoxy-Hb] in tissue has been confirmed (Hazeki and Tamura, 1988), the Beer-Lambert law can be extended to a light-scattering system like a living tissue.

NIR lights at multiple wavelengths are employed for measurement in living tissue. When NIR light at wavelength of λ_i (i = 1, 2, ...) is illuminated onto the head, the absorbance ($A\lambda_i$) is

$$A\lambda_i = \varepsilon\lambda_i[\text{oxy-Hb}]l_i + \varepsilon'\lambda_i[\text{deoxy-Hb}]l_i + \varepsilon''\lambda_i [\text{cyt. ox}] l_i + S\lambda_i, \tag{2}$$

where $\varepsilon\lambda_i$, $\varepsilon'\lambda_i$, and $\varepsilon''\lambda_i$ are molecular absorption coefficients at λ_i of oxy-Hb, deoxy-Hb, and oxidized cyt. ox. ([cyt. ox]), respectively, l_i is the path length, and $S\lambda_i$ represents absorbance due to light scattering and other chromophores at λ_i. When the conditions are changed in an activation study, assuming there is no significant change in $S\lambda_i$, the change in $A\lambda_i$ is expressed as

$$\Delta A\lambda_i = \varepsilon\lambda_i\Delta[\text{oxy-Hb}]l_i + \varepsilon'\lambda_i\Delta[\text{deoxy-Hb}]l_i. \tag{3}$$

The redox state of cyt. ox. is independent of both the mitochondrial energy state and respiratory rate (Hoshi et al., 1993a) and its redox change occurs only under extremely hypoxic conditions (Hoshi et al., 1997). Since an activation study is generally performed under normoxic conditions, the redox state of cyt. ox. is not changed (Δ[cyt. ox] = 0). By solving simultaneous equations, changes in [oxy-Hb] and [deoxy-Hb] can be calculated. Because scattering effects prevent determination of the optical path length, which differs markedly from the physical optical path length, the results are expressed in relative amounts rather than in absolute ones. Summation of changes in [oxy-Hb] and [deoxy-Hb] gives changes in [t-Hb]. The change in [t-Hb] reflects the change in blood volume within the optical field. Time resolution depends on the instrument, but in general it is about 1 second.

To carry out a study, NIR light is illuminated onto the head by a light guide and light transmitted through the brain tissue is collected by another light guide. The distance between illuminating and detecting light guides is usually in the range of 2-4 cm. Although several investigators have theoretically estimated the brain region measured by NIRS (Gratton et al., 1994; Firbank et al., 1998), it is still controversial. A recent study with time-resolved spectroscopy (TRS), in which light is input to the tissue in the form of an ultrashort pulse, and the emerging intensity is detected as a function of time with picosecond resolution, has shown that light reaches the brain tissue when the distance between two light guides is longer than 2 cm (personal communication).

II. Detection of neuronal activation by NIRS

The increase in regional cerebral blood flow (rCBF) coupled with neuronal activation exceeds the increase in oxygen consumption (Fox and Raichle, 1986). Thus, increases in [oxy-Hb] and [t-Hb] with a decrease in [deoxy-Hb] is a typical pattern showing neuronal activation in NIRS measurement (Fig. 1). When the degree of the increase in rCBF is small, however, [t-Hb] often shows no change. The behavior of the change in [deoxy-Hb] varies with each brain region and varies even in the same region during activation of brain activity, in which no change and even an increase in [deoxy-Hb] can occur. This is explained by vasodilatation caused by an increase in rCBF. That is, it is plausible that an increase in [deoxy-Hb] due to vasodilatation may cancel or exceed a decrease in [deoxy-Hb] caused by overcompensation of the flow.

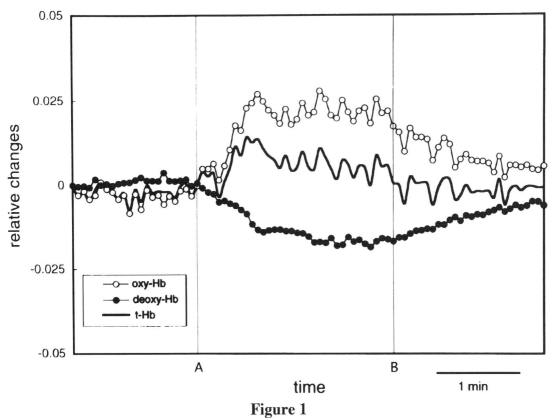

Figure 1
Changes in [oxy-Hb], [deoxy-Hb], and [t-Hb] in the left frontal region during a mental task. The digit span forward test was performed between A and B. Baselines were selected from resting state, and these values were taken as 0 for each signal. Upward (plus) and downward (minus) trends show increase and decrease in values, respectively.

Figure 2 shows changes in the hemoglobin oxygenation state in the bilateral frontal regions during solving of a mathematical problem. The subject listened to the problem, which was read out by an examiner, between A and B, and then solved it between B and C. In the left frontal region, increases in [oxy-Hb] and [t-Hb] with a decrease in [deoxy-Hb] were intermittently observed. In contrast, in the right frontal region, [t-Hb] and [oxy-Hb] first increased without accompanying a change in [deoxy-Hb], and then decreases in [t-Hb] and [oxy-Hb] were observed. This means that neuronal activity in one brain region, which differs from that in another region, varies with time. Thus, continuous measurement of hemodynamic changes is essential to map temporally dependent functional localization. In addition, the existence of right-left differences in NIRS traces provides evidence that NIRS measures changes in the brain tissue not the extracranial tissue.

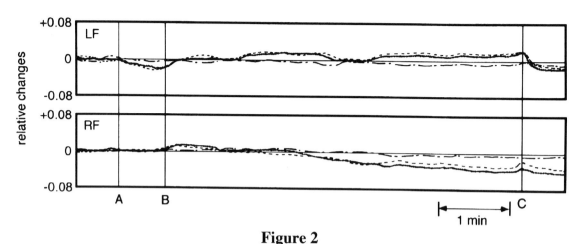

Figure 2

Changes in [oxy-Hb] (dotted line), [deoxy-Hb] (broken line), and [t-Hb] (solid line) in the bilateral frontal regions during a mental task. A mathematical problem was read out between A and B and then a subject solved it between B and C. LF, the left frontal region; RF, the right frontal region.

III. Functional NIRS

Several types of brain activity have been assessed by NIRS during finger tapping (Kleinschmidt et al., 1996), visual (Kato et al., 1993; Hoshi and Tamura, 1993b) and auditory (Hoshi and Tamura, 1993b) stimuli, and mental tasks (Hoshi and Tamura, 1997; Sakatani et al., 1998). The distinct properties of NIRS have found novel applications in functional mapping studies.

(1) Detection of sequential brain activation

Performing a mental task involves various cognitive operations. These operations are thought to be subserved by different cortical structures. A major aim of functional mapping studies is to visualize such internal operations. To achieve this aim, functional association between different areas of the brain is examined by correlational analysis of the functional mapping data. However, temporal correlation is hardly assessed by this approach. Recently, fMRI has been applied to detect the temporal dynamics of brain activation (Cohen et al., 1997), though NIRS is also useful for this purpose. In our previous study, in which the bilateral frontal regions were measured by NIRS during performance of a mental task, it was seen that activated regions moved alternatively: when in one region [oxy-Hb] that had first increased returned to the resting level, in another it started to increase (Hoshi and Tamura, 1997). These region-dependent temporal variations of brain activity probably reflect internal processes. Thus, simultaneous measurement of multiple brain regions with a multichannel NIRS instrument has the potential to detect the sequence of brain activation.

(2) Screening of brain function in neonates

Neurologically intact survival is being focused on in recent neonatal care. Early prediction of neurological prognosis is desired in clinical medicine, while methods to evaluate neonatal brain function are limited. EEG and measurement of evoked potentials (EPs) are commonly used for this purpose. However, they are not necessarily useful diagnostic tools in the neonatal period. Although fMRI enables us to assess neonatal brain function at the cortical level qualitatively (Born P et al., 1996), these studies are accompanied by greater or lesser risks, because neonates have to be transported to a specific room and be administered a medicine for sedation. In contrast, NIRS enables measurement at the bedside and even in an incubator. Since it takes a long time to detect a cortical area of interest from outside of the brain with a single-channel NIRS system, however, a multichannel NIRS system is desirable. Using a three-channel NIRS system, we could detect activation of the primary visual cortex within 15 minute in a healthy neonate during natural sleep (Fig. 3). Increases in [oxy-Hb], [t-Hb] and [deoxy-Hb] were observed about 1.5 cm above the inion during photic stimulation, while no changes or decreases in any NIRS parameters were observed at other occipital regions. Other types of brain function can be assessed in a similar way. Thus, multichannel NIRS imaging is useful for screening of brain functions in neonates.

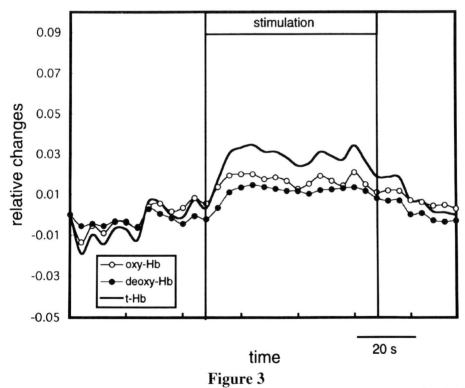

Figure 3

Changes in [oxy-Hb], [deoxy-Hb], and [t-Hb] in one neonate showing a response to visual stimulation. A stimulator was located 30 cm from the neonate's face and produced continuous flashes at 10 Hz.

(3) NIRS studies on mental and neurodegenerative disorders

Since patients with mental and neurodegenerative disorders in general have less patience with task-related anxiety, fear, pain, and physical discomfort than healthy subjects, it is often difficult to perform PET and fMRI studies on these patients. However, NIRS study can be done in a clinical office and it does not require such precise restriction of the head motion as PET and fMRI studies do, which helps get patients to participate in a prolonged study. NIRS has been applied to patients with various disorders such as schizophrenia (Okada et al., 1994) and Alzheimer's disease (Hock et al., 1997). These studies have reported that hemodynamic responses of patients differed from those of healthy subjects. For example, in response to a mirror drawing task, healthy subjects showed increases in [oxy-Hb] in the bilateral frontal regions (Okada et al., 1993), while schizophrenics showed various patterns of changes which never appeared in healthy volunteers, suggesting defective interhemispheric integration (Okada et al., 1994). These patterns disappeared when symptoms were improved by medication. Thus, NIRS study is useful for not only elucidating pathophysiology but also for judging the efficacy of therapy.

IV. Problems of NIRS imaging

(1) Quantitation of NIRS data

Quantitation by continuous intensity measurement has been difficult, because optical path length cannot be estimated. By the use of time-resolved instruments, however, we can estimate the mean optical path length and calculate absolute values of changes. Quantitation enables us to compare the degrees of changes between subjects and/or between regions in each subject, which is essential for the optical topography of brain activity.

site	L (cm)	wavelength	PL (cm)	PL/L
site 1	2.8	λ_1	12.014	4.291
		λ_2	11.968	4.274
		λ_3	11.307	4.038
site 2	2.5	λ_1	15.483	6.193
		λ_2	15.462	6.185
		λ_3	14.808	5.923
site 3	3.1	λ_1	14.981	4.833
		λ_2	14.952	4.823
		λ_3	14.358	4.632
site 4	3	λ_1	18.752	6.251
		λ_2	18.442	6.147
		λ_3	17.632	5.877
site 5	3.1	λ_1	17.325	5.589
		λ_2	17.324	5.588
		λ_3	16.312	5.262
site 6	2.8	λ_1	17.268	6.167
		λ_2	16.743	5.98
		λ_3	15.706	5.609

Table1
The separation between illuminating and detecting light guides (L) and the mean optical path length (PL) measured at the forehead. λ_1, 780nm; λ_2, 805nm; λ_3, 930nm

The mean optical path length (PL) is longer than the physical optical path length (L, the separation between illuminating and detecting light guides). Table 1 shows L, PL, and PL/L measured at three wavelengths in 6 different sites of the forehead in one healthy adult. The illuminating and detecting light guides were placed vertically and 6 sites were measured every 3 cm of the distance from the right (site 1) to the left side of the forehead. It should be noted that PL/L varied with each measurement site. This means that even though the separation between each pair of light guides (the physical optical path length) is the same, mean optical path lengths are different from each other. This is probably related to differences in scattering coefficients between brain regions. Thus, the mean optical path length has to be measured at all measurement sites in each subject.

(2) Estimation of the illuminated size of the brain tissue

It is important to estimate the illuminated size of the brain tissue not only for interpretation of NIRS data but also for developing optical topography. As to the sites at which light guides were placed, these can be identified by the use of structural MRI. In contrast, the depth of light penetration has to be estimated mathematically, which requires the determination of true absorption and scattering coefficients of skin, bone, and brain tissues. By the use of time-resolved instruments, these parameters can be measured in the human brain.

(3) Fluctuations in the hemoglobin oxygenation state during the resting period

The hemoglobin oxygenation state fluctuates during the resting period (Fig. 4). The magnitude, direction, and interval between changes in [oxy-Hb], [t-Hb] and [deoxy-Hb] vary with the time course, and these variations differ from subject to subject. These fluctuations are not related to alterations in either the heart rate or systemic blood pressure. The fluctuations are unlikely to be a result of alterations in either systemic circulation or movement artifacts since there are differences in the temporal patterns of the fluctuations even between two adjacent brain regions. Simultaneous measurements by EEG and NIRS suggested that spontaneous neuronal activity was responsible for the fluctuations (Hoshi et al., 1998). Cerebral blood flow velocity (CBFV) measured by transcranial Doppler ultrasound also shows such oscillations (Lindgaard et al., 1987) and small artery oscillations are thought to be the underlying mechanism (Diehl et al., 1991), although the exact cause of the small

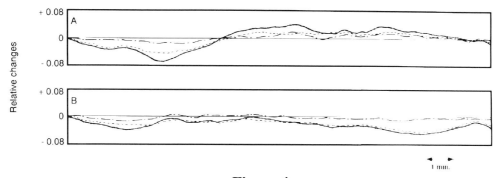

Figure 4
Changes in [oxy-Hb] (dotted line), [deoxy-Hb] (broken line), and [t-Hb] (solid line) in the frontal and occipital regions during resting period. Baselines were selected from starting point of measurement. A, the left frontal region; B, the left occipital region.

artery oscillations is still unknown. Since frequencies and characteristics of the fluctuation in the hemoglobin oxygenation state are similar to those of the CBFV oscillations, fluctuations in the hemoglobin oxygenation state might be result of vasomotor responses to spontaneous neuronal activity. The degree of changes in the hemoglobin oxygenation state caused by a mental task is sometimes within this resting variation range. Thus, taking account of these fluctuations is essential for interpretation of NIRS signals.

V. Future of NIRS imaging

Recent technical advances in NIRS have made it possible to continuously monitor hemodynamic changes in freely moving subjects like portable ECG and EEG systems do. This technique, for example, has been employed to measure hemodynamic changes in the muscles of an athlete during running. Optical signals are converted to electrical signals in the portable NIRS instrument attached to the subject and then these signals can be electrically transferred to a computer system in another place. Applying this technique to infants and children, it might be possible to examine cognitive functions while they are playing.

A time-resolved 64-channel optical tomographic imaging method has been developed (Eda et al., 1999). The system has three picosecond-pulsed light sources (semi-conductor laser diodes) and 64 fiber bundles connected to 64 corresponding time-resolved detection systems. Each detection channel consists of an optical attenuator, a fast photomultiplier, and a time-correlated single-photon counting circuit that contains a miniaturized constant fraction discriminator (CFD), time-to-amplitude converter (TAC) module, and a signal acquisition unit with an A/D converter. The performance and potentiality of the imaging system have been examined by imaging reconstruction from the measured data using solid phantoms. The reconstructed images have reasonable spatial resolution and provide quantitative information about the absorption coefficients of the phantoms. Thus, time-resolved quantitative optical imaging of the human brain will be possible in the near future.

References

Born P, Rostrup E, Leth H, Peiterson B, Lou HC. Change of visually induced cortical activation patterns during development. Lancet 1996; 347: 543.

Cohen JD, Perlstein WM, Braver TS, Nystrom LE, Noll DC, Jonides J, Smith EE. Temporal dynamics of brain activation during a working memory task. Nature 1997; 386: 604-8.

Diehl RR, Diehl B, Sitzer M, Hennerici M. Spontaneous oscillation in cerebral blood flow velocity in normal humans and in patients with carotid artery disease. Neurosci Lett 1991; 127: 5-8.

Eda H, Oda I, Ito Y, Wada Y, Oikawa Y, Tsunazawa Y, Takada M, Tsuchiya Y, Yamashita Y, Oda M, Sassaroli A, Yamada Y, Tamura M. Multi-channel time-resolved optical tomographic imaging system. R S Instru (in press).

Firbank M, Okada E, Delpy DT. A theoretical study of the signal contribution of regions of the adult head to near-infrared spectroscopy studies of visual evoked responses. NeuroImage 1998; 8: 69-78.

Fox PT, Raichle ME. Focal physiological uncoupling of cerebral blood flow and oxidative metabolism during somatosensory stimulation in human subjects. Proc Natl Acad Sci USA 1986; 83: 1140-4.

Gratton G, Maier JS, Fabiani M, Mantulin WM, Gratton E. Feasibility of intracranial near-infrared optical scanning. Psychphysiology 1994; 31: 211-5.

Hazeki O, Tamura M. Quantitative analysis of hemoglobin oxygenation state of rat brain in situ by near-infrared spectrophotometry. J Appl Physiol 1988; 64: 796-802.

Hock C, Villringer K, Müller-Spahn F, Wenzel R, Heekeren H, Schuh-Hofer S, Hofmann M, Minoshima S, Schwaiger M, Dirnagl U, Villringer A. Decrease in parietal cerebral hemoglobin oxygenation during performance of a verbal fluency task in patients with Alzheimer's disease monitored by means of near-infrared spectroscopy (NIRS) - correlation with simultaneous rCBF-PET measurements. Brain Res 1997; 755: 293-303.

Hoshi Y, Hazeki O, Tamura M. Oxygen dependence of redox state of copper in cytochrome oxidase in vitro. J Appl Physiol 1993a; 74: 1622-7.

Hoshi Y, Tamura M. Dynamic multichannel near-infrared optical imaging of human brain activity. J Appl Physiol 1993b; 75: 1842-6.

Hoshi Y, Hazeki O, Kakihana, Tamura M. Redox behavior of cytochrome oxidase in the rat brain measured by near-infrared spectroscopy. J Appl Physiol 1997; 83: 1842-48.

Hoshi Y, Tamura M. Near-infrared optical detection of sequential brain activation in the prefrontal cortex during mental tasks. NeuroImage 1997; 5: 292-7.

Hoshi Y, Kosaka S, Xie Y, Kohri S, Tamura M. Relationship between fluctuations in the cerebral hemoglobin oxygenation state and neuronal activity under resting conditions in man. Neurosci Lett 245; 1998: 147-50.

Kato T, Kamei A, Takashima S, Ozaki T. Human visual cortical function during photic stimulation monitoring by means of near-infrared spectroscopy. J Cereb Blood Flow Metabol 1993; 13: 516-20.

Kleinschmidt A, Obrig H, Requardt M, Merboldt KD, Dirnagl U, Villringer A, Frahm J. Simultaneous recording of cerebral blood oxygenation changes during human brain activation by magnetic resonance imaging and near-infared spectroscopy. J Cereb Blood Flow Metabol 1996; 16: 817-26.

Lindegaard KF, Lundar T, Wiberg J, Sjøberg D, Aaslid R, Nornes H. Valiations in middle cerebral artery blood flow investigated with noninvasive transcranial blood velocity measurements. Stroke 1987; 18: 1025-30.

Okada F, Tokumitsu Y, Hoshi Y, Mamoru T. Gender- and handedness-related differences of forebrain oxygenation and hemodynamics. Brain Res 1993; 601: 337-42.

Okada F, Tokumitsu Y, Hoshi Y, Tamura M. Impaired interhemispheric integration in brain oxygenation and hemodynamics in schizophrenia. Eur Arch Psychiatry Clin Neurosci 1994; 244: 17-25.

Roy CW, Sherrington CS. On the regulation of the blood supply of the brain. J Physiol (London) 1890;11:85-108.

Sakatani K, Xie Y, Lichty W, Li S, Zuo H. Language-activated cerebral blood oxygenation and hemodynamic changes of the left prefrontal cortex in poststroke aphasic patients. a near-infrared spectroscopy study. Stroke 1998; 29: 1299-304.

Sokoloff L. Relationships among local functional activity energy metabolism and blood flow in the central nervous system. Fed Proc 1981; 40: 2311-16.

T. Nakada (Ed.)
Integrated Human Brain Science: Theory, Method Application (Music)
© 2000 Elsevier Science B.V. All rights reserved

Chapter II.15

Clinical Application of Near-infrared Spectroscopy (NIRS) to Cerebrovascular Disorders

Satoshi Kuroda[1] and Kiyohiro Houkin

Department of Neurosurgery, Hokkaido University Graduate School of Medicine

Introduction

Near-infrared spectroscopy (NIRS) has recently been accepted as a useful tool for non-invasive monitoring of the brain oxygen metabolisms. The principle of NIRS is based on the facts that NIR lights can easily penetrate the tissues including the bone, and that oxy-hemoglobin, deoxy-hemoglobin, and cytochrome oxidase have specific NIR absorption. Using these optical properties, it has been reported that near-infrared spectroscopy (NIRS) could observe the changes in the oxygenation state of hemoglobin and redox state of cytochrome oxidase in the tissues continuously and non-invasively. For cerebral tissue, the changes in the oxygenation state of hemoglobin and redox state of cytochrome oxidase have been measured, where the former gave information on the circulatory system, and the latter the tissue oxygenation state of intracellular space.

The authors have applied NIRS to the clinical monitoring in the patients with cerebrovascular disease for these 8 years, and describe their experiences in this chapter.

How Can NIRS Detect Tissue Ischemia? – Experimental Studies

Basic principles of NIRS are precisely discussed in the chapter written by Hoshi in this book. As mentioned above, NIRS can detect the signal of hemoglobin and cytochrome c oxidase in the brain. First, the following should be noted in assessing the findings on NIRS. NIRS can detect the signals from the brain when the NIR light source and receiver are put on the scalp with the interval of at least 2.7 cm. The observed changes on NIRS, however, include the signals from both the scalp and the brain (McCormick et al. 1992; Kuroda et al.

---

[1] Correspondence: Satoshi Kuroda, MD, DMSc., Department of Neurosurgery, Hokkaido University Graduate School of Medicine, North 15 West 7, Kita-ku, Sapporo 060-8638, Japan, Tel: (81)-11-716-1161, Fax: (81)-11-706-7878, e-mail: skuroda@med.hokudai.ac.jp

1996b). Therefore, it should be careful to evaluate the NIR signals obtained from the optodes on the scalp. The signals of hemoglobin on NIRS reflect mainly the change in the oxygenation state of hemoglobin in the mixed cerebral venous blood (Tamura et al. 1988). Therefore, a decrease in the concentration of oxy-hemoglobin ([oxy-Hb]) and an increase in the concentration of deoxy-hemoglobin ([deoxy-Hb]) suggest a decrease in oxygenated blood supply and/or an increase in cerebral oxygen metabolism. In addition, previous studies concluded that the changes in the concentration of total hemoglobin ([total Hb]) are identical to those in cerebral blood volume (CBV) (Pryds et al. 1990). NIRS can non-invasively observe hypoxic hypoxia by detecting a gradual decrease in [oxy-Hb] and increase in [deoxy-Hb] in both animals and human (Hazeki and Tamura, 1988; McCormick et al. 1991). Further hypoxia causes blood pressure reduction, leading to a decrease in [total Hb], which represents a reduction of cerebral blood volume (CBV) (Fig. 1). In contrast, ischemic hypoxia gives rise to a rapid decrease of [oxy-Hb] and [total Hb], and increase of [deoxy-Hb]. Restoration of blood flow also causes a rapid recovery of these parameters (Fig. 2).

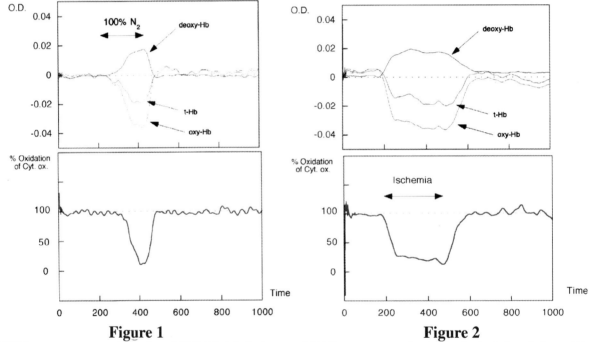

Figure 1
NIRS responses to a short-time 100% N_2 inhalation in gerbil brain. [oxy-Hb] and [total Hb] slowly decreased, and then rapidly reduced in parallel to blood flow reduction. Once reduction of cytochrome oxidase started, cytochrome oxidase was rapidly and completely reduced. Reoxygenation by air inhalation rapidly and completely resolved NIRS changes.

Figure 2
NIRS responses to 5-min forebrain ischemia in gerbil brain. Reduction of cytochrome oxidase was observed immediately after onset of ischemia. Recirculation resolved NIRS changes, followed by secondary reduction of [oxy-Hb] and [total Hb].

Second, the authors describe the importance to monitor the redox state of cytochrome oxidase during cerebral ischemia using NIRS. Cytochrome oxidase is the terminal enzyme of mitochondrial electron transport chain, and catalyses the transfer of electron from ferrocytochrome c to molecular oxygen to form water (Wong-Riley 1989). It includes two heme proteins (a and a_3) and two copper molecules (Cu_A and Cu_B). It can utilize approximately 95 % of oxygen in living organisms, playing a very important role in oxidative phosphorylation (Wikström 1981). Cytochrome oxidase has a broad absorption band around

830 nm (Beinert 1980). The calibration study showed that the oxygen affinity of cytochrome oxidase is extremely high, and that the oxygen concentrations for the half-maximal reduction (p50) of Cu_A is 0.08 µM and is completely independent of the energy state (Hoshi et al. 1993a; Matsunaga et al. 1998). The results obtained from heme a + a_3 are quite different, because the p50 of heme is 0.16 µM in state 3 respiration and 0.08 µM in state 4, dependent on its energy state (Hoshi et al. 1993a). These findings mean that the degree of the redox state of copper depends only on the tissue oxygen concentration. Thus, it can be easily accepted that the occurrence of its reduction, even if it is partial, indicates the occurrence of severe hypoxia leading to cellular energy failure. Experimental and clinical studies have shown that the redox state of copper in cytochrome oxidase is a valuable "*in vivo* marker" exhibiting intracellular oxygen concentration (Hazeki and Tamura 1989, Hoshi et al. 1993a and b).

To assess the utility of NIRS to monitor cerebral ischemia, the authors compared dynamic changes of NIRS with those in high-energy metabolites during and following forebrain ischemia of gerbil (Kuroda, 1995a). Especially, they aimed to clarify the relationship between redox state of cytochrome oxidase and bioenergetic state during cerebral ischemia. Mongolian gerbil has no vascular connection between the carotid and vertebrobasilar systems. Therefore, a brief-time occlusion of the bilateral common carotid arteries leads to a marked reduction (less than 5% of the control) of cerebral blood flow (CBF). Using the algorhism described by Hazeki and Tamura (1989), the authors monitored NIRS through the intact skull during and following 5-min forebrain ischemia of gerbil. High-energy metabolites, ATP, ADP and AMP were measured in separate series of the animals, using high performance liquid chromatography (HPLC) technique. Bilateral carotid occlusion gave rise to a rapid decrease in [oxy-Hb] and [total Hb], and an increase in [deoxy-Hb]. Copper of cytochrome oxidase also started to be reduced quickly (Fig. 2). ATP concentration showed no significant changes 15 seconds after the onset of ischemia, whereas approximately 20 % of copper were already reduced. ATP started to decrease significantly 30 seconds after the onset of ischemia, when about 60 % of copper was already reduced (Fig. 3). As reported previously, 5 min of ischemia gave rise to almost complete loss of ATP (1.9 % of its preischemic value), at which time point 90 % of copper was also reduced. These results indicated that the copper signal in non-invasive NIRS could be a very sensitive and quick parameter to detect dense ischemia which would lead to energy failure. Even partial reduction of the copper would be at risk of energy failure and electrical dysfunction. Similar results have been indicated in the ischemic heart and brain (Wiernsperger et al. 1981; Parsons et al. 1990; Li et al. 1997), although algorhisms for measuring redox state of cytochrome oxidase are still a matter of debate (Cooper and Springett 1997).

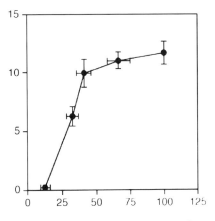

Figure 3
Relationship between % oxidation of cytochrome oxidase and the cortical level of ATP during first 5-min forebrain ischemia. Note that cortical ATP content did not decrease until about half of cytochrome oxidase was reduced.

Clinical Application of NIRS to Cerebrovascular Disease

Nowadays, several NIRS equipments are available for clinical use, although their algorithms for NIRS measurements were different with each other. For these 7 years, the authors have employed OM-100, OM-110 (Shimadzu Co., Kyoto, Japan), and HEO-200 (OMRON CO., Tokyo, Japan) to monitor the hemodynamic and metabolic changes of the brain during surgery or at bedside for the patients with cerebrovascular disorders (Kuroda et al. 1996a, b, c).

OM-100 and OM-110 are run by the different algorithms, both of which were developed by Prof. Tamura and his colleagues at Hokkaido University (Hazeki and Tamura 1988, 1989). OM-100 device can monitor the changes of [oxy-Hb], [deoxy-Hb], and [total Hb] in the brain, using 3 different wavelength lights (780, 805 and 803 nm) through laser diodes. Following is the algorithm used in the device:

$$\Delta[\text{oxy-Hb}] = -3.0 \, \Delta \, A805 + 3.0 \, \Delta \, A830$$

$$\Delta[\text{deoxy-Hb}] = 1.6 \, \Delta \, A780 - 2.8 \, \Delta \, A805 + 1.2 \, \Delta \, A830$$

$$\Delta[\text{total Hb}] = \Delta[\text{oxy-Hb}] + \Delta[\text{deoxy-Hb}]$$

On the other hands, OM-110 device can also monitor the redox changes of cytochrome oxidase (Cyt. ox.) in the brain mitochondria, as well as the changes of [oxy-Hb], [deoxy-Hb], and [total Hb] in the brain, using 4 filtered wavelength lights (700, 730, 750 and 805 nm). Following is the algorithm applied in the device:

$$\Delta[\text{oxy-Hb}] = -0.912 \, \Delta \, A700 - 2.130 \, \Delta \, A730 + 3.040 \, \Delta \, A750$$

$$\Delta[\text{deoxy-Hb}] = 0.743 \, \Delta \, A700 - 1.610 \, \Delta \, A730 + 0.868 \, \Delta \, A750$$

$$\Delta[\text{total Hb}] = \Delta[\text{oxy-Hb}] + \Delta[\text{deoxy-Hb}]$$

$$\Delta[\text{Cyt. ox.}] = -1.530 \, \Delta \, A700 + 0.768 \, \Delta \, A730 + 1.760 \, \Delta \, A750 - 1.0 \, \Delta \, A805$$

HEO-200 device, which are run by the independent algorithm reported by Shiga et al. (1995), can also measure the changes of Hb oxygenation and [total Hb] in the brain, using 2 wavelength lights (760 and 850 nm through laser diodes. The principle of measurement is based on the two-wavelength method.

These apparatus are very useful to monitor the relative changes of cerebral hemodynamics and oxygen metabolism with good time resolution in every second. The characteristics allow us a real-time monitoring of the hemodynamic and metabolic changes in the brain. Especially, HEO-200 is very compact (size 85 x 160 x 42 mm: weight 350g) and is useful for intraoperative and bedside measurements (Fig. 4).

The authors have applied NIRS for patients with cerebrovascular disorders, for mainly two purposes. First, they aimed to monitor the changes in cerebral hemodynamics and oxygen metabolism during carotid artery occlusion in order to avoid serious neurological complications resulting from cerebral ischemia. Electrophysiological parameters such as EEG and somatosensory evoked potential (SEP) have widely been employed as useful tools for the same purpose (Sundt et al. 1981; Guerit et al. 1997). That is, experimental studies have proven that cerebral cortex has the ischemic threshold for their electrical activities, which rapidly deteriorate or disappear when cerebral blood flow decreases to approximately 35 % or less of the control value. However, they cannot demonstrate the underlying changes of cerebral hemodynamics and oxygen metabolism, until blood flow declines to the level which would lead to electrical dysfunction of the brain (Branston et al. 1974, Astrup et al. 1977). EEG can monitor the whole areas of the cerebral cortexes, but requires certain time to analyze the changes. It is easier to analyze the

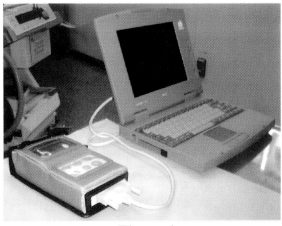

Figure 4
Photograph of portable two-wavelength NIRS apparatus, HEO-200 (Omron Co., Kyoto, Japan)

changes of latency and amplitude of N_{20} on SEP, but SEP has the disadvantage that they monitor only the specific tract and cortex for somatosensory stimuli. Transcranial Doppler sonography (TCD) has also been employed to directly measure the velocity of the middle cerebral artery, but is not available in certain subgroup of patients with narrow temporal window. On the other hands, NIRS have been expected to detect the hemodynamic and metabolic changes in the illuminated frontal lobes continuously, when optodes are put on the forehead.

Second, they aimed to non-invasively assess pathophysiological conditions of the ischemic brain in the patients with cerebrovascular diseases. For example, carotid compression test under NIRS measurement was performed to evaluate the role of the contralateral carotid artery as collaterals in the patients with carotid artery occlusion. These examinations are quite easy and possible in the bedside. They do not require any tracer such as contrast material or radioisotope. Some of the following data presented have previously been described elsewhere (Kuroda et al. 1996a, b, c, 1999)

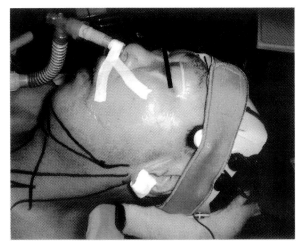

Figure 5
The patient under multi-brain monitoring for CEA, including SEP, TCD, and NIRS

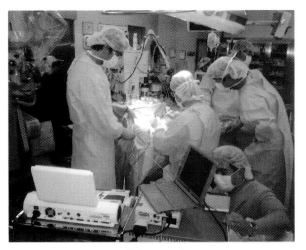

Figure 6
Intraoperative view of CEA under multi-brain monitoring.

1) Carotid endarterectomy (CEA)

It has been known that CEA could resolve hemodynamically significant stenosis and/or source of thromboemboli due to atheromatous plaque, improving the outcome of the patients (NASCET Groups 1991). It, however, is very important to avoid ischemic neurological complications because surgical procedures of CEA always require temporary clamping of the carotid artery. For this purpose, internal shunting has been employed to shorten the duration of carotid clamping, keeping blood flow to the brain during CEA.

The authors have applied NIRS for totally 40 CEA. It was possible to obtain stable near-infrared (NIR) signals in all patients. To compare the findings on NIRS with those on other monitoring modalities, SEP and TCD were simultaneously employed in all and recent 9 patients, respectively (Figs. 5 and 6). As mentioned above, TCD was not accessible in 2 out of 9 patients (22 %) due to narrow temporal window. After anesthesia induction, the optodes for NIRS were placed on the ipsilateral forehead with 35-mm intervals, and were fixed with an elastic band. Then, the electrodes for SEP and the probes for TCD were also placed on the head (Fig. 5). After exposure of the carotid bifurcation, the external carotid artery (ECA) was occluded. Following the clamp of the ECA, NIRS showed a decrease in [oxy-Hb] and [total Hb] and an increase in [deoxy-Hb] in most cases, indicating that a part of NIRS signals include those from scalp (Kuroda et al. 1996, McCormick et al. 1992). Redox state of cytochrome oxidase, however, did not change after clamping of the external carotid artery (Fig. 11). It strongly suggested that signals of cytochrome oxidase come only from the illuminated brain. After stabilization of NIR signals, the common and internal carotid arteries were clamped. Then, arteriotomy was performed, and a three-way internal shunt tube was inserted in 36 out of 40 cases. In other 4 patients, subsequent procedures were performed without insertion of internal shunt tube, because the operated ICA had too small diameter for shunting. The duration of carotid cross clamping was 5 to 10 minutes, when internal shunting was performed. During cross clamping of the carotid artery, blood pressure was kept similar to that during patients' ordinary life. The atheromatous plaque was carefully dissected and removed under operative microscope. The arteriotomy was closed and the internal shunt tube was removed.

Changes of Hb oxygenation during cross clamping of the carotid artery could be divided into 2 patterns.

In 23 cases (Group 1), cross clamping of the carotid artery transiently induced a decrease in [oxy-Hb] and [total Hb] and an increase in [deoxy-Hb]. Subsequently, these parameters gradually returned to the original levels during the carotid clamping. Recovery of [oxy-Hb] and [total Hb] was more prominent than that of [deoxy-Hb] (Fig. 7). Carotid clamping caused almost no changes of these parameters in some cases (Fig. 9). Using OM-110 device (Shimadzu Co., Japan), the authors also monitored redox state of cytochrome oxidase in 6 of these 23 cases, and no change was observed through the carotid clamping (Fig. 9). Intraoperative SEP monitoring revealed no or only a mild decrease in N_{20} amplitude. N_{20} amplitude during carotid clamping ranged from 80 % to 100 % of the control on SEP. Using TCD, velocity of the middle cerebral artery (MCA) was continuously monitored in 5 patients. Carotid clamping leaded to no or moderate decrease of mean velocity of the ipsilateral MCA to 50 % of the control. As described above, 4 patients underwent CEA without internal shunt tube, because the caliber of the ICA was too small to insert it. No ischemic complication, however, was observed after surgery in all 13 cases.

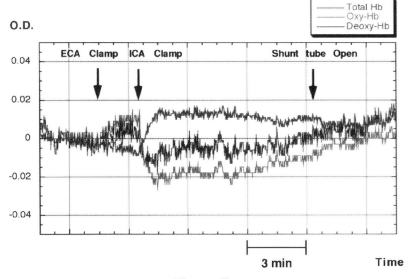

Figure 7

NIRS monitoring in a 65-year-old patient during CEA (OM-100). Cross-clamping of the carotid artery caused a decrease in [oxy-Hb] and an increase in [deoxy-Hb], followed by a gradual recovery to the control level. Transient decrease in [total Hb] was also observed.

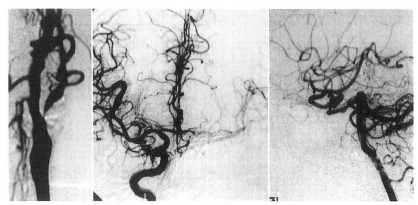

Figure 8

Cerebral angiograms of a 68-year-old patient. Left carotid angiogram (*left*) showed severe stenosis of the left internal carotid artery. Well-developed collateral circulation through the anterior and posterior communicating arteries was identified on right carotid (*middle*) and vertebral angiograms (*right*).

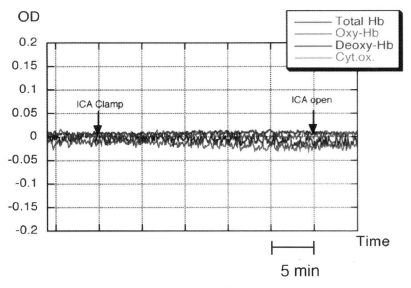

Figure 9

NIRS monitoring in a 68-year-old patient during CEA (OM-110). Cross-clamping of the left carotid artery caused no significant change in NIRS parameters because of well-developed collaterals through the circle of Willis (Fig. 8)

Illustrative Case (Group 1)

The 68-year-old male developed weakness of the right extremities and was admitted to our hospital. Neurological examination on admission revealed mild dysarthria and right hemiparesis, resolving 3 days later. MRI showed no cerebral infarct, but carotid angiograms demonstrated severe stenosis of the left ICA associated with good cross-filling through the anterior and posterior communicating arteries (Fig. 8). He underwent CEA for the left stenotic ICA. Clamping of the ICA leaded to no significant changes in the parameters on Hb. Redox state of cytochrome oxidase also showed no significant change during the carotid clamping (Fig. 9). N_{20} amplitude did not decrease on SEP. So, internal shunting was not performed, and following endarterectomy was completed. Postoperative course was uneventful, and he was discharged without any neurological deficit.

On the other hands, cross-clamping of the carotid artery leaded to a persistent decrease in [oxy-Hb] and increase in [deoxy-Hb] in other 17 cases (Group 2). Consecutive decrease in [total Hb] was also observed in 13 patients. These changes lasted until the internal shunt tube restored blood flow in all Group 2 cases (Fig. 11). In addition, [oxy-Hb] and [total Hb] increased to higher levels than the original immediately after the opening of the shunt tube, and gradually decreased to the original level in five Group 2 patients, suggesting post-ischemic hyperemia (data not shown).

The change in the redox state of cytochrome oxidase was monitored in 5 of thes 17 cases. In these 5 patients, the carotid clamping caused an immediate reduction of cytochrome oxidase (an increase in absorbance on NIRS), which lasted during the carotid clamping. Redox state of cytochrome oxidase returned to the original level just after the shunt tube restored blood flow (Fig. 11).

In Group 2 patients, SEP monitoring demonstrated a marked reduction of the N_{20} amplitude immediately after the carotid clamping, and its amplitude decreased to much lower level ranging from 10 % to 60 % of the control (Fig. 12). Using TCD, the flow velocity of the ipsilateral MCA was continuously monitored in 4 patients. Carotid clamping caused a marked decrease in its mean velocity to less than 50 % of the control.

Taken these findings observed in Group 2 patients together, it is assumed that carotid artery clamping leaded to a marked reduction of blood flow, causing cerebral hypoxia and electrical failure, due to poor collateral circulation. As described, shunting procedure resolved the parameters on NIRS, TCD and SEP in all Group 2 patients. All but one patient did not neurologically deteriorated after surgery, probably becasuse internal shunting could prevent tissue damages due to prolonged ischemia. Only one patient developed left hemiparesis lasting for 12 hours following surgery for the right ICA stenosis. Postoperative MRI revealed no infarction, and he was discharged without any deficit. Transient, but dense ischemia during carotid artery clamping was most likely to cause transient neurological deficits.

NIRS monitoring could also detect a deterioration of cerebral hemodynamics caused by the sudden obstruction of the shunt tube, which lasted for 1 minute in one Group 2 patient. SEP could not detect this sudden accident immediately.

These results strongly suggest that the redox behavior of cytochrome oxidase detected by NIRS is a useful parameter to monitor oxygen concentration in the brain during carotid occlusion, and that the serial changes of Hb oxygenation can also provide a valuable information on brain oxygen metabolism during ischemia.

Illustrative Case (Group 2)

The 63-year-old male developed weakness of the right extremities and was admitted to our hospital. Neurological examination on admission revealed mild dysarthria and right hemiparesis. MRI showed only a small cerebral infarct in the left centrum semiovale. Carotid angiograms demonstrated severe stenosis of the left ICA and occlusion of the right ICA (Fig. 10). He underwent CEA for the left stenotic ICA. Although clamping of the ECA gave rise to a partial decrease of [oxy-Hb] and [total Hb], it did not cause any change in redox state of cytochrome oxidase. Clamping of the ICA, however, leaded to a continuous decrease of [oxy-Hb] and [total Hb], and increase of [deoxy-Hb]. Cytochrome oxidase was also reduced immediately after the carotid clamping. These findings persisted during the carotid clamping (Fig. 11). N_{20} amplitude gradually decreased to about 50 % of the control level (Fig. 12). Internal shunting resolved these parameters on NIRS and SEP. Postoperative course was uneventful, and he was discharged without any neurological deterioration.

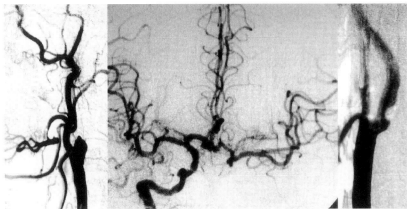

Figure 10

Cerebral angiograms of a 63-year-old patient. Left carotid angiogram (*left*) showed complete occlusion of the left internal carotid artery. Well-developed collateral circulation through the anterior communicating artery was identified on right carotid (*middle*) angiogram. Right carotid angiogram also showed significant stenosis of the right internal carotid artery (*right*).

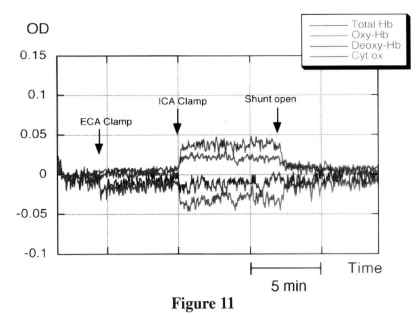

Figure 11

NIRS monitoring in a 63-year-old patient during CEA (OM-110). Cross-clamping of the right carotid artery immediately gave rise to a decrease in [oxy-Hb] and [total Hb] and an increase in [deoxy-Hb] as well as the reduction of cytochrome oxidase, because of occlusion of the contralateral internal carotid artery (Fig. 10).

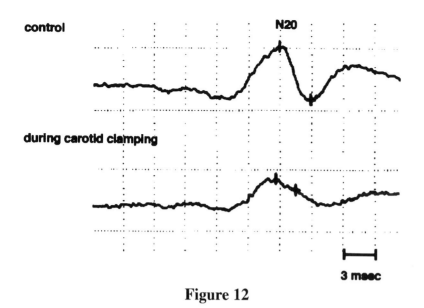

SEP monitoring in a 63-year-old patient during CEA. Note the marked reduction of N_{20} amplitude during cross-clamping of the carotid artery.

Figure 12

2) Common carotid artery - subclavian artery (CCA-SCA) bypass

Subclavian artery stenosis often causes vertebrobasilar insufficiency (VBI), including subclavian steal syndrome. Surgical treatments for subclavian artery stenosis includes transposition of VA to the CCA and CCA-subclavian artery bypass using the graft. It is essential to transiently occlude the CCA for both techniques. To avoid ischemic neurological deficits due to carotid occlusion, the authors applied NIRS monitoring for 2 patients who underwent CCA-subclavian artery bypass using the graft. One patient showed no significant changes of Hb oxygenation and [total Hb] for 25 min of CCA clamping, suggesting the presence of well-developed collateral flow.

Another patient showed a distinct deterioration of Hb oxygenation due to CCA clamping. Therefore, systolic blood pressure was raised from 120 mmHg to 160 mmHg, leading to a partial recovery of Hb oxygenation during CCA clamping (Fig. 10). No neurological symptom developed after surgery (Fig. 13).

3) Carotid balloon occlusion test

Temporary carotid artery occlusion has been frequently carried out in the patients with certain complex aneuryms of the internal carotid artery (ICA) in order to reduce the risk of aneurysm rupture during surgery. Surgical sacrifice of the ICA also remains the treatment of choice in the patients with giant or large ICA aneurysms. Therefore, it has been well known that balloon occlusion test of the ICA is very helpful to predict if ischemic symptoms would develop during or after surgery (Matsuda et al. 1988; Chen and Chen 1995). Previously, safety of carotid occlusion was judged on the basis of neurological symptoms, regional cerebral blood flow (rCBF), EEG, and stump pressure during transient carotid occlusion (Miller et al. 1977). We have applied NIRS monitoring to carotid balloon occlusion for the first time (Kuroda et al. 1996; see also Kaminogo et al. 1999).

Six patients underwent carotid balloon occlusion test under NIRS monitoring, because of large or giant ICA aneurysm in 5 patients and pseudoaneurysm of the ICA due to cavernous sinus aspergillosis in one. Another patient with severe ICA stenosis underwent NIRS monitoring during percutaneous transluminal angioplasty (PTA) using balloon catheter. NIRS monitoring did not disturb the neuroradiological procedures (Fig. 14).

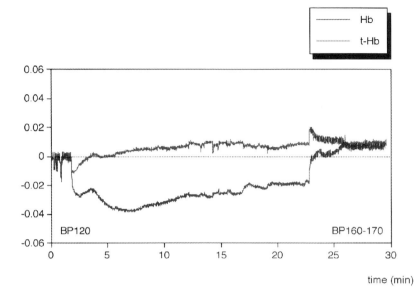

Figure 13

NIRS monitoring in a 55-year-old patient who underwent common carotid artery to subclavian artery bypass. Clamping of the common carotid artery leaded to a rapid and continuous deterioration of Hb oxygenation in the ipsilateral frontal lobe. Hb oxygenation gradually recovered in reponse to blood pressure elevation from 120 mmHg to 160-170 mmHg. NIRS parameters recovered to the control level just after re-opening of the common carotid artery.

After insertion of balloon catheter, the ICA was temporarily occluded for 1 to 15 minutes. Duration of temporary occlusion was determined by neurological status of the patients. The authors examined patients' consciousness, speech, and motor function through the test. Regional cerebral blood flow (rCBF) was simultaneously measured using ^{99m}Tc-hexamethyl-propylene amine oxime (HMPAO) and single photon emission computed tomography (SPECT), with a dose of 740 MBq (20 mCi), in 4 out of 7 patients. ^{99m}Tc-HMPAO, a lipophilic tracer, is rapidly converted to a hydrophilic form in the brain tissue, and is retained over hours. It was administered intravenously 2 to 3 min after carotid balloon occlusion

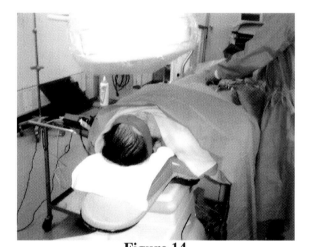

Figure 14

Balloon occlusion test of the internal carotid artery under NIRS monitoring. NIRS apparatus does not disturb interventional procedure.

(Matsuda et al. 1988). Immediately after the procedures, a brain SPECT study was carried, because the distribution of ^{99m}Tc-HMPAO is fixed in the brain in proportion to rCBF within 2-3 minutes following injection.

NIRS could clearly detect the changes of cerebral hemodynamics and metabolism in the illuminated brain, because the balloon interrupted the blood flow of the ICA, but not the ECA. Carotid balloon occlusion caused only a transient deterioration of the cerebral oxygenation state in 4 out of 7 patients. Cerebral oxygenation state judged by NIRS returned to the original levels within 1 or 2 minutes. In these patients, no remarkable change could be observed in their neurological status as well as in rCBF.

In contrast, a continuous decrease in [oxy-Hb] and [total Hb], and an increase in [deoxy-Hb] were observed during carotid balloon occlusion in other 3 patients. The authors present two illustrative cases who showed marked deterioration of cerebral hemodynamics and metabolism on NIRS, associated with temporary neurological deficits. It should be noted that the presence of cross flow through the anterior communicating artery on angiogram was not a reliable factor to determine the safety of carotid artery occlusion (see below).

Illustrative Case no. 1

The 64-year old female developed double vision on right lateral gaze, and was admitted to our hospital. Neurological examinations revealed right abducens nerve palsy. CT scans and MRI showed a round mass in the right cavernous sinus. Carotid angiograms revealed a partially thrombosed, giant aneurysm of the right ICA (Fig. 15). Carotid balloon occlusion test was planned to assess the safety of proximal ICA occlusion. Left carotid angiograms under balloon occlusion of the right ICA showed the presence of cross-flow through the anterior communicating artery (not shown). However, she developed consciousness disturbance, dysarthic speech and left hemiparesis 3 min after carotid balloon occlusion. Her neurological symptoms rapidly resolved following deflation of the balloon. Carotid balloon occlusion immediately caused a rapid and constant deoxygenation of intracerebral Hb. Deflation of the balloon induced a rapid oxygenation of Hb, which was followed by a transient hyperoxygenation of the brain (Fig. 16). Brain SPECT also revealed marked reduction of CBF up to 60% of the control in the right hemisphere (Fig. 17). These results revealed that proximal ligation of the right ICA would highly cause ischemic neurological deficits. Subsequently, she underwent proximal ligation of the right ICA following external carotid artery to middle cerebral artery using radial artery graft (Houkin et al. 1999). Postoperative course was uneventful and her neurological symptoms resolved 2 weeks after surgery.

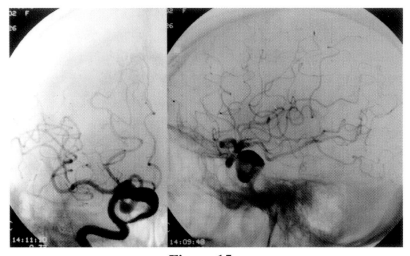

Right carotid angiogram of a 64-year-old patient. Note the partially thrombosed giant aneurysm of the cavernous portion of the internal carotid artery.

Figure 15

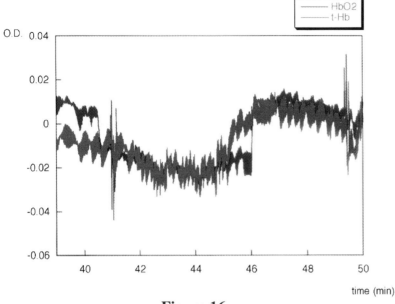

NIRS monitoring in a 64-year-old patient during balloon occlusion test of the internal carotid artery. Balloon inflation immediately caused a continuous deterioration of Hb oxygenation. Total concentration of Hb also decreased. Balloon deflation resolved these changes.

Figure 16

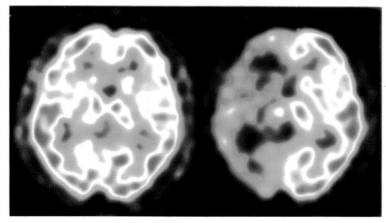

Regional cerebral blood flow (rCBF) mapping at resting state (*left*) and during balloon occlusion test (*right*). Note a marked reduction of rCBF in the right cerebral hemisphere during balloon inflation.

Figure 17

Illustrative Case no. 2

The 60-year old female developed double vision on left lateral gaze, and was admitted to our hospital. Neurological examinations revealed left abducens nerve palsy. CT scans and MRI showed a round mass in the left cavernous sinus. Carotid angiograms revealed a giant aneurysm of the left ICA (Fig. 18 *right*). Carotid balloon occlusion test was planned to assess the safety of proximal ICA occlusion. Left carotid angiograms under balloon occlusion of the left ICA showed the presence of cross-flow through the anterior communicating artery (Fig. 18 *left*). She did not show any neurological deterioration during carotid balloon occlusion test. Carotid balloon occlusion immediately caused only a transient deoxygenation of intracerebral Hb and decrease of [total Hb]. These parameters returned to the original level within 2 minutes (Fig. 19). Brain SPECT revealed no significant change of rCBF (Fig. 20). These results revealed that proximal occlusion of the left ICA would possibly be safe. Subsequently, she underwent proximal balloon occlusion of the left ICA. Postoperative course was uneventful and her neurological symptoms resolved 3 weeks after ICA balloon occlusion.

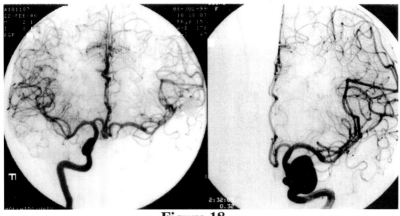

Carotid angiograms of a 60-year-old patient with a giant aneurysm of the left internal carotid artery (*right*). Right carotid angiogram during balloon occlusion of the left internal carotid artery showed good cross flow through the anterior communicating artery (*left*).

Figure 18

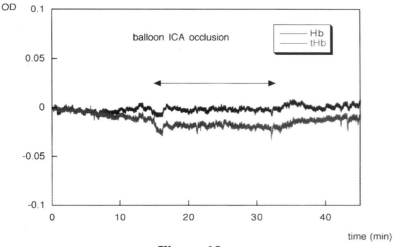

NIRS monitoring in a 60-year-old patient during balloon occlusion test of the internal carotid artery. Balloon inflation caused only a tansient deterioration of Hb oxygenation and decrease of total Hb.

Figure 19

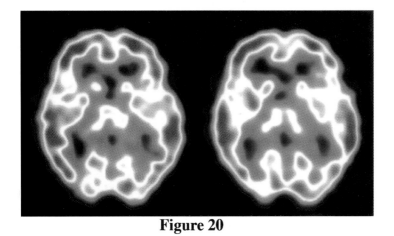

Regional cerebral blood flow (rCBF) mapping at resting state (*left*) and during balloon occlusion test (*right*). Note no significant change of rCBF in the left cerebral hemisphere during balloon inflation

Figure 20

4) Carotid compression test

Carotid artery occlusion is one of risk factors for development of ischemic stroke in the patients with poor collateral circulation. Oxygen extraction fraction and cerebrovascular reactivity to CO_2 or acetazolamide have been evaluated to predict their outcomes (Grubb et al. 1999, Vorstrup et al. 1986, Kuroda et al. 1993). The authors applied NIRS to assess the role of the contralateral ICA as the collateral circulation in 5 patients with ICA occlusion. Optodes for NIRS were placed on the ipsilateral forehead with a 35-mm source-receiver interval, and the contralateral carotid artery was manually compressed. Manual compression of the contralateral ICA caused a persistent decrease of [oxy-Hb] and [total Hb] and increase of [deoxy-Hb] in 3 patients who had well-developed collateral circulation through the anterior communicating artery. Following the release of manual compression, [oxy-Hb] transiently increased to higher levels than the original, and gradually decreased the original level. [Deoxy-Hb] also decreased to the lower levels than the original level, and gradually returned to the original one. These findings became more pronounce by repeated compression of the contralateral ICA. The results strongly indicated that the contralateral ICA could play an important role as the collateral pathway, keeping cerebral hemodynamics normal in these patients. On the other hands, NIRS monitoring showed no significant changes due to manual compression of the contralateral ICA in 2 patients who had no collateral circulation via the anterior communicating artery. Similar study were reported previously (Ferrari et al. 1987).

5) Hyperventilation Test in Moyamoya Disease

Moyamoya disease is a specific disorder, which involves both children and adults. Pediatric patients with moyamoya disease develop transient ischemic attack or stroke due to reduced cerebral perfusion reserve after hyperventilation such as running and crying. Although it is still unclear why hyperventilation causes ischemic attacks in moyamoya disease, hyperventilation gives rise to "re-build up" phenomenon after the end of hyperventilation on EEG, which is a pathognomotic finding in moyamoya disease. "Re-build up phenomenon often occurs simultaneously with transient ischemic attack. Previous studies suggested that "re-build up" phenomenon is related to a disturbed vasodilatory capacity (Kuroda et al. 1995b), decreased blood flow (Kazumata et al. 1996) and enhanced hypoxia after hyperventilation (Kameyama et al. 1986).

The authors applied NIRS to the pediatric patients with moyamoya disease in order to clarify the mechanism of "re-build up" phenomenon. Preliminary results were reported previously (Kuroda et al. 1996c). The authors monitored cerebral oxygenation state on NIRS as well as EEG during and following 3- to 4-minute hyperventilation. The optodes were placed on the forehead with the same fashion as other studies.

Hyperventilation caused a gradual decrease of [oxy-Hb] and [total Hb], and "build up" phenomenon which is the common finding in children. After the end of hyperventilation, these parameters on NIRS recovered to the control level in the patients who did not show "re-build up" phenomenon. However, NIRS revealed a further decrease of [oxy-Hb] and [total Hb], and a partial reduction of cytochrome oxidase 1 to 3 minutes after the end of hyperventilation in the patients who had "re-build up" phenomenon on EEG. These findings on NIRS were tightly related to the occurrence of "re-build up" phenomenon. The patients developed headache or transient ischemic attack during the period of "re-build up" phenomenon on EEG.

Concluding Remarks

As shown above, these results clearly showed that NIRS could detect real-time changes of cerebral oxygenation state non-invasively and continuously, and that NIRS has potential roles to avoid serious complications due to cerebral ischemia during carotid surgery and to clarify the mechanisms of various cerebrovascular disorders. Further progresses on the basic properties of lights in the tissue will make it more standard to monitor oxygen metabolism of the brain by NIRS.

References

Astrup J, Symon L, Branston NM, Lassen NA: Cortical evoked potential and extracellular K^+ and H^+ at critical levels of brain ischemia. Stroke 8:51-57, 1977

Beinert H, Shaw RW, Hansen RE, Hartzell CR: Studies on the origin of the near-infrared (800-900 nm) absorption of cytochrome c oxidase. Biochim Biophys Acta 591:458-470, 1980

Branston NM, Symon L, Crockard HA, Pasztor E: Relationship between the cortical evoked potential and local cortical blood flow following acute middle cerebral artery occlusion in the baboon. Exp Neurol 45:195-208, 1974

Chen HJ, Chen HY: Use of Tc-99m HMPAO brain SPECT to evaluate cerebral collateral circulation during Matas test. Clin Nucl Med 20:346-351, 1995

Cooper CE, Springett R: Measurement of cytochrome oxidase and mitochondrial energetics by near-infrared spectroscopy. Philos Trans R Soc Lond B Biol Sci 352(1354):669-676, 1997

Ferrari M, Zanette E, Sideri G, Giannini I, Fieschi C, Carpi A: Effects of carotid compression, as assessed by near infrared spectroscopy, upon cerebral blood volume and hemoglobin oxygen saturation. J R Soc Med 80:83-87, 1987

Guerit JM, Witdoeckt C, de Tourtchaninoff M, Ghariani S, Matta A, Dion R, Verhelst R: Somatosensory evoked potential monitoring in carotid surgery. I. Relationships between qualitative SEP alterations and intraoperative events. Electroencephalogr Clin Neurophysiol 104:459-469, 1997

Grubb RL Jr, Derdeyn CP, Fritsch SM, Carpenter DA, Yundt KD, Videen TO, Spitznagel EL, Powers WJ: Importance of hemodynamic factors in the prognosis of symptomatic carotid occlusion. JAMA 280:1055-1060, 1998

Hazeki O, Tamura M: Quantitative analysis of hemoglobin oxygenation state of rat brain in situ by near-infrared spectrophotometry. J Appl Physiol 64(2):796-802, 1988

Hazeki O, Tamura M: Near infrared quadruple wavelength spectrophotometry of the rat head. Adv Exp Med Biol 248: 63-69, 1989

Hoshi Y, Hazaki O, Tamura M: Oxygen dependence of redox state of copper in cytochrome oxidase in vitro. J Appl Physiol, 74: 1622-1627, 1993a

Hoshi Y, Tamura M: Dynamic changes in cerebral oxygenation in chemically induced seizures in rats: Study by near-infrared spectrophotometry. Brain Res 603:215-221, 1993b

Kameyama M, Shirane R, Tsurumi Y, Takahashi A, Fujiwara S, Suzuki J, Ito M, Ido T: Evaluation of cerebral blood flow and metabolism in childhood moyamoya disease: an investigation into "re-build-up" on EEG by positron CT. Childs Nerv Syst 2:130-133, 1986

Kaminogo M, Ochi M, Onizuka M, Takahata H, Shibata S: An additional monitoring of regional cerebral oxygen saturation to HMPAO SPECT study during balloon test occlusion. Stroke 30:407-413, 1999

Kazumata K, Kuroda S, Houkin K, Abe H, Mitumori K: Regional cerebral hemodynamics during re-build-up phenomenon in childhood moyamoya disease. An analysis using 99mTc-HMPAO SPECT. Childs Nerv Syst 12:161-165, 1996

Kuroda S, Kamiyama H, Abe H, Houkin K, Isobe M, Mitsumori K: Acetazolamide test in detecting reduced cerebral perfusion reserve and predicting long-term prognosis in patients with internal carotid artery occlusion. Neurosurgery 32:912-919, 1993

Kuroda S: Near-infrared monitoring of cerebral oxygenation during cerebral ischemia (in Japanese). Hokkaido Igaku Zasshi 70:401-411, 1995a

Kuroda S, Kamiyama H, Isobe M, Houkin K, Abe H, Mitsumori K: Cerebral hemodynamics and "re-build-up" phenomenon on electroencephalogram in children with moyamoya disease. Childs Nerv Syst 11:214-219, 1995b

Kuroda S, Houkin K, Abe H, Tamura M: Cerebral hemodynamic changes during carotid balloon occlusion monitored by near-infrared monitoring. Neurol Med Chir (Tokyo) 36: 78-86, 1996a

Kuroda S, Houkin K, Abe H, Hoshi Y, Tamura M: Near-infrared monitoring of cerebral oxygenation state during carotid endarterectomy. Surg Neurol 45: 450-458, 1996b

Kuroda S, Houkin K, Hoshi Y, Kazumata K, Abe H, Tamura M: Cerebral hypoxia after hyperventilation causes "re-build up" phenomenon and TIA in childhood moyamoya disease; Study by near-infrared spectroscopy. Child's Nerv Syst 12: 448-453, 1996c

Kuroda S, Houkin K, Kobayashi T, Yasuda H, Ushikoshi S, Nakayama N, Abe H: Clinical Application of Portable Near Infrared Oximetry (HEO-200) to Neurosurgical Monitoring (in Japanese). No Shinkei Geka 1999 (in press)

Li JY, Ueda H, Seiyama A, Nakano M, Matsumoto M, Yanagihara T: A near-infrared spectroscopic study of cerebral ischemia and ischemic tolerance in gerbils. Stroke 28:1451-1457, 1997

Matsuda H, Higashi S, Asli IN, Eftekhari M, Esmaili J, Seki H, Tsuji S, Oba H, Imai K, Terada H: Evaluation of cerebral collateral circulation by technetium-99m HM-PAO brain SPECT during Matas test: report of three cases. J Nucl Med 29:1724-1729, 1988

Matsunaga A, Nomura Y, Kuroda S, Tamura M, Nishihira J, Yoshimura N: Energy-dependent redox state of heme a + a3 and copper of cytochrome oxidase in perfused rat brain in situ. Am J Physiol 275(4 Pt 1): C1022-30, 1998

Matsuda H, Higashi S, Asli IN, Eftekhari M, Esmaili J, Seki H, Tsuji S, Oba H, Imai K, Terada H, Sumiya H, Hisada K: Evaluation of cerebral collateral circulation by Technetium-99m HM-PAO brain SPECT during Matas test: Report of three cases. J Nucl Med 29: 1724-1729, 1988

McCormick PW, Stewart M, Goetting MG, Balakrishnan G: Regional cerebrovascular oxygen saturation measured by optical spectroscopy in humans. Stroke 22:596-602, 1991

McCormick PW, Stewart M, Lewis G, Dujovny M, Ausman JI: Intracerebral penetration of infrared light. Technical note. J Neurosurg 76(2):315-8, 1992

Miller JD, Jawak K, Jennet B: Safety of carotid ligation and its role in the management of intracranial aneurysms. *J Neurol Neurosurg Psychiatry* 40: 64-72, 1977

Spencer MP, Thomas GI, Moehring MA: Relation between middle cerebral artery blood flow velocity and stump pressure during carotid endarterectomy. *Stroke* 23: 1439-1445, 1992

North American Symptomatic Carotid Endarterectomy Trial Collaborators: Beneficial effect of carotid endarterectomy in symptomatic patients with high-grade carotid stenosis. N Engl J Med 325:445-453, 1991

Parsons WJ, Rembert JC, Bauman RP, Greenfield JC Jr, Piantadosi CA: Dynamic mechanisms of cardiac oxygenation during brief ischemia and reperfusion. Am J Physiol 259(5 Pt 2):H1477-85, 1990

Pryds O, Greisen G, Skov LL, Friis-Hansen B: Carbon dioxide-related chnages in cerebral blood volume and cerebral blood flow in mechanically ventilated preterm neonates: comparison of near infrared spectrophotometry and ^{133}Xe clearance. Pediatr Res 27: 445-449, 1990

Shiga T, Tanabe K, Nakase Y, Shida T, Chance B: Development of a portable tissue oximeter using near infra-red spectroscopy. Med Biol Eng Comput 33:622-626, 1995

Sundt T, Sharbrough F, Piepgras D, KEarns T, Messick J, O'Fallon W: Correlations of cerebral blood flow and electroencephalographic changes during carotid endarterectomy. *Mayo Clin Proc* 56: 533-543, 1981

Tamura M, Hazeki O, Nioka S, Chance B, Smith DS: The simultaneous measurements of tissue oxygen concentration and energy state by near-infrared and nuclear magnetic resonance spectroscopy. Adv Exp Biol 222: 359-362, 1988

Vorstrup S, Brun B, Lassen NA: Evaluation of the cerebral vasodilatory capacity by the acetazolamide test before EC-IC bypass surgery in patients with occlusion of the internal carotid artery. Stroke 17:1291-1298, 1998

Wiernsperger N, Sylvia AL, Jöbsis FF: Incomplete transient ischemia: a non-destructive evaluation of in vivo cerebral metabolism and hemodynamics in rat brain. Stroke 12:864-868, 1981

Wikström M: Energy-dependent reversal of the cytochrome oxidase reaction. Proc Natl Acad Sci U S A 78:4051-4054, 1981

Wong-Riley MT: Cytochrome oxidase: an endogenous metabolic marker for neuronal activity. Trends Neurosci 12:94-101, 1989

Chapter II.16

fMRI guided Near-infrared Spectroscopy:
An example for Parkinson's Disease

Ingrid L. Kwee[a1] and Tsutomu Nakada[a,b]

[a]*Department of Neurology, University of California, Davis, VANCHCS*
[b]*Department of Integrated Neuroscience, Brain Research Institute, University of Niigata*

Introduction

A conservative estimate indicates that over 500,000 individuals in the United States have idiopathic Parkinson's disease (PD). In the early 21st century, this number is expected to increase three to fourfold reflecting the explosive growth in the geriatric population in developed countries. Several million people in the US, primarily elderly, will by then suffer from PD. PD is generally categorized as a degenerative movement disorder, the main feature of which is neuronal death within the zona compacta of the substantia nigra. The resultant decline in the dopamine pool of the neo-striatum is considered to be the main pathophysiologic process giving rise to the characteristic symptoms of bradykinesia, rigidity, tremor and loss of postural reflexes. Until recently, treatment in PD has targeted primarily the motor dysfunctions and less attention has been paid to cognitive aspects, such as working memory dysfunction, in part due to lack of consensus on the very nature of the dysfunction and underlying pathogenetic mechanisms. However, it has become more evident that PD is in fact more than a "movement" disorder and certain of the cognitive/behavioral abnormalities may be attributed to the primary pathological processes of PD (Mayeux et al., 1990, Cooper et al., 1991, Levin and Katzen, 1995). It is now essential for investigators to actively address the cognitive aspects of PD for scientific reasons as well as for devising treatment and improving patient quality of life.

Among the various cognitive dysfunctions described in PD, prefrontal dysfunction, especially that of the dorsolateral prefrontal cortex (DLPF), related to working memory represents the most consistent one (Baddeley, 1992, Jacobs et al., 1995, Owen et al., 1997). Considering the massive connections between the frontal association cortex and the basal ganglia and ventral tegmental area, it is plausible that working memory dysfunction may

---

[1] Correspondence: Ingrid L. Kwee, M.D. Department of Neurology, University of California, Davis, VANCHCS, 150 Muir Road, Martinez, CA 954553, Tel: (1)-925-370-4734, Fax: (1)-925-299-4897, e-mail ilkwee@ucdavis.edu

reflect a primary, rather than secondary pathological process in PD. Although neuropsychological batteries support a prefrontal dysfunction in PD, technical difficulties have precluded its more precise delineation. Conventional behavioral batteries designed to study prefrontal function uniformly require subject motor response. Quantitative assessments are often dependent on reaction time. Such tests are in principle not suitable for use in PD patients whose primary motor dysfunction such as bradykinesia, rigidity, or tremor may adversely influence the results.

Recent advancements in blood oxygenation level dependent (BOLD) contrast functional magnetic resonance imaging (fMRI) have revolutionized the field of cognitive neurology (D'Esposito et al., 1998). It is now possible to perform objective analysis of higher function on an individual patient basis (Nakada et al., 1998a, 1998b, Kwee et al., 1999). One draw back of fMRI is its technical complexity. Highly reproducible analysis suitable for clinical application requires extensively optimized instrumentation. It is impractical and prohibitively costly to perform fMRI on PD patients as part of routine clinical work up. To overcome this limitation while still benefiting from the obvious advantages of fMRI, we introduce here a new technique of fMRI guided near infrared spectroscopy (NIRS). The technique is a hybrid technique based on the complementary application of high-field fMRI and portable NIRS (Figure 1). The former provides accurate activation maps of selected sets of complex paradigms, while the latter provides a convenient bedside tool for evaluating such activation. The technique is applied to the evaluation of dorsolateral prefrontal (DLPF) activation reflecting working memory in response to presented complex cognitive paradigms. The study reports effectiveness of this new technique in PD patients providing entirely objective measures the results of which are not dependent on the motor abilities of the patient.

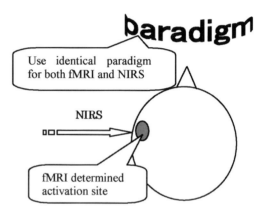

Figure 1
Concept of fMRI guided NIRS

Materials and Methods

High-Field fMRI

Ten normal volunteers, ages 65-75 years, were imaged according to the human research guidelines of the Internal Review Board of the University of Niigata. Subjects viewed a screen where visual tasks were presented. Each session consisted of multiple 30-second epochs in boxcar configuration.

A Signa 3 T (GE Medical System, Waukesha, Wisconsin, USA) imaging system with a superconductive magnet operating at 3.0 Tesla (Magnex, Abingdon, Oxon, UK) was used to perform all imaging studies. Gradient echo echo-planar images (GE-EPI) were obtained using the following parameter settings: FOV 40 cm x 20 cm; matrix 128 x 64; slice thickness 5 mm; inter-slice gap 2.5 mm; TR 1 sec. Spatial resolution was approximately 3 mm x 3 mm x 5 mm. Sessions which had brain motion exceeding 0.6 mm were re-performed to avoid so-called fictitious activation due to pixel misalignment. fMRI time series data consisting of consecutive EPI images for each slice were analyzed utilizing SPM96 (the Wellcome Department of Cognitive Neurology).

The data were smoothed using a 3-mm full width at half maximum (FWHM) kernel. Statistical analysis was performed using a delayed (6 seconds) boxcar model function in the context of the general linear model as employed by SPM96. To minimize effects of physiological noise, a high pass filter and global normalization were applied within the design matrix. Specific effects were tested by applying appropriate linear contrasts to the parameter estimates for each condition, resulting in a t statistic for each and every voxel. These t statistics, which were transformed to Z statistics, constitute an activation map. These images were interpreted by referring to the probabilistic behavior of a Gaussian field. fMRI images were presented with contrast between the two conditions specified and show activated areas which conformed to statistical criteria of significance ($p < 0.01$).

NIRS

Informed consent as approved by the Institutional Review Board of the University of California, Davis was obtained from all ten cognitively intact Parkinson's subjects, Hoehn & Yahr stage II=III, ages 63-75 years, undergoing NIRS studies detailed below.

The FDA approved Invos 4100 (Somanetics, Troy MI) clinical cerebral oximeter consisted of an electronic computer display, connecting cables and two flexible adhesive probes containing miniature light-emitting diodes (LED) and light detectors was utilized. The LEDs use two wavelengths of near-infrared light, centered at 730 and 810 nm, to measure the ratio of oxyhemoglobin to deoxyhemoglobin in the field beneath the oximeter probe, thus providing an index of changes in brain hemoglobin oxygen saturation (Dujovny et al., 1994).

Adhesive sensors will be applied onto each subject's forehead bilaterally, taking care to position the sensors so that the LED is located in the FP2 lead of the international EEG 10-20 system. This placement results in the optical detectors being approximately 5 cm lateral from the midline of the forehead. Subject will be then be asked to view a colored computer monitor, which delivers the cognitive tasks previously validated to elicit DLPF activation. In addition to the n-back task, three categories of tasks based on the Wechsler Adult Intelligence Scale, Third Edition (WAIS-III) are presented, namely, picture completion (missing object), matrix reasoning (which symbol comes next) and picture arrangement (story board).

Baseline hemoglobin oxygen relative saturation index (rSO2) and Blood Volume readings are obtained during 5-minute pre- and post- test intervals in which the subject is requested to remain silent while viewing a nature scene on the monitor. Each task is also preceded and followed by a similar 1-minute rest interval designed to allow rSO_2 and Blood Volume levels to return to baseline. Subjects are asked to focus on the monitor and remain silent throughout the study other than to give a simple verbal indication with the completion of each task. It is not essential that subjects give the correct answer.

Simultaneous with the appearance of each task on the monitor, an event mark is entered on the Invos unit and the subject is asked for the solution. The moment the subject signals completion, a second event mark is entered and the 1-minute rest interval begins. A total of 20 tasks are presented with varying degrees of difficulty ranging from easy to moderately difficult. If the subject is unable to solve a task within two minutes time, the experimenter moves on to the next task. Depending on the subjects' response time the entire study requires approximately 50 to 70 minutes to complete.

Data is output digitally from the Invos unit to a laptop computer as a text file, which is later processed using Microsoft Excel. The data includes rSO_2 and Blood Volume levels recorded at 1-second intervals as well as task marks and date and time information. This data is then plotted onto separate graphs for rSO_2 and Blood Volume vs. Time together with task start and stop marks.

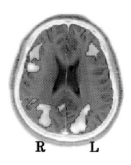

Figure 2

Results

Representative fMRI is shown in Figure 2. Consistent activation in normal subjects were observed in DLPF and intraparietal (IP) bilaterally. Corresponding fMRI guided NIRS is shown in Figure 3. All ten normal volunteers and ten Parkinson's patients without cognitive dysfunction showed DLPF activation detected by fMRI guided NIRS.

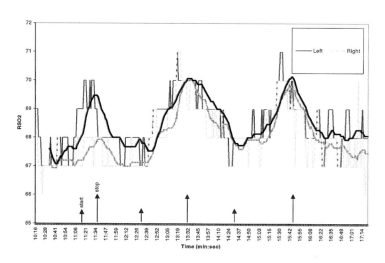

Figure 3
fMRI guided NIRS

Relative saturation index of regional hemoglobin (rSO_2) shows typical activation corresponding to task performance. Short arrows indicate the time when the tasks are presented to the subject, while long arrows indicate the time when the subject signals completion.

Discussion

Near-infrared spectroscopy relies on the measurement of absorption of light at specific wavelengths to distinguish the relative quantities of chromophores in living tissue. A natural "window" exists in human tissue between the wavelengths of 690-880 nm where the absorption by water and C-H bonds is insignificant and near-infrared light can penetrate to a depth of several centimeters. Because certain molecules exhibit marked differences in spectral absorption when they are oxygenated (hemoglobin, myoglobin and cytochrome aa$_3$), measurement of changes in oxygenation *in vivo* is possible. The Beer-Lambert Law describes how the concentration of a chromophore can be quantified using light which travels in a specific path L:

$$\ln OD = C\mu aL$$

where OD is the optical density caused by attenuation of light as it traverses through tissue from source to detector, C is the concentration of a chromophore in the illuminated region, μ_a is the absorption coefficient of the chromophore, and L is the exact path length of light in tissue. Correction for light scattering makes use of the concept of "mean photon path" through tissue between light source and detector which establishes the most likely path a photon will take.

Considering the concentration of chromophores other than hemoglobin is very small, changes in optical attenuation at two wavelengths λ_1 and λ_2 on the frontal skull is primarily influenced by changes in oxygenated and reduced hemoglobin and the mean path of light <L> using this modification of the Beer-Lambert Law:

$$\Delta \ln OD\lambda 1 = [(\Delta CHb \times \mu aHb\lambda 1) + (\Delta CHbO2 \times \mu aHbO2\lambda 1)] \times <L\lambda 1>$$
$$\Delta \ln OD\lambda 2 = [(\Delta CHb \times \mu aHb\lambda 2) + (\Delta CHbO2 \times \mu aHbO2\lambda 2)] \times <L\lambda 2>$$

Thus, assuming path length is approximately equal at both wavelengths, two wavelengths of near-infrared light can quantify changes in oxyhemoglobin and reduced hemoglobin and also changes in total hemoglobin which is their sum. Knowing these variables also allows the calculation of changes in oxygen saturation, however, these variables are not sufficient to calculate absolute oxygen saturation of blood directly (Mancini et al., 1994, Hock et al., 1995, Levy et al., 1999).

Extracranial contamination remains a potential concern in the measurement of cerebral oxygenation (Liu et al., 1995). The proposed method of fMRI guided NIRS can effectively ameliorate this problem. In general, the use of multiple detectors can effectively suppress extracerebral oxygenation and perfusion events provided they occur over a specific range of blood volume changes. Definite causes of inaccuracy occur when measurements are attempted over hematomas or when venous engorgement of the scalp is present (Gopinath et al., 1993). The excessive absorption caused by the pooled blood prevents a significant portion of light from entering the brain, thereby shifting the measurement from cerebral blood to extracerebral blood. In addition to suppression of extracranial oxygenation changes, multiple detectors can make significant improvements in accuracy.

Dementia and depression represent the two most commonly recognized non-motoric abnormalities in PD (Figure 4). Selective impairment of cognitive function without dementia or depression is also found to be common in PD. These are believed to be manifestations of functional disturbances of single or multiple parallel frontal subcortical circuitries. Representative connections include: (1) dorsolateral prefrontal cortex to dorsolateral caudate to lateral globus pallidus to ventral anterior and medial dorsal thalamus to prefrontal cortex circuit; (2) lateral orbital cortex to ventromedial caudate to medial globus pallidus to ventral anterior and medial dorsal thalamus to orbital cortex circuit; and (3) anterior cingulate to nucleus accumbens to globus pallidus to medial dorsal thalamus to cingulate circuit. Although by no means settled, since cognitive impairment may appear early in PD, it is thought to be the result of primary dysfunction of subcortical neurons in these circuitries. In general, the prefrontal syndrome is characterized by slowed information processing, difficulty with visuospatial relations, and impaired executive functions such as planning, sequencing and innovation; the orbitofrontal syndrome by inappropriate behavior, changed personality and irritability; the cingulate syndrome by abulia, anergia, anhedonia, and timidity. The orbitofrontal and cingulate syndromes may simulate depression but are remarkably free of the association with the feeling of guilt, remorse, or despair.

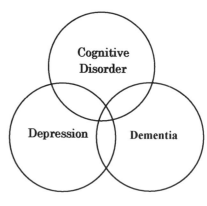

Figure 4

Unfortunately, to date, there has been no entirely objective bedside testing tool developed to evaluate these cognitive dysfunction, primarily because of technical difficulties. Conventional behavioral batteries designed to study prefrontal function require varyingly elaborate subject motor response. Quantitative assessments are often dependent on so-called "reaction time". They are in principle not well suited for use in PD patients whose motor dysfunction, such as bradykinesia, rigidity, or tremor, could adversely influence test results. The current study clearly demonstrated that objective measures of DLPF function can readily be performed utilizing fMRI guided NIRS.

Acknowledgement

The study was supported by grants from Ministry of Education (Japan) and Department of Veterans Affairs (USA). The study was presented in part at the XIII International Congress on Parkinson's Disease, Vancouver, Canada.

References

Baddeley AD: Working Memory. Science 1992;255:556-559.
Cooper J, Sagar HJ, Jorden N, Harvey NS, Sullivan EV: Cognitive impairment in early, untreated Parkinson's disease and its relationship to motor disability. Brain 1991;114:2095-2122.
D'Esposito M, Aguirre GK, Zarahn E, Ballard D, Shin RK, Lease J: Functional MRI studies of spatial and nonspatial working memory. Cogn Brain Res 1998;7:1-13.

Dujovny M, Slavin KV, Cui W, Lewis G, Ausman JI: Somanetics INVOS 3100 Cerebral Oximeter: Instrumentation, technique, and technology. Neurosurg 1994;34:935-936.

Gopinath SP, Robertson CS, Grossman RG, Chance B: Near-infrared spectroscopic localization of intracranial hematomas. J Neurosurg 1993;79:43-47.

Hock C, Müller-Spahn F, Schuh-Hofer S, Hofmann M, Dirnagl U, Villringer A. Age dependency of changes in cerebral hemoglobin oxygenation during brain activation: a near-infrared spectroscopy study. J Cereb Blood Flow Metab 1995;15:1103-1108.

Jacobs DM, Marder K, Côté LJ, Sano M, Stern Y, Mayeux R: Neuropsychological characteristics of preclinical dementia in Parkinson's disease. Neurology 1995;45:1691-1696.

Kwee IL, Fujii Y, Matsuzawa H, and Nakada T. Perceptual processing of stereopsis in human: High-field (3.0T) functional MRI study. Neurology 1999;53:1599-1601.

Levin BE, Katzen HL: Early cognitive changes and nondementing behavioral abnormalities in Parkinson's disease. Adv Neurol 1995;65:85-95.

Levy WJ, Carpenter J, Fairman RM, Golden MA, Zager E: The calibration and validation of a phase-modulated near-infrared cerebral oximeter. J Clin Monit 1999;15:103-108.

Liu H, Chance B, Hielscher AH, Jacques SL, Tittel FK: Influence of blood vessels on the measurement of hemoglobin oxygenation as determined by time-resolved reflectance spectroscopy. Med. Phys. 1995;22:1209-12017.

Mancini DM, Bolinger L, Li H, Kendrick K, Chance B, Wilson JR: Validation of near-infrared spectroscopy in humans. J Appl Physiol 1994;77:2740-2747.

Mayeux R, Chen J, Mirabello E, Marder K, Bell K, Dooneief G, Cote L, Stern Y: An estimate of the incidence of dementia in idiopathic Parkinson's disease. Neurology 1990;40:1513-1517.

Nakada, T, Fujii Y, Suzuki K, Kwee IL: "Musical Brain" revealed by high-field(3tesla) functional MRI. Neuroreport 1998;9:3853-3856.

Nakada, T, Fujii Y, Suzuki K, Kwee IL: High-field (3.0T) functional MRI sequential epoch analysis: an example for motion control analysis. Neurosci Res 1998;32:355-362.

Owen AM, Iddon JL, Hodges JR, Summers BA, Robbins TW: Spatial and non-spatial working memory at different stages of Parkinson's disease. Neuropsychologia 1997;35:519-532.

Section III

Neuroscience of Music

This chapter represents the proceedings of the International Symposium on the Neuroscience of Music 1999 (ISNM '99), Niigata, Japan, October 15-17, 1999.

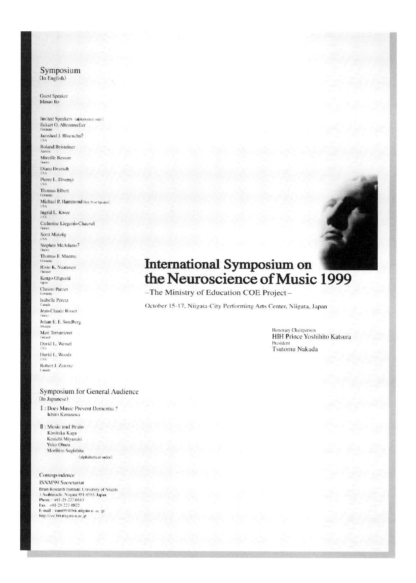

Welcome to Japan and Niigata, Host City of the International Symposium on the Neuroscience of Music. The unique program of this Symposium brings together world scientists in the field of music and musicians for the purpose of achieving a better understanding of the brain cognitive processing of music. I hope you share my enthusiasm for this exciting opportunity to experience, over the next two and a half days, the joining of art and science together in furthering our knowledge of how the brain works.

yoshihito

Honorary Chairperson
HIH Prince Yoshihito Katsura

Chapter III.1

Perception of Musical Sound : Simulacra and Illusions

Jean-Claude Risset [1]

Laboratoire de Mécanique et d'Acoustique, CNRS

Computer synthesis permits for the first time to manufacture sounds directly from blueprints of their physical structure. To produce computer music, one must provide a synthesis program, such as Music V (Mathews, 1969) or CSound (Boulanger, 2000), with a precise and thorough description of the physical parameters of the desired sounds. The auditory effect is experienced by listening to the synthesis results. This permits to relate the auditory effect and the physical structure. An understanding of this psychoacoustic relation is necessary for mastering the musical possibilities of digital sound synthesis. Thus Chowning, Mathews and myself, trying since the 60s to take advantage of this new way to create musical material, developed new psychoacoustic knowledge and knowhow.

I would like to report here some findings concerning the perception of musical sound. The exploration of the musical resources of computer music has led to the following conclusion : the perceived attributes are in more complex relation with the physical parameters of the sound than is generally assumed. This conclusion will be illustrated here by 20 sound examples, presenting notably musical simulacra and auditory illusions.

Pitch and frequency ratios

Even at the level of individual sounds, the extraction of subjective attributes such as pitch or timbre is often elaborate and happens at a high level. This is confirmed by other data : tonotopia is not a mapping of pitches, and pitch can sometimes be inferred only through comparisons of the firings from the two auditory nerves.

Pitch intervals relate with frequency rations : a ratio close to 3/2 will be heard as an interval of a fifth. However sound example # 1 shows that above 5000 Hz, the interval between two tones becomes practically impossible to recognize or qualify. This breakdown in our capacity to evaluate intervals explains why the higher limit of the tessitura of musical instruments is below 5000 Hz. It can be attributed to the collapse of temporal cues for the

---

[1] Correspondence: Jean-Claude Risset, Laboratoire de Mécanique et d'Acoustique, CNRS, 13402 Marseille Cedex 9, France, Tel: (33)-4-91-16-40-78, Fax: (33)-4-91-22-08-75; e-mail: jcrisset@lma.cnrs-mrs.fr

evaluation of pitch : even with volley-like mechanisms, the neural firings cannot follow the high rate of periodicity.

Adding sounds with neighbouring frequencies

In sound example # 2, one first hears a periodic tone of frequency 55 Hz, with 10 harmonics of equal amplitude. This is heard as a sustained low A. Then one hears the result of mixing this tone with seven similar tones of neighbouring frequencies : 55 + 1/20 Hz, 55 + 2/20 Hz, 55 + 3/20 Hz, ... , 55 + 7/20 Hz. Since the frequency difference is so small - 1/20th Hz, corresponding to a 1/60th of a semi-tone - one would expect to hear either a kind of cluster or an almost unchanged similar tone. Yet one hears a pattern of harmonics waxing and waning at different rates.

This is understandable if one remembers that

$$\cos 2\pi f_1 t + \cos 2\pi f_2 t = 2\cos\{2\pi(f_1+f_2)/2t\}\cos\{2\pi(f_1-f_2)/2t\}.$$

This mathematical identity is the key to the phenomenon of beats : if one adds two sinusoidal components with frequencies f_1 and f_2 when the frequencies are very close, $(f_1-f_2)/2$ becomes a very low frequency, so that the superposition will be heard as a tone with frequency $(f_1+f_2)/2$ modulated at a rate $(f_1-f_2)/2$. For instance, if the frequency difference is 2 Hz, the amplitude will go up and down twice per second. Here the difference is 1/20th Hz, so the amplitude of the fundamental will wax and wane in 20 s - but the period of the amplitude change will be 20/2 second for the 2d harmonic, 20/3 second for the 3d harmonic, ..., 20/10 second for the 10th harmonic. Thus the different harmonics wax and wane at a rate proportional to their rank. In the example, there are not only 2 tones, but 8 tones with frequencies in arithmetic progression : this makes the pattern much clearer by turning the amplitude maxima into sharp peaks - a well-known effect in multiple-source interferometry.

Thus it can be hard to predict the auditory result, even for relatively simple sound structures like those used in this example. However such effects can be musically useful. Clearly, the high precision for specifying the physical parameters of a synthetic sound can give rise potentially interesting effects.

Simulation of musical instruments

The simulation of instruments demonstrates that the spectrum is far from being the sole cue for identifying the timbre of a musical instrument : hearing tends to rely on cues that are more robust and less likely to be distorted in the propagation between source and listener.

Sound example # 3 shows that it is not sufficient to describe an instrumental tone by specifying a single frequency spectrum, an attack transient and a decay. Tones synthesized from this recipe do not evoke the instrument : hence the description is not aurally relevant.

In the case of brass tones, the main cue is a relation between spectrum and loudness : the louder the tone, the richer the spectrum (Cf. Risset & Mathews, 1969; Risset & Wessel, 1999). This is illustrated by sound example # 4. Such a feature allows the listener to recognize whether the tone has been blown loudly or softly, regardless of the distance between instrumentalist and listener.

With other instruments, the cue(s) can be different. As Mathews showed, the spectrum of the tone produced by a bowed string changes in a complex way as a function of frequency, so that the vibrato cannot be mimicked by a simple frequency modulation : to sound realistic, it must be accompanied by a synchronous spectral modulation (cf. sound example # 5).

Hearing appreciates the "gait" of a sound : an accurate rendering is not always critical. In the two latter cases, the main cue for the timbre is not an invariant parameter, but a relation between the spectrum and another parameter, which is better preserved during the propagation that the spectrum itself. The singing voice can be evoked by playing on still another aspect of the sound. One chooses a tone with a spectrum resembling one that could be emitted by a vocal tract. As can be heard on example # 6, by introducing an irregular "vibrato" (frequency modulation) on such a tone, one endows the tones with a vocal quality - one can have the impression of a male or a female voice, although it does not originate in a body. This process was originally introduced by Chowning and McNabb. Chowning has shown that synchronous frequency fluctuations are critical to make frequency components "cohere" and be fused by the listener as a single auditory entity, distinct from components which undergo different fluctuations - even if their average frequencies coincide.

It is easy to endow sounds with a percussive quality by imposing a sharp attack followed by a decay. Beyond this overall amplitude behavior, specifics in spectral evolutions or frequency patterns often suggest details about the way the sounds were produced - e.g., a membrane with varying tension, a certain material of the struck surface (skin with or without snares, wood, metal). This can be heard in sound example # 7 - even the example, realized with additive synthesis, did not attempt in any way a physical modeling.

The realization of instrumental simulacra ensures that the model has captured the features of the sound that are relevant for the ear - a genuine analysis by synthesis. It strongly suggests that hearing is highly sensitive to features that can help to infer through which mechanical action the sound was produced.

The computer imitation of instrumental sounds is not only a foolproof test of the aural relevance of their description : it can have musical interest, since it allows to control and modify those sounds produced in novel ways.

Thus, in sound example # 8, one can hear a bell-like tone constructed so as to have an ambiguous pitch which can be matched by two different frequencies. A soprano (Irène Jarsky) sing a high pitch. A carefully contrived bell-like tones begins on the same pitch, but the lower resonance is echoed by a lower tone from the soprano.

In sound example # 9, one hears a set of bells, each of which is composed like a chord, with a prescribed inner harmony. In the second part of the example, one hears textures with the same inner harmonies : each bell has been turned into a fluid tone by turning the percussive envelope which modulates the amplitude of each frequency component into a smooth bell-shaped curve. Individual sound objects - "bells" - are liquefied while keeping the same frequency content, like a substance changing physical state : a novel possibility of interest to composers.

Auditory paradoxes and illusions

Visual illusions have been around for a long time. The precision of computer synthesis permits to contrive the sounds so as to produce auditory illusions or paradoxes. These effects evidence idiosyncrasies of auditory perception. As the physiologist Purkinje wrote more than a century ago, "Illusions, errors of the senses, are truths of perception."

Pitch is a subjective attribute of sound that relates to frequency, but this relation is more complex than one generally believes. I have produced examples where pitch varies in an unintuitive way.

As demonstrated in example 11, doubling the frequencies of the components may result in lowering the pitch of the sound (Risset, 1971). This effect is obtained with sounds made up of frequency components in stretched octave relation. Thus, when frequencies are doubled, each component is replaced by a component a little lower in frequency, except at the extremities. This can be viewed as a proximity effect. Example 11 shows in a striking way that frequency relations are not necessarily mapped into similar pitch relations. One might object that such an *ad hoc* physical structure is very unlikely to be encountered with acoustically-produced sounds. However the fact is that many inharmonic sounds - such as those of bells and gongs - are subject to similar effects : transposing them in frequencies can give rise to surprising pitch transpositions (cf. Terhardt, 1974).

A similar effect can be demonstrated for rhythm. Example 12 shows that doubling the speed of a tape recorder reproducing a sequence of beats may result in slowing down the perceived tempo (Risset, 1986).

Several authors have proposed that pitch was a compound attribute, with two concurring components: tonal pitch, the component implied in the notion of pitch class (C, D, ...), probably originating from neurral autocorrelation analysis in the brain, and spectral pitch, correlated with the centroid of the spectrum (hence related to timbral brilliances) and thus to the position of maximum stimulation of the basilar membrane. By synthesizing tones made of octave components, one can divorce variations of spectral pitch and tonal pitch. These variations appear to be best detected respectively by the left ear and the right ear (Cf. Charbonneau & Risset, 1975).

Synthetic tones can be made to go up (or down) indefinitely in pitch (Shepard, 1964, Risset, 1971) : example 12 presents (a portion of) an endless glissando downwards - only tonal pitch varies here. In example 13, a tone goes up in pitch, without octave jump, while staying a E flat : only spectral pitch varies in this case. In example 14, the sound seems to go down in pitch, yet it is much higher at the end than where it started. Tonal and spectral pitch are varying in opposite directions. After an octave variation, the pitch class is back to what it was originally : on the contrary, spectral variations accumulate. Thus, in example 14, one might say that pitch goes down locally - in the direction of tonal pitch - but up globally - in the direction of spectral pitch.

Again, as demonstrated in examples 15, 16 and 19, similar behaviors to those of pitch in examples 12, 13 and 14 can be demonstrated for rhythm. One can notice that a cyclic pseudo-acceleration is dynamogenic, whereas a pseudo-slowing-down is rather depressing, even though these rhythmic variations do not lead anywhere in the long run.

In stereo listening, the direction of a source can be inferred on the basis of differences between both the levels and the arrival times of the signals received in each ear. The diffraction of the sound around the head modifies the spectrum and yields subtle cues which are called for to evaluate the height of the source or to distinguish between front and back (cf. Blauert, 1983).

Hearing appreciates the surrounding space by attending to the echoes and reverberation that have been superimposed to the original sound : it does so in an elaborate way, as evidenced by the so-called precedence effect (sometimes called Haas effect), whereby echoes tend to be fused with the original sound without smearing the localization. Now one can alter a sound by adding electroacoustically-produced echoes and reverberation, thus suggesting this sound is immersed in a "virtual" acoustic space - one can also give the impression that the sound source is spreading (as in example 17) or moving (as in example 18, 19 and 20) in this virtual space.

The latter example uses the recipe described by Chowning, who has produced powerful illusions of fast sound movements (1972) with a 4-channel system by carefully controlling the evolution in time of the following parameters : the amplitude in each of the four channels, the direct to reverberant ratio, and a pitch transposition mimicking a Döpler effect. These illusions evidence some of the cues our hearing resorts to in order to appreciate the space around us, the distance of the sound sources and their motion. The Döpler effect - the apparent change of frequency of a vibration when its source moves with respect to the observer - strongly suggests the movement of sources. In example 20, which does not attempt to reproduce any Döpler effect precisely, frequency variations reinforce the illusion of movement.

The cues for evaluating the distance of sound sources are of great practical importance for the listener : they are not limited to the overall loudness. Turning down the amplification in a radio receiver does not cause the sound to recede in the distance : the sound just becomes softer. As Chowning demonstrated, the impression of a receding sound source can be conveyed by adding reverberation : if the direct sound diminishes in level while the reverberation is kept constant, one has the impression that the source is going away from the listener.

Ecological perception : auditory scene analysis

The above findings imply that the brain implements elaborate strategies to extract from the impinging sounds reliable and useful information about the environment. Perceptual processing parses the auditory signal into entities which are assigned to different hypothetical sound sources : Bregman has proposed to call such processing "auditory scene analysis". To identify the hypothetical sources, features of the sound that are fragile and easily distorted in the course of sound propagation are either disregarded (e.g., phase relations between the harmonics of periodic sounds), or interpreted to yield additional information (e.g., Döpler frequency changes due to the movements of sound sources). Hearing has developed a high sensitivity to frequency aspects, which are very robust during wave propagation. The ear is very good at evaluating both the loudness of the source and its distance, and it attempts to get from the overall sound a feel for both the excitation that made it appear - the dynamic, agogic aspects - and the response of the excited vibrating system - the stable, structural aspects. Many such capabilities of the auditory system favor survival, and they appear to have developed through evolution in a world where sound sources are produced by mechanical vibrations. These processes continue to function somewhat mistakenly with electrically produced sounds, and sounds produced electronically tend to have a stronger identity if they can be interpreted by the listener as originating from a recognizable acoustic process such as hitting, scraping or blowing.

Music is an activity involving body and mind, rationality and emotion. Schlaug has shown recently that the practice of a musical instrument at a young age reinforces the structures responsible for communication between the two hemispheres of the brain. Music takes advantage of the fine capabilities of hearing developed to help survival, and it calls for them to perform in a gratuitous way : it can offer case studies of the brain functions and skills.

Summary

The exploration of the musical resources of computer music has led to the following conclusion : the perceived attributes are in more complex relation with the physical parameters of the sound than is generally assumed. This is illustrated by sound examples of musical simulacra and illusions. These findings imply that the brain implements strategies to extract from the impinging sounds reliable and useful information about the environment. Music takes advantage of these capabilities and calls for them to perform in a gratuitous way : it can offer case studies of the brain functions and skills.

References

Bever, T., & Chiarello, R.J., Cerebral dominance in musicians and nonmusicians, Science, 1974, 185, 537-539.
Blauert, J., Spatial hearing, M.I.T. Press, Cambridge, Mass., 1983
Boomsliter,P.C., & Creel, W., Hearing with ears instead of instruments, J. Audio Engin. Soc., 1961, 18, 407-412.
Boulanger, R., ed., The C-Sound Book, M.I.T. Press, Cambridge, Mass., to be published in 2000.
Braus, I. Retracing one's steps: an overview of pith circularity and Shepard tones in European music- 1550-1990. Music Perception, 1995, 12 (3), 323-351.
Bregman, A. S. Auditory scene analysis - the perceptual organization of sound. Cambridge, Mass: M.I.T. Press, 1990.
Brown, G.J., & Cooke, M. Computational auditory scene analysis. Comput. Speech Language (UK), 1994, 8 (4), 297-336.
Brown, G.J., & Cooke, M. Perceptual grouping of musical sounds : a computational model. Journal of New Music research (Netherlands), 1994, 23 (2), 107-132.
Cadoz, C., Luciani, A., & Florens, J.L. CORDIS-ANIMA: a modeling and simulation system for sound and image synthesis - the general formalism. Computer Music Journal, 1993, 17 (1), 19-29.
Carterette, E.C. & Friedman, M.P., Handbook of Perception, vol. IV: Hearing, Academic Press, 1978.
Charbonneau, G., & Risset, J.C., Circularité de hauteur sonore. C.R. Acad. Sci. Paris, Série D, 1973, 281, 163-166.
Charbonneau, G., & Risset, J.C., Différences entre oreille droite et oreille gauche pour la perception de la hauteur des sons. C.R. Acad. Sci. Paris, Série D, 1975, 281, 163-166.
Chowning, J.M., The simulation of moving sound sources, Journal of the Audio Engineering Society, 1971, 19, pp. 2-6.
Chowning, J.M., The synthesis of complex audio spectra by means of frequency modulation, Journal of the Audio Engineering Soc., 1973, 21, 526-534 (réédité dans Roads & Strawn, 1985).

Chowning, J.M., Computer synthesis of the singing voice, in Sound generation in winds, strings, computers, Royal Swedish Academy of Music, Stockholm, 1980, 4- 13.
Deutsch, D., editor, 1999, The psychology of music, Academic Press.
Gibson, J.J. The senses considered as perceptual systems. Boston, Houghton Mifflin, 1966.
Grey, J.M. An exploration of musical timbre, Doctoral Dissertation, Stanford, Calif., 1975.
Guttman, N., & Pruzansky, S., Lower limits of pitch and musical speech, J. of Speech and Hearing Research ;1962, 5, 207-214.
Kubovy, M., & Pomerantz, J.R., editors, Perceptual organization, Erlbaum, Hillsdale, N.J., 1981
Lewis, E.R., Speculations about noise and the evolution of vertebrate hearing, Hearing Research, 25,1987, pp. 83-90.
Bigand, E, ed. Thinking in sound: the cognitive psychology of human audition. Oxford: Clarendon Press, 1993.
McAdams, S., & Bregman, A.S., Hearing musical streams, Computer Music Journal, 1979, 3 (4), pp. 26-43.
Marr, D., Vision, Freeman, San Francisco, 1982.
Mathews, M.V., The technology of computer music, M.I.T. Press, Cambridge, Mass.,1969.
Mathews, M.V., & Kohut, J.,, Electronic simulation of violin resonances, Journ. of the Acoust. Soc. of America 53, 1620-1626 1973.
Mathews, M.V., & Pierce, J.R., ed. Current Directions in Computer Music Research (with a compact disk of sound examples), M.I.T. Press, Cambridge, Mass., 1989.
Music Perception, Univ. of California Press, 1984, 1 (3) (dedicated to Helmholtz).
Nilssone, A., & Sundberg, J., Differences in ability of musicians and nonmusicians to judge emotional state from the fundamental frequency of voice samples, Music perception, 1985, 2 (4), 507-516.
Pierce, J.R. The science of musical sound. San Francisco: Freeman/Scientific American (with sound examples on disk), 1983 .
Plomp, R., Aspects of tone sensation, Academic Press, 1976.
Risset, J. C., Sur certains aspects fonctionnels de l'audition, Annales des Télécommunications, 1968, 23, 91-120.
Risset, J. C. , An introductory catalog of computer synthesized sounds (with sound examples), Bell Laboratories, Murray Hill, N.J., 1969. Reprinted in Wergo 2033-2: The historical CD of digital sound synthesis, 1969, pp. 88-254.
Risset, J. C., Paradoxes de hauteur: le concept de hauteur sonore n'est pas le méme pour tout le monde, Actes du 7e Congrès International d'Acoustique, Budapest, 1971.
Risset, J. C., Sons, Encyclopedia Universalis 13, pp. 168-171 (également dans la nouvelle édition), 1973.
Risset, J. C., Musical Acoustics, in E.C. Carterette & M.P. Friedman, 1978, Handbook of Perception, vol. IV: Hearing, Academic Press, 1978, 521-563.
Risset, J. C., Pitch and rhythm paradoxes: Comments on "Auditory paradox based on a fractal waveform", Journal of the Acoustic Society of America, 1986, 80, pp. 961-962.
Risset, J. C., & Mathews, M. V. Analysis of musical instrument tones. Physics Today, 1969, 22, No. 2, 23-30.
Risset, J. C. & Wessel, D.L., Exploration of timbre by analysis and synthesis, in Deutsch, The Psychology of Music, Academic Press, , 1999, 113-169.
Schlaug, G., Jäncke, L. Huang, Y., Staiger, J.F., Steinmetz, U. Increased corpus callosum size in musicians. Neuropsychologia, 1995, 33, 1047-1055.

Sethares, W.A., Tuning, timbre, spectrum, scale, Springer-Verlag, London (with a compact disc of sound examples), 1998.

Shepard, R.N.,, Circularity of relative pitch, Journ. of the Acoust. Soc. of America 1964, 36, 2346-2353.

Shepard, R.N., Psychophysical complementarity. In M. Kubovy & J.R. Pomerantz, Perceptual organization, Erlbaum Ass., Hillsdale, N.J., 1981.

Slaney, M., Lyon, R., & Naar, D. Auditory model inversion for sound separation, Proceedings of 1994 ICASSP, Adelaide, Australia, 1994, vol. II, 77-80.

Sloboda, J., The musical mind - the cognitive psychology of music, Clarendon Press, Oxford, England, 1985.

Sloboda, J., J. W. Davidson, M.J.A. Howe, & D.G. Moore, The role of practice in the development of expert musical performance. British Journal of Psychology, 1996, 87, 287-309.

Sundberg, J., Nord, L., Carlsson, R. Music, language, speech and brain, McMillan, 1991.

Terhardt, E., Calculating virtual pitch, Hearing research, 1979, 1, 155-182.

Van Noorden, L., Temporal coherence in the perception of tone sequences. Eindhoven, Holland: Instituut voor Perceptie Onderzoek, 1975.

Wessel, D.L. Timbre space as a musical control structure, Computer Music Journal, 1979, 3 (2), 45-52.

Wessel, D.L. & Risset, J.C., Les illusions auditives, Universalia (Encyclopedia Universalis), 1979, 167-171.

Description of sound examples

Sound example # 1 (10 s)

First, two melodic intervals which are both perceived as a fifth. The component frequencies are 440 Hz/660 Hz (A/E) and 2500 Hz/ 3750 Hz (D#/A#)- a ratio of 3/2 between the first and second tone. Then, again two melodic intervals : while the first one is still perceived as a fifth, the second one corresponds to an upward jump, but it is hard to describe it in comparison to a fifth, a third or any other musical interval. The component frequencies here are 880 Hz/1320 Hz (A/E) and 5000/7500 Hz - still a ratio of 3/2 between the first and second tone : however neither the two last pitches nor the last interval can be identified. Above 5000 Hz, it becomes impossible to appreciate pitch intervals with precision.

Sound example # 2 (23 s)

First, one hears a periodic tone of frequency 55 Hz, with 10 harmonics of equal amplitude. This is hears as a sustained low A. Then one hears the result of mixing this tone with seven similar tones of neighbouring frequencies : 55 + 1/20 Hz, 55 + 2/20 Hz, 55 + 3/20 Hz, ... , 55 + 7/20 Hz. Although one would expect to hear either a kind of cluster or a very similar tone, since the frequency difference is so small - corresponding to a 1/60th of a semi-tone - one hears a pattern of harmonics waxing and waning at different rates.

Sound example # 3 (6 s)

The four tones heard are synthetic copies of a brass instrument from a recipe which specifies a single frequency spectrum, an attack transient and a decay. Tones synthesized from this description (that of a trumpet as found in a classic Musical Acoustics treatise) do not evoke the instrument : hence the description is not aurally relevant.

Sound example # 4 (11 s)

One hears various brass-like tones (the last one, somewhat tuba-like, together with a cymbal-like sound) synthesized at Bell Laboratories in 1964 and 1965 from the characteristic law : the louder the tone, the richer the spectrum.

Sound example # 5 (5 s)

This is a stylized evocation of bowed-string tones (separated by a soft bassoon-like tone), obtained with the MusicV program by making the spectrum fluctuate in synchrony with the pitch.

Sound example # 6 (14 s)

Two overlapping tones are presented. Initially their identity is unclear. Introducing an irregular "vibrato" (frequency modulation) endows the tones with a vocal quality : they can then evoke a male voice (for the first tone) and a female voice (for the second tone).

Sound example # 7 (29 s)

This presents several sounds evoking percussion. The synthesis processes are described in examples # 400, 410, 411, 420, and 440 of the computer sound catalog (Risset, 1969). The sounds may give hints on the way the sounds were produced mechanically - percussion on a membrane with varying tension, on a skin surface with or without snares, on metal ... However all sounds were synthesized by additive synthesis - with no actual matter or percussion involved in the process.

Sound example # 8 (11 s)

One hears a soprano (Irène Jarsky) sing a high pitch (B flat). A carefully contrived bell-like tones begins on the same pitch, but this tone has ambiguous pitch with a lower resonance which is echoed by a second tone (D) from the soprano.

Sound example # 9 (23 s)

One first hears a set of bells, each of which is composed like a chord, with a prescribed inner harmony. Then one hears fluid textures with the same harmonies : each bell has been turned into a fluid tone by turning the percussive envelope modulating the amplitude of each frequency component into a smooth bell-shaped curve.

Sound example # 10 (7 s)

Two tones are presented. Although most listener judge the second one lower in pitch than the first one (by about a semi-tone), the second tone is obtained from the first one by doubling the component frequencies. The frequencies of the components are separated by a little more than an octave (frequencies in Hz such as 49.6, 102.4, 211.2, 435.2, 896, 1843.2, 3788.8). The amount of octave stretching is not critical : if the stretching is very small, it will yield a descent of a smaller pitch interval, but with tones which sound very ordinary and "normal".

Sound example # 11 (23 s)

Two patterns of beats, A and B, are compared - one hears A-B, then again A-B. Most listeners judge B a little slower than A. However B is obtained from A by doubling the speed of THE tape recorder - doubling all rates as well as all frequencies.

Sound example # 12 (46 s)

This sound glides downwards. However the physical parameters go back to their original value after a descent of one octave. Hence the sound is cyclic, it could be pursued as long as one wishes : pitch goes down indefinitely. Only the so-called tonal pitch varies.

Sound example # 13 (5 s)

This sound glides upwards, without octave jumps, while staying a B flat. Only the so-called spectral pitch varies.

Sound example # 14 (19 s)

This sound glides down the scale : however it is much higher in pitch at the end of the example. Tonal pitch and spectral pitch vary in opposite directions.

Sound example # 15 (1 mn 04 s)

In this cyclic example, the tones seem to go down indefinitely and to slow down indefinitely.

Sound example # 16 (41 s)

Here, the pitch goes up the scale, but it is much lower at the end; the beat seems to speed up, but it is much slower at the end.

Sound example # 17 (30 s)

A singing voice (that of soprano Irène Jarsky) recorded in a relatively dead environment. Then strong echoes are added, suggesting extension in space.

Sound example # 18 (5 s)

A speaking voice (saying twice "Ils se meuvent hors de l'espace") modified so as to give the impression of spatial movement.

Sound example # 19   (38 s)

Here, the pitch goes down the scale, but it is much faster at the end; the beat seems to slow down, but it is much faster at the end. In addition, the sound source seems to rotate in space - this apparent rotation is faster at the end of the example.

Sound example # 20 (46 s)

Above a sustained low tone, a high melodic line seems to rotate in space. (This is an excerpt of the author's composition *Songes*).

Chapter III.2

Primitive Intelligence at the Sensory Level in Audition

Risto Näätänen[1] and Mari Tervaniemi

Cognitive Brain Research Unit, Department of Psychology, University of Helsinki

In many information-processing models, the role of the preattentive auditory processes was restricted to feature analysis and to the transient storing of sensory data. Subsequent processing was thought to require focused attention engaging limited centrally allocated capacities. Thus, one could expect from this description of auditory stimulus processing that the processes organizing complex auditory scenes evident in our perception might occur beyond the sensory brain areas. Although models of later processing stages postulate sensory-specific subsystems within the centrally managed resources (e.g., Baddeley, 1986), many of these models (e.g., Treisman, 1982) still maintain the requirement of attention for all but the most elementary sensory-processing functions. In recent years, however, some researchers suggested that pre-attentive processes in audition might have a much more extensive role than previously assumed (Bregman, 1990, Näätänen, 1992). In this more recent view, the auditory scene is pre-attentively monitored, with its structure and regularities being maintained in a neural model representing the current event in the context of the immediate past independently of the direction of focused attention (Näätänen & Winkler, 1999).

Neurophysiological evidence supporting this hypothesis largely originates from investigations using the mismatch negativity (MMN; Näätänen et al. 1978; for a review, see Näätänen 1995). This component of the auditory event-related potential (ERP) is elicited by any infrequent discriminable change in some repetitive aspect of auditory stimulation (for illustrations, see Tervaniemi, this volume). Recent results show, however, that MMN is also elicited by violations of complex auditory regularities. This suggests that MMN is based on the neural traces of the well-structured auditory scene, being elicited when an incoming stimulus violates of these complex regularities. Consequently, in addition to reflecting sound discrimination, the MMN also indexes a number of higher cognitive processes occurring at the sensory level in audition. This is remarkable in particular in view of the fact that MMN is elicited even in the absence of attention. For example, MMN has been recorded in unconscious coma patients a few days before

---

[1] Correspondence: Prof. Risto Näätänen, Cognitive Brain Research Unit, Department of Psychology, P.O. Box 13 FIN-00014, University of Helsinki, Finland, Tel: (358)-9-1912-3445, Fax: (358)-9-1912-2924, e-mail: risto.naatanen@helsinki.fi

the recovery of their consciousness (Kane et al. 1993). Therefore, MMN can indeed be used to probe the auditory scene at the pre-attentive level.

The main neural generators of the MMN are located in the supratemporal plane (for a review, see Alho 1995). These generators loci differ between simple and complex musical sounds (Alho et al., 1996) and between phonetic and musical sounds (Tervaniemi et al., 1999)(for a review, see Tervaniemi, this volume).

The purpose of the present article is to review the different types of cognitive achievements of central auditory processing which occur independently of the subject's attentional focus. They are suggested to be as follows:

Auditory stream formation and segregation. In everyday situations, acoustic information from several sound sources arrives at the two ears simultaneously or nearly simultaneously. An important function of the central auditory system, therefore, is to segregate the concurrent sounds from each other and to attribute them to their original sources. This segregation occurs in the basis of, for example, the pitch and the location of the origin of the sounds, or the timbral characteristics of a speaker's voice. The process of grouping those sounds that belong together and segregating those that derive from different sources was termed auditory stream segregation by Bregman (1990).

Auditory stream segregation takes place, at least mainly, in the auditory cortex in the early, pre-attentive stages of auditory processing. This was demonstrated by Sussman et al. (1999) who, using the MMN as the measure of cortical change detection, showed that the segregation of high and low tones to separate streams can occur without attention directed to the sounds. The authors manipulated the high and low sounds so that the separate regularities could only be detected after the tones were segregated to two separate streams. MMN was then elicited separately by deviants within each stream but not when all tones were perceived as forming a single stream. It was, therefore, concluded that auditory stream segregation occurs in an early preattentive stage of auditory processing preceding the stage of (automatic) MMN generation (see also Shinozaki et al., 2000).

Temporal and sequential auditory grouping. Sussman et al. (1998) recently demonstrated pre-attentive auditory grouping within a single sound source. In separate stimulus blocks, they presented a repetitive sequence of 5 tones (AAAAB) at a fast (SOA 100 ms) or slow (SOA 1.3 sec) pace to subjects ignoring the tones. MMN was elicited by the infrequent B tone when the SOA was long but not when it was short. This suggests that the 5 tones were preattentively grouped together as a single repeating pattern when presented at the fast rate and therefore the B tone became part of the regularity (or standard) represented by the sensory-memory traces formed, and, therefore, no MMN was elicited. In contrast, with the slow presentation rate, the perceptual unit of stimulation was a single tone and therefore the B tone was detected by the sensory-memory system as a deviant with respect to the repeating single A tone.

Short-term extrapolatory sound traces. Tervaniemi et al. (1994) showed that an MMN can also be elicited when an unattended tone sequence, regularly descending in pitch, is interrupted by an ascending tone (back to the preceding tone frequency), and even by a tone repetition. These data suggest that the auditory cortex may form extrapolatory traces on the basis of the regularities or trends detected in the auditory past. These traces represent auditory events of the immediate future as such as they should occur on the basis of the trend or regularity detected (Näätänen 1992). A corresponding negativity to an abrupt deviation in sound location change was recorded by Winkler et al. (in preparation), who presented reading subjects with a regular virtual sound movement (produced by free-field stimulation).

Discrimination of short-term abstract sound patterns. Saarinen et al. (1992) presented tone pairs to their subjects reading a book. The standard pairs were such that the second tone was higher in frequency than the first tone (i.e., ascending pairs), whereas the deviant pairs were descending ones. The first tones of all tone pairs, standards and deviants alike, randomly occurred at 5 different frequency levels, there thus being no physically identical repeating standard stimulus (tone pair). Instead, the common standard-stimulus feature was a higher-order ("abstract") one, the *direction* of the tone pair. An MMN was nevertheless elicited by the descending tone pairs, indicating that the automatically formed memory traces represent no particular tone pair with certain fixed acoustic parameters but rather the abstraction "ascending pair". In addition to the direction of the tone interval, also the interval size is automatically encoded in the auditory cortex (Paavilainen et al., 1999). This was evidenced by the MMN elicitation by smaller or larger intervals among minor seventh standard intervals (all intervals were randomly presented at 10 frequency levels). Furthermore, the cortical interval-representation may also be formed on the basis of converging information from both ears as indicated by the MMN elicitation when the two successive sounds of an interval are delivered to separate ears (Paavilainen et al., 1998).

Most importantly, Paavilainen et al. (1995) demonstrated that these results indeed represented achievements of preattentive processing rather than those of post-perceptual cognitive operations. The subject was instructed to detect deviant pairs within the input to one ear while stimulus-pair sequences were concurrently presented to each ear. An MMN was elicited by deviant pairs even in the ignored ear, however only when this was the right ear, suggesting that it might be the left hemisphere that is specialized in performing such a pattern-change detection task.

Long-term sound traces. Importantly, the MMN does not only reflect short-term sensory traces but also long-term ones. Using a complex spectro-temporal stimulus pattern with occasional frequency deviations in one of its 8 segments, Näätänen et al. (1993) found no MMN in several of their subjects in an ignore condition in the beginning of the experimental session. Consistent with this, in the discrimination task, these subjects´ performance was at the chance level. In the subsequent passive condition, however, some MMN could be recorded and now the subject showed improved performance in the subsequent discrimination task. The development of this training effect depended on attention (discrimination task), however, as passive repetition lasting even hours resulted in no long-term effects.

This learning result appears to indicate a gradual refinement or sharpening of the informational contents of the sound traces developed during the experiment: the better is the informational quality of the representation, the easier it is to discriminate stimuli deviating from it (Näätänen & Alho, 1997). These long-lasting traces probably serve as recognition traces for the corresponding stimuli in auditory perception, explaining, for instance, the fact that we can immediately recognize a large number of different speakers in the phone.

Long-term abstract sound-pattern traces. The existence of language-specific memory traces was demonstrated by Näätänen et al. (1997). They found that Finnish subjects' MMN to an occasional replacement of the vowel /e/ shared by the Finnish and Estonian languages was larger when the deviant sound was a vowel in Finnish (/ö/) than when it was not but was, instead, a vowel in Estonian (/õ/). In contrast, both deviant sounds elicited an enhanced MMN in Estonian subjects for whom both of these deviant sounds were vowels. Magnetoencephalographic (MEG) recordings of the magnetic equivalent (MMNm) of MMN located the origin of this enhanced MMN of Finnish subjects to the Finnish vowel /ö/ to the left auditory cortex. A considerably smaller MMN was generated in parallel in the right auditory cortex. A very small MMN was also generated in the auditory cortices of both hemispheres when the Estonian /õ/ was the deviant stimulus. These small MMNs were obviously elicited because of the (mere) acoustic deviance. The authors proposed that the development of such traces is a necessary prerequisite for the correct perception of speech, i.e., that these traces serve as pattern-recognition traces for the corresponding phonemes of the spoken language. More recently, Cheour et al. (1998) found evidence for learning of native-language phoneme categories to take place between 6 and 12 months of age.

Speech-sound recognition cannot, of course, be based on the representations of acoustically constant, or nearly constant, stimuli, as different speakers have acoustically very different voices (e.g., in pitch and timbre) and as phonemes also vary acoustically according to the word context (allophonic variation). Therefore, phoneme (syllable and word) recognition must be based on such long-term traces that can accommodate, or normalize, this variation. Consequently, these traces represent no acoustic sound feature as such but rather some invariant code which is shared by different speakers (and word contexts) when the same phoneme is perceived as the same irrespective of wide acoustic variation.

The human brain's ability to automatically detect a pattern change in an acoustically varying stimulation was demonstrated with non-linguistic stimuli, in addition to the Saarinen et al. (1992) study reviewed above. Also Tervaniemi et al. (submitted) found that a short behavioral task facilitated automatic pattern-change detection in a group of musicians. This was indicated by MMN elicited by a frequency change in 5-tone frequency patterns randomly occurring at 12 different frequency levels, the subjects' attention being directed elsewhere. In addition, Paavilainen et al.'s (1999) subjects showed an MMN to changes in the frequency ratio between two consecutive tones, the first randomly occurring at 12 different frequency levels. Very importantly, this MMN to ratio violation was also obtained when the two tones were presented in parallel, forming a complex tone. This result demonstrates the existence of such neuronal populations that might subserve phoneme perception in the presence of wide acoustic variation (see Näätänen in press).

Table 1
Perceptual-cognitive functions of the central auditory system

PROCESS OUTCOME	FUNCTION
SHORT-TERM SENSORY-MEMORY TRACE OF A SOUND (CENTRAL SOUND REPRESENTATION)	SOUND PERCEPTION AND SHORT-TERM SENSORY MEMORY
AUDITORY CHANGE SIGNAL	AUDITORY CHANGE DETECTION
SORTING OF MULTI-CHANNEL AUDITORY INPUT INTO SOURCES	AUDITORY STREAM FORMATION AND SEGREGATION
TEMPORAL/SEQUENTIAL AUDITORY GROUPING	SOUND-OBJECT FORMATION
SHORT-TERM EXPLORATORY SOUND TRACE	PREATTENTIVE SOUND ANTICIPATION
SHORT-TERM ABSTRACT SOUND-PATTERN TRACE	DISCRIMINATION OF SHORT-TERM ABSTRACT SOUND PATTERNS
LONG-TERM SOUND TRACE	LONG-TERM SOUND RECOGNITION AND DISCRIMINATION (e.g., FAMILIAR VOICES AND ENVIRONMENTAL SOUNDS)
LONG-TERM ABSTRACT SOUND-PATTERN TRACE	SOUND-CATEGORY RECOGNITION AND DISCRIMINATION (e.g., SPEECH PERCEPTION)
REPRESENTATION OF RELATION BETWEEN TWO ATTRIBUTES	DETECTION OF RELATION BETWEEN TWO ATTRIBUTES

References

Alho K. Cortical generators of mismatch negativity (MMN) and its magnetic counterpart (MMNm)elicited by sound changes. *Ear Hear* 1995; 95:93-96.

Alho K, Tervaniemi M, Huotilainen M, Lavikainen J, Tiitinen H, Ilmoniemi RJ, Knuutila J, Näätänen R. Processing of complex sounds in the human auditory cortex as revealed by magnetic brain responses. *Psychophysiology* 1996; 33: 369–375.

Baddeley A. *Working memory.* Oxford University Press, Oxford, England, 1986.

Bregman AS. *Auditory scene analysis: The perceptual organization of sound.* MIT press, Cambridge, MA, 1990.

Cheour M, Caponiene R, Lehtokoski A, Luuk A, Allik J, Alho K, Näätänen R. Development of language spesific phoneme representations in the infant brain. *Nature Neurosci* 1998; 1: 351-354.

Kane NM, Curry SH, Butler SR, Cummins BH. Electrophysiological indicators of awakening from coma. *Lancet* 1993; 341: 688

Näätänen R. *Attention and brain function.* Erlbaum, Hillsdale, NJ, 1992.

Näätänen R. The Mismatch negativity: A powerful tool for cognitive neuroscience. Ear Hear 1995; 16: 6-18.

Näätänen R, Alho K. Higher-order processes in auditory change detection. *Trends in Cogn Neurosci* 1997; 1: 44-45.

Näätänen R, Gaillard AWK, Mäntysalo S. Early attention effect on evoked potential reinterpreted. *Acta Psychologica* 1978; 42: 313-329.

Näätänen R, Lehtokoski A, Lennes M, Cheour M, Huotilainen M, Iivonen A, Vainio M, Alku P, Iloniemi RJ, Luuk A, Allik J, Sinkkonen J, Alho K. Language-spesific phoneme representations revealed by electric and magnetic brain responses. *Nature* 1997; 385: 432-434.

Näätänen R, Schröger E, Karakas S, Tervaniemi M, Paavilainen P. Development of memory trace for a complex sound in the human brain. *NeuroReport* 1993; 4: 503-506.

Näätänen R, Winkler I. The concept of auditory stimulus representation in cognitive neuroscience. *Psychol Bul* 1999; 125: 826-859.

Paavilainen P, Jarmillo M, Näätänen R. Binaural information can converge in abstract memory traces. *Psychophysiology* 1998; 35: 483-487.

Paavilainen P, Jarmillo M, Näätänen R, Winkler I. Neuronal populations in the human brain extracting invariant relationships from acoustic variance. *Neurosci Lett* 1999; 265: 179-182.

Paavilainen P, Saarinen J, Tervaniemi M, Näätänen R. Mismatch negativity to changes in abstract sound features during dichotic listening. *J Psychophysiol* 1995; 19: 607-610.

Saarinen J. Paavilainen P, Schröger E. Tervaniemi M, Näätänen R. Representation of abstract stimulus attributes in human brain. *NeuroReport* 1992; 3: 1149-1151.

Shinozaki N, Yabe H, Sato Y, Sutoh T, Hiruma T, Nashida T, Kaneko S. Mismatch negativity (MMN) reveals sound grouping in the human brain. NeuroReport 2000; 11: 1597-1602.

Sussman E, Gomes H, Nousak JM, Ritter W, Vaughan HG Jr. Feature conjunctions and auditory sensory memory. *Brain Res* 1998; 18: 95-102.

Sussman E, Ritter W, Vaughan HG Jr. An investigation of the auditory streaming effect using event-related brain potentials. *Psychopsysiology* 1999; 36: 22-34.

Tervaniemi M, Kujala A, Alho K, Virtanen J, Ilmoniemi RJ, Näätänen R. Functional specialization of the human auditory cortex in processing phonetic and musical sounds: A magnetoencephalographic study. *NeuroImage* 1999; 9: 330–336.

Tervaniemi M, Maury S, Näätänen R. Neural representations of abstract stimulus features in the human brain as reflected by the mismatch negativity. *NeuroReport* 1994; 5: 844-846

Tervaniemi M, Schröger E, Saher M, Näätänen R. (submitted).Effects of spectral complexity and sound duration on automatic complex-sound pitch processing in humans – a mismatch negativity.

Tervaniemi M. (submitted). Automatic processing of musical information as evidenced by EEG and MEG recordings. Proceeding of the International Symposium for Neuroscience of Music. Niigata, Japan, Oct 15-17, 1999.

Treisman AM. Perceptual grouping and attention in visual search for features and for objects. *J Exp Psychol: Human Percept Perf* 1982; 8: 194-214.

Chapter III.3

Grouping and Differentiation
Two Main Principles in the Performance of Music

Johan Sundberg[1]

Department of Speech, Music and Hearing, KTH

Introduction

Music receives important characteristics from the performers. This is evident from many observations, e.g. the notorious experience that musically quite pathologic performances emerge, when the musician is replaced by a computer that *verbatim* converts the score into the corresponding sound sequences. The belief that this should be at all possible is remarkable, as it assumes that the "real" music is the one defined by the score, thus failing to realize that the "real" music is the sounding music.

In any event, the performers' contributions are implicitly well known in terms such as articulation, phrasing, and the performers' commitment and musicality. Explicit descriptions of what these terms mean have started to emerge from music performance research. The purpose of the present paper is to review some results from our own research in this area and to discuss them from the point of view of music communication.

Method

During the last decades much research has been devoted to music performance (Gabrielsson, 1999). While analysis-by-measurements has been the most frequently used method, we have used the analysis-by-synthesis strategy, illustrated in Figure 1. It implies that the computer reads the score, converted into a music file, and performs it on a synthesizer. In this conversion a set of context-dependent performance rules introduce deviations from the music score, in terms of changes of duration, amplitude and vibrato characteristics of tones, and insertion of accents and micropauses.

The basic strategy of analysis-by-synthesis is the testing of hypotheses. Our basic

---
[1] Correspondence: Johan Sundberg, Department of Speech, Music and Hearing, KTH, Stockholm, Sweden, Tel: (46)-8-790-78-73, Fax: (46)-8-790-78-54, e-mail: pjohan@speech.kth.se

hypothesis has been that performers' deviations from the score are not random but can be accounted for in terms of context dependent rules. Furthermore, hypothetical rules are formulated which are tested by applying them to musical examples. No occasional exceptions to rules are accepted.

In formulating or modifying each hypothetical rule we have been guided by comments and recommendations of an expert listener, professor Lars Frydén, professional violinist and teacher of performance. The experimental situation is similar to the classical teacher-pupil interaction in educating musicians; when the teacher has listened to the student's rendering of a piece, he instructs her/him how to change the performance in order to improve it.

The outcome of the research procedure is a performance rule system, *Director Musices*. It can be said to constitute a generative grammar of music performance in the sense that it automatically generates performances according to a set of well-defined rules, thereby exhaustively defining the rule contexts and describing how they modify the acoustic shape of the performance. On the other hand, it is not strictly generative in the classical sense, as it does not pretend to define *all possible* professional performances of music.

Detailed accounts of our performance grammar have been published elsewhere (see e. g., Friberg 1995a). The rules contain two aspects, one defining the context of the *target* tones, to which the rule will be applied, and one defining the *quantity* of the effect (e. g., the amount of tone lengthening) that the rule will induce on its target tone. Thus, the magnitude of the effect of a specific rule can be continuously varied from nil to wildly exaggerated. By choosing a negative value for the quantity parameter, a rule can be inverted, such that, e.g., a lengthening is replaced by a shortening of the target notes.

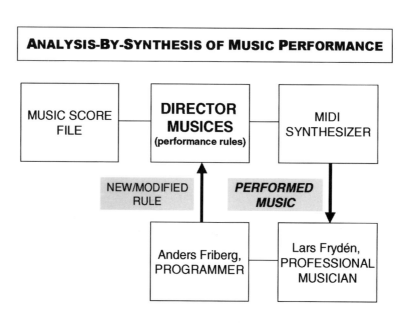

Figure 1
Schematic illustration of the analysis-by-synthesis strategy.

Rules

The rules can be grouped into different categories depending on the function that they appear to serve in music communication, Figure 2. One group of rules seems to *increase the differences between different tone categories or classes*, such as note values, pitches, and intervals. Another type of rules seems to *enhance the grouping of tones that belong together*, such as melodical gestures, subphrases, and phrases. A third type of rules organizes *tuning and timing within ensembles* and a fourth type seems to merely reflect characteristics of *playing dialects* or *playing styles*. Some rules appear to serve more than one of these purposes.

Differentiation rules

There are two types of differentiation rules. One seems one related to the music listener's categorical perception of pitch and tone duration, and the other to listeners' expectancies.

Duration contrast increases the difference between long and short notes by shortening the short tones and reducing their sound level, Figure 3 (sound example 1). *Duration contrast articulation* can be used for inserting micropauses after tones, the duration of which is inversely proportional to tone's duration. This rule is particularly useful in some types of Baroque music.

High – Loud and *High – Sharp* increase the difference between high and low pitches by sharpening the intonation of high tones and by increasing their sound level.

Melodic Charge is applicable to tonal music. It increases the difference between the various tones of the scale. It adds duration and sound level and, if applicable, also increases the vibrato extent of a tone to an extent proportional to its distance, along the circle of fifths, from the root of the prevailing chord. The melodic charge values of scale tones are shown in Figure 4a and the effect of the rule on an example is illustrated in Figure 4b (sound example 2).

The rule also sharpens the intonation during the first part of tones to an extent depending on the tone's melodic charge; tones falling along the right side of the circle of fifths are sharpened in proportion to their melodic charge, and similarly tones located on its left side are flattened, Figure 5. This tuning according to the tone's melodic charge has bearings in the differentiation of major and minor melodic intervals; major seconds, thirds, sixths and septimas are slightly widened and their minor variants are narrowed. This can be seen as an application of the differentiation principle.

Harmonic Charge is a rule that is applicable to tonal music, enhancing the difference between chords. It is defined as the weighted sum of the chord tones' melodic charge computed with the root of the tonic in the "noon" position, see Figure 6a. The rule introduces a crescendo combined with a rallentando when a chord with a greater harmonic charge is approaching and vice versa, Figure 6b (sound example 3).

Emphasis rules seem to respond to listeners' expectancies. These expectancies depend on the musical context. For example, given the harmonic environment of a C major chord, the pitch of C or G are expectable, while tones like D# and C# are unexpected, or remarkable. Obviously the *Melodic charge* rule responds to this expectancy. Also other rules seem to serve this same purpose. *Harmonic Charge* has a similar function, although in the harmonic domain.

Director Musices performance rules

Differentiation rules	Elements		Effect
Duration-contrast	dr sl	Note values	The longer, the longer and louder and vice versa
Duration-contrast articulation	Dro	Note values	The shorter note the longer micropause
High-loud	Sl	Pitch classes	The higher the pitch, the louder
High-sharp	F0	Pitch classes	The higher pitch, the sharper
Melodic charge	sl dr va	Pitch classes	Emphasizes scale tones remote from the current chord
Harmonic charge	sl dr	Chord classes	Emphasizes chords remote from the current key
Chromatic-charge	dr sl	Emphasis	Emphasizes note sequences close in pitch in atonal music

Grouping rules

Inegales	dr	Verse feet	Long-short patterns of consecutive eighth notes (swing)
Faster-uphill	dr	Gesture	Shortens notes in ascending melodic motion
Leap-tone-duration	dr	Gesture	Shortens note initiating ascending leap; lengthens target note in descending leap
Leap-articulation	dro	Gesture	Inserts micropauses in leaps
Punctuation	dr dro	Gesture	Identifies musical gestures, lengthens their final note and adds a micropause
Phrase-articulation	dro dr	Phrase	Adds micropause at phrase and subphrase boundaries; lengthens phrase final note
Phrase-arch	dr sl	Phrase	Arch-like tempo and sound level curve for phrases: starting slow/soft, then faster/louder, and slowing/diminuendo towards end
Final-ritard	dr	Piece	Inserts ritardando modeled from stopping runners at the end of the piece

Ensemble rules

Ensemble timing	dr	Ensemble	Creates a common time table in ensembles
Ensemble-swing	dr	Verse feet	Models timing and swing ratios proportional to tempo in ensembles
Ensemble tuning	dr	Ensemble	Creates a common tuning strategy in ensembles
Mixed-intonation	F0	Interval classes	Combines melodic and harmonic intonation

Other

Repetition-articulation-dro	dro	Technicality	Inserts micropause in tone repetition
Social- care-duration	dr	Technicality	Increases duration of extremely short notes
Double-duration	dr	Dialect	Decrease duration contrast for notes with duration ratio 2:1
Offbeat sound level	Sl swing	Dialect	Increase sound level at offbeats

Figure 2
Rules contained in the Director Musices generative performance grammar.

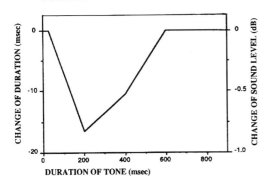

Figure 3
Relation between nominal tone duration and shortening according to the Duration contrast rule.

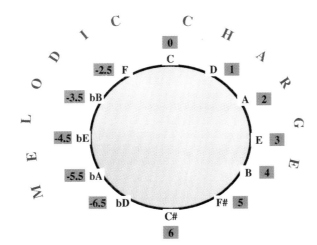

Figure 4
a. Melodic charge values for the various scale tones in the context of a C major chord. b. Effects of the Melodic Charge rule on the tone durations, sound levels and vibrato extent when applied to the theme of the first Kyrie in J S Bach's b minor mass.

FINE TUNING

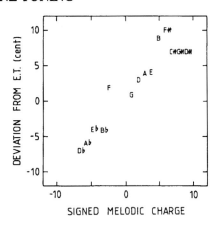

Figure 5
Effect of the Melodic Charge rule on fine tuning.

a.

HARMONIC CHARGE OF CHORDS

Tonic (C E G)	0
Dominant (G B D)	2.0
Subdominant (F A C)	1.6
Double dominant (D F# A)	4.0
Double subdominant (Bb D F)	2.7
Relative of Tonic (A C E)	1.3
Relative of Dominant (E G B)	5.0
Relative of Subdominant (D F A)	1.7
Dominant of Relative of Tonic (E G# B)	6.5
Dominant of Relative of Dominant (B D# F#)	7.0
Dominant of Relative of Subdominant (A C# E)	5.4

b.

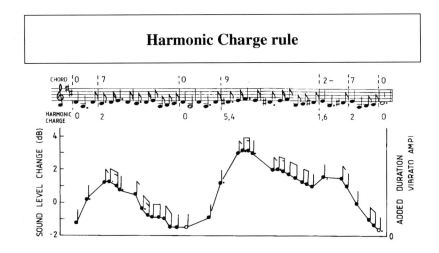

Figure 6
a. Harmonic charge values for some chords. The examples refer to a C major tonality. Harmonic charge is defined as a weighted sum of the chord tones' melodic charge values, computed with the root of the tonic in the "noon" position. **b.** Effects of the Harmonic Charge rule when applied to the second theme of the first movement of Franz Schubert's Unfinished symphony.

Chromatic charge is a rule applicable to atonal music. After all notes have been transposed into the same octave, the chromatic charge is computed as the inverted running mean distance, in semitones, of five consecutive notes. According to the rule an increase of chromatic charge is accompanied by crescendo and rallentando, and vice versa.

In singing, also linguistic factors in texts may affect an attentive listener's expectancies substantially. Text words that turn the linguistic content and/or change the emotional ambiance are said to be focused. Such words can be said to be unexpected within their linguistic context and tend to be emphasized in expressive performances (Sundberg, 1999). While in the rule system pitch and harmony events that are remarkable for musical reasons are identified and emphasized automatically, events that are remarkable for linguistic reasons have to be marked manually.

Grouping rules

Inégales induces a long – short pattern in sequences of eighth notes by transporting some duration from the note in stressed position to the subsequent note in unstressed position.

Faster uphill shortens tones that form an ascending melodic sequence and thus tends to group them into melodical gestures.

Leap tone duration shortens tones initiating an ascending melodic leap and lengthens the target note of a descending leap. Its effect is comparable to the one of *Faster uphill* in that it packs tones that belong together more densely in time.

Leap articulation introduces micropauses in leaps, the duration of which increases with the size of the leap. In music where leaps are used to mark boundaries between melodic gestures or other structural elements, the rule has the effect of grouping these elements. In cases of split melodic lines it differentiates the two lines.

Punctuation (Friberg & al., 1998) introduces micropauses at the end of melodical gestures, i. e., small groups of tones, mostly less than 7, that are perceived as belonging together. More specifically, the final part of the final tone of such groups is replaced by a short pause. The rule automatically identifies melodical gestures by means of a set of subrules which first identify potential endings in terms of, e. g., appoggiatura, pitch jumps, and a disruption of the melodical line or a change of its direction. A second set of subrules makes a selection among these candidates. Finally, a third set of subrules insert micropauses after these endings and decides their magnitudes.

Phrase articulation (Friberg, Sundberg & Frydén, 1987) and *Phrase arch* (Todd, 1985; Friberg, 1995b) are rules that are both triggered by markings of phrase and subphrase endings. These markings are introduced manually in the music file. The former rule lengthens the final note of a subphrase and inserts a micropause after it. The latter rule adds an arch-like tempo curve by introducing an accelerando at the beginning of subphrases and a rallentando at the end, Figure 7 (sound examples 4a-d). The lengthening of the final note can be adjusted. The sound level is modeled accordingly, such that it varies in proportion to the tempo, slow tempo being performed softer.

Final ritard introduces a ritard toward the end of a piece. Its tempo curve was derived from the velocity decrease of stopping runners, Figure 8 (Friberg & Sundberg, 1999).

Ensemble rules

Timing and tuning within ensembles are handled by a special set of rules. Timing is taken care of by *Ensemble timing*; all rules that affect tone duration are applied to one single melodic line, the synchronization voice. In every moment this voice consists of the shortest note with the highest melodic charge appearing in the score at that moment. The durational rules are then applied to the synchronization voice, and onsets of other tones in the ensemble are synchronized with those of the synchronization voice. *Ensemble swing* is applicable to jazz and is similar to *Inegales*. The resulting effect is often referred to as swing. The duration of the short beat is about 100 ms but longer at tempi slower than 150 bpm.

Ensemble tuning and *Mixed intonation* apply the fine tuning of melodic intervals according to the melodic charge rule at tone onsets, so that the melodic intervals are tuned according to the data shown in Figure 5 above. For tones longer than 400 ms, however, the intonation starts, after 120 ms, to approach just tuning at a rate of 5 cent/s. In this way beats of mistuned consonant intervals are gradually reduced.

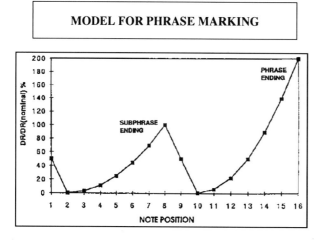

Figure 7
Effects of the rule Phrase Arch on the tempo, plotted as performed-to-nominal-duration ratio, when applied to the subphrase and phrase levels.

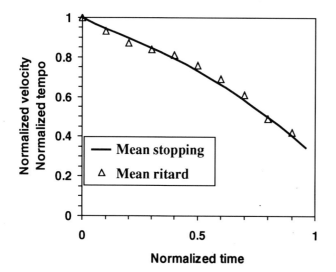

Figure 8
Comparison between the tempo decrease in final ritards in music performances according to Sundberg & Verrillo (1980) (triangles) and the mean decrease of velocity of stopping runners (solid curve). From Friberg & Sundberg (1999).

Playing dialects

Tone repetition introduces a micropause at the end of repeated tones. *Social duration care* is an output constraint ensuring that tones do not get overly shortened.

Double duration operates on sequences of 2:1 durational contexts. It takes some duration from the longer tone and adds it to the following short note. The effect of this rule is thus opposite to that of the *durational contrast* rule described above. *The shorter, the longer* is the inverse of *The shorter the shorter*, as it lengthens rather than shortens short notes and creates an espressivo effect in some types of music.

Off beat sound level adds some extra sound level to tones that appear in unstressed bar positions. This rule is sometimes useful in jazz styles

This group of rules is less coherent than the ones described above. It is possible that in the future their roles in music communication can be understood more accurately, in some cases possibly in relation to analogies between music and motion.

Verification

An obvious method of verifying the grammar is to compare data from real performances with those generated by the system. Figure 9 shows one example of such a comparison concerning the fine tuning of scale tones. Performance data were quoted from Garbusov's (1948) measurements on three violinists' performances of the same piece, and the predictions were based on the *Melodic Charge* rule. In the real performances the mean departures from the equally tempered tuning were similar, but clearly greater than those resulting from the performance grammar.

Our performance grammar assumes that performers' deviations from the music score can be described in terms of a set of rules; it seemed reasonable to first test the explanatory power of the simplest possible description. One way of testing this assumption is to find out to what extent different performances of the same piece can be generated by Director Musices. Figure 10 shows such a comparison of tone durations between real and rule generated performances of Robert Schumann's *Träumerei*. The rule-generated performances were obtained by iterating rule quantity values so as to match the given performances, however, keeping the same quantity for each rule throughout the piece (Friberg 1995b). As can be seen in the figure the agreement is, by and large, quite good. This indicates that the rules are capable of generating essential aspects of the durational deviations in different artistic performances of the same piece. For specific notes, however, substantial discrepancies do occur, although they are mostly limited to the amount, rather than to the direction of the deviation. One salient discrepancy is caused by the performers' lengthening of penultimate note in phrases. Interestingly, penultimate note lengthening does not seem to occur at the end of musical gestures; it may be relevant, that unlike the penultimate note of melodical gestures, phrase-final notes are mostly long already in the score (Drake, 1993).

The discrepancies between real and rule-generated performances demonstrate that the Director Musices grammar is as yet incomplete and/or that the quantity of the various rules is dependent on musical context, such that it is varied during the performance of a piece.

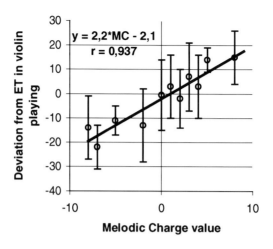

Figure 9
Comparison of departures from equally tempered tuning observed by Garbusov (1948) in three violinists' performances of the same piece and generated by Melodic Charge rule.

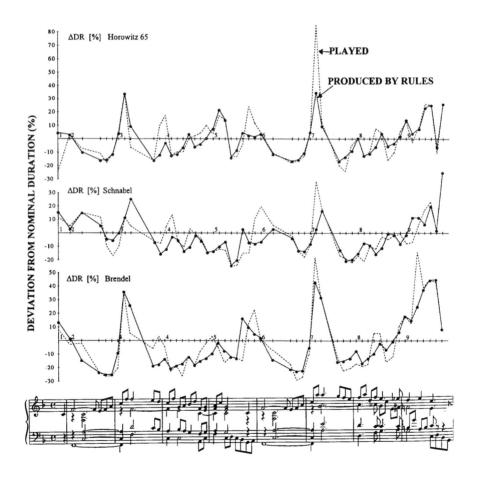

Figure 10
Deviations form nominal durations in three performances of an excerpt of Robert Schumann's Träumerei by three pianists (dotted curves, measurements by Repp, 1992) and produced by the Director Musices program by chosing different quantity values. (From Friberg, 1995b)

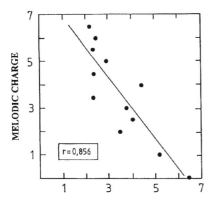

Figure 11
Relation between melodic charge and Krumhansl's (1990) mean probe tone ratings that reflect how well the different scale tones appear to serve as a continuation of a diatonic major scale.

Another method to test the performance grammar has been to ask professional musicians to adjust the magnitudes of various rules by repeated listening to a set of music examples, whereby zero magnitude, i. e., switching off the rule, was an available choice. The results from such experiments have shown that, on average, the musicians' preference was often close to one, two or three times the smallest perceptible effect of the rule (Friberg, 1995a). Thus, the musicians confirmed that most rules improved the quality of the performance of the examples tried in the test.

The Melodic Charge rule adds emphasis to remarkable tones by increasing their duration, loudness and vibrato extent. This remarkableness is calculated from the tone's distance, on the circle of fifths, from the root of the chord. As mentioned, Melodic Charge is likely to be related to a listener's expectancies, such that remarkable tones are unexpected. This suggests that Melodic Charge be related to the probe tone rating values that Carol Krumhansl obtained in experiments, where subjects were asked to rate the goodness of continuation of the different scale tones when presented after a complete scale (Krumhansl 1990). Figure 11 compares these probe tone ratings with the Melodic Charge values, showing a correlation of .856. This correlation is interesting, given the fact that probe tone ratings and melodic charge values were derived from entirely different experiments, viz. listening tests and analysis-by-synthesis of musical performance.

Limitations

Our performance grammar obviously has limitations. One is that the grammar is general rather than tailored to fit a specific instrument. Different versions of the grammar should be developed for different instruments so as to take advantage of the individual instrument's expressive potentials.

Another limitation is that almost all rules are triggered by a particular musical context as specified by the score, and in many cases this context is short, as pointed out by e.g., Oosten (1993) and Shaffer (1995). Furthermore, the rules react almost exclusively on music structure as given by the score, while in reality, a performer would also take other factors into account, such as the emotional characters of the piece. On the other hand, different performances of a piece can be obtained by varying the quantity parameter of the various rules, as demonstrated above. It remains a question for future research to explore the limit of this variability, e.g., to what extent different performance styles can be matched by the rule system.

The performance grammar is also capable of generating performances of differing emotional characters (Bresin & Friberg, 1998). Also here, the quantity parameter plays a key role. This implies that different emotional characters can be derived from the musical structure. It was, however, necessary to complement the score information by two factors normally not given explicitly in the score, namely, tempo and mean overall loudness. It does not seem principally excluded that, to some extent, also some of this information can be derived automatically from the score in the future.

A third limitation is that the rules are deterministic in the sense that each rule is applied with the same quantity throughout a piece, every time its context description is fulfilled. This does not appear as a realistic model of a performing musician. Rather, the quantity parameter would vary depending on factors not yet implemented in the grammar. For example, a player would hardly play a section of a piece in exactly the same way the second time it appears in a composition. To solve problems of these types, a system of meta-rules is needed that identifies larger musical structures and vary the quantity parameter of rules within a piece. Also, it seems likely that performers occasionally may cheat listeners' performance expectations, established from listening to other performances of the same piece.

Basic Performance Principles

Our results have suggested that much of musicians' deviations from the nominal description represented by the score can be explained as the application of two basic overall principles, differentiation and grouping. These deviations are often referred to as expressive deviations. Our results suggest that they are also meaningful, so the term *meaningful deviations* would be more appropriate.

As musicians apparently assist the listener in these tasks of differentiation and grouping, it seems reasonable to assume that they respond to global demands for music communication. These demands may originate either from musical tradition or from the human perceptual system.

Some support for the latter alternative is offered by the fact that the principles of differentiation and grouping can be found also in the other type of inter human communication by acoustic signals, viz. speech. The most apparent analogue is grouping. It is well known that the structure of a spoken sentence is reflected in prosody. Recently Fant and Kruckenberg (personal communication) found interesting examples of *symbolic pitch patterning;* certain pitch characteristics reflected the parsing of a sentence into different structural elements, such that speakers tended to mark boundaries between different structural elements by pitch shifts. Another example is the typical repetition of a specific pitch pattern for each member of a series of repetitions, as illustrated in Figure 12. Such parallels between music performance and speech prosody speaks to the paramount importance of marking structure in communication by acoustic signals.

Also the principle of differentiating categories can be observed in speech. It is well known that speech sounds are perceived categorically. In Swedish, most vowels appear in two versions, long and short. In addition to duration, being the primary differentiation cue, secondary differentiation cues are added in terms of different formant frequencies and, under some conditions, also the length of the following consonant. The intrinsic pitch of vowels adds pitch differences between some vowels, that would facilitate differentiation. Consonants is another example. The main characteristics of consonants are formant frequency patterns. However, in addition to this main cue, secondary cues are often introduced, e.g., in terms of pitch gestures. For example, pitch drops tend to accompany the pronunciation of voiced stops. Although such drops can be explained by acoustic/aerodynamic factors, they nevertheless should facilitate the differentiation between consonant categories.

The special case of differentiation between expected and unexpected events are obviously found very frequently in speech, where expectancies play an important role in communication. Indeed, speech is very hard to understand, if emphasis in not added to the less expected parts of a message.

The principles of differentiation and grouping appear also outside the areas of music and speech. For example, the color difference between traffic lights is enhanced by their different locations along a light ramp. Uniforms are used to enhance the differentiation between different categories of, e.g., airplane staff. Gender difference is enhanced by differences in fur coating in many animals and in clothing style among humans. The various parts of a house facade, such as roof, walls, and basement are often differentiated by different surface structures and colors, offering an example of a combination of differentiation and grouping principles.

These parallels between music performance and extramusical communication suggest that the principles of differentiation and grouping are essential to communication in general. The reason for this may be a question of potential interest to neuroscientists.

Why are differentiation and grouping needed? Differentiation should facilitate the listener's decoding of the message, and should hence make communication more robust. Grouping in music is mostly evident also when it is not marked, for example in a beginners' way of playing, and also in speech, as when young children read aloud without understanding what they are reading. On the other hand, more than one grouping is often possible in music, and musicians may chose different groupings of the same piece. In speech different groupings can carry different meanings.

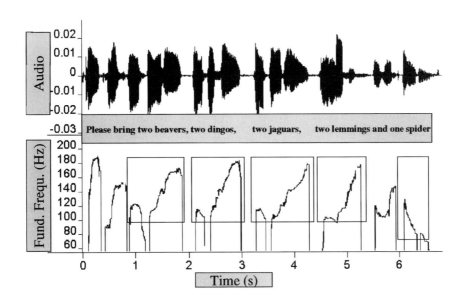

Figure 12
Example symbolic pitch patterning marking juxtapositions in a reading of the sentence "Please bring two beavers, two dingos, two jaguars, two lemmings, and one spider".

Code

It is also interesting to contemplate the codes used to facilitate differentiation and grouping in music. In many cases the code is the same as in speech. An emphasis marker frequently used in speech is duration; prominent words are lengthened (Fant & Kruckenberg, 1999). Melodic charge uses the same code. Another emphasis marker common to speech and music is delayed arrival; an emphasized syllable is often delayed (Sundberg, 1999); lengthening of the penultimate note uses the same code. In speech, sounds that belong to the same group are packed more densely in time (Fant, Kruckenberg & Liljencrants, forthcoming), and boundaries between such groups are marked by special signs, such as pauses and final lengthening. The Faster Uphill rule can be seen as a parallel, when ascending scales form a melodical gesture, and the Phrase Articulation and Phrase Arch rules mark phrase endings in a similar manner.

Thus, in many cases speakers and musicians use similar codes to facilitate emphasis and grouping. In some cases the code seems to originate from other areas than communication by sound as in the case of final ritards mentioned above (Friberg & Sundberg, 1999). The shape of the tempo curve for the final ritard was copied from the velocity decrease of stopping runners, and can therefore be regarded as an icon; the form of the sign reminds of its significance. The final ritard seems to represent a tangible example of musical semiotics.

Almost 100% of the population in this world appreciate listening to music. Music performers' use of codes for grouping and differentiation, that listeners are familiar with from their extramusical experience, may contribute to the explanation why music is appreciated by almost anyone.

Conclusions

The Director Musices generative grammar for music performance specifies certain deviations that musicians typically make from the nominal description in the score and defines the contexts in which they occur. The grammar suggests that the deviations serve two main purposes in music communication, differentiation and grouping. The differentiation rules would facilitate the listeners' sorting of tones according to pitch and note value categories and add emphasis to unexpected tones. The grouping rules would facilitate the listener's identification of structural elements such as melodical gestures, subphrases, and phrases. Thus, a musically acceptable performance seems to facilitate musical perception and cognition. The principles of differentiation and grouping appear also in other types of interhuman communication, and may therefore be relevant to the neurophysiological aspects of perception and cognition in general. The acoustic code used for differentiation and grouping in music performance shows striking similarities with that used in speech and may reflect neurophysiological demands on communication.

Acknowledgements

The author acknowledges the invaluable, inspiring co-operation in this long term research with his two friends and the department colleagues, dr Anders Friberg, researcher, programmer and pianist, and dr h.c. Lars Frydén, professional violinist and music teacher. Throughout his research career, but particularly in writing this article, the author has profited greatly from discussions with professor Gunnar Fant. The article is partly based on the author's presentation at the Meeting of the Society of Systematic and Comparative Musicology in Berlin, September -97.

References

Bresin R & Friberg A (1998) "Emotional expression in music performance: synthesis and decoding", TMH-QPSR (Speech Music and Hearing Quarterly Progress and Status Report) 3-4/1998, Stockholm, 85-94.

Drake C (1993) "Perceptual and performed accents in musical sequences", Bulletin of the Psychonomic Society 31, 107-110

Fant G & Kruckenberg A (1999) "Prominence correlates in Swedish prosody", paper given at the International Conference of Phonetic Sciences, San Fransisco.

Fant G, Kruckenberg A & Liljencrants J (forthcoming) "Acoustic-phonetic analysis of prominence in Swedish", manuscript to be published.

Friberg A, Bresin R, Frydén L and Sundberg J (1998) "Musical punctuation on the microlevel: Automatic identification and performance of small melodic units", Journal of New Music Research, 1998, Vol. 27, No. 3, 271-292

Friberg A, Sundberg J. (1999) "Does music performance allude to locomotion? A model of final ritardandi derived from measurements of stopping runners", Selected research article, *J Acoust Soc Amer* 105, 1469-1484.

Friberg, A. 1985a. "A Quantitative Rule System for Musical Performance". *diss. KTH, Stockholm.*

Friberg, A. 1995b. "Matching the rule parameters of Phrase Arch to performances of "Träumerei": a preliminary study", in A Friberg & J Sundberg, eds, *Proc of the KTH Symposium on Grammars for Music Performance 37-44.* Department of Speech Music Hearing, KTH, Stockholm.

Friberg, A., Sundberg, J. & Frydén, L. (1987). "How to terminate a phrase. An analysis-by-synthesis experiment on the perceptual aspect of music performance", in A. Gabrielsson (ed.), Action and Perception in Rhythm and Music, Stockholm: Royal Swedish Academy of Music, Publication No. 55, pp. 49-55.

Gabrielsson A (1999) "The performance of music". Chapter 14 in D Deutsch, ed: *The Psychology of Music*, 2nd ed, San Diego: Academic Press.

Garbuzov N (1948) *Zonnaja priroda zvukovysotnogo slucha*, : (Akademija Nauk SSSR Moscow, in Russian).

Krumhansl, C. (1990). *Cognitive Foundation of Musical Pitch*, New York: Oxford University Press.

van Oosten P (1993) "Critical study of Sundberg's rules for expression in the performance of melodies", *Contemporary Music Review* 9, 267-274.

Repp, B. 1992. "Diversity and commonality in music performance: An analysis of timing microstructure in Schumann's "Träumerei", *J Acoust Soc Amer 92:2546-2568.*

Shaffer LH (1995) "Musical performance and Interpretation", *Psychology of Music* 23, 17-38.

Sundberg J, (1999) "Emotive transforms", paper given at the symposium *Speech Communication and Language Development, in honor of Björn Lindblom*, Stockholm University, June 17-19.

Sundberg J & Verrillo V (1980) "On the anatomy of the ritard: A study of timing in music", J Acoust Soc Amer 68, 772-779.

Todd, N. (1985). "A model of expressive timing in tonal music", *Music Perception* 3:33-58.

Chapter III.4

Horizontal and Vertical Programming in Musical Performance

Pierre L. Divenyi [1]

Speech and Hearing Research, Department of Veterans Affairs

Musical performance, especially performance of music of the classical Western type, represents one of the most complex and demanding human activities. Performance is complex: it involves flawlessly recreating, traditionally from memory and in front of a generally sophisticated audience, a certain composition. The act of re-creating should be done not only with absolute deference to the composer's intention (as it is best understood, or at least surmised, by historical musicology and practicing tradition) but also with a freedom of expression and individualism that should instill a life in the piece as if it were created at the moment. This act of creation makes the art form of music stand out among other forms of art: the strict real-time requirement differentiates it from theater, and the requirement of absolute faithfulness to the printed text from dance. As to the physical exercise aspect of the performance, although its demands often match those of the most perfect, spectacular, and daring circus act, it differs from acrobatics by virtue of stemming from a need of artistic expression. And this expression is driven by the dual (and seemingly contradictory) exigencies of being at once maximally faithful to the composer's thought process (who, because he is likely to be no longer alive, cannot offer his help) and, at the same time, also of unquestionably reflecting the performer's individual cognitive, esthetic, and emotional spheres.

In order for a performer to succeed recreating the piece of his/her (or, more typically, the artist management's) choice, the performer's brain must be a specialized device that modern technology cannot even dream of, let alone design and build. Just as a computer, the performer must possess and rely on a program that he/she will have to execute without a single snag. However, unlike the computer's, the performer's program thrives on the unexpected, the odd, intelligently and sensibly woven into the fabric of a well-specified and stylistically often "frozen" style. Despite diligent efforts (see e.g., (Todd 1989)), artificial intelligence – fuzzy logic notwithstanding – has yet to conquer the mastery of emulating a composer or a performer happily engaged in breaking the rules. So, what sort of a brain do we

---

[1] Correspondence: Pierre L. Divenyi, Speech and Hearing Research, Department of Veterans Affairs, Martinez, CA, U.S.A., Tel: (1)-925-370-6745, Fax: (1)-925-228-5738, e-mail: pdivenyi@marva4.ebire.org

possess, what brain is capable of fulfilling all the requirements imposed by musical performance, in real time? This question has been of great interest to neurologists and neuroscientists for a very long time – at least since it has been plainly shown that a seasoned performer's brain looks different from the norm even to the naked eye (Somogyi 1930) – and was the subject of increasingly intense scientific investigation throughout the 20th century. Aspects of this research are well illustrated in the different chapters of the present volume, each depicting specialized and specific functions of the brain that make musical composition, performance, or listening possible. The real-time nature of music imposes serious constraints on much of neuroscientific investigations of musical activity: apart from the areas of timbre and harmony,[i] all aspects of music involve stretches of events happening over time. The question is: how long a time span does an investigator of music, especially musical performance, need to worry about? The urgency of this question is undeniable because of the orders-of-magnitude discrepancy between temporal ranges investigated in neuroscience and those encountered in a musical composition. True enough, recent advances in technology and statistics may extend the time window of the analysis of brain activity elicited by music: as Makeig[ii] has demonstrated, neurophysiological responses to rhythmic structures as long as 30 s can now be reliably measured. The fact that musically meaningful conclusions can be inferred from cortical activity recorded over such a long time span is truly remarkable. Nevertheless, the present chapter will argue that much more needs to be understood about the brain before we could form more than merely an approximate idea of how a 30-min piece is performed. These arguments will be presented from the biased point of view of a single performer playing Western classical music, but it is our firm belief that the same processes are active in, and the same requirements stand for, musical performance regardless of the number of performers and across diverse styles.

The performance program

While a performer may have various initial motivations to learn and perform a piece, we will assume that the major one among them, and the only one considered on these pages, is to recreate the given piece with the desire that his/her interpretation should approach to the greatest extent the supposed idea of the composer when he wrote the composition. Naturally, the performer has no way of precisely knowing what the composer (often a composer who has been dead for a long time) had in mind, not even when he/she is thoroughly familiar with all the historical, musicological, and music-theoretical background of the piece. Nonetheless, there is one crucial question that, in addition to or aside of his/her knowledge of the background, the performer has to work sometimes hard to find a satisfactory answer to: why did the composer write the piece? What were the intellectual challenges, esthetic cravings, emotional drives of the composer that nothing besides carving out the composition could solve, satisfy, appease? Of course, there is no unique answer to this question and, for the most part, different performers are bound to come up with answers that are not only different but

---

[i] Even harmony, i.e., the field of simultaneously sounding tones also called musical objects or events, lives in time because the listener (and presumably the composer) is influenced by what happened before. And, except for the most rule-bound compositions, such as some following a strict serial style, a tonal center inevitably emerges suggested by foregoing harmonies, a center against which subsequent musical events are evaluated.

[ii] Scott Makeig, personal communication

could also be at times widely divergent. Yet, every performer must embark on a journey, trying to retrace the composer's footsteps, as he/she becomes more and more familiar with a given composition – a process that, for many, represents a lifelong trek.[iii]

Such a journey is not without adventures. The adventure most apparent to the outside is the plain execution: the notes written must be played correctly, no matter what it takes. However, for even a moderately seasoned performer, the technical execution will come as a fringe benefit of learning to penetrate the depths of the piece. A more subtle, although almost as obvious, adventure is the discovery of the ways the composer bent the rules of his days' (or his own) style, where he broke them, where and how he is deceiving our expectations and, thereby, creates meaning. As Leonard Meyer said, "...[musical] meaning arises when a[n] ... expectation is delayed or blocked" (1957) The composer knew well, and very accurately, just how much information per unit time he could convey, i.e., how often he could deceive us, in order to create meaning: increasing entropy beyond a certain degree would actually decrease the information transmitted – random music, just like white noise, is ultimately uninformative. Thus, often a very simple motif can form the seed of a musical piece and perhaps nobody was able to spin a nothing-of-an-idea into magnificent edifices better than Beethoven, as illustrated by the descending pairs of E-flat major chord thirds at the beginning of the piano sonata op. 7, or by the repeated notes and descending thirds in the first theme of his fifth symphony. Simple material yields better and more numerous opportunities to deceive the listener's expectations, and it is the performer's task to display and exploit those opportunities.

There are definite consequences that the meaning-bearing instances will imprint on the performance, the most noticeable of which will be that the meter, i.e., the strict time measure or beat, will seldom coincide with the rhythmic flow. If an exact, implacable beat is our reference, then sometimes the music has to hurry and at other times it has to give, just in order for it to sound rhythmic – as it has been demonstrated by Sundberg ((Sundberg, Friberg et al. 1991) and in the chapter of this book).

Horizontal programming: the question of time unit

As we pointed out, music is a temporal process. In actuality, nothing in music is more important than its flow, its continuous evolution. But what is the unit of temporal organization in music, the unit which the performer must use to give a horizontal structure to the composition he/she is about to recreate? How short or how long is this unit? Obviously, the performer needs to look farther than the beat – generally very much farther ahead. To illustrate just what happens when a performer attempts to choose units that are too brief, the reader should listen to the first four audio tracks reproducing the first theme of the first movement of Mozart's Sonata in a minor (K. 310, see the printed music in Figure 1). Audio Track 1 demonstrates a beat-by-beat rendition of the theme, while Track 2 presents a two-beat-based, Track 3 a one-measure-based, and Track 4 a two-measure-based structuring. However, it should become clear to any listener that not even two-measure units are

---

[iii] It is extremely instructive to listen to an artist's different recordings of the same piece made over a longer period of time. For example, Claudio Arrau's 1930-40 rendition of a certain Beethoven piano sonata is very different from his 1960-80 recordings of the same piece. Although the listener is able to find, in both, undeniable signs of a pianistic genius and a great musician, the simplicity and serenity of the later recordings bespeak the thought and experimentation Arrau must have gone through during the several decades long dialog with the composition. Or with the composer?

Figure 2
Beginning of the B-major Fugue from the 1744 Second Volume of the Well-Tempered Clavier by J. S. Bach (BWV 892). The subject's first entry is illustrated in Track 7 of the audio examples. All entries of the fugue's subject are indicated by arrows. The first countersubject entry in measure 5 is indicated by the broken arrow (Track 8). Note that the second entry of the subject (by the tenor voice) in measure 5 is above the countersubject (by the bass voice), except for the third note in measure 7, where the bass jumps over the tenor. Maintaining a clear voice leading in such instances of voice crossings is one of the challenges of contrapuntal performance. The exposition of the fugue (Track 9) ends with the subject entry in F-sharp in measure 27 and, in measure 28, the second countersubject is introduced by the soprano voice (broken arrow). The beginning of the second countersubject exposition can be heard on Track 10. Note that the steady eight-note progression of the second countersubject is maintained throughout the rest of the fugue, except for two emphasized chords in the coda (Track 11, not shown in the figure).

sufficiently long to convey actually where Mozart wanted to get to with the first theme. To show the way in which the theme is spun forth – modulating from a minor to the key of the second theme, C-major, by way of the dominant G major (flirting with c minor, of which it is also the dominant) – Track 5 presents the whole first-theme area. Mozart is seen to draw a direct line, although not necessarily a straight one, from the beginning to the end of the first theme, but he ends the first theme area with a column, rather than a period, a column that is expected to lead to the second theme. We see that Mozart was careful to invent a first theme difficult to forget, he took care to present it twice at the onset, and he kept its rhythmic bounce at the G-major closure of the first-theme area, so that we remember the theme all too well throughout the development, which uses essentially no other thematic material. Track 6 reproduces the development where the first theme traverses some real drama; the chromatic 16th-note slide into the recapitulation makes the re-appearance of the first theme sound, in its exact initial form, maximally inevitable.

Panel a: First theme of the first movement of the a-minor Piano Sonata (K. 310) by W. A. Mozart. The ▲ symbol under each of the beats (in the first two measures) signifies the one-beat unit adopted in Track 1 of the audio examples; ▲ that introduces the C-major second theme.

Figure 1

The peripeties Mozart subjects the first theme to, or the first theme Mozart, would not sound meaningful without the performer planning them, constructing them, from the moment on he/she begins playing or, actually, long before the moment. The line from the beginning to the end of the piece must be present in the performance program, unbroken and implacable. How long is this program? In our example, we are talking of a performance program for the first movement of a Mozart sonata, a composition that lasts about eight minutes and the program must contain the whole piece. The program of the whole piece, however, is not made up from adroitly juxtaposing smaller entities, such as motifs, measures, melodies, chords. Rather, details of the performance involving those smaller entities must derive from the entire program of the piece – the devil is in the whole.

Vertical programming: the question of ordered streams

Programming is not restricted to planning the performance sequentially – one would be tempted to say, linearly. The same hierarchical ordering that proceeds from the whole toward the details is valid for vertical programming of the performance of a piece. Here, the term "vertical" refers to simultaneous execution of sequential entities. Loosely, the sequential entities may be considered individual *streams*, as defined by Bregman (Bregman and Campbell 1971; Bregman 1991), although whether or not the separate sequential entities written by the composer will be actually segregated by the listener's auditory system into separate auditory streams, when played simultaneously, is not the primary objective of the performer. Nevertheless, we must trust the composer: when he intended us to hear simultaneous sequential entities as segregated streams, he did write them far enough apart in pitch and close enough to each other in time, in order to satisfy van Noorden's dual segregation requirement (1975). It was this way of handling simultaneous sequences, undoubtedly one of the countless manifestations of his genius, that made Bach so admirably successful when writing three-four-five-voice fugues for keyboard, at times for instruments (such as the harpsichord) that had a very rapid decay after attack. Although the fugues are much easier to perform on a modern piano than on a harpsichord, or even on Bach's clavichord, it is quite challenging for the performer to keep the voices distinct while still achieving a balance – a balance that suggests the kind of unforced dialog between separate voices that constitutes the essence of counterpoint.

As an example, the B-major fugue from the second volume of Bach's Well-Tempered Clavier (Figure 2) shows how difficult Bach made the emergence of segregated voices, starting with the choice of a countersubject (Track 8) that, at one point, jumped over the subject (Track 7), as the fugue's exposition, with the five entries of the subject, shows (Track 9). But Bach does abandon the countersubject immediately after the exposition and introduces a more linearly flowing one (Track 10), possibly to make easier his task of building the whole cathedral of the fugue. That this second countersubject eminently well suits this architectural purpose is clear when listening to the measures modulating back to the final B-major appearance of the subject followed by the coda (Track 11). Here again, just as in the Mozart sonata example, programming the whole fugue as a single entity is an absolute necessity for the performer. Were he/she unable, or unwilling, to accomplish the task of a builder, prior to as well as during performance, neither the contrapuntal dialog, nor the architectonic qualities of the fugue would be conveyed.

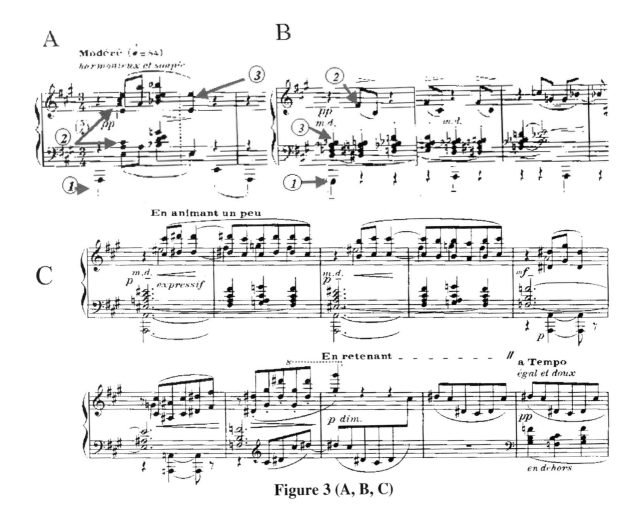

Figure 3 (A, B, C)

Figure 3 (D)

Excerpts from the Prelude "...Les sons et les parfums tournent dans l'air du soir" from the First Book of Preludes by C. Debussy. Panel a: beginning measures with three clearly separate simultaneous textures indicated by arrows and circled numbers. The three textures are also heard sequentially on Track 12. Panel b: end of first phrase with three textures (see the arrows and the three circled numbers), also on Track 13. The complete first phrase can be heard on Track 14. Panel c: Second phrase (also on Track 15). Panel d: Third phrase. Note that this phrase repeats the beginning but modulates from the original A major by a half tone, i.e., into A-flat major. It will go back to A major only after the harp-like descending arpeggios, at the end of the phrase. The third phrase can also be heard on Track 16.

On a piano, however, counterpoint can take several disguises. One of these is the contrast between simultaneous, sequentially presented textures and/or timbres. Perhaps the foremost master of composing such textures is Debussy. In the fourth piece from his first book of Preludes ("...Les sons et les parfums tournent dans l'air du soir" ["The sounds and fragrances revolve in the evening air", quote from a Ch. Baudelaire poem), three vertical textures can be identified in the first phrase (Track 12, Figure 3a) and three in the second (Track 13, Figure 3b). Although some of the textures contain more than one note, such as the four-note chord progression (suggesting a quartet of French horns) in the third one played on Track 13, the separate textures are audible even when they are put together vertically. The complete first phrase can be heard on Track 14 (Figure 3a-b) and the second on Track 15 (Figure 3c).

However, the textures change, sometimes kaleidoscopically, sometimes abruptly, as the piece goes on, as can be heard in the third example (Track 16, Figure 3d) showing what Debussy does to the four-note main motif using harmonic, and textural, variations. It is all these various textures that, when understood and programmed by the performer, can create the enormous timbral variety, that (according to the opinion of his contemporaries) Debussy was able to tease out of the piano when he played.

Conclusion

A program for the performance of a musical composition is, actually, a blueprint for the real-time execution of hierarchically ordered horizontal and vertical structures. The role of hierarchical ordering implies that the structures to be created during performance become larger and larger until, on final count, the program becomes a single entity, with the neatly ordered sequential and simultaneous structures telescoped into one meta-object, timeless and spaceless. Performance, therefore, only represents the unfolding this object. The process of unfolding thus becomes one of creation, not unlike the one that Mozart is told to have gone through just before writing down a new composition, when "he felt to hold the whole piece in his hand like an apple."

To accomplish such a feat represents a considerable challenge for the mechanism responsible for the development and the execution of the program. This is to say, the brain must be able to order, and keep the order amongst, sometimes lengthy and horizontally/vertically complex structures, such as the exposition of the first movement of Beethoven's Sonata op. 53 is (Track 17). The most demanding of the requirements is the one of temporal organization: although simple regurgitation of sequentially adjacent structures would reduce the length of the necessary time window, it would prevent the performer from "saying something." For the act of saying something, and hoping to say what the composer's message is, no time window shorter than the composition itself would suffice and the brain must be able to hold that window somewhere, somehow. Where and how, however, will yet have to be uncovered. But is not this what neuroscience is for?

Acknowledgments

The research on which the present chapter is based was supported by a grant from the Natinal Instituts of Health and by the Veterans Affairs Medical Research. The author wishes to thank Dean Michael Hammond, Dr. Tsutomu Nakada, and Dr. Ingrid Kwee for many helpful discussions, and Dr. David L. Wessel for kindly letting the audio examples be recorded at the Center for New Music and Audio Technology of the Department of Music, University of California, Berkeley. The author also would like to ask the reader to adopt an indulgent attitude when listening to the audio examples, noting that they were realized by a non-professional performer (the author) without the benefit of help by a recording engineer.

References

Bregman, A. S. (1991). Auditory scene analysis. Cambridge, Mass., Bradford Books (MIT Press).

Bregman, A. S. and J. L. Campbell (1971). "Primary auditory stream segregation and perception of order in rapid sequences of tones." Journal of Experimental Psychology **89**: 242-249.

Meyer, L. B. (1957). "Meaning in music and information theory." Journal of Aesthetics & Art Criticism **15**: 412-424.

Somogyi, I. (1930). "Adatok a zenei kepesseg agyilokalizaciojahoz." Orvosi Hetilap **74**: 77-82 (in Hungarian).

Sundberg, J., A. Friberg, et al. (1991). Common secrets of musicians and listeners: An analysis-by-synthesis study of musical performance. Representing musical structure. P. Howell, R. West and I. Cross. London, Academic Press: 161-200.

Todd, P. (1989). A sequential network design for musical applications. Proceedings of the 1988 Connectionist Models Summer School. San Mateo, CA, USA, Morgan Kaufmann, Inc: 76-84.

van Noorden, L. P. A. S. (1975). Temporal coherence in the perception of tone sequences, Unpublished doctoral dissertation, Technische Hoogschool, Eindhoven (the Netherlands).

Chapter III.5

Automatic Processing of Musical Information as Evidenced by EEG and MEG Recordings

Mari Tervaniemi[1]

Cognitive Brain Research Unit, Department of Psychology, University of Helsinki

Introduction

By recording the event-related potential (ERP), all stages of the neural sound processing can be probed with a millisecond accuracy (Regan 1989), starting from from the physical stimulus features (brain stem and middle-latency potentials; Picton 1980, Celesia 1976) ending up to cognitive memory- and attention-related processes (e.g., Näätänen 1992). Due to easy accessibility and relatively long tradition of cognitive ERPs especially in auditory psychophysiology, several ERP components have been used to clarify the neural basis of music perception. Among the first components under interest were the P3 (peak latency around 300 ms from the tone onset; Verleger 1990; Crummer et al., 1994; Janata 1995) and Late Positive Component (LPC, also termed P300 and P600; peak latency at 600-700 ms; Besson et al. 1994, 1998; Besson & Faita 1996; Patel et al. 1998). These components reflect the degree of expectancy violation in musical context but always under subjects' attentional control.

However, the importance of focussed attention in music perception was questioned about ten years ago (for a review, see Tervaniemi 1999). These first studies on automatic neural encoding of musical material employed the mismatch negativity (MMN) component of ERPs. The MMN is evoked by an infrequently presented auditory stimuli ("deviant") differing from the frequently-occurring stimuli ("standard") in one or several physical or abstract parameters (Figure 1) (Näätänen 1992; Näätänen & Winkler 1999). It reflects the discrepancy between the neural code formed by the standard sound and that of the deviant infrequent sound. The MMN can be recorded even when the subject is performing a task unrelated to the stimulation under interest such as reading a book or playing a computer game.

Multiple studies have shown that the MMN parameters closely correlate with several

---

[1] Correspondence: Dr. Mari Tervaniemi, Cognitive Brain Research Unit, Department of Psychology, P.O. Box 13 FIN-00014, University of Helsinki, Finland Tel: (358)-9-1912-3408, Fax (358)-9-1912-2924, e-mail: mari.tervaniemi@helsinki.fi.

indicators of subjects' perceptual accuracy as determined in a separate experimental session. For instance, the MMN amplitude and latency correlate with subjects' behavioral performance as determined by musicality tests (Lang et al., 1990; Tervaniemi et al., 1997) and by hit rates and reaction times (e.g., Tiitinen et al., 1994; Koelsch et al., 1999; Kraus et al., 1996). In addition, the degree of perceptual similarity between different musical instrument timbres highly correlates with the MMN amplitude (Toiviainen et al. 1998). This correspondence between the MMN parameters and perceptual accuracy imply that pre-attentive neural functions determine the accuracy of the subsequent attentive processes (Novak et al., 1990).

Thus, the MMN can be used to probe the level of perceptual auditory abilities without being contaminated by, for instance, subjects' motivational involvement or their vigilance. By now, particular interest these features received during past years among clinicians who need to evaluate the integrity of neurocognitive functions in patients without contaminating the evaluation by motivational or attentional factors (for a recent review, see Näätänen in press). For instance, it has been shown that automatic discriminative functions of dyslexic adults are worse than those of control subjects (Schulte-Körne et al. 1998; Baldeweg et al. 1999; Kujala et al. in press) and that aphasic patients are selectively impaired in detecting sound changes in right-ear stimulation (Ilvonen et al. submitted). In addition, patients suffering from Alzheimer's disease have selectively impaired capacity in maintaining sound information when sounds are delivered with relatively slow stimulation rate (Pekkonen et al. 1994).

In the following, MMN recordings investigating the neural basis of musical sound processing and musical abilities will be introduced.

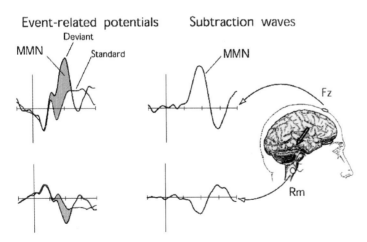

Figure 1
Left panel: The event-related potential elicited by standard (thin line) and deviant (thick line) sounds. The sounds were spectrally rich and differed in pitch by 2.5%. The blue area indicates the MMN. The x-axis displays time in milliseconds and the y-axis the amplitude of the response in microvolts. Right panel: The subtraction wave obtained by subtracting the ERP elicited by the standard sound from that of the deviant sound. Data from Tervaniemi et al. (submitted b).

Generators of the MMN

MMN is generated mainly in primary auditory cortex or in its immediate vicinity. This has been evidenced by magnetoencephalogram (MEG) (e.g., Hari et al. 1984; Tiitinen et al. 1993;

Winkler et al. 1995), intracranial recordings (Kropotov et al. 1995), positron emission tomopgraphy (Tervaniemi et al., submitted a), and functional magnetic resonance imaging (Opitz et al. 1999 a, b). In electric recordings this is reflected by polarity reversal above the Sylvian fissure from the fronto-central negative maximum to positivity in mastoid electrodes (when the nose reference is used) (see Figure 1). Additional generators have been found in the frontal (Giard et al. 1990; Rinne et al., submitted) and parietal (Levänen et al. 1996) lobes (for a review, see Alho 1995). Within the auditory cortex, MMN may have multiple generators since its scalp topography differs as a function of the parameter in which the deviance occurs (e.g., intensity, duration, frequency; Giard et al. 1995). This suggests that neural codes for separate sound parameters are spatially distinct.

The human cerebral cortex is functionally specialized to sensory-specific and association areas. In addition, some of the sensory-specific areas are known to be functionally organized according to the stimulus complexity (e.g., Hubel 1988). The auditory cortex is tonotopically organized as a function of spectral pitch (Pantev et al. 1995; Tiitinen et al. 1993; Yamamoto et al. 1992) and the temporal (periodicity) aspect of pitch (Langner et al. 1997; see also Pantev et al. 1989). These notions led us to investigate whether auditory cortices might also be spatially organized according to the stimulus complexity and/or stimulus structure. To this end, two MEG experiments were conducted.

In the first study, the MMNm (magnetic counterpart of MMN) was recorded to three types of musical stimuli while the subjects were instructed to read a self-selected book (Alho et al. 1996). In each condition, the stimuli had an identical frequency change embedded in them but that change was presented in a different context in separate conditions: among single sinusoidal tones, parallel chords, and sequential chord patterns. A frequency change elicited the MMN with all these three sounds without latency or amplitude differences between the sound types. However, comparisons of the frequency MMN generators (modeled as equivalent current dipoles) revealed that the MMN generator was, on the average, 1 cm deeper (more medial) in the auditory cortex with sequential and parallel chords than with the single tones. In other words, the MMN source was different for complex than for simple sounds. This suggests that even within the auditory cortex, there are spatially separable areas specialized in encoding spectrally/temporally complex information.

The second MEG experiment was conducted to determine whether the auditory cortex is functionally specialized in pre-attentive encoding of phonetic vs. musical information (Tervaniemi et al. 1999). The subjects, while watching a silent movie, were presented with frequent and infrequent phonemes (/e/ vs. /o/) or chords (A major vs. A minor). These phonetic and musical stimuli were matched in complexity as well as in the magnitude of the frequency change embedded in them. It was found that in both hemispheres, the source of the MMNm elicited by the infrequent sounds among phonemes and chords was located posteriorly to the source of the earlier P1m component which is known to be generated at the primary auditory areas. In addition, the MMNm source for a phoneme change was located superiorly to that of the chord change. These data thus indicate that there are distinct cortical areas specialized in representing phonetic and musical sounds in both hemispheres. In other words, the functional specialization does not only cover stimulus complexity (sinusoidal vs. chord; Alho et al. 1996) but also the informational content (phonetic vs. musical, Tervaniemi et al. 1999). However, this specialization is not necessarily present prior to memory-related processing as suggested by the dissociation between P1m and MMNm data (Tervaniemi et al. 1999).

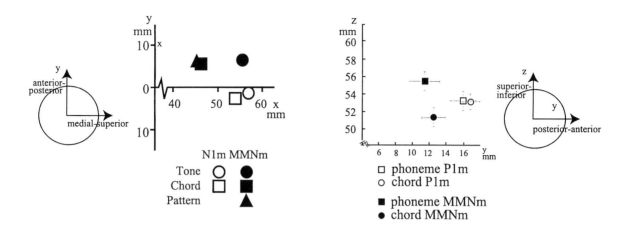

Figure 2

Left panel: The MMNm generator loci (as modelled by equivalent current dipoles) for identical frequency change when presented among single sounds (black circle), parallel chords (black square), and in sequential chords (black triangle). The head is illustrated from above. The MMNm was generated more medially within the auditory cortex when the frequency change was presented among chords than when it was among single sounds. The N1m generator loci (white circle and square) did not differ between different stimuli. Figure adapted from Alho et al. (1996). Right panel: The MMNm generator loci (as modelled by equivalent current dipoles) for phoneme change (black square) and for chord change (black circle). The head is illustrated laterally from the right side. The MMNm was generated more superiorly when the change occurred in the phonemes than when it occurred in the chords. The P1m generator loci (white circle and square) did not differ between phonemes and chords. Figure adapted from Tervaniemi et al. (1999).

Effects of stimulus structure and familiarity on pitch discrimination

Preliminary behavioral evidence suggested that the pitch of spectrally rich sounds is easier to discriminate than that of sinusoidal tones consisting of only one frequency component. First, just noticeable difference in frequency discrimination was slightly smaller with synthetic vowels than for pure tones (Flanagan and Saslow 1958). Second, the tuning of musical instruments (and thus also pitch discrimination) is easier relative to instruments with rich harmonic spectra than relative to musical instruments with more pure sound quality (Sundberg 1991). Third, pitch discrimination was more accurate when stimuli were square waves consisting of several harmonic partials than when stimuli were sinusoidal tones or had only two harmonic partials (Sidtis 1980). These notions led us to study whether pitch-discrimination of spectrally rich sounds is facilitated even when subjects concentrate on a task unrelated to the sounds.

In the first study, the MMN elicited by 2.5%, 5%, and 10% frequency change was recorded from subjects who were asked to read a book of their own choice and to ignore the sound stimulation (Tervaniemi et al. 2000). The sounds were (in separate blocks) pure tones with only one sinusoidal frequency component or consisted of three harmonic partials. In a separate condition, the subjects were asked to indicate detection of pitch change by a button press. The results show that the MMN was elicited with a larger amplitude and shorter latency by frequency changes in spectrally rich tones than by change in pure tones. Furthermore, the subjects' behavioral responses were more accurate for spectrally rich tones than for sinusoidal tones. Interestingly, no effects of the size of frequency change or interaction between frequency change and sound structure (suggesting the facilitation caused by harmonic partials being more pronounce, for instance, with smaller frequency deviations) were observed.

The facilitation of pitch discrimination caused by harmonic partials could be explained in acoustical terms since spectrally rich tones carry more spectral as well as temporal information than sinusoidal tones (Moore 1989). In spectral terms, pitch perception results from the firing of tonotopical organization of the auditory system (see e.g., Terhardt 1974). With regard to the pitch of harmonically complex sounds, this theory assumes that a complex sound is reduced into its harmonic components in the cochlea and, further, that higher auditory mechanisms encode the pitch by recognizing the pattern of resolved harmonics. If the frequency encoded by each resolved partial is subsequently encoded by the cortical memory trace (reflected by MMN), each harmonic partial might elicit its separate MMN. Thereafter these parallel MMNs could be summed together, causing more accurate pitch-discrimination performance. In temporal terms, pitch perception results from the property of single cells and neuronal nets to respond to a tone in a time-locked manner. Thus, the pitch of complex sound is encoded by higher auditory mechanisms that determine the inter-spike interval of neural activation in the cochlea. According to recent results, pitch discrimination can even be facilitated by increasing temporal information if it is accumulated across a sufficiently long time period: a pitch-MMN in the missing-fundamental sounds was elicited when they were of 500-ms duration (Winkler et al. 1997). However, when sounds were of 150-ms duration, pitch change did not elicit the MMN and it was also difficult to detect in a separate discrimination condition. This suggests that even when pitch of a relatively short harmonic sound cannot be determined because its fundamental frequency is missing, it is possible with prolonged sound samples.

Consequently, the second study was conducted to compare the relative effects of the spectral complexity and sound duration on automatic pitch discrimination (Tervaniemi et al. submitted b). The MMN component was recorded from reading subjects to a small (2.5%) pitch change in six conditions: The pitch change was presented in pure tones with only one sinusoidal frequency component (500 Hz) and in spectrally rich tones with three (500...1500 Hz) and five (500...2500 Hz) harmonic partials. Stimulus durations of 100 ms and 250 ms were employed. The results show that the MMN was elicited with larger amplitude by change in spectrally rich tones, be them with three or five partials, than by change in pure tones. The sound duration did not significantly affect the MMN parameters. It may thus be concluded that increased spectral rather than temporal information facilitate pitch discrimination with spectrally rich sounds.

The third alternative explanation for facilitated pitch processing with spectrally rich sounds could be the difference in the stimulus familiarity. That is, since our auditory system has extensive experience with spectrally rich stimulation like speech and music even prior to birth, it reacts more vigorously to changes in spectrally complex information than to changes in tones consisting only of the fundamental frequency.

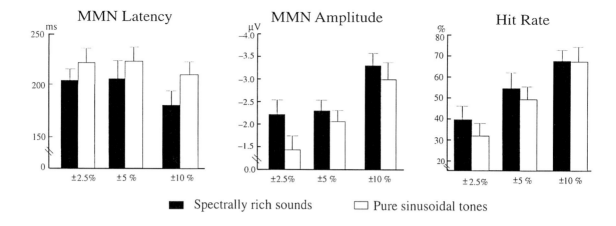

Figure 3
Left panel: The MMN amplitude for 2.5%, 5%, and 10% frequency change among spectrally rich (black bar) and sinusoidal (white bar) tones. Middle panel: The MMN latency for 2.5%, 5%, and 10% frequency change among spectrally rich (black bar) and sinusoidal (white bar) tones. Right panel: The hit rate for 2.5%, 5%, and 10% frequency change among spectrally rich (black bar) and sinusoidal (white bar) tones. Figure adapted from Tervaniemi et al. (2000).

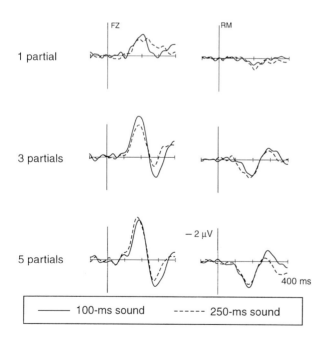

Figure 4
The MMN elicited by a 2.5% pitch change when the sounds were pure tones with only one sinusoidal frequency component (top) and in spectrally rich tones with three (middle) and five (bottom) harmonic partials. Continuous line denotes the MMN elicited by 100-ms sound and dashed line the MMN elicited by 250-ms. Figure adapted from Tervaniemi et al. (submitted b).

The aspect of sound familiarity was investigated by employing stimulation differing from studies described above. Instead of spectrally complex sounds, temporally complex sound patters of about half a second in duration were constructed and a frequency change of identical magnitude was presented among them (Brattico et al. submitted). In the Scale condition, these patterns consisted of pure tones embedded in a five-tone pattern with tonal musical relationships (i.e. A major scale). In the Compressed condition, the tones of this five-tone pattern were without musical relationships (non-musical compressed whole-tone steps). In addition, the same frequency change was presented in the Single-tone condition among isolated sinusoidal tones. The frequency deviant in the Scale condition changed the mode of the pattern from A major to A minor while in the other two conditions it represented no musically meaningful change. To determine the influences of musical expertise, the subjects were either musicians (N=10) or non-musicians (N=10).

The MMN was elicited in all conditions and both subject groups, being, in general, of larger amplitude in the conditions in which sounds were temporally complex than when they were single tones. In addition, the MMN parameters differentiated musicians and non-musicians. First, the MMN latency was shorter in all conditions in musicians than in non-musicians. Second, the MMN in non-musicians was larger in amplitude in the Scale condition than in the Compressed or Single-tone conditions. This suggests that the non-musicians more readily processed the musically meaningful sound change in familiar context (or, alternatively, simple frequency ratio intervals over complex ratios; see Cuddy 1971; Dewar et al. 1977; Schellenberg and Trehub 1994). In contrast, in musicians, no difference in MMN amplitude between the Scale and Compressed conditions was observed, MMN being smallest in the Single-tone condition. It could thus be speculated that the auditory system in musicians is tuned to process structurally complex sound events and, further, that the familiarity or complexity of sound intervals is of smaller importance. In addition, the shorter MMN latency in musicians than in non-musicians implies that in musicians, auditory processing is facilitated in a generalized manner.

Musical expertise and automatic pitch processing

As seen above, musical expertise may be reflected even at the automatic level of sound processing. The superiority of musicians', and, especially, violin players' auditory system to process slight pitch changes was recently addressed. There the standard stimulation consisted of 3-part major chords (Koelsch et al. 1999). The deviant stimulus was the same chord as the standard stimulus, except that the middle tone of the chord was marginally mistuned (<1%). This stimulation was presented to subjects while they were reading a book (the first and third blocks of the experiment) and while they were asked to detect the deviant chords (the second block). During the reading task, the deviants elicited the MMN only in musicians. During the discrimination task, non-musicians detected, on the average, about 15% and musicians about 85% of the deviants. ERPs recorded during the discrimination task showed that in Musicians, a significant MMN was followed by an additional negative ("N2b") and a subsequent positive deflection ("P3b"). This N2b-P3 complex reflects higher cognitive processes concerned with the conscious detection and evaluation of deviants. Non-musicians had a small MMN without subsequent N2b or P3. The third block was presented to see whether intermediate attentive task facilitated subjects' automatic pitch processing. This, however, was not the case - musicians showed a MMN which did not differ from their MMN of the first block and non-musicians showed no MMN. These results demonstrate that highly trained musicians automatically detected differences in auditory information which were undetectable for non-musicians and, further, that these automatic functions were not modified by attentional manipulations. Interestingly, however, the MMN evoked by a small (<1%) or large (10%) pitch change in sinusoidal tones did not differentiate violinists and non-musicians. This suggests that musical training facilitates pre-attentive cortical auditory processing especially when the sound structure is musically relevant and when the stimulus change is of musically relevant magnitude.

Conclusions

The data reviewed above suggest that auditory information with musical relevance is automatically encoded by the human central auditory system. In fact, the pitch of spectrally or temporally rich sounds is processed even more vigorously than that of pure tones. Musical expertise is reflected in these automatic auditory functions in a broad manner. However, the present evidence, leaving the direction of causality between the primary auditory functions and musical expertise unanswered, invites us to further investigations.

Acknowledgements

The studies summarized above were supported by the Academy of Finland and by the University of Helsinki. This review was written while the author was a visiting research fellow at the Department of pschology, University of Leipzig, supported by the Deutscher Akademischer Austauschdienst. The author thanks Dipl.-Psych. Stefan Koelsch for his comments on the manuscript.

References

Alho K. Cerebral generators of mismatch negativity (MMN) and its magnetic counterpart (MMNm) elicited by sound changes. Ear Hear 1995; 16: 3-50.

Alho K, Tervaniemi M, Huotilainen M, Lavikainen J, Tiitinen H, Ilmoniemi RJ, Knuutila J, Näätänen R. Processing of complex sounds in the human auditory cortex as revealed by magnetic brain responses. Psychophysiology 1996; 33: 369-375.

Baldeweg T, Richardson A, Watkins S, Foale C, Gruzelier, J. Impaired auditory frequency discrimination in dyslexia detected with mismatch evoked potentials. Annals Neurol 1999; 45: 495-503.

Besson M, Faita, F. An event-related potential (ERP) study of musical expectancy: Comparison between musicians and non-musicians. J Exp Psychol: Human Percept Perf 1996; 21: 1278-1296.

Besson M, Faita F, Peretz I, Bonnel AM, Requin, J. Singing in the brain: independence of lyrics and tunes. Psychol Sci 1998; 9: 494-498.

Besson M, Faita F, Requin, J. Brain waves associated with musical incongruities differ for musicians and non-musicians. Neurosci Lett 1994; 168: 101-105.

Brattico E, Tervaniemi M, Näätänen R. Context effects on pitch perception according to the degree of musical expertise - Evidence from brain recordings, submitted.

Celesia GG. Organization of auditory cortical areas in man. Brain 1976; 99: 403-414.

Crummer GC, Walton JP, Wayman JW, Hantz EC, Frisina RD. Neural processing of musical timbre by musicians, nonmusicians, and musicians possessing absolute pitch. J Acoust Soc Am 1994; 95: 2720-2727.

Cuddy LL. Absolute judgement of musically-related pure tones. Canadian J Psychol 1971; 25: 42-55.

Dewar K, Cuddy LL, Mewhort DJK. Recognition memory for single tones with and without context. J Exp Psychol 1977; 3: 60-67.

Flanagan JL, Saslow MG. Pitch discrimination for synthetic vowels. J Acoust Soc Am 1958; 30: 435-442.

Giard MH, Lavikainen J, Reinikainen K, Perrin F, Bertrand O, ThÇvenet M, Pernier J, Näätänen R. Separate representation of stimulus frequency, intensity and duration in auditory sensory memory: An event-related potential and dipole-model analysis. J Cogn Neurosci 1995; 7: 133-143.

Giard MH, Perrin F, Pernier J, Bouchet P. Brain generators implicated in the processing of auditory stimulus deviance: A topographic event-related potential study. Psychophysiology 1990; 27: 627-640.

Hari R, Hämäläinen M, Ilmoniemi R, Kaukoranta E, Reinikainen K, Salminen J, Alho K, Näätänen R, Sams M. Responses of the primary auditory cortex to pitch changes in a sequence of tone pips: neuromagnetic recordings in man. Neurosci Lett 1984; 50: 127-132.

Hubel DH. Eye, Brain, and Vision. Scientific American Library, New York 1988.

Ilvonen TM, Kujala T, Tervaniemi M, Salonen O, Näätänen R, Pekkonen E. Mismatch negativity (MMN) indicates deficits in pre-attentive auditory discrimination after left hemisphere stroke. submitted.

Janata P. ERP measures assay the degree of expectancy violation of harmonic contexts in music. J Cogn Neurosci 1995; 7: 153-164.

Koelsch S, Schröger E, Tervaniemi M. Superior attentive and pre-attentive auditory processing in musicians. NeuroReport 1999; 10: 1309-1313.

Kraus N, McGee T, Carrell TD, Zecker SG, Nicol TG, Koch DB. Auditory neurophysiologic responses and discrimination deficits in children with learning problems. Science 1996; 273: 971-973.

Kropotov JD, Näätänen R, Sevostianov AV, Alho K, Reinikainen K, Kropotova OV. Mismatch negativity to auditory stimulus change recorded directly from the human temporal cortex. Psychophysiology 1995; 32: 418-422.

Kujala T, Myllyviita K, Tervaniemi M, Alho K, Kallio J, Näätänen R. Basic auditory dysfunction in dyslexia as pinpointed by brain-activity measurements. Psychophysiology, in press.

Lang AH, Nyrke T, Ek M, Aaltonen O, Raimo I, Näätänen R. Pitch discrimination performance and auditive event-related potentials. In Brunia CHM, Gaillard AWK and Kok A. (Eds.), Psychophysiological Brain Research, Vol. 1. Tilburg University Press, Tilburg, 1990; pp. 294-298.

Langner G, Sams M, Heil P, Schulze H. Frequency and periodicity are represented in orthogonal maps in the human auditory cortex: Evidence from magnetoencephalography. J Comparat Physiol A 1997; 181: 665-676.

Levänen S, Ahonen A, Hari R, McEvoy L, Sams M. Deviant auditory stimuli activate human left and right auditory cortex differently. Cerebral Cortex 1996; 6: 288-296.

Näätänen R. Attention and Brain Function. Hillsdale: Lawrence Erlbaum Associates, 1992.

Näätänen R. (Ed.) Mismatch negativity and its clinical applications. Special Issue. Audiology & Neuro-otology, in press.

Novak GP, Ritter W Vaughan, HG Jr., Wiznitzer ML. Differentiation of negative event-related potentials in an auditory discrimination task. Electroencephal clin Neurophysiol 1990; 75: 255-275.

Opitz B, Mecklinger A, von Cramon DY, Kruggel F. Combining electrophysiological and hemodynamic measures of the auditory oddball. Psychophysiology 1999a; 36: 142-147.

Opitz B, Mecklinger A, Friederici AD, von Cramon DY. The functional neuroanatomy of novelty processing: integrating ERP and fMRI results. Cerebral Cortex 1999b; 9: 379-391.

Pantev C, Oostenveld R, Engelien A, Ross B, Roberts LE, Hoke M. Increased auditory cortical representation in musicians. Nature 1998; 392: 811-814.

Pantev C, Bertrand O, Eulitz C, Verkindt C, Hampson S, Schruierer G, Elbert T. Specific tonotopic organizations of different areas of the human auditory cortex revealed by simultaneous magnetic and electric recordings. Electroencephal clin Neurophysiol 1995; 94: 26-40.

Patel AD, Gibson E, Ratner J, Besson M, Holcomb PJ. Processing syntactic relations in language and music: an event-related potential study. J Cogn Neurosci 1998; 10: 717-733.

Pekkonen E, Jousmäki V, Könönen M, Reinikainen K, Partanen J. Auditory sensory memory impairment in Alzheimer's disease: an event-related potential study. NeuroReport 1994; 5: 2537-2540.

Picton TW. The use of human event-related potentials in psychology. In: Martin I, Venables PH, editors. Techniques in Psychophysiology. New York, Wiley, 1980: 357-395.

Regan D. Human Brain Electrophysiology: Evoked Potentials and Evoked Magnetic Fields in Science and Medicine. New York: Elsevier, 1989.

Rinne T, Alho K, Ilmoniemi RJ, Virtanen J, Näätänen R. Separate time behaviors of the temporal and frontal MMN sources, submitted.

Schellenberg EG, Trehub SE. Frequency ratios and the discrimination of pure tone sequences. Perception Psychophys 1994; 56: 472-478.

Schulte-Körne G, Deimel W, Bartling J, Remschmidt H. Auditory processing and dyslexia: evidence for a specific speech processing deficit. NeuroReport 1998; 9: 337-340.

Sidtis J. On the nature of the cortical function underlying right hemisphere auditory perception. Neuropsychol 1980; 18: 321-330.

Sundberg J. The Science of Musical Sounds. Academic Press, San Diego, 1991.

Tervaniemi M. Pre-attentive processing of musical information in the human brain, J New Music Research 1999; 28: 237-245.

Tervaniemi M, Ilvonen T, Karma K, Alho K, Näätänen R. The musical brain: Brain waves reveal the neurophysiological basis of musicality in human subjects. Neurosci Lett 1997; 226: 1-4.

Tervaniemi M, Ilvonen T, Sinkkonen J, Kujala A, Alho K, Huotilainen M, Näätänen R. Harmonic partials facilitate pitch discrimination in humans: Electrophysiological and behavioral evidence. Neurosci Lett 2000; 279: 29-32.

Tervaniemi M, Kujala A, Alho K, Virtanen J, Ilmoniemi RJ, Näätänen R. Functional specialization of the human auditory cortex in processing phonetic and musical sounds: A magnetoencephalographic study. NeuroImage 1999; 9: 330-336.

Tervaniemi M, Medvedev SV, Alho K, Pakhomov SV, Roudas MS, van Zuijen TL, Näätänen R. Lateralized automatic auditory processing of phonetic versus musical information: a PET study, submitted a.

Tervaniemi, M, Schröger E, Saher M, Näätänen R. Effects of spectral complexity and sound duration in complex-sound pitch processing in humans- a mismatch negativity study, submitted b.

Tiitinen H, Alho K, Huotilainen M, Ilmoniemi RJ, Simola J, Näätänen R. Tonotopic auditory cortex and the magneto encephalographic (MEG) equivalent of the mismatch negativity. Psychophysiology 1993; 30: 537-540.

Tiitinen H, May P, Reinikainen K, Näätänen R. Attentive novelty detection in humans is governed by pre-attentive sensory memory. Nature 1994; 372: 90-92.

Toiviainen P, Tervaniemi M, Louhivuori J, Saher M, Huotilainen M, Näätänen R. Timbre similarity: convergence of neural, behavioral, and computational approaches. Music Perception 1998; 16: 223-241.

Verleger R. P3-evoking wrong notes: unexpected, awaited, or arousing? Internat J Neurosci 1990; 55: 171-179.

Winkler I, Tervaniemi M, Huotilainen M, Ilmoniemi R, Ahonen A, Salonen O, Standertskjöld-Nordenstam C-G, Näätänen R. From objective to subjective: sound representation in the human auditory cortex. NeuroReport 1995; 6: 2317-2320.

Winkler I, Tervaniemi M, Näätänen R. Two separate codes for missing fundamental pitch in the auditory cortex. J Acoust Soc Am 1997; 102: 1072–1082.

Yamamoto T, Uemura T, Llinas R. Tonotopic organization of human auditory cortex revealed by multi-channel SQUID system. Acta Otolaryngol 1992; 112: 210-214.

Chapter III.6

Music as a Second Language

Tsutomu Nakada [1]

Department of Integrated Neuroscience, Brain Research Institute, University of Niigata

Introduction

The final goal of neuroscience is the elucidation of how the human brain works. In this context, a detailed understanding of the neuronal mechanisms of language processing is one of the main objectives. Music and language are cognitive traits uniquely developed in humans. Both are distinct tools of communication and expression, and may be utilized simultaneously, as in song. Similar to language, humans have a natural ability for acquiring the rules of music after exposure to only a few examples and generalizing these for the purposes of composition and performance. As in language, literacy in music is a skill which requires specific and often painstaking educational effort in order to attain proficiency. Given that music and language represent different functional aspects of the brain, studies identifying the neuroanatomic substrates involved in the cortical processing of music may lead to a clearer understanding of language processing.

In spite of extensive work by various investigators, many technical difficulties have hampered these efforts from reaching conclusive data regarding the cortical processes of music. The introduction of functional magnetic resonance imaging (fMRI), especially high field fMRI (HF-fMRI), has drastically improved the opportunities for investigating human higher cortical function. The synergistic effects of the high signal to noise ratio and susceptibility effects of a high field system yield highly reproducible, high resolution functional activation maps for each individual subject. Such individual based studies obviate the necessity of applying artificially established standardization of brain anatomy (Kwee et al., 1999, Fujii et al, 1998, Nakada et al., 1998a, b,). As a result, activation maps obtained by functionally related, modality unrelated, independently performed paradigms (e.g., languages vs. music) can effectively be compared to determine functionally overlapping, spatially discrete areas in the brain. Such an approach would be difficult, if not impossible, using other types of activation studies such as H_2O^{15} positron emission tomography (PET) and fMRI performed on conventional systems. HF-fMRI provides hitherto unobtainable opportunities for performing multiple subject analysis based on

---

[1] Correspondence: Tsutomu Nakada, M.D., Ph.D., Department of Integrated Neuroscience, Brain Research Institute, University of Niigata, Niigata 951-8585, Japan, Tel: (81)-25-227-0677, Fax: (81)-25-227-0821, e-mail tnakada@bri.niigata-u.ac.jp

individually defined activation maps. Such individual based studies are inherently less artifact prone than studies based on single composite activation maps derived from multiple subjects and/or sessions employed in other modalities of activation studies. This article deals with a review of such studies performed in our laboratory in an attempt to clarify the essential identity of music and language.

Methods

Subjects

Normal volunteers were imaged according to the human research guidelines of the Internal Review Board of the University of Niigata. Informed consent was obtained from all subjects. All subjects were native Japanese speakers, literate in English and music, and right-handed. Stated handedness was confirmed using the Edinburgh inventory (Oldfield, 1971).

HF-fMRI

A General Electric (Waukesha, Wisconsin, USA) Signa-3.0 T system equipped with an Advanced NMR (ANMR) EPI module was used to perform all fMRI studies. Each session consisted of 30 second epochs configured in the box car alternate sequence. Gradient echo echo-planar images (GE-EPI) were obtained using the following parameter settings: FOV 40 cm x 20 cm; matrix 128 x 64; slice thickness 5 mm; inter-slice gap 2.5 mm; TR 1 sec. Spatial resolution was approximately 3 mm x 3 mm x 5 mm. Sessions which had brain motion exceeding 0.6 mm were re-performed to avoid so-called fictitious activation resulting from pixel misalignment. fMRI time series data consisting of consecutive EPI images for each slice were analyzed using SPM96 (the Wellcome Department of Cognitive Neurology) (Friston et al., 1994). The data were smoothed using a 5-mm full width at half maximum (FWHM) kernel. Statistical analysis was performed using a delayed (6 seconds) boxcar model function in the context of the general linear model as employed by SPM96. To minimize effects of physiological noise, a high pass filter and global normalization were applied within the design matrix. Specific effects were tested by applying appropriate linear contrasts to the parameter estimates for each condition resulting in a t statistic for each and every voxel. These t statistics, which were transformed to Z statistics, constituted an activation map. These images were interpreted by referring to the probabilistic behavior of a Gaussian field. As pointed out by several investigators, such univariate transformation did not transform a t field into a Gaussian field unless the degree of freedom of t statistics was reasonably high (Worsley, 1994). The condition was dealt with on the assumption that the activation map based on Z statistics was a reasonable lattice representation of an underlying continuous Gaussian field which was conformed using a spatial smoothing process, a 5-mm FWHM kernel described above. fMRI images were presented with contrast between the two conditions specified and showed activated areas which conformed to statistical criteria of significance ($p < 0.001$). Data were analyzed and reproducibility was confirmed for each subject individually. Anatomical identification of activated areas was performed individually by mapping these areas onto the subject's own anatomical images obtained with identical coordinates.

Experimental

The leading models of cortical language processing consider it to involve combinations of spatially discrete cortical functional subunits (Figure 1). This localization based approach, exemplified by brain activation studies using H_2O^{15} positron emission tomography (PET) or fMRI, has been fruitful in providing substantial insight into the cortical processing of language. According to recent models of cortical language processing, visually presented language symbols are first decoded within the occipital cortex, and subsequently meet the cortical processes resulting from auditory language processing within the temporal lobe and its vicinity. This area, lesions of which produce Wernicke's aphasia, is defined as the subunit where the process of visual language comprehension (reading) first converges with that of auditory language comprehension (listening). This definition, therefore, can be effectively utilized to specifically define the paradigms.

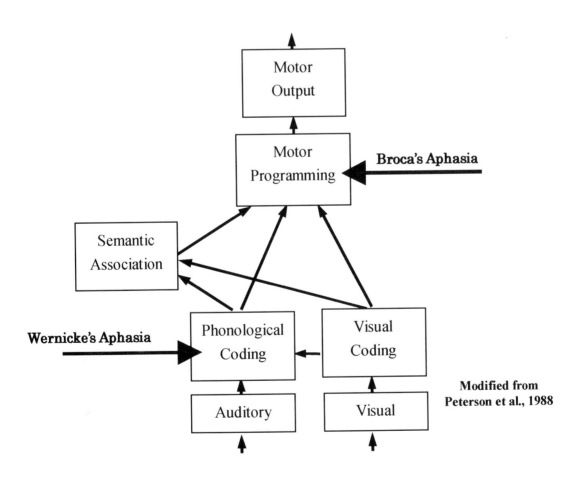

Figure 1

Results

Decoding the symbols

Decoding processes of visually presented phrases of music and language (sight-reading) were investigated utilizing music score and Japanese text (contrast: non-specific visual stimuli). A representative activation map is shown in Figure 2 with an example of the paradigm utilized in the study. A schematic summary of activation maps is shown in Figure 3. Cross sectional figures, indicated by coordinate numbers correspond to figures of the Talairach-Tournoux atlas. The statistics are summarized in Table 1.

Table 1

	left-FG	left-LG	right-LG	left ITS	right-TOS
Japanese	8/10	3/10	2/10	10/10	0/10
Musical Score	9/10	10/10	9/10	0/10	10/10

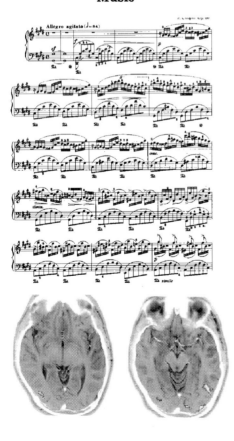
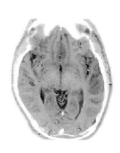
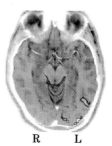

Figure 2

The study demonstrates that the left fusiform gyrus (FG) represents the primary area for the reading decoding process, regardless of the specific modality of the symbolic system. The cortex flanking the posterior part of the left inferior temporal sulcus (ITS) is most specific for reading kanji. The cortex flanking the right transverse occipital sulcus (TOS) is specific for music literacy (Nakada et al, 1998b, Nakada, 2000).

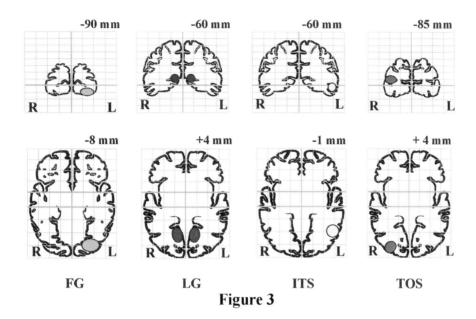

Figure 3

Auditory Conversion

Auditory conversion of visually presented symbols of music and language was investigated by obtaining functional maps associated with reading music notes vs. Japanese alphabet, kana (contrast: non-specific character). Twenty subjects literate in both music and Japanese all showed virtually identical activation of the left auditory association area. A representative activation map is shown in Figure 4 with an example of the paradigm utilized.

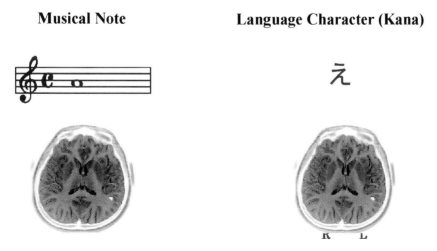

Figure 4

Discussion

In "Syntactic Structures" published in 1957, Noam Chomsky presented a revolutionary concept in linguistics that language function cannot be explained according to the laws of stimulus and response. Instead, speech is a creative process based on a universal grammar inherent to human brain and shared by all natural languages. Human children acquire their language skills from the given environment as a means of conveying their intelligence. Non-human primates do not have the capability of "acquiring language" although they do have the capability of articulation. Heinrich Schenker, a German musician, described a similar concept in his treatise on musicology more than twenty years prior to Chomsky's Syntactic Structures (Sloboda, 1985). It is not known whether or not Chomsky's concept originated from that of Schenker's. Nevertheless, time by time, music has been treated as a form of language.

Clinical reports indicate that the neuroanatomic structures underlying music and language processing overlap to high degree, but are nonetheless separable. The majority of patients who suffer from Wernicke's aphasia shows some degree of amusia during their recovery phase. In contrast, the case of the Russian composer Vissarion Shebalin, who continued to be able to compose highly complex music even after having been rendered aphasic, is a dramatic example of the virtually complete separation of musical capability from language in this individual.

Whether music is a right hemisphere or left hemisphere function oversimplifies the neuroscience of music. Nevertheless, clinical reports have indicated that the right hemisphere is more specifically related to musical function. Those patients who develop disorders of music associated with left hemispheric lesions almost always have accompanying language disorders. On the other hand, virtually all patients with musical disorders unaccompanied by noticeable language deficits have right hemispheric lesions. Investigations on the musical abilities of each hemisphere using the Wada test have previously shown that barbiturate injection into the right hemisphere can produce severe deficits in melody recognition (Bogen & Gordon, 1971). However, barbiturate injection into either hemisphere cannot produce total block of rhythmic capability (Robinson & Solomon, 1974).

The HF-fMRI studies presented here are consistent with the published literature and yet has provided further insight into the cortical processing of music. Literacy in music shares a large portion of cortical decoding processes with language, but recruits additional cortical areas, especially the cortex flanking the right transverse occipital sulcus (TOS). Both music notes and language characters activate Wernicke's area. Compared to language, music has a rather mathematical structure allowing for simpler simulation. The neuroscience of music opens a new window towards a more comprehensive understanding of human cognitive functions, especially those related to language.

Acknowledgement

The study was supported by grants from the Ministry of Education (Japan).

References

Bogen JE, Gordon HW. Musical tests for functional lateralization with intracarotid amobarbital Nature 230:524-, 1971

Chomsky N. Syntactic Structures Mouton, The Hague, 1957

Kwee IL, Fujii Y, Matsuzawa H, and Nakada T: Perceptual processing of stereopsis in human: High-field (3.0T) functional MRI study. Neurology 1999;53:1599-1601.

Friston KJ, Jezzard P, Turner R. The analysis of functional MRI time series. Hum Brain Mapping 1:153-171, 1994

Fujii Y, Nakayama N, Nakada T. High-resolution T_2 reversed MRI on high-field system. J Neurosurg 1998;89:492-495.

Nakada T, Fujii Y, Suzuki K, Kwee IL. High-field (3.0T) functional MRI sequential epoch analysis: an example for motion control analysis. Neurosci Res 1998a;32:355-362.

Nakada T, Fujii Y, Suzuki K, Kwee IL. 'Musical brain' revealed by high-field (3 Tesla) functional MRI. NeuroReport 1998b;9:3853-3856.

Nakada T.: Functional MRI of literacy. In Kato T. (ed.), *Frontiers of the Mechanism of Memory and Dementia*, Elsevier, Amsterdam, 2000, pp. 21-24.

Sloboda JA. The Musical Mind. Oxford University Press, Oxford, 1985.

Oldfield RC. The assessment and analysis of handedness: The Edinburgh inventory. Neuropsychologia 1971;9:97-113.

Petersen SE, Fox PT, Posner MI, Mintun M, Raichle ME Positron emission tomographic studies of the cortical anatomy of single word processing. Nature 1988;331:585-589.

Robinson G, Solomon DJ. Rhythm is processed by the speech hemisphere. J Exp Psychol 102:508-511, 1974

Worsley KJ Local maxima and the expected Euler characteristic of excursion sets of X^2, F and t fields. Adv Appl Probab 1994;26:13-41.

T. Nakada (Ed.)
Integrated Human Brain Science: Theory, Method Application (Music)
© 2000 Elsevier Science B.V. All rights reserved

Chapter III.7

The Neural Basis of Musical Processes

Robert J. Zatorre[1]

Montreal Neurological Institute, McGill University

Introduction

Music is among the most complex of human cognitive functions, and therefore of significant interest in achieving a fuller understanding of human cognitive neuroscience. Yet, it is only recently that the brain mechanisms associated with music have been the focus of systematic study. In this paper I will briefly summarize some studies from our laboratory aimed at exploring brain correlates of musical function, emphasizing the use of recent brain imaging techniques, notably positron emission tomography (PET). Functional Magnetic Rresonance Imaging (fMRI) has also begun to be used to study musical perception. This technique has numerous advantages over PET, particularly because of its improved spatial and temporal resolution. Until recently its application to auditory studies was complicated by the loud acoustic artifact which is inherent to fMRI; however, recent technical advances have provided a partial solution to this problem (Hall et al., 1999; Belin et al., 1999) and future work is very likely to yield important new insights.

The development of functional imaging in general represents a major advance for cognitive neuroscience, as it permits for the first time a relatively direct way to investigate changes in cerebral activity patterns as a function of specific task performance in normal subjects. The basic idea behind the application of PET relevant to the studies to be described below is that a short-lived radioactive tracer (oxygen-15) is used to measure cerebral blood flow (CBF) during a 60-second period. A scan is then reconstructed which represents a three-dimensional map of the CBF distribution in the entire brain during that time, with a spatial resolution on the order of 14 to 18 mm. Typically, several such scans are obtained in each of several individual subjects. Data from a group of subjects may then be averaged, after appropriate stereotaxic normalization is applied to correct for differences in brain size, shape, and orientation. Averaged CBF data from a given condition may then be compared to another condition by

---

[1] Correspondence: Robert J. Zatorre, Ph.D., Montreal Neurological Institute, 3801 University St., Montreal, Quebec, H3A 2B4, Tel: (1)-514-398-8903, Fax: (1)-514-398-1338, e-mail: md37@musica.mcgill.ca

superimposition of the relevant scans, and application of a pixel-by-pixel subtraction algorithm which detects significantly different areas of CBF in one condition as compared to another (Worsley et al., 1992). The assumption is that the difference image reflects areas of cerebral activity specifically related to the task in question, relative to the baseline condition which typically represents an attempt to control for certain aspects of the task. Note that, unlike ERP or MEG methods, in which the location of a dipole must be modelled, and hence can only be inferred, PET provides a way of reconstructing the distribution of CBF which is a fairly accurate reflection of the true activity concentration in the brain. A final aspect of the functional imaging work presented here is that structural MRIs are also obtained for each subject, and are co-registered to the PET images, thus allowing for accurate anatomical localization of CBF changes.

Melodic Processing

Over the past several years our laboratory has attempted to explore the neural correlates of pitch and melody processes, not only to localize the systems responsible, but also to address whether music may rely on specialized neural operations distinct from those used for speech or other auditory processes. A number of experiments performed over the years have examined the effects of unilateral temporal-lobe damage to various musically-relevant functions. One conclusion to emerge from research in our own laboratory and that of others is that the right temporal cortex plays a particularly important role for processing of pitch and of spectral pattern information (Divenyi & Robinson, 1989; Milner 1962; Robin et al., 1990; Samson & Zatorre, 1994; Sidtis & Volpe, 1988; Zatorre, 1988; Zatorre & Samson, 1981). Despite this importance of neural systems within the right cerebral hemisphere, additional evidence from the study of patients with bilateral temporal lesions clearly indicates that there are important contributions to musical processes from the left hemisphere as well (Peretz et al., 1994). Even unilateral left superior temporal lesions can, in some instances, result in mild or moderate melodic processing deficits (Zatorre, 1985; Samson & Zatorre, 1988), particularly for recognition memory tasks (Samson & Zatorre, 1992).

The above studies formed the backdrop for a functional imaging study (Zatorre et al., 1994) whose goal was to better understand the neural basis for perception of melodic patterns and for retention of pitch information in working memory. Our goal was to test two specific hypotheses: (1) that perceiving a novel tonal melody would entail neuronal processing in both left and right superior temporal regions, with a possibly greater contribution from the right; and (2) that right frontal-lobe mechanisms would be engaged when subjects make specific judgments that require retention of pitch over a filled interval (Zatorre and Samson, 1991).

We tested twelve normal subjects without formal musical training using two classes of stimuli: noise bursts and melodies. The noise bursts were constructed so as to approximate the acoustic characteristics of the melodies in terms of number, duration, inter-stimulus presentation rate, intensity, and onset/offset shape. Sixteen different eight-note tonal melodies were also prepared, all identical in their rhythmic configuration, with the aim of allowing pitch judgments of either the first two notes, or the first and last notes.

Four separate conditions were run during each of the four scanning periods. During the first condition, termed the "noise" condition, subjects listened to the series of noise bursts described above, and after each "noise melody" depressed a key to control for motor activity. In the second condition, termed "passive melodies," the subjects were presented with each of the sixteen tonal melodies, and depressed a key after each one, as before. No overt judgments were required, but subjects were instructed to listen carefully to each melody. In the third condition, the "2-note" pitch comparison, subjects listened to the same melodies as before, but this time were instructed to determine whether the pitch of the second note was higher or lower than that of the first note. Finally, in the "first/last" pitch judgment, subjects were asked to compare the pitch of the first and last notes, ignoring the notes in between, and to respond as before, according to whether the pitch rose or fell. Subjects kept their eyes shut throughout the scanning period.

The experiment was set up to permit specific comparisons, accomplished via subtraction of relevant conditions. The first comparison, passive melodies minus noise, permits examination of the cerebral regions specifically active during listening to novel tonal melodies, as opposed to the activation that might be present with any auditory stimulus with similar acoustic characteristics. The principal result indicated a significant CBF increase in the right superior temporal gyrus, anterior to the primary auditory cortex. A much weaker CBF increase was also visible within the left superior temporal gyrus. In addition, and unexpectedly, a significant focus was also identified in the fusiform gyrus of the right hemisphere, within area 19. The finding of an activation within the right superior temporal gyrus while listening to melodies fits in well with our prediction, and likely reflects the specialization of neuronal networks within the right secondary auditory cortices for perceptual analysis of tonal information, consistent with the human lesion evidence reviewed above (Milner, 1962, Zatorre, 1985; 1988). Although subjects were listening "passively," it is evident that they would be extracting perceptual information during this phase, and the CBF changes we observed nostly likely reflect these automatically engaged processes. The weak activity in the left superior temporal area may indicate the additional but perhaps less important or less consistent participation of left temporal cortices in melody processes, which is also suggested by the evidence from lesion studies (Peretz et al., 1994).

Note that, in this subtraction, no CBF increase was present in the primary auditory cortices beyond that elicited in the control condition. This result is explained by the control condition: by using acoustically matched noise bursts, nonspecific auditory processing can be dissociated from that uniquely elicited by listening to melodies. We previously demonstrated that similar noise bursts result in primary auditory cortical stimulation when contrasted to a silent condition (Zatorre et al., 1992). These findings, together with findings from prior PET studies using speech sounds or tones (e.g. Démonet et al., 1992) point to differential activation of primary vs. secondary auditory areas within the superior temporal gyrus, according to the nature of the processing elicited by a given stimulus. Although the noise stimuli proved successful in demonstrating the intended dissociation, caution must be still exercised in interpreting the results, for the noise bursts are clearly not physically identical to the melodic sounds. For example, the noise stimuli contain no periodicity, whereas the tones do; their spectral composition also is quite different. It remains to be established, therefore, which specific features of the melodies may lead to the observed pattern of activation.

The second and third comparisons in this study both used the passive melody condition as the baseline, so that any activation seen represents neural responses beyond those already present during initial listening to the same stimulus materials. Subtraction of the passive conditin from the 2-note condition resulted in significant activation within the right frontal lobe, as predicted. The first/last-passive melodies subtraction yielded a number of cortical and subcortical activation sites in both hemispheres. Among the more relevant results were CBF increases within the right frontal lobe, consistent with the predictions, including a focus in area 47/11 identical to that observed in the 2-note condition. Of particular interest was an area of significant CBF increase within area 21 of the right temporal lobe, indicating that this condition resulted in greater activity within the right auditory association cortex than already present during passive listening to melodies.

The pattern of results from these conditions implicates frontal-lobe mechanisms in effecting pitch comparisons, as had been predicted, with a particularly important contribution from right-frontal regions. In the first/last minus passive melody comparison, we observed a greater number of separate foci of CBF change over a wider swath of cortical and subcortical territory than was evident in the lower memory-load condition of judging the first two notes; this finding perhaps reflects the complexity and increased cognitive demands of the task, which was also manifested in increased error rate and slower reaction times. Although we are not in a position to interpret all of these foci, we may speculate that the numerous frontal-lobe sites observed might be associated with successful performance of distinct aspects of the task (such as for example, maintenance of pitch information in working memory, monitoring the presentation of the tones and their temporal order, directing the appropriate pitch comparison, etc.).

Putting these together with the lesion literature discussed earlier, a preliminary outline of a model to describe the neural substrates associated with pitch processing may be suggested. We may speculate that the primary auditory cortex is chiefly involved in early stages of processing (which might include computation of such signal parameters as pitch, duration, intensity, and spatial location), whereas more complex feature extraction, involving temporally distributed patterns of stimulation, is performed via populations of neurons within the secondary cortices. Neuronal systems located in both temporal lobes likely participate in higher-order perceptual analysis of melodies, but those on the right seem to be particularly important, perhaps because they are specialized to extract the features that are most relevant for melodic stimuli (including, for example, invariant pitch-interval relationships, and spectral characteristics important for pitch and timbre perception).

In both pitch judgment conditions we observed significant CBF increases within the right frontal cortex. Only in the first/last comparison, however, did we observe an additional CBF increase in the right temporal lobe, beyond that seen in passive listening. We interpret this result, together with the right frontal activation, as evidence that the high memory load imposed by the first/last task engaged a specialized auditory working memory system, and that this system is instatiated in the brain via interaction of inferior frontal and superior temporal cortices in the right cerebral hemisphere (Marin & Perry, 1999). This conclusion would be in accord with our earlier study (Zatorre & Samson, 1991), in which deficits in pitch retention were observed after right frontal and/or temporal-lobe lesions.

Musical Imagery

Many people, musically trained or not, report a strong subjective experience of being able to imagine music or musical attributes in the absence of real sound input. But subjective reports are of limited use to assess the characteristics of cognitive representations in a scientifically rigorous manner. Therefore, in recent years, psychologists have tried to find more objective means of evaluating the nature of imagery processes. Much of this research has concentrated on the visual domain, and has yielded the conclusion that visual imagery processes operate with similar characteristics to perceptual processes (see Farah, 1988, for a review). This view leads to the hypothesis that perception and imagery may share at least partially, the same neural substrate.

In our laboratory we have examined this hypothesis within the context of musical imagery, using both a behavioral lesion approach (Zatorre & Halpern, 1993), and via PET functional imaging (Zatorre et al., 1996). We adapted a paradigm originally developed by Halpern (1988), in which musically untrained subjects compared the pitch of two lyrics from a familiar, imagined song. (For instance, is the pitch corresponding to "sleigh" higher or lower than that of "snow" in the song "Jingle Bells"?)

In the first neuropsychological study (Zatorre & Halpern, 1993), we examined whether auditory imagery and perception may share similar neural mechanisms by presenting a modification of the tune scanning task to patients having undergone right or left temporal-lobe excision for the relief of intractable epilepsy. A perceptual version of the task was devised in which the listener made pitch judgments while actually hearing the song. As well, subjects participated in an imagery condition in which judgments of pitch were made to imagined tunes indexed by the lyrics. The results of that study were very clear and striking. While all subjects did better on the perception task compared to imagery, patients with left-temporal excisions showed no deficits whatsoever relative to normal controls, whereas those with damage to the right temporal area were significantly worse than the other groups on both tasks, and by about the same amount on each task. We concluded that structures in the right temporal lobe were crucial for successful performance of both imagery and perception tasks, suggesting the same kind of neuroanatomical parallelism (and by extension functional parallelism) shown by Farah (1988), Kosslyn et al. (1993), and others for visual imagery and perception.

PET methodology allows us to study the neural processes of normal subjects with greater anatomical precision compared to many other physiological techniques, including lesion studies. We therefore designed an experiment to investigate the putative similarity between perceptual and imagery mechanisms. We presented three tasks to a group of 12 normal participants: a visual baseline condition and two active tasks, one termed "perception," the other "imagery." The latter two were similar to those used by Zatorre and Halpern (1993): Two words from a familiar tune were presented on a screen, and the task was to decide if the pitch corresponding to the second word was higher or lower than the pitch corresponding to the first word. In the perceptual task, participants actually heard the song being sung, while in the imagery task they carried out the task with no auditory input. In the basleine task subjects viewed the words and performed a visual length judgment. By subtracting the activation in the visual baseline from both the perception and imagery tasks, we should, in principle, eliminate cerebral activity related to nonspecific processes shared by the two tasks, such as reading words on a screen, making a forced-choice decision, pressing a response key, etc. Thus, any CBF changes still remaining must be due to the unique demands of listening to a tune or imagining it, and making a pitch comparison.

The most important findings from these subtractions were that for nearly every region demonstrating CBF change in one condition, there was a corresponding CBF peak in the other condition, often within a few millimeters. The similarity in CBF distribution across the two conditions supports the idea that the two processes share a similar neural substrate.

Not surprisingly, highly significant CBF increases were found within the superior temporal gyrus bilaterally when subjects were processing the auditory stimuli for the perceptual task, as compared to the baseline task, in which no auditory stimulation was provided. More interesting is the finding that regions within the superior temporal gyrus were also activated, albeit at a much weaker level, when subjects imagined hearing the stimulus, again as compared to the baseline condition. Note that this latter subtraction entails two entirely silent conditions, so that positive CBF changes in the superior temporal gyri (associative auditory cortices) cannot be due to any external stimulation, but are most likely attributable to endogenous processing.

It is of interest to note that the temporal-lobe activation in the perceptual-baseline comparison incorporated primary auditory cortex and extended well into association cortical regions along most of the length of the superior temporal cortices. In contrast, this was not the case for the imagery-baseline comparison: CBF increases in that case occurred exclusively in association cortex (and were of lower relative magnitude). This distinction may be important, and supports the idea that primary sensory regions are responsible for extracting stimulus features from the environment, whereas secondary regions are involved in higher-order processes, which might include the internal representation of complex familiar stimuli.

Activation of the supplementary motor (SMA) area was also observed, and is of particular interest, given its role in motor processes. This region has consistently shown CBF increases during various types of motor tasks, including situations in which a motor task is only imagined, rather than overtly executed (Rao et al. 1993; Wise et al., 1991). The finding of SMA activation may therefore imply that the SMA is part of a substrate for both overt and covert vocalization, and therefore supports the idea that imagery for songs includes not only an auditory component ("hearing the song in one's head"), probably related to temporal cortical activity, but also a subvocal component ("singing to oneself"), reflected in SMA activity.

Because Zatorre et al. (1996) observed bilateral STG activity in their imagery condition, their finding could not support the conclusion derived from prior studies for a more important contribution of right STG areas in tonal processing. This was particularly notable given that Zatorre and Halpern (1993) had earlier shown that right temporal-lobe excision resulted in deficits on a similar imagery task whereas left temporal lesions did not. The bilateral STG activity may simply reflect the fact that the stimuli used contained both verbal and tonal information, however. In order to test this idea, a further PET study was carried out, involving strictly nonverbal materials (Halpern & Zatorre, 1999). In this study subjects were presented with a tonal cue, consisting of the first few notes of a familiar tune devoid of verbal content (e.g., the first four notes of Beethoven's fifth symphony), and asked to generate, using imagery, the remainder of the tune. This task was contrasted to one in which subjects heard a similar set of notes that did not correspond to any known tune. The results indicated significant activity in the right STG, as predicted, in a region posterior to HG quite close to the sites associated with other tonal processing studies (Zatorre et al., 1994, 1998; Perry et al., 1999), thus confirming the role of this lateralized region in the re-evocation of tonal patterns in the absence of acoustic input, consistent with the behavioral-lesion data (Zatorre & Halpern, 1993). This study also revealed SMA activity, as observed previously, as well as a large CBF response in the right inferior frontal cortex, perhaps associated with retrieval of the tune form long-term memory.

Taking the findings of the PET studies together with the behavioral lesion study of imagery (Zatorre & Halpern, 1993), we conclude that there is good evidence that perception and imagery share partially overlapping neural mechanisms, and that these include superior temporal auditory regions, as well as motor areas.

Absolute Pitch

Another recent application of functional imaging to study musical processing concerns the phenomenon of absolute pitch (AP), which refers to the ability to name the pitch of a musical tone or to sing a note on demand without reference to other sounds. Anatomical correlates of this ability have been identified using structural imaging: the region of the planum temporale appears to show a more exaggerated leftward asymmetry among AP possessors (Schlaug et al., 1995; Zatorre et al., 1998). However, the functional significance of this observation remained to be established. Zatorre et al. (1998) used PET to examine CBF patterns in subjects with AP and in control musicians without AP. In one condition they listened to pairs of musical tones without explicit instructions; in another condition, they judged whether the musical interval formed by the tones within a pair was major or minor. Each condition was compared to a baseline in which acoustically matched noise bursts were presented.

The results indicated that listening to musical tones resulted in similar patterns of increased CBF bilaterally in auditory cortical areas in both groups, including a region of the right STG, in a comparable location to that identified by Zatorre et al. (1994). Thus, AP would not appear to involve differences at the level of the initial stages of perceptual analysis. It is also of interest to note that occipital areas showed increased CBF in both subject groups, as noted in several previous studies. In addition to the similar location of most activation foci, the AP group but not the control group demonstrated activation of the left posterior dorsolateral frontal cortex in this simple listening condition. However, a similar pattern of left dorsolateral frontal activity was also observed in both groups when they made relative pitch judgments of intervals, such as minor or major. This portion of the frontal cortex has been implicated in conditional associative learning of sensory stimuli (Petrides et al., 1993). AP may be characterized as the ability to retrieve an arbitrary association between a stimulus attribute (the pitch of a sound) and a verbal label, which may be considered as a form of conditional associative learning. Thus, the CBF increase in the dorsolateral frontal cortex among the AP group but not the RP group while listening to tones may reflect the engagement of an associative mechanism. The finding of similar activity in the this area in both groups during the interval classification task may be explained since this task requires that the tonal stimuli be labelled (i.e., the verbal association "major" or "minor"). The interval labels are themselves learned associations, but both AP and RP subjects would have access to them by virtue of their musical training.

In addition to the differential recruitment of the posterior DLF cortex across the two groups, a further dissociation is apparent in this study, in that the right inferior frontal region (Brodmann area 47/11) -- active in both groups in the simple listening condition -- shows no CBF change in the AP group in the interval judgment condition. This finding is consistent with data presented earlier indicating that this region participates in maintenance of pitch information in auditory tonal working memory. Subjects without AP may use tonal working memory in both tasks, but AP possessors would not need access to this mechanism for interval classification because they are able to classify each note within the interval by name. Thus, rather than compute the size of the musical interval based on its sound, which would require maintaining pitch in working memory, AP possessors may simply obtain the correct response by knowing what the individual notes are within the interval. This conclusion is concordant with the reported absence of the P300 evoked electrical component during interval classification among AP subjects (Hantz et al., 1992), which has been interpreted as reflecting the AP subjects' use of a long-term memory representation to accomplish the interval labelling task, rather than needing to update working memory on every trial. Taken together, the findings suggest that AP may not be associated with a unique pattern of cerebral activity, but rather may depend on the recruitment of a specialized network involved in the retrieval and manipulation of verbal-tonal associations.

Music and Emotion

Traditionally, cognitive neuroscience has dealt with processes such as perception, memory, reasoning, and so forth, while leaving aside emotion. In the past decade this situation has changed dramatically, with emotion becoming the focus of intense scientific inquiry. Music can play a particularly important role in this respect, since there is such an obvious and important affective component. In this section, a study will be presented (Blood et al., 1999) that used PET to examine the neural correlates of music-induced emotion.

The ability of musical stimuli to elicit emotion is particularly intriguing since, unlike most other stimuli which evoke emotion, such as smell, taste, or facial expression, music has no obvious intrinsic biological or survival value. One usually thinks of music and emotion in the context of the positive emotions that people typically experience, and which is the reason why most people listen to music in the first place. However, the scientific study of this phenomenon is complicated by the fact that music preference tends to be highly individual (just ask any teenager and any parent for an example of disagreement). To obviate this problem, we opted to study the negative affective reactions elicited by dissonance, which appear to be relatively consistent and stable. Listeners who have been exposed to the Western tonal idiom typically respond readily to dissonance, even in the absence of formal musical training. This phenomenon presumably reflects the fact that listeners have internalized the tonal rules of music in their culture, and react to violations of these rules (Krumhansl, 1990).

Few neuropsychological data are available concerning the affective component of musical processing. A recent case study of a patient with amusia, however, has suggested that perceptual and emotional analysis of music may be dissociated: paralleling work in visual face processing, Peretz et al. (1998) have documented that judgments of affective content of a melody (happy vs. sad) could be carried out by an amusic patient who was unable to identify or recognize a melody. Blood et al. performed their study, in part, to investigate whether such dissociations could be understood in terms of distinct neural mechanisms engaged by musically induced affect, and to establish their functional anatomy.

Blood et al. used a novel melody which was made to sound more or less consonant or dissonant by varying the harmonic structure of its accompanying chords. PET scans were obtained while subjects with no more than amateur musical training listened to six versions of this stimulus, designed based on pilot studies to spontaneously elicit a continuum of pleasant to very unpleasant emotional responses. Regression analysis was used to correlate CBF with degree of dissonance. Subjects also rated the emotional quality of the music, using a rating scale with eight pairs of adjectives. The hypothesis was that the variations in affective quality of the stimuli would correlate with CBF changes in regions involved in emotional processes.

The results demonstrated that CBF changes in specific paralimbic and neocortical areas, known from prior studies to play a role in affective processing, correlated with increasing dissonance or consonance. These included the right parahippocampal gyrus and right precuneus, which increased CBF with increasing dissonance; and bilateral orbitofrontal, medial subcallosal cingulate and right frontal polar regions which increased CBF as a function of increasing consonance. Furthermore, regional covariation analyses established that as CBF increased in the areas related to consonance, they decreased in areas related to dissonance, and vice-versa, implying a reciprocal functional relationship across these sets of regions. Dissonance/consonance was also associated with certain positive or negative subjective emotional ratings, suggesting that the regions in question are involved specifically in response to these emotions, rather than other emotions which did not change as a function of dissonance. The paralimbic and neocortical regions identified are distinct from areas of secondary auditory cortex which emerged in control subtractions. Activity in these auditory cortical areas likely represents operations related to processing the perceptual aspects of consonance and dissonance.

The orbitofrontal cortex, subcallosal cingulate, and frontal polar regions have all been implicated in emotional processing by numerous previous studies (Lane et al., 1997; Rolls et al., 1994; Dias et al., 1996). Damasio (1996) has suggested that the ventromedial portion of prefrontal cortices may be involved in making judgments about stimuli based on their emotional valence. Finally, the subcallosal cingulate region has been shown to exhibit decreased baseline rCBF in depressed patients compared to normals (Drevets et al., 1997); the present data are in agreement with this finding, since activity in the subcallosal region decreased with unpleasantness, while it increased with pleasantness of the stimuli.

In conclusion, this study demonstrates that musical dissonance can elicit affective reactions, which in turn are related to separable neural circuits in paralimbic and neocortical areas associated with positive and negative emotion. The findings thus begin to characterize the neural basis for emotional responses to music. Interestingly, these regions differ from those which are active during non-emotional components of music perception, as reviewed above. The findings of this study therefore not only begin to define a neural network associated specifically with emotional responses to music, but also demonstrate dissociations between perceptual and affective processing.

Conclusion

The studies outlined here illustrate an approach to how the complex neural processes involved in musical cognition can be studied systematically. The converging evidence provided by the complimentary approaches of lesion and functional imaging appear to be quite powerful, and we are thus optimistic about being able to provide more complete models of this intriguing aspect of cognition in the future.

Acknowledgements

The research reported in this paper was supported by the Medical Research Council of Canada and by the McDonnell-Pew program in Cognitive Neuroscience. For further resources, including audio examples of stimuli used, visit: www.zlab.mcgill.ca.

References

Blood AJ, Zatorre RJ, Bermudez P, Evans AC. Emotional responses to pleasant and unpleasant music correlate with activity in paralimbic brain regions. Nature Neuroscience 1999; 2: 382-387.

Damasio AR. The somatic marker hypothesis and the possible functions of the prefrontal cortex. Philos. Trans. R. Soc. Lond. B Biol. Sci. 1996; 351: 1346, 1413-1420.

Démonet JF, Chollet F, Ramsay S, Cardebat D, Nespoulous JL, Wise R, Rascol A, Frackowiak R. The anatomy of phonological and semantic processing in normal subjects. Brain 1992; 115: 1753-1768.

Dias R, Robbins TW, Roberts AC. Dissociation in prefrontal cortex of affective and attentional shifts. Nature 1996; 380: 69-72.

Divenyi PL, Robinson AJ. Nonlinguistic auditory capabilities in aphasia. Brain Lang 1989; 37: 290-326.

Drevets WC, Price JL, Simpson JR Jr, Todd RD, Reich T, Vannier M, Raichle ME. Subgenual prefrontal cortex abnormalities in mood disorders. Nature 1997; 386: 824-827.

Farah MJ Is visual imagery really visual? Overlooked evidence from neuropsychology. Psychol Rev 1988; 95: 307-317.

Halpern AR Mental scanning in auditory imagery for tunes. J Exp Psychol: Learn Mem Cognit 1988; 14: 434-443.

Halpern AR, Zatorre RJ. When that tune runs through your head: A PET investigation of auditory imagery for familiar melodies. Cerebral Cortex 1999; 9: 697-704.

Hantz EC, Crummer GC, Wayman JW, Walton JP, Frisina RD. Music Percept 1992; 10: 25-42.

Kosslyn SM, Alpert NM, Thompson WL, Maljkovic V, Weise SB, Chabris CF, Hamilton SE, Rauch SL, Buonanno FS. Visual mental imagery activates topographically organized visual cortex: PET investigations. J Cog Neurosci 1993: 5; 263-287.

Krumhansl CL. Cognitive Foundations of Musical Pitch; Oxford Psychology Series. No. 17, Oxford University Press: New York, 1990.

Lane RD, Reiman EM, Bradley MM, Lang PJ; Ahern GL, Davidson RJ, Schwartz GE. Neuroanatomical correlates of pleasant and unpleasant emotion. Neuropsychologia 1997; 35: 1437-1444.

Marin OSM, Perry DW. Neurological aspects of music perception and performance. In: The Psychology of Music, 2nd Edition In: Deutsch D, editor, New York: Academic Press, 1999.

Milner B. Laterality effects in audition. In: Interhemispheric Relations and Cerebral Dominance. In: Mountcastle VB, editor, Baltimore: Johns Hopkins Press, 1962: 177-195.

Perry DW, Zatorre RJ, Petrides M, Alivisatos B, Meyer E, Evans AC. Localization of cerebral activity during simple singing. NeuroReport, 1999; 10: 3979-3984.

Peretz I, Gagnon L, Bouchard B. Music and emotion: perceptual determinants, immediacy and isolation after brain damage. Cognition 1998; 68: 111-141.

Peretz I, Kolinsky R, Tramo M, Labrecque R, Hublet C, Demeurisse G, Belleville S. Functional dissociations following bilateral lesions of auditory cortex. Brain 1994; 117: 1283-1301.

Petrides M, Alivisatos B, Evans AC, Meyer E. Proceedings of the National Academy of Sciences (U.S.A.) 1993; 90: 873-877.

Rao SM, Binder JR, Bandettini PA, Hammeke TA, Yetkin FZ, Jesmanowicz A, Lisk LM, Morris GL, Mueller WM, Estkowski LD, Wong EC, Haughton VM, Hyde JS. Functional magnetic resonance imaging of complex human movements. Neurology, 1993; 43: 2311-2318.

Robin DA, Tranel D, Damasio H. Auditory perception of temporal and spectral events in patients with focal left and right cerebral lesions. Brain Lang 1990; 39: 539-555.

Rolls ET, Hornak J, Wade D, McGrath J. Emotion-related learning in patients with social and emotional changes associated with frontal lobe damage. J Neurol Neurosurg Psychiatry 1994; 57: 1518-1524.

Samson S, Zatorre RJ. Discrimination of melodic and harmonic stimuli after unilateral cerebral excisions. Brain Cognit 1988; 7: 348-360.

Samson S, Zatorre RJ. Contribution of the right temporal lobe to musical timbre discrimination. Neuropsychologia 1994; 32: 231-240.

Schlaug G, Jäncke L, Huang Y, Steinmetz H. In vivo evidence of structural brain asymmetry in musicians. Science 1995; 267: 699-701.

Sidtis JJ, Volpe BT. Selective loss of complex-pitch or speech discrimination after unilateral lesion. Brain Lang 1998; 34: 235-245.

Wise RJ, Chollet F, Hadar U, Friston K, Hoffner E, Frackowiak R. Distribution of cortical neural networks involved in word comprehension and word retrieval. Brain 1991; 114: 1803-1817.

Worsley KJ, Evans AC, Marrett S, Neelin P. A three-dimensional statistical analysis for CBF activation studies in human brain. J Cereb Blood Flow Metab 1992; 12: 900-918.

Zatorre RJ. Discrimination and recognition of tonal melodies after unilateral cerebral excisions. Neuropsychologia 1985; 23: 31-41.

Zatorre RJ. Pitch perception of complex tones and human temporal-lobe function. J Acous Soc Amer 1999; 84: 566-572.

Zatorre RJ, Evans AC, Meyer E, Gjedde A. Lateralization of phonetic and pitch processing in speech perception. Science 1992; 256: 846-849.

Zatorre RJ, Evans AC, Meyer E. Neural mechanisms underlying melodic perception and memory for pitch. J Neurosci 1994; 14: 1908-1919.

Zatorre RJ, Halpern AR. Effect of unilateral temporal-lobe excision on auditory perception and imagery. Neuropsychologia 1993; 31: 221-232.

Zatorre RJ, Halpern AR, Perry DW, Meyer E, Evans AC. Hearing in the mind's ear: A PET investigation of musical imagery and perception. J Cognit Neurosci 1996; 8: 29-46.

Zatorre RJ, Perry DW, Beckett CA, Westbury CF, Evans AC. Functional anatomy of musical processing in listeners with absolute pitch and relative pitch. Proceedings of the National Academy of Sciences (U.S.A.) 1998; 95: 3172-3177.

Zatorre RJ, Samson S. Role of the right temporal neocortex in retention of pitch in auditory short-term memory. Brain 1991; 114: 2403-2417.

Chapter III.8

The Relationship between Musical Pitch and Temporal Responses of the Auditory Nerve Fibers

Kengo Ohgushi[1] and Yukiko Ano

Faculty of Music, Kyoto City University of Arts

Introduction

A musical scale is a succession of notes arranged in ascending or descending order. Each note on a musical scale has a specific note name or pitch name and these names are the same in all octaves. In music theory, the frequency ratio relations between any two notes on a musical scale within an octave are maintained in other octaves.

However, in the high-frequency range, the frequency relations among tones within one octave are perceived to be slightly shifted upward. For example, woodwind players suggest that performed tones by the piccolo often sound lower in pitch than what should be. Further, the octave enlargement phenomenon is pronounced in the high-frequency range. Therefore, we made two experiments in the high-frequency range investigating both the production and perception aspects: 1) measuring the fundamental frequencies of performed tones on the piccolo, 2) measuring the frequencies of tones corresponding to each pitch name by absolute pitch possessors.

Fundamental Frequency Measurement of Performed Tones on the Piccolo

Piccolo players (in general flutists play the piccolo in the orchestra) often have difficulty in playing tones with a correct pitch in the high-frequency range. The piccolo is about one-half the length of the flute, and it sounds an octave higher. They suggest from their experience that the performed fundamental frequency is often higher than the theoretical value, when

---

[1] Correspondence: Kengo Ohgushi, Ph.D., Faculty of Music, Kyoto City Universiy of Arts, 13-6 Kutsukake-cho, Nishikyo-ku, Kyoto 610-1197, Japan, Tel: (81)-75-332-0701, Fax: (81)-75-332-0709

measuring the fundamental frequency of the tone by an electronic counter which they often have with them. We decided to measure systematically the fundamental frequency of performed pitch to confirm the experiential result and to obtain the quantitative data.

Method

Nine flutists (6 professionals and 3 music students) produced tones consisting of C major scale tones, from C6 to C8, on their own piccolo. The standard tone A4 (442 Hz) was presented to the players using a piano. The flutists played long succcessive tones of 2-3 s without vibrato. The frequency measurement was carried out by measuring the periods of the waveform. Stable portions, where frequency fluctuation were small, were used for the measurement.

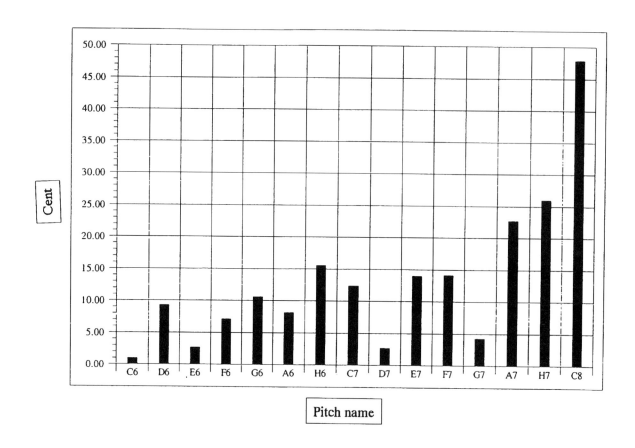

Figure 1
The fundamental frequency of the performed tone as a function of pitch name. Nine flutists produced tones consisting of C major scale tones, from C6 to C8, on their own piccolo. The ordinate shows the frequency deviation in cents, of each frequency from the theoretical value.

Results

Fig.1 shows the frequency of the performed tone as a function of pitch name. The abscissa represents the pitch name from C6 to C8. The ordinate shows the frequency deviation, expressed in cents, of each frequency from the theoretical value, where the thoretical value is calculated according to the equal tempered scale with A4 of 442 Hz.

As seen in Fig.1, there are rather large individual differences in frequency. Fig.2 shows the average data over 7 players, excluding two players whose frequencies were the highest and lowest. The ordinate shows the average deviation from the theoretical value. This figure shows that all tones were produced higher in frequency than the theoretical value and that this tendency is striking in higher tones. We were able to confirm the experiential fact that the electornic tuner's indication does not fit the perceived pitch in the high-frequency range.

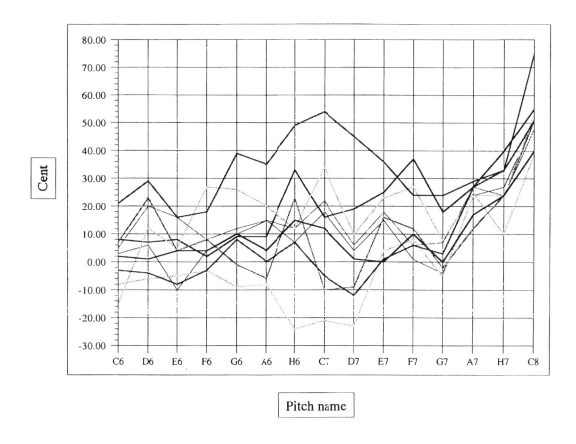

Figure 2
The average frequency deviation over seven flutists, excluding two players whose frequencies were the highest and the lowest.

Pitch Judgment of High-Frequency Pure Tones by Absolute Pitch Possessors

The perceptual octave is known to be slightly larger than the physical octave, which is a factor of two. This tendency is more pronounced in the high-frequency range. These facts make us expect that a high-frequency tone induces a higher pitch than would be predicted from the conventional musical scale. However, the perceived pitch of tones has never previously been measured without a reference tone. To measure the perceived pitch of tones without any references, we asked absolute pitch possessors to serve as subjects. Absolute pitch possessors can identify the pitch of tones by a specific note name. They can also judge whether a given musical tone deviates from the standard pitch or not. The results shown in Fig.1 and Fig.2 suggest that the frequency corresponding to each pitch name for pure tones may deviate from the theoretical value in the high-frequency range.

Method

Thirty-one pure tone stimuli for each pitch name were generated for 15 pitch names from C6 to C8 and for A4. The experimental method of the experiment used here was the "method of limits". Ten subjects with absolute pitch were presented a tone changing in frequency by 2.3 cents on every step. They judged whether the tone was higher than, equal to, or lower than a given pitch name. The points of subjective equality (PSE) were calculated from these data.

Results

The individual difference are less than what might be expected. For example, the PSE for the A4 tone ranged from 440.3 Hz to 441.5 Hz (5 cent width) over ten subjects. Further, the PSE for the C8 tone ranged from 4225 Hz to 4242 Hz (7 cent width). Fig.3 shows the average data over ten subjects. Comparison between Fig.2 and Fig.3 revealed that the trend of the pitch deviations is similar. Both figures show greater deviations in the high-frequency range. However, the amount of the deviation in Fig.3 is generally smaller than that in Fig.2.

Discussion

Relation between frequency and musical pitch

As mentioned above, the frequency ratios between any two notes on a musical scale within an octave should be maintained in other octaves. However, the results of the frequency measurement of performed pitch on the piccolo showed on average rather large upward deviations from the theoretical values in the high-frequency range, although there were considerable individual differences. Furthermore, the results of the listening experiments for the absolute pitch possessors also showed upward deviations although they were smaller. This shows that the normal relation between frequency and musical pitch no longer holds in the high-frequency range.

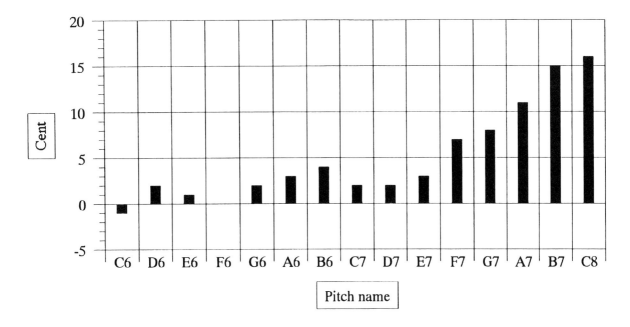

Figure 3
The frequencies corresponding to each pitch name judged by ten absolute pitch possessors.

Physiological origin of upward pitch shift in the high-frequency range

An acoustic stimulus is transformed in the innner ear into neural impulses which are conducted through the auditory nerve fibers. The auditory nerve fibers discharge impulses corresponding to peaks of the stimulating waveform for low frequency tone. The temporal distribution of these impulses can be represented in the form of an interspike interval histogram. Fig.4 shows an example of such histogram for impulses from a single auditory nerve fiber of a squirrel monkey when a pure tone of 1000 Hz was presented (Rose et al.1967). The abscissa represents the interspike interval with a binwidth of 0.1 ms and the ordinate indicates the number of intervals in each bin. This histogram shows that the interspike interval distribution is polymodal, and that each modal value corresponds approximately to the period of the tone and its integral multiples. Let μi be the i-fold value of the period of tone divided by the ith modal value. It was observed that the value of μ1 was not exactly 1.0, but depends on the frequency of the tone. I calculated μ1 value from the data for four auditory nerve fibers published by Rose et al. (1967, 1968). The calculated result is shown in Fig.5. The abscissa shows the frequency of the sinusoidal stimulus and the ordinate indicates the value of μ1. This figure shows that the values of μ1 depend clearly on the frequency of the tone.

Neural responses to an acoustic stimulus are considered to carry two types of information; by spatial distribution (or tonotopic organization) of firing fibers and by temporal distribution of neural impulses. The interspike interval corresponding to successive peaks of the stimulating waveform is an important cue for pitch perception. The octave enlargement phenomenon was explained by the fact that µl values become smaller in the high-frequency range (Ohgushi 1983). The fact that the value of µl lowers with increasing frequency also suggests that the musical pitch shifts upward in the high-frequency range.

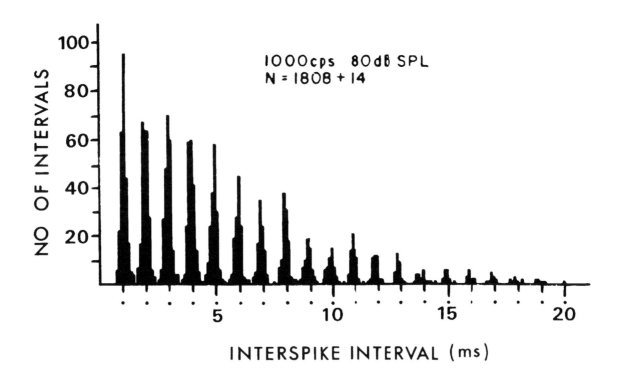

Figure 4
An example of an interspike interval histogram of impulses from a single auditory nerve fibers of a squirrel monkey when a pure tone of 1000 Hz was presented (Rose et al.1967).

Difference between performed pitch and perceived pure tone pitch

In general, the performed pitch was higher than the perceived pitch of pure tones in the high-frequency range. How can we account for this difference? In the performance experiment, the flutists played each tone as a pitch in the C major scale. They did not play each tone independently, but played under a melodic context. In octave matching experiments, the melodic octave is known to be larger than the harmonic octave (Ohgushi,1983). The melodic context may have influenced the experimental data. On the other hand, the perceived pitch experiment for pure tones was carried out without any melodic context. This difference may be one of the reason why the performed pitch was higher than the perceived pitch.

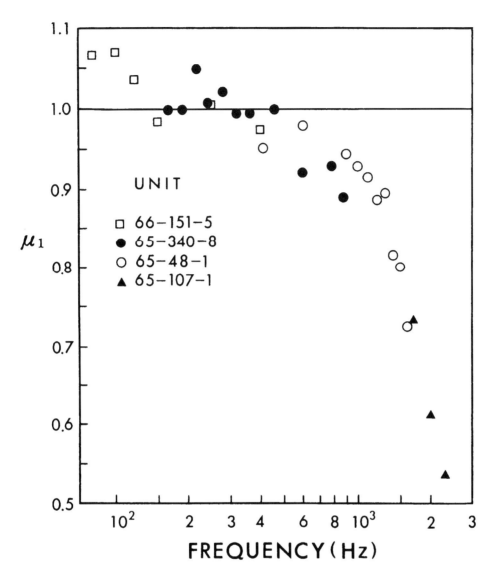

Figure 5
The calculated value $\mu 1$. $\mu 1$ is the value of the period of tone divided by the first modal value (Ohgushi,1983).

Relation to tuning curve for pianos

It is known that in pianos tones higher than C6 are tuned slightly deviated from equal temperament. The amount of the deviation is 1-2 cents for C6, 10 cents for C7 and 30 cents for C8. The inharmonicity in piano strings is considered to be the main reason why pianos are "stretch-tuned"(Rossing,1990). However, the tendency is almost the same as the data in Fig.3, where there was no problem of inharmonicity because we used pure tones. In our experiment, however, the upward frequency shift in the high-frequency note is more pronounced for the performed pitch than the perceived pitch. This suggests that deviation from equal temperament in the high-frequency range is due to the process of transmission in the auditory nervous system as well as the inharmonicity in piano strings.

References

Ohgushi,K. On the role of spatial and temporal cues in the perception of the pitch of complex tones. J.Acoust Soc.Am 1978; 64: 764-771.

Ohgushi,K. The origin of tonality and a possible explanation of the octave enlargement phenomenon. J.Acoust Soc.Am 1983; 73: 1694-1700.

Rose,JE. Brugge,JF. Anderson DJ. and Hind,JE. Phase-locked response to Low-frequency tones in single auditory nerve fibers of the squirrel monkey. J.Neurophysiol 1967; 30: 769-793.

Rose,JE. Brugge,JF. Anderson,DJ. and Hind,JE. Patterns of activity in single auditory nerve fibers of the squirrel monkey. In: de Euck, AVS and Knight,J. Editors. Hearing Mechanisms in Vertebrates. London: Churchill,1968; 144-168.

Rossing,TR. The science of sound 2nd edition. Addison-Wesley Publishing Company,1990; 290-292.

Ward,WD. Subjective musical pitch. J.Acoust Soc.Am 1954; 26: 369-380.

T. Nakada (Ed.)
Integrated Human Brain Science: Theory, Method Application (Music)
© 2000 Elsevier Science B.V. All rights reserved

Chapter III.9

What Can the Brain Tell Us about the Specificity of Language and Music Processing ?

Mireille Besson [1]

Center for Research in Cognitive Neuroscience, CNRS

Introduction

In this chapter, I will address the question of the specificity of language processing by comparing the computations realized by the brain to process some aspects of language with those involved when processing some aspects of music. It should be noted that, in the general considerations as well as in the specific experiments described below, I will only consider classical Western music. I will start by reviewing two fundamental arguments that have been used in favor of the specificity of language processing. These arguments are mainly issued from the theory of generative grammar, first developed by Chomsky (1957), but have been called into question by recent alternative linguistic theories (Benveniste, 1966; Fauconnier, 1997; Fuchs, 1997; Langacker, 1986; Robert, 1997; Talmy, in press). I will then rapidly describe the different levels of processing involved in language perception and comprehension as well as in music perception. Some words will also be said about the brain imaging methods that have been used (Event-Related brain Potentials and functional Magnetic Resonance Imaging), before summarizing the results of several experiments directly aimed at the comparison between different aspects of language and music processing.

The Arguments in Favor of the Specificity of Language Processing

Several models in psycholinguistic research have been proposed to explain how an acoustic signal or a chain of characters is identified and transformed so as to access meaning.

---

[1] Correspondence: Mireille Besson, Ph.D., Center for Research in Cognitive Neuroscience, 31, Chemin Joseph Aiguier, 13402 Marseille Cedex 20, France, Tel: (33)-4-91-16-43-05, Fax: (33)-4-91-77-49-69

Let's consider natural speech. The first stage would be to segment the speech signal into significant phonological units, the phonemes, and to assemble them to form a phonological representation. Based on this phonological representation, a lexical entry would be selected that allows to identify the spoken word. The nature of the lexical representation is still a matter of debate since it is unclear whether morphological, syntactic and semantic information are also included within the lexical representation or, are, in contrast, stored in different, specific representations. In any event, both semantic and syntactic information are available regarding the meaning of the word and its role within the sentence (noun, verb, pronoun, preposition, etc…). Accessing these semantic and syntactic attributes allows to understand the meaning of a word and its function and to build, through an integration process, a coherent representation of the utterance.

This quick description of the processes involved in understanding natural speech is not without theoretical assumptions. For instance, this description implies a serial and independent chain of computations realized by specialized sub-systems or modules, to use the terminology adopted by Fodor (1983). In his view, the linguistic system comprises several modules, each responsible for a specific aspect of language processing, phonological, prosodic, semantic, syntactic and pragmatic. However, while several models in psycholinguistics rely on such conceptions (Frazier, 1987; Gorrel, 1995, Friederici, 1998), others are, in contrast, issued from an interactive conception of information processing. In these models, the outcome of one aspect of processing is constantly influencing the outcome of the other aspects (Marslen-Wilson & Tyler, 1980; McClelland, St John & Taraban, 1989; Seidenberg, 1997). Whether the syntactic structure of an utterance is extracted before, and independently of, its semantic representation or whether both are processed in interaction and in parallel is still at the center of hot debates in the literature. Similarly, whether building a phonological representation is a pre-requisite to access a semantic representation, or whether both are accessed simultaneously is still an open question. While those are interesting and important issues, I will not discuss them further, but rather concentrate on the question of the specificity of linguistic computations.

Several authors, following Chomsky (1957), advocate the idea, central to the theory of generative grammar, that language is autonomous from other cognitive functions (Jackendoff, 1994; Pinker, 1994). Jackendoff presents two fundamental arguments in favor of this position. The first one relies on the existence of a mental grammar. The fact that every normal human being speaking his mother tongue, is able to build and create an infinite number of sentences from a finite number of words, implies that the human brain entails unconscious grammatical principles. On the basis of such principles one can decide that one sentence is grammatical while the other is not. The well-known example used by Chomsky (1957) to illustrate this point, as well as the independence of semantics and syntax, is that the sentence "Colorless green ideas sleep furiously" is grammatical while "Furiously sleep ideas green coloursless" is not. The second fundamental argument relies on an innate knowledge of the main aspects of grammar and what Jackendoff (1994) calls the "paradox of language acquisition". While normally developed children learn the basic language principles very quickly and without much effort, the most talented linguists, helped with the most powerful computers, are still unable to explain the basic principles that govern the languages of the world. These arguments are summarized in the following citation from Pinker (1994, p.18): "Language is a complex, specialized skill, which develops in the child spontaneously, without effort or formal instruction, is deployed without awareness of its underlying logic, is qualitatively the same in every individual, and is distinct from more general abilities to process information or behave intelligently". While each aspects of this citation rises numerous questions that are still at the center of important research programs, the final claim, that language is independent from other general cognitive abilities, is what I would like to focus on in the following of this chapter.

The idea that the computations involved to process linguistic information differ from those necessary to process other type of information has been called into question (Fuchs, 1997; Kail, 1997; Robert, 1997; Seidenberg, 1997). Within the sequence of computations that will be performed by a listener, for instance, to extract meaning from the basic acoustic signals, some computations are probably specific to the fact that these signals form the words of a language. Other computations, however, are probably common to other auditory signals that are organized in function of precise structural and functional rules. Language would be "an emergent property relying on the general principles of cognition and entertaining numerous homologies with other cognitive functions and, in particular, with perception" (Fuchs, 1997, p.16). This view is also advocated by Seidenberg (19970 who wrote: "Brain organization therefore constrains how language is learned, but the principles that govern the acquisition, representation and use of language are not specific to this type of knowledge" (p. 1603). In order to demonstrate that language relies on some general principles of cognition, it is necessary, first, to specify the different computations that are involved in processing language and, second, to compare these computations with those required to process information issued from other organized systems: music is an interesting candidate.

The Comparison between Language and Music

This topic was at the center of hot debates between philosophers and scientists from the 17th to the 19th century. Early on, the discussions mainly focused on the common or independent origins of language and music. Jean-Jacques Rousseau (1781) was a fervent advocate of the idea that language and music had a common origin with music appearing first: primitive languages were sung rather than spoken. Their functions were to express feelings: love, hate, angryness... While also considering that the basic function of language was to express emotions, and agreeing upon the common origins of language and music, the philosopher Herbert Spencer (1857) favored the idea that music evolved out of language. He proposed a physiological theory of language and music, following which emotions and movements are directly connected: the stronger the emotion, the larger the movement. Thus, strong emotions lead to large movements of the articulatory system and to the production of large intervals in the intonation of the voice. Rene Descartes (1618/1987) had first developed this idea. A music with a rapid tempo strongly activates the "animal spirits" that circulate between some parts of the brain and the muscles, thus inducing rhythmic movements of the body. Placing the debate within an evolutionary perspective, Darwin (1871/1981) also favored the conception that language and music have a common origin, with music evolving from the love calls during reproduction: "Musical notes and rhythm were first acquired by the male or female progenitors of mankind for the sake of charming the opposite sex" ("The descent of man", p. 336).

In contrast, musicologists of the 19th and 20th centuries seem to favor the view following which language and music evolved independently. Wallaschek (1891) hold, for instance, that music finds its origin in primary rhythmic pulsions necessary to use the energy left over after the accomplishment of elementary needs. Newman (1905) also consider that music evolved before and independently of language mainly because it is easier to produce music than language: "Man certainly expressed his feeling in pure indefinite sound long before he had learned to agree with his fellows to attach certain meanings to certain stereotyped sounds" (p. 210).

Differences between language and music

While the question of the origin of language and music may remain unsolved, a related, less controversial issue is linked with the function of language and music. Whether to charm the opposite sex or to use the left over energy when elementary needs have been fulfilled, most authors would agree that the basic function of music, and of primitive languages, is to express emotion. Music is a language that allows to translate emotional experiences into artistic forms (Nadel, 1930), to communicate with the unknown, as can be seen from the religious rituals in primal tribes. Music ensures the cohesion of the social group and, as such, plays a powerful social function in all human societies (Arom & Khalfa, 1999; Boucourechliev, 1993; Levman, 1992; Nadel, 1930). Language also plays social functions but different from music, in that language has developed to express rational thought. In the course of evolution, language lost the isomorphism between sound and meaning (e.g., onomatopees) to become symbolic, through the development of an arbitrary relationship between the acoustic features and the associated concept (Levman, 1992; Saussure, 1916). Thus, one of the main differences between language and music seems to rely on their social function. While both serve to communicate, language mainly expresses rational communication and music expresses emotional communication. In this consensus rises the discordant voice of Pinker who recently wrote: "As far as biological cause and effect are concerned, music is useless... music could vanish from our species and the rest of our lifestyle would be virtually unchanged" (p. 528). This, however, appears as an ill-posed problem, since the point is rather to explain why all cultures have developed music.

Aside from the aforementioned difference in the social function of language and music, a second important difference is linked with the problem of the semantic of music. In other words, does music have meaning? While this again is a controversial issue, it seems that several authors would agree that while words have meaning by reference to an extra-linguistic designated space, music is self-referential (Boucourechliev, 1993; Kivy, 1991; Jakobson, 1973; Meyer, 1956). As stated by Leonard Meyer, in his influential book "Emotion and meaning in music" (1956): "Music means itself. That is, one musical event ... has meaning because it points to and makes us expect another musical event" (p. 35). Thus, if one considers the meaning of a word in language as the emergent property of the arbitrary relationship between the acoustic form and the concept, music, in contrast to language has no meaning. The expressive communicative power of music is in its form, in the musical figures, rather than in the notes or chords that do not express anything by themselves.

So far, I have considered two main differences between language and music. One is linked with their social function (rational versus emotional communication) and the other, with the question of meaning. A third difference emerges from their structural organization. While language is only organized horizontally, a succession of sounds in time, music offers both an horizontal organization, the melody formed by the pitch relationship between successive sounds, and a vertical organization, the simultaneous production of two or more sounds such as in chords. Note, however, that what can be considered as a difference at one level of analysis may function as a similarity at another level. For instance, let's consider the phonems and the chords: they are both similar and different. Different based on their acoustic and functional characteristics within the linguistic or musical system. Similar because they both derive from large pitch variations combined in space and time. Most importantly, both phonems and chords acquire their functional value from the relationship that they entertain with the other linguistic or musical elements. A phonem is assembled to another to form a morphem, morphems are tied together to form words and the assembly of words creates the sentence. A chord also acquires its functional value from the relationship that ties it to the other chords: the succession of chords creates the musical theme, the musical phrase, the musical piece...

Similarities between language and music

Whatever the level considered for analysis, there are also interesting similarities between language and music. Both have been developed by all human cultures and seem specific to humans[2]. Both language and music rely on a sequential organization of sounds that unfold in time. These sound are in both cases characterized by pitch, duration, intensity and timber parameters. They are structured into separate units by variations in voice intonation, prosody, that tend to go down at the end of sentences, and by the cadence at the end of musical phrases (i.e., the classic succesion of tonic, dominant, sub-dominant and tonic chords at the end of musical phrases). Interestingly, categorical perception effects (De Witt & Samuel, 1990) and phonemic restauration effects (Warren, 1970) have been reported in music as well as in language perception.

Levels of processing in language and music

Most importantly both language and music perception rely on different levels of representation. In language, one classically differentiates the orthographic, phonological, prosodic, semantic and pragmatic levels, and the musical notation, rhythmic, melodic and harmonic levels in music. The research that I report below was precisely aimed at comparing some aspects of these different levels of processing. However, before going into the results of these experiments, I would like to make some general predictions based upon the considerations stated above. We saw that music has no meaning in the sense that music is not referenced to an extra-musical space: music is self-referential. Therefore, by comparing the semantic aspects of language with the melodic or harmonic aspects of music, results should highlight the differences rather than the similarities between language and music. In contrast, classical, baroque music is organized following the precise rules of harmony and counter-point. Harmony determines which chord is allowed in a specific musical context (vertical organization) and organizes the relationships between successive sounds (horizontal

---

[2] Without entering the debate about animal communication, it seems to be mostly directed toward survival (e.g., alarm calls) or reproductive functions (e.g., bird songs).

organization). Counterpoint, as the art to combine different melodic lines, also constraints the horizontal organization and is most beautifully illustrated in the work of Jean-Sebastien Bach (1685/1750). Therefore, syntactic rules govern the organization of both music and language. Note, however, that musical syntax seems to evolve more rapidly than linguistic syntax: there are larger differences between the works of Bach and Beethoven than between the works of Descartes and Stendhal.

Finally, concerning syntax, it is interesting to note that the argument used by Chomsky (1957) to demonstrate the existence of unconscious mental grammar, the fact that every human individual normally constituted is able to judge that the sentence "Colourless green ideas sleep furiously" is grammatical, while "Furiously sleep ideas green coulorsless" is not, can be applied to music as well. No specific musical training is indeed necessary to determine that a series of sounds that unfolds in time is musical while the same succession but in reversed order is not. Following Chomsky's logic, this would imply that the human brain contains unconscious musical principles allowing us to decide what is musical and what is not. As for language this, however, remains to be demonstrated, and is probably largely influenced by each individual specific cultural background. To conclude this short review, I would like to once again cite Jackendoff (1994) who, after having developed many arguments in favor of the specificity of language, notes: "We see then, that language is not splendidly isolated among human mental capacities. All its basic characteristics are mirrored in our ability to understand music" (p. 171). The door is thus wide open for the comparison between language and music processing.

Brain Imaging Methods

In order to address the question of the similarity of language and music processing, two brain imaging methods were used. One, the Event-Related potential (ERPs) method, offers a very good temporal resolution and, therefore, allows to study the time-course of the mental operations required to process the information of interest. The other, functional Magnetic Resonance Imaging (fMRI) offers a very good spatial resolution and allows to determine which cerebral structures, or which network of brain structures, are activated when processing the information of interest. Of course, one would hope to be able to combine the results from these two methods to understand the spatio-temporal dynamics of the brain activity that underlies behavior, and more specifically language and music processing. This remains one of the major challenges in cognitive neurosciences. I will come back to this issue at the end of this section, but, before, I will first describe in more details what are the principles underlying each method and some of the main results.

The Event-Related Potential method

The ERPs method allows to study the variations in the brain electrical activity that are time-locked with the event of interest. Each event, be it external, a sound, a light, a touch..., or internal, a decision, an expectation, a preparation to act..., induces small variations in the Electroencephalogram (EEG, Berger, 1929). This variation, the evoked potential, is however, of such small amplitude (5-15 μV) that it cannot be detected within the larger background noise of the EEG, on the order of magnitude of 100 μV.

As was first proposed by Dawson (1953), it is therefore necessary to average the evoked potentials by taking as a time origin the onset of the stimulation (e.g., the sound, the light...). While all variations that are time-locked to stimulus onset will sum up and emerge from the background EEG noise, all random fluctuations that are not time-locked will cancel out. Therefore, the results of the averaging procedure is an average ERP that reflect the variations directly evoked by the stimulus as a function of time. The ERP is composed of several components that are positive or negative relative to a baseline, that is, the mean amplitude of the brain electrical activity within the 200 ms before stimulus onset. Components also differ by their amplitude, their onset latency, and their distribution across the electrodes placed at several locations over the scalp. Moreover, components are also differently sensitive to the manipulation of experimental variables (Donchin, 1979).

Using this method, Kutas and Hillyard (1980) were able to show that the presentation of an incongruous word at the end of a sentence (e.g., "I take coffee with cream and dog") elicits a negative component with maximum amplitude around 400 ms, the N400 component (see Figure 1). Further research showed that incongruous words are not a necessary condition to elicit an N400 component. Rather, every word is associated with an N400, which amplitude varies as a function of semantic expectancy: the less a word is expected within a sentence context, the larger the N400 amplitude. Since this seminal study, a large number of experiments have been aimed at specifying the functional significance of the N400 and have used the N400 to test models issued from psycholinguistic research. It has thus been shown that N400 amplitude is modulated by factors

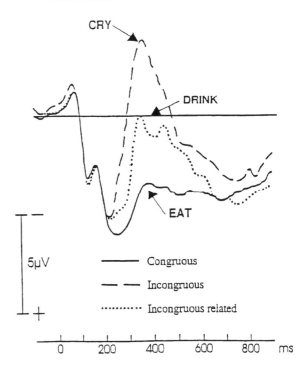

Figure 1
ERPs elicited by sentence-final words at the central recording site (Cz) for congruous and incongruous words and for incongruous words that are semantically related to the best sentence completion. The amplitude of the negative component, peaking at 400 ms post-final word onset (N400) is largest for incongruous words, intermediate for incongruous words related to the best sentence completion and smallest for congruous words. In this and subsequent figures, amplitude (μV) is represented on the ordinate, with negative voltage up, and time (ms) on the abscissa *(from Kutas & Hillyard, 1980).*

such as a word lexical frequency, repetition, contextual probability, etc... and that its occurrence is most likely linked with the integration of a word meaning within the sentence context, or more generally, the linguistic context within which it occurs.

One of the main advantages of the ERP method, as mentioned above, is its very good temporal resolution[3]. This is particularly useful when the processes of interest unfold very rapidly in time, as is the case for the computations involved in language processing. On average, a normal individual is able to understand the meaning of a word in less than 200 ms and to integrate very rapidly the meaning of different words to form a coherent representation of the utterance. One of the main disadvantages of this method, however, is the difficulty to localize the generators of the effects recorded on the scalp. As demonstrated by Helmholtz, the so-called "inverse problem", is an ill-posed problem because of an infinite number of solutions. In other words, the number of neuronal configurations that can give rise to the pattern of activity recorded on the scalp is infinite. One solution to partially solve this problem is to constraint the number of possible solutions based on the information provided by other brain imaging method with excellent spatial resolution. One such method is fMRI.

Functional Magnetic Resonance Imaging

While the ERP and MEG methods permit a direct measure of brain activity, the fMRI method is an indirect measure of brain activity through the changes in metabolic activity associated with the variations in cerebral blood flow. It has been known for a long time that local blood flow and oxygen consumption increase in the brain structures that are active. As the increase in local blood flow is larger than oxygen consumption, oxygen concentration increases in the active brain area. Linus Pauling was given one of his Nobel Prize (1954) for the discovery that oxygen concentration modify the magnetic properties of protons (he was also awarded the Peace Nobel prize in 1962). Indeed, oxyhemoglobin (i.e., the hemoglobin carrying the oxygen molecules) is less magnetic than desoxyhemoglobin (i.e., the hemoglobin that lost his oxygen molecules). Therefore, the magnetic perturbations drawn on protons are less strong when a brain area is active (i.e., the concentration in oxygen is high) than when a brain area is not active (i.e., the concentration in oxygen is low). Consequently, the magnetic resonance signal emitted by the concerted excitation of hydrogen protons is stronger when the brain area is active than when it is not. This is called the "Blood oxygenation level-dependent" signal (Kwong et al, 1992).

One major disadvantage of the fMRI method is its poor temporal resolution linked with the long latency of the hemodynamic response (i.e., the time needed for the oxygen concentration to increase in the active regions of the brain), between 4 and 6 seconds. As we saw above, this is much longer than the duration of the mental computations involved in language processing (e.g, access to the meaning of a word in less than 200 ms). To accommodate with the long latency of the hemodynamic response, the experimental designs used in fMRI research have mainly been "blocked designs". Basically, stimuli belonging to the same experimental condition are repeated in succession within a block of trials to allow time for the hemodynamic response to develop. However, research in experimental psychology has

---

[3] Magnetoencephalography (MEG), that allows to analyze the changes in the magnetic fields associated with the changes in electrical activity, also offers an excellent temporal resolution and is a direct measure of brain activity.

previously demonstrated that the informational content of a stimulus is processed differently when subjects are able to anticipate on the basis of one stimulus what the next stimulus will be. Processing strategies therefore differ from those used in so-called "mixed designs", in which stimuli belonging to different experimental conditions are mixed together and occur randomly, or pseudo-randomly, within a sequence. Recently, mixed designs have been used in fMRI research, based upon Event-related fMRI, so as to analyze the hemodynamic response to a specific event rather than to a group of events (Dale & Buckner, 1997). This permits to use the same basic designs than in experimental psychology and results can be compared. Moreover, it also allows to sort trials as a function of the response and to take into consideration only those in which the subject was correct. Finally, event-related fMRI offers an increased temporal resolution and it is now becoming possible to examine the variations in the BOLD signal associated with events appearing in close temporal succession (less than 1 second).

One last point should be considered regarding the fMRI method. This method is aimed at discovering the cerebral structures associated with specific mental functions. Consequently, experimental designs are most often based upon the theoretical model of the mental function under study. Hierarchical models of information processing have been widely used in which different stages of processing subserve different computations with the output of one stage serving as input for the next stage (Sternberg, 1969). Therefore, the assumption is that by subtracting the activity in the lower stage from the activity in the higher stage, one is able to isolate the specific mental computation that differentiate the lower and the higher processing stages. This may, however, not always be the case, specifically when considering cognitive functions such as language and music, in which the hierarchical organization, if it exists, is not well-defined. Related to this point is the choice of the control condition or baseline level. Again, one has to make sure that the control condition only differs from the experimental condition by the factor of interest, that is the computation that needs to be isolated.

Conclusion

The use of fMRI in cognitive neuroscience often relies upon the assumption of the modularity of cognitive functions. One will search for the brain areas involved in language processing or in solving mathematical problems, for instance. However, results gained with this method tend to show that a strictly localisationist conception cannot account for the complexity of the effect observed. In contrast, the idea following which distributed networks of brain structures are activated, simultaneously or in parallel, as a function of the materials to process and the task at hand, seems more realistic. The most important issue is therefore to integrate results from methods with excellent temporal resolution (e.g., ERPs) and excellent spatial resolution (e.g., fMRI) to study the spatio-temporal dynamics of the activation of these cerebral networks. This aim is actually difficult to reach both because of technical issues (it is difficult to simultaneously record electrical and hemodynamic responses) and theoretical problems (the relationships between the electrical and hemodynamic activities are not well-understood). Based on the extremely rapid progress that is being made in brain imaging research, one can hope that the technical and theoretical tools necessary to integrate the information gained from these two methods will be available shortly.

What do we learn about the specificity of language processing by using brain imaging methods?

While research on the neural substrates of the different components of auditory language processing, phonology, semantics, syntax and to a lesser extend pragmatics, has largely been developed over the past twenty years, studying the neurophysiological basis of music processing is a relatively new research topic. Thus, research on the different components of music processing has benefited from results acquired in language research and similar experimental designs are often used in both research areas. Furthermore, the direct comparison of the effects found in language and music processing elicits growing interest. The idea is that if qualitatively similar effects are found when comparing specific aspects of language, such as prosody, semantic and syntax, and specific aspects of music, such as rhythm, melody and harmony, this will provide strong arguments against the position advocated by Jackendoff (1994) and Pinker 1994), and described in the introduction, following which the computations performed to process language are specific to language. Moreover, this conclusion would be strongly supported if similar networks of brain structures are shown to be activated by specific aspects of language and music processing.

Semantic, melody and harmony

A starting point in the study of the neurophysiological basis of language processing has been the discovery of the N400 component by Kutas & Hillyard (1980). As mentioned previously, this negative component of the ERPs, peaking around 400 ms after word onset, is elicited by words that are semantically unexpected within a linguistic context, be it a sentence or a single word (e.g., "I carry my daughter in my nostrils"; "Fruit - Lion"). Results of numerous experiments have lead to consider that the N400 reflects the integration process by which the meaning of a word is integrated within the linguistic context. The first experiments that we conducted in our laboratory were aimed at determining whether a qualitatively similar ERP component will be elicited by the presentation of melodically and harmonically unexpected notes at the end of monodic familiar and unfamiliar musical phrases (Besson & Macar, 1987; Besson, Faita & Requin, 1994; Besson & Faita, 1995). Familiar musical phrases were chosen from the classical repertoire of Western occidental music from the 18th and 19th centuries. Unfamiliar musical phrases were composed for the experiment by a musician who followed the basic rules of harmony (see Figure 2). This musical material was presented to musicians, who had at least 10 years of musical training, and to non-musicians, who never listened to music except as background noise. Melodically unexpected final notes were chosen within the tonality of the musical phrases but were not the most expected ending. Harmonically unexpected final notes were chosen out of the tonality of the musical phrase and were clearly perceived as wrong notes. By using both melodically and harmonically unexpected terminal notes we hoped to establish a degree of musical incongruity.

Results clearly showed that both types of wrong notes elicit the occurrence of a Late Positive component, P600, which amplitude varies as a function of the degree of incongruity: it is larger for the most unexpected, harmonically wrong notes than for the less unexpected, melodically wrong notes (see Figure 3). Furthermore, the amplitude of the P600 is larger for

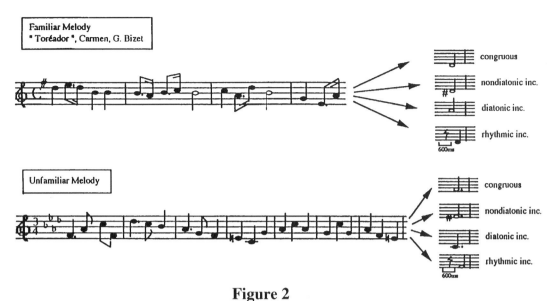

Figure 2
Examples of the stimuli used in the experiment *(from Besson & Faïta, 1995)*.

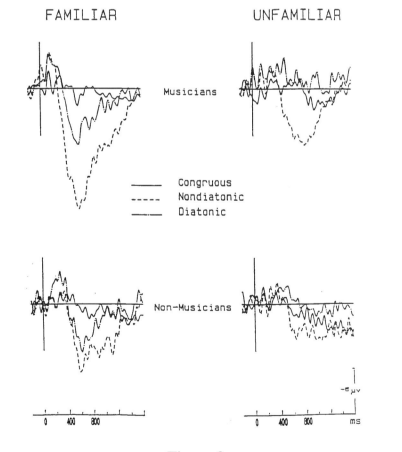

Figure 3
ERP results for musicians and non-musicians are presented separately for familiar and unfamiliar musical phrases. The vertical lines mark the onset of the final note. Results are from one typical recording site, the parietal location (Pz). The amplitude of the P600 component is larger for the nondiatonic than diatonic incongruity, for musicians than non-musicians and for familiar than unfamiliar musical phrases *(from Besson & Faïta, 1995)*.

familiar than unfamiliar musical phrases and for musicians than for non-musicians. These findings clearly demonstrate that music generates strong expectancies. In this respect, language and music share some similarities. However, they show that the processes that govern semantic expectancy, and that are reflected by the N400 component, are qualitatively different from those governing musical expectancies, that are reflected in the P600 component. Therefore, these results argue for the specificity of the processes involved in processing the semantic aspects of language. They are in line with the ideas developed in the introduction following which music has no intrinsic meaning and that, in contrast to language, music is a self-referential system that generates musical expectancy.

Semantic and harmony in opera

The "Orfeo" of Monteverdi (1604) is generally considered as the first written opera. Since this time, a recurrent question has been to determine which of the words or the music plays the most important role when we listen to songs. Authors such as Stendhal, in his "Life of Rossini" have argued that music is more important, and that "its function is to animate the words". Ethnomusicologists, such as Levman (1992), have pointed out that the lyrics are subordinate to the music in tribal songs and rituals. In contrast, Wagner considered that both aspects are intrinsically linked: "Words give rise to the music and music develops and reinforces the language", an opinion recently shared by Boulez (1966): "The text is the center and the absence of the musical piece". Richard Strauss once even composed an opera "Capriccio" to illustrate the complementarity of the words and music.

To study the processes involved when we listen to opera, and try to determine, based on scientific grounds, whether the words or the music play the most important role, we orthogonally manipulated semantic and harmonic congruity (Besson, Faita, Peretz, Bonnel & Requin, 1998). Two hundred excerpts from French opera, lasting between 8 and 20 seconds, were sung a capella by a woman singer. Each excerpt was sung in each of four experimental conditions: the final word of the excerpt was semantically congruous and sung in tune, semantically incongruous and sung in tune, semantically congruous and sung out of tune and both semantically incongruous and sung out of tune (see Figure 4).

Based on previous results (e.g., Kutas & Hillyard, 1980), it was of interest to determine whether semantically incongruous words will also elicit an N400 component when they are sung. Similarly, it was of interest to determine wether congruous words sung out of tune will also elicit a P600 component (e.g., Besson & Faita, 1995). Of most interest, was the double incongruity condition: will semantically incongruous words sung out of key elicit both an N400 and a P600 component? Will these effects be additive (i.e., equal to the sum of the effect associated with each type of incongruity alone) or interactive? To answer these questions, we recorded the ERPs associated with the final words of each excerpt from 16 professional musicians from the opera in Marseille.

To summarize, results demonstrate that sung incongruous words did elicit an N400 component, thus extending to songs results previously reported for written and spoken language (Kutas & Hillyard 1980; MacCallum et al, 1984; see Figure 5A). Moreover, words sung out of tune did elicit a P600 component, thus extending to songs results previously reported for out of tune notes (Besson & Faita, 1995; Paller et al, 1992; Verlager et al, 1992; see Figure 5B). Most interesting are the results in the double incongruity condition. They show that incongruous words sung out of tune elicit successively N400 and P600 components (see Figure 5C). Furthermore, the N400 occurred earlier than the P600 which is taken as evidence that the words are processed faster than the music. Finally, effects in the double incongruity condition were not significantly different from the sum of the effects observed in each simple incongruity condition which is taken as a strong argument for the independence (i.e., the additivity) of language and music processing. When we listen to opera, we seem to process the lyrics and the tunes separately.

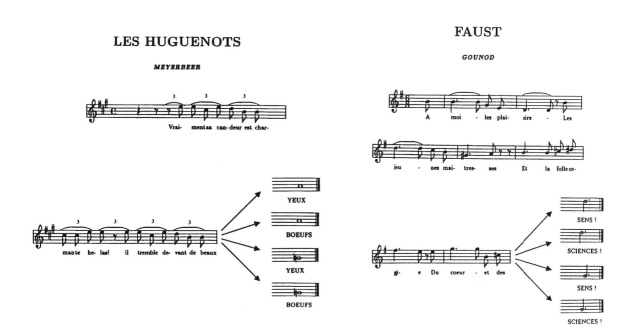

Figure 4
Example of the opera's excerpts used in the experiment. Approximate translation of the excerpts, from "Les Huguenots" (Meyerber): "Really, his naïvity is charming. However, he trembles in front of beautiful eyes", and from "Faust" (Gounod) : "For me the pleasures and young mistresses, the crazy orgy of the heart and the senses". Note that, in french, the final incongruous words "boeufs" and "sciences" rhymes with the expected completions 'yeux" and "sens". The final note of the excerpt is in or out of tune *(from Besson, Faïta, Peretz, Bonnel & Requin, 1998).*

The influence of attention

We tracked these results further by conducting another series of experiments aimed at studying the effect of attention, again with professional musicians from the opera in Marseille (Regnault & Besson, in preparation). We hypothesized that if lyrics and tunes are processed independently, listeners should be able to pay attention only to the lyrics or only to the tunes

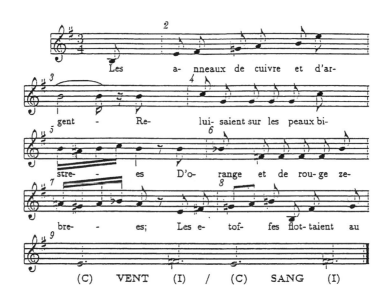

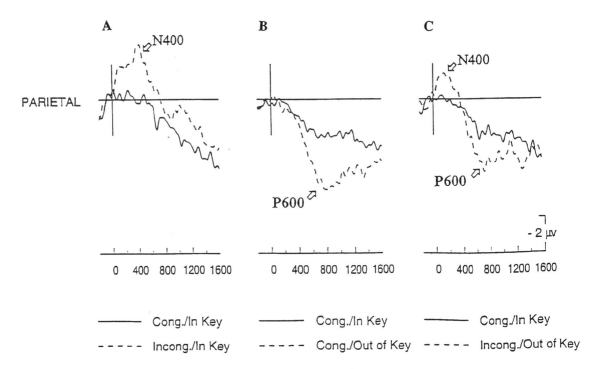

Figure 5

ERPs results averaged across 16 professional musicians and recorded from the parietal electrode (Pz). Terminal congruous words sung in key are compared to (A) semantically incongruous words sung in tune, (B) semantically congruous words sung out of tune and (C) semantically incongruous words sung out of tune. The vertical lines mark the onset of the final word of the excerpts. A large N400 component develops in the 50 - 600 ms that follow the presentation of semantically incongruous words (A). In marked contrast, a P600 develops in the 400 - 1200 ms that follow the presentation of words sung our of key (B). Most importantly, both an N400 and a P600 develop in response to the double incongruity (C; *from Besson, Faïta, Peretz, Bonnel & Requin, 1998).*

depending upon the instructions. Without going into the details of the results, an N400 component was elicited to sung incongruous words and a P600 was associated with congruous words sung out of tune, thus replicating our previous results (Besson et al, 1998). Most interestingly, the N400 to incongruous words completely vanished when participants paid attention only to the music (see Figure 6). This implies that the musicians were not processing the meaning of words, and consequently, not differentiating congruous from incongruous endings, when focusing attention on the music. Similarly, the amplitude of the P600 was significantly reduced when musicians paid only attention to the language, so that they were not able to detect that the final word was sung out of tune. Taken together these results provide strong arguments in favor of the independence of lyrics and tunes. There is some limit to such processing independence, however. Results in the double incongruity condition showed that the presence of one type of incongruity influenced the processing of the other type of incongruity. Musicians noticed that the words were sung out of tune when they were also semantically incongruous, even if their attention was focused on language.

Syntax and harmony

The rules of harmony and counterpoint have often been described as the grammar of tonal music. As syntax is used to extract the fundamental structure of an utterance by assigning different roles to different words, the rules of harmony allow to specify the different elements, notes and chords, that fulfill a specific harmonic function. Results of experiments manipulating the harmonic function of target chords have shown that violations of harmonic expectancies are associated with P600 components (Janata, 1995; Regnault, Bigand & Besson, in press). Interestingly, research on syntax using ERPs has also shown that different types of syntactic violations, such as violations of gender, word order, noun-verb agreement..., elicit a positive component, peaking around 600 ms (Hagoort, Brown & Groothusen, 1993; Osterhout & Holcomb, 1992, Friederici, 1998). Moreover, both components show a similar centro-parietal distribution over the scalp, which, together with their similar polarity and latency, seems to indicate that they reflect qualitatively similar processes.

In order to further test this hypothesis, Patel & collaborators (1998) conducted an experiment directly aimed at comparing the P600 components elicited by harmonic and syntactic violations. ERPs associated to a word within a grammatically simple, complex or incorrect sentence were compared to those associated with the presentation of a chord that belonged to the same, a nearby or a distant tonality than the one induced by the chords sequence. Results showed that independently of the basic differences between words and chords, due to the acoustic characteristics of these two types of auditory signals, the effects associated with the violation of syntactic and harmonic expectancies are not significantly different (see Figure 7). Therefore, these results raise the interesting possibility that a general cognitive process is called into play when participants are asked to process the structural aspects of an organized sequence of sounds. Finally, an early right anterior negativity develops around 300-400 ms in response to a chord belonging to a distant tonality. These results parallel those obtained in language experiments showing that a left anterior negativity is also associated with some syntactic violations (Friederici, Pfeifer & Hahne, 1993). While these two negative components show a different distribution over the scalp, with a left predominance for language and a right predominance for music, they may reflect functionally similar processes.

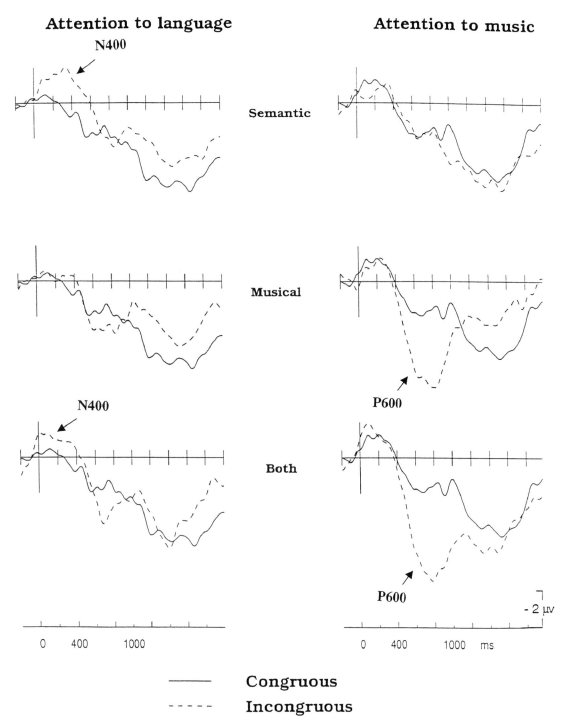

Figure 6
Overlapped are the ERPs to congruent and incongruent endings, recorded at the central recording site (Cz), when participants payed attention only to the language (left column) or only to the music (right column) of the opera's excerpts. Clearly, a large N400 effect is generated when participants focus their attention on language. This effect completely vanish when attention is focussed on music (top row). Similarly, the P600 effect is much larger when participants payed attention to music than when they payed attention to language (medium row). Finally, when words are both semantically incongruous and sung out of tune, the N400 effect is larger when participants payed attention to the language and the P600 effect is larger when they payed attention to the music (bottom row; *from Regnault & Besson, in preparation*).

Language

Music

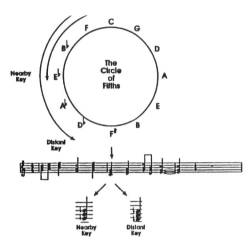

Simple: "Some of the senators had promoted an old idea of justice".

Complex: "Some of the senators endorsed promoted an old idea of justice".

Ungrammatical: "Some of the senators endorsed the promoted an old idea of justice".

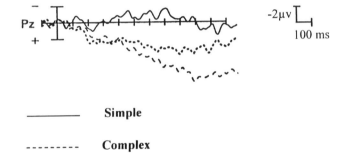

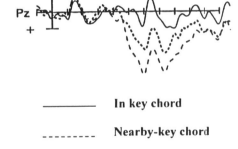

———— Simple

- - - - - - Complex

············ Ungrammatical

———— In key chord

- - - - - - Nearby-key chord

············ Distant-key chord

Figure 7

Left side: Examples of the sentences presented in the auditory language experiment. Results showed an increased positivity from the simple to ungrammatical sentences. Right side: Representation of the circle of fiths. Examples of the stimuli used in the experiment. The target chord, shown by the downward-pointing vertical arrow, is the congruous chord. The two arrows below the musical notation point to moderately incongruous (nearby key) and highly incongruous (distant key) target chords. Results also showed an increased positivity from the in key chords to the distant-key chords *(from Patel, Gibson, Ratner, Besson & Holcomb, 1998).*

Temporal structure

Spoken language, as music, is structured by acoustic events that unfold in time. Thus, based upon the temporal structure, specific events are expected at specific times. The central question addressed in the next series of experiments was to determine whether the processes involved to analyze temporal structures are similar for language and music, thereby relying on general cognitive mechanisms, or are rather different as a function of the materials to be processed, thereby favoring the specificity of temporal processing. We used both the ERP method to analyze the time course of the effects of temporal structure violations and fMRI to analyze the cerebral structures activated by temporal violations in language and music.

In previous experiments (e.g., Besson & Faita, 1995) we introduced an unexpected silence between the before to the last and the last note of a musical phrase. Results showed that a large biphasic, negative-positive, potential, the emitted potential (Sutton et al, 1967), was elicited at the time when the final note should have been presented but was not since it was delayed by 600 ms. The amplitude of this effect was similar for musicians and non-musicians, but was larger for familiar than unfamiliar melodies (see Figure 8). These findings clearly indicate that both musicians and non-musicians were able to anticipate the precise moment when the final note was to be presented, and were surprised when it was not. Moreover, knowing the melodies allowed participants to generate more precise expectancies than when melodies were unfamiliar. Therefore, these results indicate that the occurrence of an emitted potential can serve as a marker of temporal expectancy.

To further track these effects, it was of interest to determine whether results similar to music, would be found for spoken language (Besson, Faita, Czternasty & Kutas, 1997). To this aim, we presented both familiar (e.g., proverbs) and unfamiliar auditory sentences to participants. In half of sentences, final words occurred at their normal position, while in the other half, they were delayed by 600 ms. Results showed that an emitted potential, very similar to the one described for temporal ruptures in music, develops when the final word should have been presented (see Figure 9). Therefore, these results indicate that qualitatively similar processes seem to be responsible for temporal processing in language and music.

In order to strengthen this interpretation, it was important to determine whether the same brain structures are activated by the processing of temporal ruptures in language and music. As mentioned above, fMRI allows to localize brain activation with an excellent spatial resolution. Moreover, the MEG permits to localize the generators of the effects observed on the scalp more precisely than the ERP method, and also offers an excellent temporal resolution. Therefore, in collaboration with Prof. H. Heinze and his research team, we conducted three experiments, in which we presented both auditory sentences and musical phrases (Weyerts et al, in preparation). These experiments used a blocked design in which only sentences or musical phrases without temporal ruptures were presented within a block of trials, and only sentences or musical phrases with temporal ruptures were presented in another block of trials. The ERP method was used in the first experiment to replicate, within subjects, the results found previously with two different groups of subjects, and the fMRI and the MEG methods were used in the other two experiments, respectively.

Comparison of the conditions with and without temporal violations revealed a somewhat different pattern of results using the MEG and fMRI methods. MEG results indicated that the neuronal populations generating the biphasic potential recorded on the scalp using both the ERPs and the MEG methods, were most likely localized in the primary auditory cortex of both hemispheres. FMRI results showed activation in the associative auditory cortex in both hemispheres, as well as some parietal activation. These differences still need to be explained. However, the main point is that similar brain areas were activated by temporal violations in both language and music.

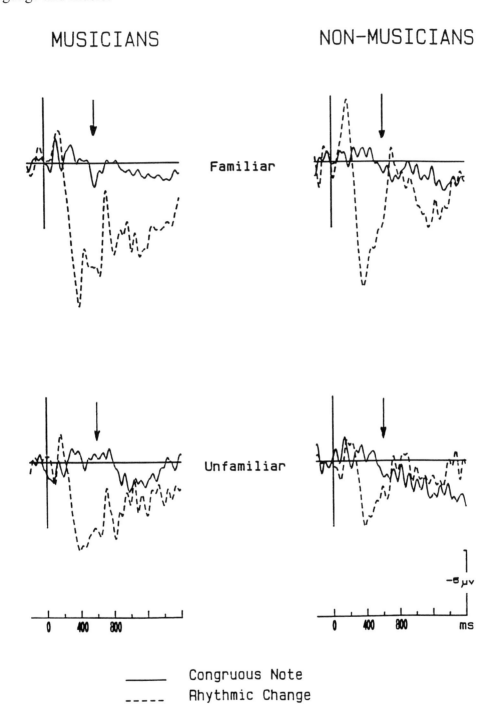

Figure 8
Overlapped are the ERPs to congruous notes and to the rhythmic incongruities ending familiar and unfamiliar musical phrases for musicians and non-musicians. Recordings are from the parietal electrode (Pz). Large emitted potentials are elicited when the final note should have been presented (vertical bar) but was delayed by 600 ms. The arrow points to the moment in time when the final note was presented *(from Besson & Faïta, 1995)*.

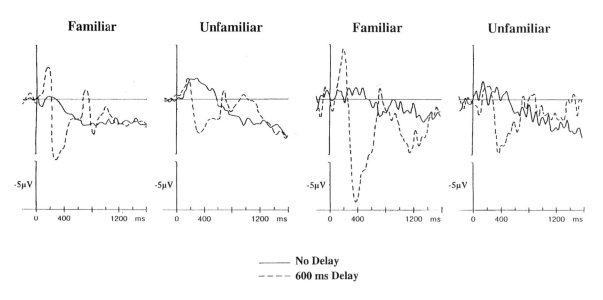

Figure 9
Comparison of the effects of temporal violations in language and music. Recordings are from the parietal electrode (Pz). Left side: Overlapped are the ERPs to congruous words and to the temporal disruptions ending familiar and unfamiliar sentences. In both language and music, large emitted potentials are elicited when the final event should have been presented (vertical bar) but was delayed by 600 ms. Note that the amplitude of the emitted potential is larger in music than in language, but that in both cases, its amplitude is larger for familiar than unfamiliar materials.

Conclusion

In this chapter I have addressed one of the most important question for human cognition, the question of the specificity of language processing. Is language an autonomous system, independent from other human cognitive abilities or does language rely on general cognitive principles? To address this question, we have conducted several experiments aimed at comparing some aspects of language processing with some aspects of music processing. We mainly used the Event-Related Potentials method, that offers an excellent temporal resolution, and therefore permits to study the time course of information processing and to determine whether the processes involved in language and music are qualitatively similar or different.

Taken together, results have shown that the semantic computations required to access the meaning of words, and their integration within a linguistic context, seem to be specific to language. Indeed, while unexpected words within a sentence context are associated with the occurrence of an N400 component, unexpected notes or chords within musical phrases elicit a P600 component. In contrast, words that are unexpected on the basis of the syntactic structure of the sentence, and chords that are unexpected as a function of the harmonic structure of the musical sequence, elicit similar effects in both cases, namely a P600 component. Earlier negative effects, the Left Anterior Negativity and the Right Anterior Negativity, that develop between 200-300 ms, have also been reported in experiments manipulating syntax and harmony, respectively. While their different scalp distribution seems to indicate that they reflect the involvement of different brain structures, more research is needed to further track their functional significance. Finally, violations of temporal structure within language and music also elicit similar effects in both cases, that is, a biphasic negative-positive complex, the emitted potential. The occurrence of the emitted potential show that in both language and music, words and notes or chords are expected to occur at specific moments in time. Therefore, when we listen to language and music we do not only expect words or chords, with specific meaning and specific function, but we also expect them to be presented on time!

The question of the specificity of language processing has broad implications for our understanding of the human cognitive architecture, and even more generally, for the fundamental problem of the relationship between structures (the different brain regions) and functions (language, music, ...). While the research reported here shed some light on some specific aspects of language processing and highlight some of the similarities and differences with music processing, more research clearly need to be conducted within this fascinating research domain. It will be of most interest to use brain imaging methods that offer complementary information about the spatio-temporal dynamics of brain activity in order to be able to pinpoint the networks of cerebral structures that are involved when we are engaged in two of the most human cognitive abilities: language and music.

References

Arom S, Khalfa J. Descartes en Afrique. In: V Gomez Pin, editor. "Descartes lo racional y lo real". Univesitat Autonoma de Barcalona, 1999.

Benveniste E. Problèmes de linguistique générale. Paris: Gallimard, 1966.

Berger H. Uber das Electrenkephalogramm das menchen. Archiv für Psychiatrie 1929; 87: 527-570.

Besson M, Faïta F. An Event-Related Potential (ERP) study of musical expectancy: Comparison of musicians with non-musicians. Journal of Experimental Psychology: Human Perception and Performance, 1995; 21: 1278-1296.

Besson M, Faïta F, Czternasty C, Kutas M. What's in a pause: event-related potential analysis of temporal disruptions in written and spoken sentences. Biological Psychology 1997; 46, 3-23.

Besson M, Faita F, Peretz I, Bonnel AM, Requin J. Singing in the brain: Independence of lyrics and tunes. Psychological Science 1998; 9: 494-498.

Besson M, Faïta F, Requin J. Brain waves associated with musical incongruity differ for musicians and non-musicians. Neuroscience Letters 1994; 168: 101-105.

Besson M. & Macar F. An event-related potentials analysis of incongruity in music and other non-linguistic contexts. Psychophysiology 1987; 24: 14-25.

Boucourechliev A. Le langage musical. Collections les chemins de la musique. Fayard, 1993.

Boulez P. Son et verbe. In Relevés d'apprenti. Paris: Le Seuil, 1966.

Chomsky N. Syntactic structures. La Haye: Mouton & Co, 1957.

Dale AM, Buckner RL. Selective averaging of rapidly presented individual trials using fMRI. Human Brain Mapping 1997; 5: 329-340.

Darwin C. *The Descent of Man*. Princeton, NJ: Princeton University Press, 1871.

Dawson GD. Cerebral responses to nerve stimulation in man. British Medical Bulletin 1950; 6: 326-329.

Descartes. Abrégé de musique. Compenduim musicae. Épiméthée. Paris: P.U.F., (1618) [1987].

DeWitt LA, Samuel AG. The role of knowledge-based expectations in music perception: Evidence from musical restoration. Journal of Experimental Psychology: General, 1990; 119: 123-144.

Donchin E. Event-Related brain potentials: a tool in the study of human information processing. In: H. Begleiter, editor, Evoked potentials and behavior. New York: Plenum, 1979: 13-75.

Fauconnier G. Mappings in thought and language. Cambridge: Cambridge University Press, 1997.

Fodor J. Modularity of mind. Cambridge, MA: MIT Press, 1983.

Frazier L. Sentence processing: A tutorial review. In: M Coltheart, editor. Attention and performance XII. Hillsdale, NJ: Erlbaum, 1987: 559-586.

Friederici AD. The neurobiology of language comprehension. In: Friederici AD, editor. Language comprehension: A biological approach. Berlin/Heidelberg/New York: Springer, 1998: 263-301.

Friederici AD, Pfeifer E, Hahne A. Event-Related brain potentials during natural speech processing: Effects of semantic, morphological and syntactic violations. Cognitive Brain Research 1993; 1: 182-192.

Fuchs C. Diversité des représentations linguistiques: Quels enjeux pour la cognition? In: Fuchs C, Robert S, editors. Diversité des langues et représentations cognitives. Ophrys, 1997: 5-24.

Gorell P. Suntax parsing. Cambridge, UK: Cambridge University Press, 1995.

Hagoort P, Brown C, Groothusen J. The syntactic positive shift as an ERP-measure of syntactic processing. Language and Cognition Processes 1993; 8: 439-483.

Jackendoff R. Patterns in the mind: Language and human nature. Basic books, 1994.

Jakobson R. Essais de linguistique générale. II. Rapports internes et externes du langage. Paris; Editions de Minuit, Arguments, 1973.

Janata P. ERP measures assay the degree of expectancy violation of harmonic contexts in music. Journal of Cognitive Neuroscience 1995; 7: 153-164.

Kail M. Variations linguistiques et contraintes cognitives dans le traitement et le développement du langage. In: C Fuchs, S Robert, editors, Diversité des langues et représentations cognitives. Ophrys, 1997: 210-226.

Kivy P. Music Alone. Philosophical Reflection on the Purely Musical Experience. Cornell Paperbacks, 1991.

Kutas M, Hillyard SA. Reading senseless sentences: Brain potentials reflect semantic incongruity. Science 1980; 207: 203-205.

Kwong KK, Belliveau JW, Chesler DA, Goldberg IE, Weisskoff RM, Poncelet BP, Kennedy DN, Hoppel BE, Cohen MS, Turner R, Cheng H-M, Brady TJ, Rosen BR. Dynamic magnetic resonance imaging of human brain activity during primary sensory stimulation. Proceedings of the National Academy of Sciences, USA, 1992; 89: 5675-5679.

Langacker RW. An introduction to cognitive grammar. Cognitive Science 1986; 10: 1-40.

Levman BG. The genesis of music and language. Ethnomusicology 1992; 36: 2, 147-170.

Marslen-Wilson WD, Tyler LK. The temporal structure of spoken language understanding. Cognition 1980; 8: 1-71.

MacCallum WC, Farmer SF, Pocock PV. The effects of physical and semantic incongruities on auditory event-related potentials. Electroencephalography and Clinical Neurophysiology 1984; 59: 477-488.

McCLelland JL, St John M, Taraban R. Sentence comprehension: A parallel distributed processing approach. Language and Cognition Processes, 1989; 4: 287-335.

Meyer L. Emotion and meaning in music. University of Chicago Press, 1956.

Nadel S. The Origin of Music. Musical Quarterly 1930; 16: 531-546.

Newman E. Herbert Spencer and the Origin of music. In: Newman, editor, Musical studies. New York: Haskell House Publishers, 1905/1969.

Osterhout L, Holcomb PJ. Event-Related brain potentials elicited by syntactic anomaly. Journal of Memory and Language 1992; 31: 785-804.

Paller KA, McCarthy G, Wood CC. Event-Related Potentials elicited by deviant endings to melodies. Psychophysiology 1992; 29: 202-206.

Patel A, Gibson E, Ratner J, Besson M, Holcomb P. Processing syntactic relations in language and music: An Event-Related Potential study. Journal of Cognitive Neurosciences 1998; 10: 717-733.

Pinker S. The language instinct: How the mind creates language. Harper Perennial, 1994

Pinker S. How the mind works. New York: W.W. Norton & Company, 1998.

Regnault P, Bigand E, Besson M. Different brain mechanisms mediate sensitivity to sensory consonance and harmonic context: Evidence from auditory Event-Related brain Potentials. Journal of Cognitive Neuroscience (in press).

Regnault P, Besson M. Singing in the brain, Part II: The effect of attention (in preparation).

Robert S. Variations des représentations linguistiques: des unités à l'énoncé. In: Fuchs C, Robert S, editors, Diversité des langues et représentations cognitives. Ophrys, 1997: 5-24.

Rousseau JJ. Essai sur l'origine des langues. Paris: Flammarion, 1781/1993.

Saussure F. Cours de linguistique générale. Paris: Payot, 1916/1971.

Seidenberg MS. Language acquisition and use: Learning and applying probabilistic constraints. Science 1997; 275: 1599-1603.

Spencer H. The origin and function of music. Fraser's Magazine 1857; 56: 396-408.

Sutton S, Braren M, Zubin J, John ER. Evoked potential correlates of stimulus uncertainty. Science1967; 150: 1187-1188.

Sternberg S. The discovery of processing stages: Extension od Donder's method. Acta Psychologica 1969; 30: 276-315.

Talmy L. The relation of grammar to cognition. Toward a cognitive Semantics. Vol. 1. MIT Press, (in press).

Verleger R. P3-evoking wrong notes: Unexpected, awaited or arousing? Journal of Neuroscience 1990; 55: 171-179.

Wallaschek R. On the Origin of Music. Mind 1891; 16: 375-386.

Warren R. Perceptual restoration of missing speech sound. Science 1970; 167: 392-393.

Weyert H, Besson M, Tempelmann C, Scholz M, Fell J, Kutas M, Woldorff MG, Heinrichs H, Scheich H, Heinze HJ. An analysis of temporal structure in language and music using ERPs, MEG and fMRI techniques (in preparation).

Chapter III.10

Attention to Pitch in Musicians and Non-Musicians: An Event-related Brain Potential Study

Thomas F. Münte[a,c,1], Wido Nager[a], Oliver Rosenthal[a], Sönke Johannes[a], and Eckart Altenmüller[b]

[a]*Neurologische Klinik, Medizinische Hochschule Hannover*
[b]*Institut für Musikermedizin und Musikphysiologie, Hochschule für Musik und Theater Hannover*
[c]*Neuropsychologie, Otto-von-Guericke-Universität Magdeburg*

Introduction

Over the last few years, a great number of publications has addressed the question, whether the brain of trained musician processes musical stimuli different from non-musicians. Many of these studies have used neuropsychological, brain-imaging and electrophysiological techniques (Altenmüller, 1989; LaBarba, Kingsberg, 1990; LaBarba *et al.*, 1989; Mazzucchi *et al.*, 1981) to answer the question, whether the involvement of the two hemispheres in the processing of music is different in musicians.

Considerably less attention has been devoted to the question, whether or not the cognitive organisation of auditory stimulus processing and/or its temporal organisation might be different in trained musicians and non-trained listeners. Given recent evidence that cortical event-related brain potentials can be used to track neural plasticity due to brain lesions or extensive training, it seems promising to search for such plasticity effects in basic attention-driven auditory stimulus processing in musicians.

The ERP effects of selective auditory attention have been studied extensively over the last two decades or so (Alho *et al.*, 1987; Alho *et al.*, 1989; Alho *et al.*, 1994; Hansen, Hillyard, 1980; Hansen, Hillyard, 1983; Hansen, Hillyard, 1988; Alain *et al.*, 1994). Paying attention to one stream of auditory stimuli, defined by the stimuli's location or pitch, while ignoring other concurrent streams gives rise to a more negative ERP waveform, termed the Nd (for negative displacement), the onset of which coincides with the exogenously evoked N1

---

[1] Correspondence: Thomas.F.Münte, Dept. of Neurology, Medizinische Hochschule Hannover, 30623 Hannover, Germany, Fax: (49)-511-5323-115, e-mail: muente.thomas@mh-hannover.de

component. As far as attention to pitch is concerned, Hansen and Hillyard (Hansen, Hillyard, 1980) could show in an early study that the Nd wave could be subdivided into an earlier phase (100-300 ms, henceforth early Nd) and a more frontally distributed later phase (300-beyond ms, henceforth late Nd). The functional correlates of the early and late Nd are still under debate. A recent study (Teder-Sälejärvi, Hillyard, 1998) suggests, however, that auditory selective attention might be organized in two stages, an early, more broadly tuned stage taking place over the first 200 ms after the stimulus followed by a later stage during which a more finely tuned stimulus selection takes place. Several reports have addressed the neural generators underlying the auditory attention effect (Giard *et al.*, 1991; Giard *et al.*, 1988). The early portion of the Nd might originate from the supratemporal plane while the late Nd seems appears over frontal areas, but probably originates from deeper sources. One attractive hypothesis, proposed by Näätänen's group, regarding the functional correlates of the attention effects in auditory ERPs has been the „attentional trace" hypothesis (Alho *et al.*, 1989). According to this hypothesis, the additional negativity may reflect a comparison process between the incoming stimuli and an attentional trace. The better a stimulus corresponds to this trace, the larger the amplitude of the negativity.

In the present investigation we studied the early and late components of the Nd during selective attention to pitch. Given that musicians are highly trained auditory discriminators we hypothesized that they should exhibit a changed auditory attention effect especially for the late Nd.

Methods

Subjects

Twelve musicians (mean age 24.8 y, 7 women), all enrolled as students of the Hannover School of Music, gave informed consent to participate. All of the musicians had had extensive musical training since early childhood and were practicing music at a professional level. Four of the musicians said to have absolute pitch.

Twelve non-musicians (mean age 23.6 y, 3 women) were recruited from the student population at Hannover Medical School. None of the non-musicians had any formal training in music or experience with a musical instrument.

None of the subjects had a history of neurological or psychiatric diseases.

Stimuli and Procedure

Stimuli were computer-generated brief sine-wave tone-pips. Two channels of information, defined by the pitch of the stimuli, were presented via a speaker standing in front of the subjects. Three different conditions were created on the basis the pitch separation between channels:

50 Hz condition   800 Hz ./. 850 Hz
100 Hz condition   800 Hz ./. 900 Hz
500 Hz condition   800 Hz ./. 1300 Hz

Within each channel 90 % of the stimuli were of 60 ms duration (standard stimuli) while 10 % of the stimuli were of slightly longer duration (80-120 ms depending on performance of the subject) that served as the target stimuli.

For any given run subjects were instructed to selectively attend to one of the channels (high tones or low tones) and to answer target stimuli in the attended channel by a speeded button response.

Stimuli were presented with a randomized interstimulus-interval between xx and xx ms at about 60 dB(SL) under the control of a micro-computer. Before the experiment proper subjects were familiarized with the procedure and their target discrimination was tested. The duration of the target stimuli was adjusted such that a target detection rate of about 70 % was achieved.

For each condition 2 runs, each lasting about 5 min, were recorded. Subjects were tested in a single session while seated in an easy chair in a dimly lit experimental chamber. Instructions were given to minimize movements and eye-blinks during the experiment. The total experiment took an average of 150 min to complete including the placement of the electrodes.

Recording and Analysis

EEG was recorded from all 19 standard scalp-sites of the 10/20 system (Jasper, 1958) plus ten additional intermediate sites using tin electrodes mounted in an electrode cap with reference electrodes placed at the mastoid processes. Biosignals were recorded against one of the mastoid processes and were rereferenced off-line to the algebraical mean of the activity at both mastoid processes. Additional electrodes were affixed at the right external canthus and at the right lower orbital ridge to monitor eye movements for later off-line rejection. The biosignals were amplified with a bandpass from 0.01 to 70 Hz, digitized at 250 points per second and stored on magnetic disk. After artifact rejection by an automated procedure using an individualized amplitude criterion on the eye-channel and the frontopolar channels, ERPs were averaged for 1024 epochs including a 100 ms prestimulus interval. To ensure that only ERPs to physically identical stimuli were compared, measurements were taken exclusively on the ERPs to the 800 Hz stimuli. The waveforms were quantified by mean-amplitude measures in time windows 250 to 300 ms and 600 to 800 ms. These data were subjected to repeated measures analyses of variance. Since auditory attention effects are maximum over fronto-central scalp-regions measurements were taken on a set of fronto-central channels (F3/4, C3/4, FC1/2, FC5/6). Additional analyses addressing the distribution of the attention effect in the difference waves were carried out on the whole electrode set after normalization of the data according to the method of McCarthy and Wood (McCarthy, Wood, 1985). Statistical comparisons were conducted using an ANOVA design with group (musician vs. non-musician) as between subjects factor, and pitch difference (50, 100, 500 Hz), attention (attended vs. unattended) and electrode site as within subject factors. The Huyhn-Feldt correction for inhomogeneities of covariance was used whenever applicable. Where appropriate, post-hoc comparisons were carried out.

Results

Behavioral Results

The overall hit rate of the musicians (75.5 % SD 10.6) was better than that of the controls (63.8, SD 18.9; $F_{(1,22)}=4.14$, p=0.054). Performance in the three different pitch separation condition was similar (50 Hz: 67.9, SD 15.9; 100 Hz: 70.2, SD 17.4; 500 Hz: 70.8, SD 16.1; $F_{(2,44)}=1.95$, n.s.).

Musicians reacted slower than control subjects (569 ms, SD 62 vs. 502 ms, SD 59, $F_{(1,22)}=8.37$, p< 0.009). There was no interaction between pitch separation condition and group ($F_{(2,44)}=0.15$, n.s.).

Waveform Morphology

Grand average event-related potentials are shown for both groups for the frontal midline electrode and the three different pitch separation conditions (fig. 1). Waveforms are characterized by an initial positivity at around 100 ms followed by a negativity peaking at about 180 ms (N1). In the 500 Hz pitch separation condition the onset of the attention effect appears to coincide with the upward flank of the N1. The attention effect extends to the end of the recording epoch in the musicians, while in the control subjects the attention effect lasts up to only about 600 ms in the 100 Hz and 500 Hz conditions. The time-course of the attention effect can be illustrated by the computation of difference waves subtracting the ERPs to the unattended stimuli from those to the attended stimuli (fig. 2). It becomes apparent that (1) the Nd-wave in the musicians outlasts that of the control subjects and (2) the early and late phases of the Nd wave are clearly separated in the musicians.

To enhance signal to noise ratio for the topographical comparisons, difference waves from all three pitch separation conditions are collapsed and shown for all 29 scalp sites in figure 3. Over the entire scalp, difference waveforms reflecting the differential neural activity related to attention to pitch did not differ between control subjects and trained musicians up to about 270 ms. For the later portion of the waveform especially beyond 500 ms the musicians exhibit an extended negativity with a maximum over frontal scalp-sites.

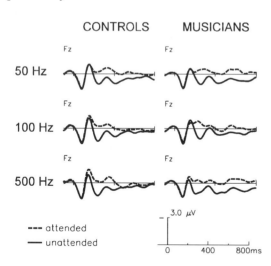

Figure 1
Group averages for the musicians and controls for the frontal midline site. From about 150 ms onwards the ERPs to the attended tones are characterized by a more negative waveform in both groups. This effect is more short-lived in the control group, however.

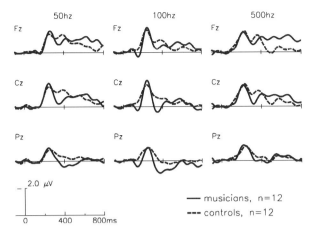

Figure 2
Comparison of the difference waves in the three pitch separation conditions. A long-lasting frontal negativity is visible in the musician group.

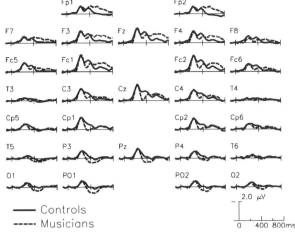

Figure 3
Comparison of the difference waves (collapsed over pitch separation conditions) for the entire electrode set. While the early attention effects are indistinguishable between musicians and controls, the late attention effect in the musicians is more frontally distributed

The distribution of the early and late attention effects can also be derived from the spline-interpolated isovoltage maps in fig. 4a. While the center of gravity of the distribution of the early attention effect is virtually identical, the late Nd is considerable more frontally distributed. In terms of absolute position, the maximum of the late attention effect is shifted frontally by about 3.5 cm. The more anterior distribution of the late attention effect is also visible in fig. 4b. For this graph, the mean amplitude of the difference wave in the 600-700 ms time window was normalized to mean of 1.

The waveforms for the target stimuli are depicted in the left column of fig. 5. The most striking difference between musicians and non-musicians is the more pronounced negativity in the 150-300 ms range in the control group. This negativity corresponds to the well-known mismatch negativity. Over parietal areas, on the other hand, the musicians showed a slichtly larger P3 component. In the unattended channel as well deviant stimuli were associated with a larger mismatch negativity in the control group.

Statistical Results

A highly significant main effect of attention was observed for both, early (250-300 ms, $F_{(1,22)}= 58.14$, $p < 0.0001$) and late (500-750 ms, $F_{(1,22)}=19.28$, $p < 0.0002$) time windows. While the size of the attention effect was virtually identical during the early time-window (group x attention interaction $F_{(1,22)}=0.01$, n.s.), the larger late Nd wave in the musicians was reflected in a group x attention interaction for the late time window ($F_{(1,22)}=4.42$, $p < 0.05$).

To test for differences in distribution of the attention effects measurements were taken on all 29 electrodes and entered into an ANOVA after rescaling (McCarthy, Wood, 1985). For

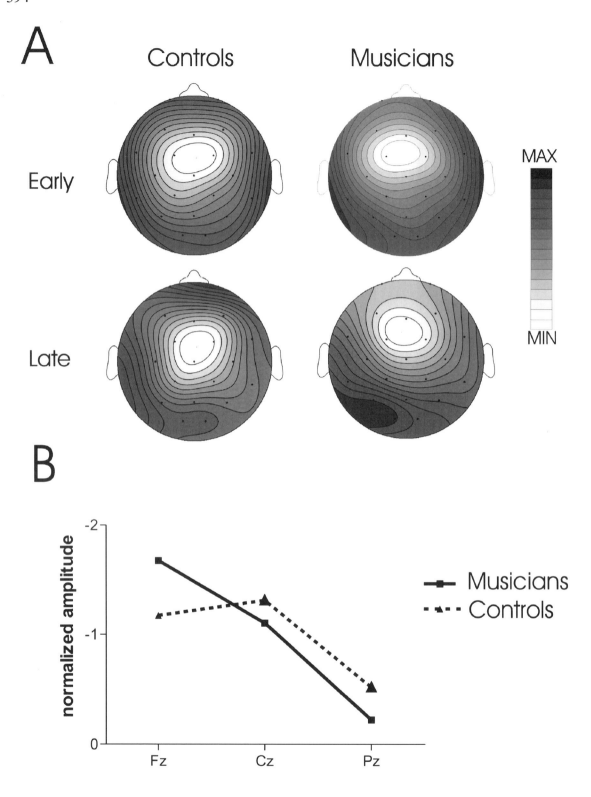

Figure 4

A Isovoltage maps obtained by spline interpolation. The upper panel shows the distribution of the attention effect in an early (250-300 ms) time-window, while the lower time-window shows the distribution during the 500-750 ms interval. The late effect is considerably more frontally distributed in the musician group. **B** Normalized mean amplitudes during the late time window for three midline channels. Again, a more frontal distribution in the musicians is apparent.

the early time window, no group x electrode-site interaction was present ($F_{(28,616)}=0.46$, n.s.) indicating a similar distribution. For the late time-window on the other hand a significantly different distribution of attention effects in musicians and non-musicians was revealed ($F_{(28,616)} = 2.46$, $p<0.0001$, after Huynh-Feldt correction: $p<0.04$). For the target ERPs, the frontal negativity was quantified for the set of frontal channels time-window 200-300 ms). The main effect of group showed a trend towards significance ($F_{(1,22)}=4.21$, $p=0.052$). The parietal P3 was quantified in the time window 500-700 ms (P3, P4, Pz). Again, only a trend towards significance was attained ($F_{(1,22)}=2.91$, $p=0.102$, n.s.).

For the deviant stimuli in the unattended channel, the mismatch negativity tended to be larger for the control group ($F_{(1,22)}=3.95$, $p=0.059$).

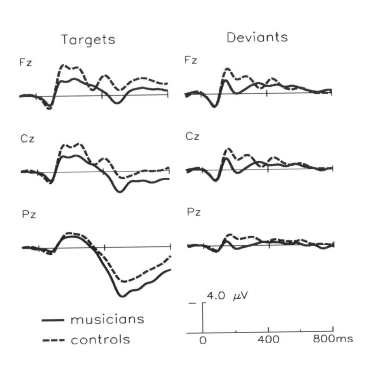

Figure 5

ERPs from the targets (left column) and deviant stimuli in the unattended channel (right column). The control subjects exhibit a sizeable mismatch negativity compared to a smaller negativity in the musicians for both, target and deviant stimuli. This contrasts with the findings for the P3-component to the target stimuli, which appears to be larger for the musicians.

Discussion

The present study aimed at delineating possible differences in electrophysiological concomitants of selective auditory attention in musicians and non-musicians. The study was successful in that several important differences emerged for the two subject groups that can be tentatively interpreted in the light of what is known about the specific components involved.

The first difference between musicians and control subjects was an extended late Nd wave in the former group that was absent or much smaller in the latter. This extended negativity that furthermore had a more frontal distribution can be derived from figs. 3 and 4. Differences between early and late auditory attention effects have been observed in several circumstances. An early interpretation of the late Nd (also termed processing negativity) was that it indicated a match between an attentional trace, i.e. a template against which incoming stimuli can be matched, and the eliciting stimulus (Alho *et al.*, 1987; Alho *et al.*, 1989). In the present circumstances, this would indicate that musicians achieve a better fit between the attentional trace and incoming stimuli, possibly because their attentional trace is more finely tuned and precise.

Another possible interpretation of the prolonged attention effect in musicians can be deduced from a recent study on the spatial gradient of auditory attention (Teder-Sälejärvi, Hillyard, 1998). In this study, it was shown that stimuli from locations immediately adjacent to the attended location showed an Nd effect. Moreover, these stimuli had a larger late Nd effect than the attended stimuli themselves. This finding indicates that the late Nd might reflect the extent and intensity of analysis alloted to the stimulus. Thus, it can be speculated that musicians as highly trained auditory observers analyze incoming auditory stimuli more thoroughly than less experienced listeners.

It is interesting to note that the distributions of the late Nd were different for the two groups. This apparent shift in distribution might correspond to the fact that musicians pose much greater demands upon their auditory system and thus might recruit additional brain areas. For the somatosensory system there are now numerous reports indicating plastic brain organization including evidence from string players (Elbert *et al.*, 1995). Several papers have addressed the localization of auditory attention effects (Giard *et al.*, 1991; Giard *et al.*, 1988) suggesting that the later attention-related components emanate from frontal cortex. This corresponds with the current finding of a shifted topographic maximum in musicians. Clearly, it would be a worthwhile exercise to attempt a source localization on the present data.

A further difference between musicians and non-musicians can be found in the target and deviant ERPs. Here, control subjects showed a greater negativity over anterior scalp sites corresponding to the mismatch negativity. This effect contrasts with recent findings (Koelsch *et al.*, 1999; Tervaniemi *et al.*, 1997) of larger mismatch negativities in musicians. These studies, however, used a passive listening task with subjects reading a book. Moreover, the deviants were defined by changes in the temporal structure of complex auditory stimuli. Nevertheless, it is noteworthy that under the present circumstances musicians show a smaller mismatch negativity.

To summarize, attention effects to pitch are different in musicians in both their timing and topography suggesting a more thorough analysis of auditory input of these highly trained listeners possibly involving additional brain regions. Further studies are needed to obtain a more precise localization of these effects, to compare the effects of attention to pitch with those of attention to space, and to assess the musicians' treatment of deviant and target stimuli more deeply.

Summary

Event-related brain potentials were recorded in a group of 12 highly trained musicians and a group of musically naive control subjects while they were attending to streams of auditory stimuli defined by their pitch. In different runs three different pitch separations (50, 100, 500 Hz) between attended and unattended stimulus channels were used. Both groups showed an attention effect in the ERP in the form of a more negative going ERP for the attended stimuli. This early part of this attention effect up to about 270 ms was indistinguishable between the two groups. The later part of the attention effect, however, was more pronounced and more frontally distributed in the musicians. This can be taken as a sign of a more thorough analysis of attended auditory input by the musicians. By contrast, the ERP sign of stimulus deviance, the mismatch negativity, was smaller in the musicians for both attended and unattended deviants.

Acknowledgments

The technical assistance of J. Kilian is gratefully acknowledged. This work was supported by grants from the DFG awarded to EA and TFM.

References

Alain A, Woods DL, Ogawa KH (1994) Brain indices of automatic pattern processing. N*euroreport*, 6, 140-144.

Alho K, Donauer N, Paavilainen P, Reinikainen K, Sams M, Näätänen R (1987) Stimulus selection during auditory spatial attention as expressed by event-related potentials. *Biol Psychol*, 24, 153-162.

Alho K, Sams M, Paavilainen P, Reinikainen K, Näätänen R (1989) Event-related brain potentials reflecting processing of relevant and irrelevant stimuli during selective listening. *Psychophysiology*, 26, 514-528.

Alho K, Teder W, Lavikainen J, Näätänen R (1994) Strongly focused attention and auditory event-related potentials. *Biol Psychol*, 38, 73-90.

Altenmüller E (1989) Cortical dc-potentials as electrophysiological correlates of hemispheric dominance of higher cognitive functions. *Int J Neurosci*, 47, 1-14.

Elbert T, Pantev C, Wienbruch C, Rockstroh B, Taub E (1995) Increased cortical representation of the fingers of the left hand in string players. *Science*, 270, 305-307.

Giard MH, Perrin F, Pernier J (1991) Scalp topographies dissociate attentional erp components during auditory information processing. *Acta Otolaryngol Suppl (stockh)*, 491, 168-174.

Giard MH, Perrin F, Pernier J, Peronnet F (1988) Several attention-related wave forms in auditory areas: a topographic study. *Electroencephalogr Clin Neurophysiol*, 69, 371-384.

Hansen JC, Hillyard SA (1980) Endogenous brain potentials associated with selective auditory attention. *Electroencephalogr Clin Neurophysiol*, 49, 277-290.

Hansen JC, Hillyard SA (1983) Selective attention to multidimensional auditory stimuli. *J Exp Psychol Hum Percept Perform*, 9, 1-19.

Hansen JC, Hillyard SA (1988) Temporal dynamics of human auditory selective attention. *Psychophysiology*, 25, 316-329.

Jasper HH (1958) The ten-twenty system of the international federation. *Electroencephalogr Clin Neurophysiol*, 10, 371-375.

Koelsch S, Schröger E, Tervaniemi M (1999) Superior pre-attentive auditory processing in musicians. *Neuroreport*, 10, 1309-1313.

Labarba RC, Kingsberg SA (1990) Cerebral lateralization of familiar and unfamiliar music perception in nonmusicians: a dual task approach. *Cortex*, 26, 567-574.

Labarba RC, Kingsberg SA, Martin PA, Pellegrin K (1989) Cerebral lateralization of music perception in the dual task paradigm: unfamiliar melody recognition in sinistrals. *Neuropsychologia*, 27, 247-250.

Mazzucchi A, Parma M, Cattelani R (1981) Hemispheric dominance in the perception of tonal sequences in relation to sex, musical competence and handedness. *Cortex*, 17, 291-302.

Mccarthy G, Wood CC (1985) Scalp distributions of event-related potentials: an ambiguity associated with analysis of variance models. *Electroencephalogr Clin Neurophysiol*, 62, 203-208.

Teder-Sälejärvi WA, Hillyard SA (1998) The gradient of spatial auditory attention in free field: an event- related potential study. *Percept Psychophys*, 60, 1228-1242.

Tervaniemi M, Ilvonen T, Karma K, Alho K, Näätänen R (1997) The musical brain: brain waves reveal the neurophysiological basis of musicality in human subjects. *Neurosci lett*, 226, 1-4.

Chapter III.11

Cortical Activation Patterns During Perception and Imagination of Rhythm in Professional Musicians: A DC-Potential Study

Eckart Altenmüller[a],[1], Roland Beisteiner[b], Wilfried Lang[b], Gerald Lindinger[b], and Lüder Deecke[b]

[a] *Institute for Music Physiology and Performing Arts Medicine*
[b] *Department of Clinical Neurology, University of Vienna*

Introduction

Rhythm as a fundamental element of music is defined as the serial relation of durations between different acoustical events in a train of sounds, i.e. rhythm represents a serial durational pattern (Jones 1987). In contrast to rhythm, metre involves a temporal invariance in terms of the regular recurrence of pulses or beats marking off equal durational units, which can be organised as measures (Smith and Cuddy 1989). Metre therefore represents a more complex acoustical "Gestalt", since its perception and production requires information on sound intensity (accented and unaccented events) and on periodicity of rhythmical events. Perception and creation of metre is a prerequisite of the musician's ability to make music "swing".

The cerebral mechanisms underlying the processing of rhythm and metre are largely unknown. Lesion studies in patients with disorders of the sense of rhythm reveal conflicting results concerning the cerebral structures predominantly involved in rhythm processing. In his pioneering work on amusias, Henschen (1920) cites a case of isolated preservation of the sense of rhythm after a left-hemispheric lesion but a loss of all melodic abilities. In contrast, Mavlov (1980) reported of a professional musician who developed an isolated but supramodal defect in the recognition and reproduction of rhythms with relatively preserved melody processing following a stroke of the left parietotemporal region. Moreover, the ability to perceive musical time structures such as rhythm and metre seems to be dissociable: I. Peretz

---

[1] Correspondence: Eckart Altenmüller, Institut für Musikphysiologie und Musiker-Medizin der Hochschule für Musik und Theater Hannover Plathnerstr, 35 D-30175 Hannover, Tel: (49)-511-3100-552, Fax: (49)-511-3100-557, e-mail: altenmueller@hmt-hannover.de

(1990) found in a group of patients with unilateral right- or left-hemispheric brain-damage spared metric judgement in the presence of disrupted rhythmic discrimination, irrespective of whether the right or left hemisphere was lesioned. In a recent investigation in 65 patients who had undergone unilateral temporal cortectomy for the relief of intractable epilepsy, this dissociation between rhythmic and metric judgement was confirmed (Liégois-Chauvel et al. 1998). For metre processing, a critical involvement of the anterior part of the superior temporal gyrus was found, whereas rhythm processing seemed to rely more on the posterior parts of the right superior temporal gyrus. In a very elegant multimodal auditory and visual paradigm, Penhune et al. (1999) were able to demonstrate in chronic epileptic patients a modality specific dissociation with isolated involvement of the anterior secondary auditory areas in the right temporal lobe during processing of acoustically presented time structures. It should be kept in mind, however, that the generalisation of the results remains difficult, since a long history of epileptic seizures and pathological neuronal activity leads to remarkable changes in neuronal networks due to compensatory long-term cortical plasticity (Sutula et al. 1989, Vargha-Khadem et al. 1991).

To circumvent the problem of long term cortical reorganisation, we investigated 20 patients during a very early time window 5 to 10 days post lesion after having suffered from small unilateral cerebrovascular cortical lesions (Schuppert et al. 2000). Using a discrimination paradigm similar to the test published by Peretz (1990), processing of acoustically presented rhythms and metres was assessed. Detailed analysis of the individual patterns of neuropsychological deficits revealed a hierarchical organisation, with an initial right-hemisphere recognition of metre followed by identification of rhythm via left-hemisphere subsystems. In addition, individual aspects of musicality and musical behaviour as well as musical knowledge contributed to the formation of neuronal subsystems underlying the perception of musical time-structures.

With respect to the functional anatomy of rhythm processing in normal subjects, three new studies exist. Patel et al. (1997) investigated with PET-technique brain activation during regularly and irregularly presented series of tones. They found a pronounced involvement of the left frontal Broca-region and concluded that language processing and rhythm processing might be closely related to each other. Contradictory results emerged from a PET-study by Penhune et al. (1998), testing the perception and reproduction of regular isochronous or complex novel time structures in both, the auditory and visual modality. Auditory perception of rhythm produced an activation of the right planum temporale. Since the design of the study required additionally production of rhythms, activation of the somatosensory cortices and of the cerebellar hemispheres was demonstrated. In a recent fMRI-work, processing of simple (1:2, 1:3, 1:4) and complex (1:2.5, 1:3.5) time relationships was compared (Sakai et al.1999). Simple rhythmical relationships yielded an activation of left prefrontal and parietal brain areas, whereas complex relationships were processed in right prefrontal, premotor and parietal regions. Both conditions produced an additional activation of the cerebellum, the simple task mainly in the anterior lobe, the more complex in the posterior lobe. The latter finding fits nicely into the theories of a cerebellar timekeeping module (Ivry 1993). Although not discussed by the authors, the results can be interpreted in the light of Lerdhal and Jackendoff's two component model (1983). According to their model, rhythm- and metre-sense rely on two different cognitive operations which may be processed in different hemispheres: processing of rhythm requires a left hemispheric "local"-level serial cognitive operation, processing of metre a right hemispheric "global"-level, holistic strategy linked to grouping or chunking mechanisms. Applied to the fMRI-experiment, simple rhythmic relationships can be understood as "local tasks" since they are accessible for analytical and sequential "counting" strategies. In contrast, this processing mode is not accessible for perception of the complex stimuli, which therefore had to be perceived in a holistic way as auditory "gestalt". Shifts from local to global task solving strategies or vice versa might be of particular importance with respect to effects of expertise and music education on brain activity during music processing. The influence of music education and of different - procedural versus explicit - teaching strategies on brain activation patterns during melodic processing had been demonstrated previously (Altenmüller et al. 1997).

Summarising the results of lesion- and brain imaging studies, a puzzling and in many instances contradictory variety of findings emerged. This might be due to the different experimental paradigms, to different measurement methods and to poorly defined groups of subjects with respect to their musical knowledge. The present study attempts to contribute to the clarification of the neuronal substrates of the processing of musical time structures. In order to assess cortical activation patterns during rhythm processing, the topographical distribution of sustained surface negative DC-potential shifts was recorded using scalp electrodes. Since these DC-potentials reflect activation of the underlying cortex (Altenmüller and Gerloff 1998), their local distribution reveals task specific patterns which correspond to the brain structures specifically involved in processing of the respective task (Altenmüller 1989). It is a general problem of brain activation studies investigating complex mental functions, that multiple, almost simultaneous perceptive and cognitive processes occur. Although the subtraction paradigms (for a review see Frackowiack et al.1997) are designed to circumvent this problem, correlation of brain activation patterns to specific mental events may remain difficult and equivocal. To solve this problem, an experimental condition was designed which allowed the separation of mental processing of rhythms from auditory perception: subjects were required to mentally imagine rhythms without actual acoustic stimulation. There is evidence that mental imagery of physical percepts relies on an activation of the same brain structures which are involved in the actual processing of such physical percepts. This has been demonstrated in the visual and visuo-spatial modality (Uhl et al. 1990) and in the auditory modality for the imagination of spoken language (Altenmüller et al. 1993, Thomas et al. 1997). During music processing, we applied the mental imagery technique for an investigation of cortical activation patterns related to analytic and creative processing of melodic structures (Beisteiner et al. 1994).

Methods

Subjects

With respect to findings indicating that cortical activation patterns during melodic and harmonic processing may be influenced by musical expertise (Altenmüller 1986, Pantev et al. 1998, Schlaug et al. 1995), only trained professional musicians were included in the present study: 18 right-handed students of music at the Viennese conservatory (10 females, 8 males) aged between 17 and 27 years (mean age 24,3) were tested. Handedness was assessed using the modified Edinburgh handedness questionnaire (Oldfield 1971). Musical expertise was assured, since all subjects had passed the competitive entrance examination to the conservatory. They either played a melody-instrument (violin, viola, violoncello, clarinet) or sang professionally. Since piano-playing was scheduled for all students at the conservatory, subjects were additionally trained in the fundamentals of piano playing.

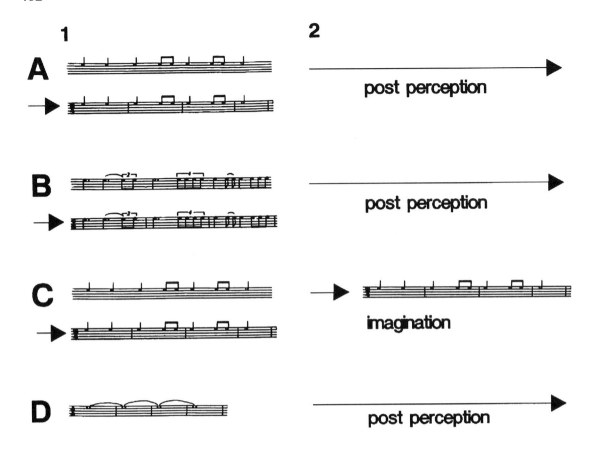

Figure 1
Stimuli and tasks. On the left side, examples of the four different stimuli and the imposed tasks during the "perception"-period (1) are displayed (notation tenor-clef, pitch a' duration 4 sec.), on the right, the respective "post perception" periods and - in condition C- the "imagination" period are symbolised (2). In condition A simple rhythms, in condition B complex rhythms were delivered. Subjects were instructed to mentally create a metre on line to those stimuli (symbolised by bars in the arrow-marked examples beyond the stimuli). In condition C, a simple rhythm was given and the subjects had to create a metre on line and to reiterate this rhythm mentally in the second period. Condition D consisted in a sustained, temporally unstructured tone of four sec. duration, designed to control effects of unspecific acoustic stimulation.

Material and tasks

Four different experimental conditions were investigated: subjects either had to listen to a simple rhythm (condition A) or to a complex rhythm (condition B), or had to imagine a simple rhythm after listening to it (condition C). A "non-rhythm" condition consisted in listening to a sustained note without any rhythmical characteristics (condition D). The stimuli were composed as follows (examples given in Fig. 1):

<u>Simple rhythm</u> patterns (condition A and C) consisted of sequences of notes with "regular" temporal intervals. The temporal sequences of subsequent notes were composed of 1:2, 2:1, 1:3, 3:1, 1:4, 4:1 ratios, or, - expressed in musical terms – of classical time-values of notes such as crochets, quavers, triplets and semiquavers. There were no variations in pitch.

Complex rhythm patterns (condition B) consisted of sequences of notes with "irregular" temporal intervals. Their temporal intervals consisted of more complex ratios such as 4:5, 4:7, 5:13 etc. In musical notation, these time values demand ligatures and dotted notes.

"Non-rhythm" condition (condition D) consisted of a note of 440 Hz frequency lasting 4 sec.. This type of stimulus was designed to control effects of unspecific acoustic stimulation. All stimuli lasted 4 s and were produced by the sound channels of an Atari Mega ST2 computer. The base frequency of the rhythms and notes was set at 440 Hz, the intensity at 60 Dezibel (spl). No accents occured.

The tasks were divided into two 4s periods. During the first period ("perception"-period), subjects were asked to listen attentively to the acoustical material. In order to attract and maintain attention and to stimulate processing of rhythms, subjects were instructed to create a metre "on-line", i. e. while listening. During the subsequent time-period, subjects either had to perform no specific task ("post-perception"-period in conditions A, B, D) or they had to imagine the simple rhythm by mental reiteration ("imagination"-period in condition C).

Experimental procedure

Subjects were seated comfortably 2 m in front of the Atari-speaker. Volume was kept constant over all experiments. To prevent artefacts in EEG-recordings, subjects were requested to fix their gaze on a small circle in front of them and to avoid body movements or vocalisations while performing the tasks.

Subjects began each trial by pressing a button as soon as they were ready. At that instant, a tape recorder delivered a command concerning the forthcoming task. This was "listen" for the simple rhythm (A), complex rhythm (B) and the non-rhythm condition (D), and "imagine" with the imagination condition (C). After the end of the tape recorded instruction (end of the 6 sec. pre-trigger period in Fig. 2), a break of 3 sec. occurred, during which the baseline ("0" in Fig. 2) was assessed. The baseline-period was followed by the 4 sec. perception-period ("1" in Fig. 2). Subsequently to the perception-period, subjects were required either to wait without any specified cognitive task for a period of about 6 sec. (post-perception-period, "2" in Fig. 2) or to mentally reiterate the rhythm (imagination-period, "2" in Fig. 2).

Every condition comprised 40 trials. The total of 160 trials was presented in random order. After the experiment (total duration about 3 h) subjects had to fill in a questionnaire concerning the judgement of task-difficulty, self-evaluation of performance and task-solving strategies (verbal vs. holistic, non verbal strategies). Participation in the experiment was honoured with 35 US $.

Data acquisition

DC-potentials were recorded from 10 electrodes positioned according to the 10/20 system over left and right frontal (F3, F4), frontobasal (F7, F8), anterior temporal (T3, T4), posterior temporal (T5, T6) and parietal (P3, P4) brain regions. Linked earlobe electrodes served as a reference. Impedance was reduced to less than 1 kOhm. To optimise stability of DC-recordings, an advanced DC-recording technique was used (5). To control artefacts arising from eye-movements, the electrooculogram (EOG) was obtained (medial upper versus lateral lower right orbital rim). The frequency band of amplification ranged from DC to 35 Hz. Sampling rate was 100 Hz using a 12 bit analogue-to-digital converter (MicroVAX II).

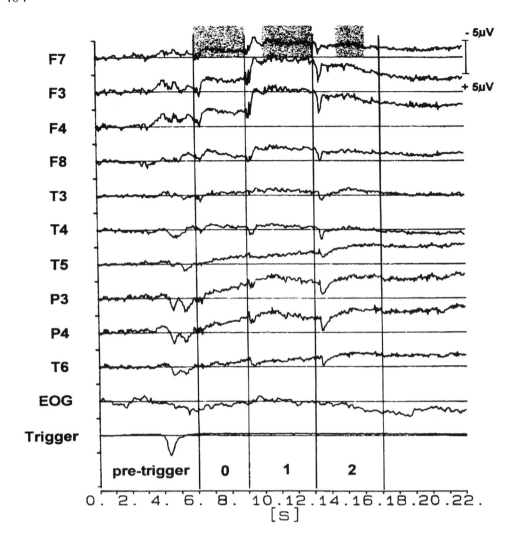

Figure 2
Grand average over all subjects in condition C (simple rhythms and subsequent imagination). Electrode positions are labelled according to the 10/20 system, EOG and trigger are shown. According to the EEG-conventions, negativation of DC-potentials is pointing upwards (scale: 10 uVs). During the pre-trigger-period, subsequent to the button press (marked in the trigger trace) the information on the forthcoming task is given. Afterwards, the 3 sec. baseline-period (Period 0, sec. 6-9) is started. During the following 4 sec. (sec. 9-13) the acoustic stimuli were presented (period 1). After the perception period, the imagination period (sec. 13-17) was assigned. Time intervals assigned for data quantification are marked as grey areas in the top trace (F7): mean amplitudes during sec. 10 to 13 of the perception period and during sec. 14.5 to 16 of the processing period (period 2) were compared to the mean amplitudes during the baseline period (sec. 6-9). Acoustic stimulation yields a pronounced negativation over left frontal (F3), right frontal (F4), left fronto-temporal (F7) and right fronto-temporal (F8) and parietal brain regions (P3, P4). The subsequent imagination of the rhythm causes an activation especially over the left and right posterior temporal (T5, T6) and parietal (P3, P4) brain regions.

Data analysis

Each data sampling epoch lasted 22 seconds. Off line, trials contaminated by artefacts were excluded from further analysis. For each condition, about 30 artefact-free trials were averaged per subject. Based on the individual averages, grand-averages across all subjects were derived.

For selected time intervals, mean amplitudes of DC-shifts were calculated (see Fig. 2, upper trace, marked zones). Analysis of the perception period was done between sec. 10-13. The first second of the perception period was excluded from analysis because unspecific EEG-components, related to arousal and orientation occur in this time interval. Analysis of the imagination or post-perception-period was done between sec. 14.5 and 16. The latter interval was selected to avoid overlap of DC-potentials with the off-effect of the last note of the perception period and to assure task-specific activation. The baseline was taken from the period prior to the perception period (sec. 6-9).

Statistical analysis

Data were subjected to repeated measures ANOVA. For a strict separation of material effects and task effects, effects of material were only assessed for the conditions 1 (simple rhythms), 2 (complex rhythms) and 4 (non-rhythm condition), whereas effects of the task were exclusively tested for the conditions 1 (simple rhythms, perception) and 3 (simple rhythms, imagination).

As within-subjects factors Material (three levels: simple rhythms, complex rhythms and non-rhythm), Task (2 levels: perception, imagination), Period (2 levels: perception period, post-perception or imagination period), and Electrode (11 levels according to the electrode-positions) were tested. Influences of material or task on the hemispheric lateralization were assessed introducing an additional within-subjects factor Hemisphere (2 levels: right-hemispheric vs. left-hemispheric electrode-positions).

In order to detect differences in topographic distribution rather than in amplitude, in a second step data were normalised according to the procedure proposed by McCarthy and Wood (1985) and ANOVA was repeated on these normalised values.

For correction of violations of the sphericity assumption, the Huynh-Feldt epsilon was used. Only those variables, which after correction showed significant ($p < 0.01$) main effects or significant interaction terms were further analysed using the appropriate T-Test. Since this procedure applies multiple T-Tests, the p-values were Bonferroni-adjusted.

Group statistics on hemispheric lateralization were assessed by calculating the preponderance of negativity over left- or right-hemispheric homologous brain-areas in each individual. The data were dichotomised (left-hemispheric mean vs. right-hemispheric mean) and the group data were tested as Bernoulli trials for departure from equal probability (Dixon and Mood 1946).

Results

General course of potentials

The general course of the DC-potential shifts is shown in Fig. 2. Prior to pressing the button (sec. 1-4), a slow negative shift occurred in frontal electrode-positions (F3, F4). The button press - starting the delivery of the instructions - yielded a biphasic positive wave of about 2 sec. duration predominantly over parietal brain regions. During the ensuing 3 sec baseline period, preparatory activation occurred as a plateau-like negativation of small amplitude over frontal brain regions and as a ramp-like increase in negativity over parietal and posterior temporal brain regions. The subsequent acoustic stimulation caused a plateau-like, high amplitude negative DC-potential shift mainly over bilateral frontal and parietal, but not over temporal areas. During the imagination of rhythms, the activation decreased in frontal electrode-positions, but over temporal and parietal brain regions of both hemispheres, a sustained long-lasting negative shift was recorded, which exceeded the activation during the preceding perception period in amplitude.

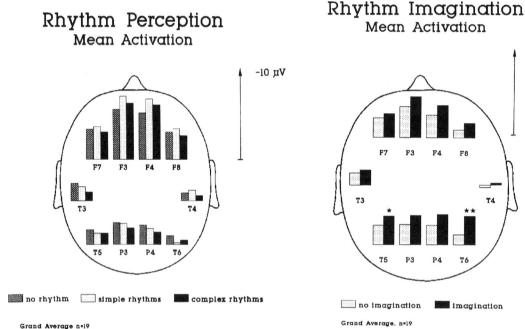

Figure 3
Topographical display of mean amplitudes during acoustic stimulation. Maximal amplitudes occur over frontal brain areas. No significant differences between simple and complex rhythms and between the non-rhythm condition can be observed

Figure 4
Topographical display of mean amplitudes during the imagination period. Imagination of rhythm causes in all electrode positions higher activation, statistically significant over right (**: $p < 0.01$) and over left (*: $p < 0.05$) posterior temporal brain regions.

For the perception period, the mean values of DC-potential amplitudes are visualised topographically in Fig. 3 for the conditions A, B and D: Maximal activation occurred over bilateral frontal (F3, F4) and lower frontal (F7, F8) brain regions, irrespectively from the acoustic material. The mean amplitudes during the imagination period of condition C, - compared to the post-perception period of condition A - are displayed in Fig. 4: imagination of simple rhythms yielded higher amplitudes in all electrode-positions when compared to the post-perception period. Maximal amplitudes were obtained bilaterally over frontal brain regions.

Material and task effects tested with ANOVA

During the perception period, DC-potential shifts showed no significant material effects, i. e. the activation patterns did not differ with the type of perceived stimulus (material x electrode interaction p = 0. 992). When retesting with normalised data, again no stimulus dependent differences in topographical distribution could be demonstrated (material x electrode interaction p = 0. 986).

Subsequently, the effect of task, i.e. perception vs. imagination was tested. Since there was a significant task x period interaction (p = 0.0009) and electrode x period interaction (p = 0.00001), ANOVA was performed on task x electrode per period. There was no significant task-effect during the perception period (p = 0.643), independently whether the task was to reiterate mentally the rhythm or to listen and to create a metre "on-line". In contrast, during the imagination period, there was a significant task-effect (p = 0.0001). Imagination of a simple rhythm caused higher amplitudes at the electrode positions T6 (p = 0.011) and T5 (p = 0.052). When retesting with normalised data, this effect remained significant (T6: p = 0.042, T5: p = 0.043). In frontal electrodes, there was a tendency of higher activation during imagination (F3: p = 0.064, F4: p = 0.094).

When testing with the within-subjects factor "hemisphere", no significant effect of task or material on hemispheric lateralization could be detected (interaction-terms hemisphere x material: p = 0.943, hemisphere x task: p = 0.341).

Group statistics on hemispheric lateralization

The group statistics on percentages of left- or right-hemispheric lateralization are given in Tab. 1. Over frontal brain areas, left- or right-hemispheric lateralization were equally probable. Over temporal areas, there was a tendency to left-hemispheric lateralization in all three perception conditions. During the imagination period no significant lateralization effect over temporal brain areas emerged. Over fronto-temporal areas (F7/8), left-hemispheric lateralization occurred more frequently with imagination (p = 0.05). However, there was no significant difference in lateralization during the imagination of a simple rhythm when compared to the post perception period of condition A.

Results of the questionnaires

12 of the subjects judged the perception and finding of metre in complex rhythms to be the most difficult task, three judged imagination of rhythm more difficult, three found all tasks difficult and only one subjects reported that she found all tasks easy to solve. The self-evaluation of percentage of correctly solved tasks varied between 60% and 100% in the perception task and between 80% and 100% in the imagination task. 11 of the subjects answered the question on task-solving strategy with "no special strategy, just feeling" or with "don't know", 7 reported on an occasional analytic strategy using inner speech to denote the rhythmic patterns. But none of these subjects used this way of processing predominantly.

PERCENTAGE OF LEFT-HEMISPHERIC LATERALIZATION IN GROUP STATISTICS (n= 19)

Perception	F7/F8	F3/F4	T3/T4	T5/T6	P3/P4
Simple r.	53%	47%	74%*	68%	58%
Complex r.	63%	52%	68%	74%*	68%
Imagination	58%	53%	63%	63%	74%*
No rhythm	42%	53%	79%*	58%	58%
Post-period					
Simple r.	79%*	74%*	84%*	63%	58%
Imagination					
Simple r.	74%*	68%	68%	47%	53%

Table 1
Group statistics on percentages of left-hemispheric lateralization in corresponding electrode positions. Significant lateralization is marked with * ($p = < 0.05$)

Discussion

To our knowledge, this is the first study on cortical activation patterns during imagination of acoustically presented rhythms. For proper evaluation of our results it must be emphasised that musical experience and expertise may influence processing strategies and - as a consequence - brain activation (Altenmüller 1986, 1989, Altenmüller et al. 1997, Koelsch et al 1999, Pantev et al. 1998, Tervaniemi et al. 1997). To control this parameter, the present investigation was restricted to a relatively homogeneous group of professionally trained musicians. A generalisation of the present findings on musically naive subjects should not be undertaken. The main results can be summarised as follows:

1.) Acoustic stimulation produces a sustained cortical activation especially over frontal brain regions, irrespective of whether the stimuli contain a simple, a complex or no temporal structure.

2.) Imagination of rhythm is accompanied by an activation of bilateral posterior temporal brain regions.

3.) The processing of rhythms and sustained notes tends to cause in the majority of musicians a left-hemispheric lateralization over temporal brain areas.

Sustained cortical activation during acoustic stimulation masks material and task effects

In the present study, acoustic stimulation yielded in all stimulus conditions a widespread, long lasting activation of bilateral frontal and parietal cortical regions. This "perstimulatory" DC-potential was described by Koehler and Wegener (1955) for the first time and was investigated systematically by David et al. (1969) and Keidel (1971). They could demonstrate that these DC-potential shifts are intramodally specific, depend on physical parameters such as loudness and frequency, and are generated in the primary and secondary auditory cortex. Since these cortical areas are located in the sylvian planes, electrical dipoles arising during activation are oriented tangentially to the temporal scull and therefore produce maximal amplitudes in midline and near-midline electrode positions (i.e. F3, F4, P3, P4) at the convexity of the scull (Lutzenberger et al. 1987). The surprising negative result that neither the stimulus material nor the imposed cognitive task did influence the cortical activation patterns during stimulation can be explained by a masking phenomenon: since already processing of simple stimuli such as the temporally unstructured notes in the "non-rhythm" condition produced widespread high-amplitude negative DC-potential-shifts, other smaller effects, related to rhythm processing or to the cognitive task imposed probably were masked by the predominant activation of the primary and secondary auditory cortices. The only way to rule out this interference is a modified stimulation mode, e.g. presenting the rhythms in the visual or tactile modality. However, results gained from visual or tactile stimulation cannot per se be generalised to the acoustic modality, since modality specific disturbances of rhythmic subfunctions with relative preservation in the visual modality, but loss in the auditory modality seem to occur after localised lesions (Fries and Swihart 1990, Penhune et al. 1999).

Imagination of rhythms and the posterior temporal lobes

In order to separate perceptive processes from cognitive processes, the imagination condition was included in the present experiment. During imagination of a rhythm, masking phenomena as discussed above were not expected to play a role. In agreement with this hypothesis, the findings revealed a clear effect related to the imposed cognitive task: Imagination of a rhythm by mental reiteration of a previously heard rhythmical sequence caused a bilateral activation at the electrode positions T5 and T6. These electrodes are located over the posterior part of the superior temporal gyrus at the transition to the angular gyrus and the parietal lobule (Homan et al. 1987). Convergent evidence for an outstanding role of the posterior temporo-parietal brain regions in rhythm processing is provided by several observations. Mavlov's case (1980) developed a disturbance of recognition and reproduction of rhythms following a lesion in the left parietotemporal cortex. Creutzfeldt and Ojeman (1989) recorded single unit activity from the posterior superior temporal gyrus during open brain surgery while stimulating with rhythmical rock music and found an increase in neuronal discharge in both hemispheres.

Although these findings are based on the physical perception of acoustical events, similar brain areas might be involved in the generation of mental images of such physical percepts. Evidence for a parallelism between the processing of physical percepts and the creation of mental images of these percepts has been provided in the visual modality: imagination of a visuo-spatial task yielded maximal activation of the parieto-occipital cortex whereas the imagination of faces caused maximal activation at occipito-temporal regions (Uhl et al. 1990).

Imagination of a route-finding task activated the superior occipital, the posterior inferior temporal and the posterior superior parietal cortex (Roland and Friberg 1985). In the latter study, the generation of mental images was accompanied by an additional activation of the superior prefrontal cortex which - in a generalised manner - was attributed to the temporal organisation of thinking. The outstanding role of the prefrontal cortex for the temporal organisation of behaviour and cognition has been proved in numerous experiments (for a review see the pioneering work of Ingvar 1985). Correspondingly in the present study, a tendency to more pronounced activation of the frontal cortices during imagination of rhythms has been found. In summary, as visual imagination activates modality specific primary, secondary, and higher order cortices, a similar activation mode may be expected for imagination in the acoustic modality. Additionally, maintenance and control of the temporal structure of thinking and imagination may cause a supra-modal activation of the prefrontal cortex. At present it remains open, whether the involvement of temporo-parietal areas is due to an inter-modal transformation of time-structures from the auditory modality into the visual modality. Since our highly trained professional musicians were able to represent music as scores, we can speculate that these musicians might have additionally used visual cues for the imagination of time-structures.

Which hemisphere is superior in processing of rhythm?

During perception and imagination of rhythms, there was no clear preponderance of one hemisphere. In group statistics, the majority of subjects exhibited a left temporal lateralization during rhythm perception, but the lateralization effects were weak and did not differ significantly from the "non-rhythm" condition. When reviewing the classical lesion studies, a similar result emerges: in stroke patients, impaired rhythm processing occurs slightly more often following lesions of the left hemisphere than of the right hemisphere (Brust 1980, Henschen 1920, Uvstedt 1937, Wertheim and Botez 1961). However, in these studies processing of rhythm and processing of metrical structures were not assessed separately. Since Lerdhal and Jackendoff's two component model (1983) predicts a dissociation of left hemispheric rhythm and right hemispheric metre processing, this variable has to be taken into account. It should be mentioned in this context, that up to now lesion studies investigating separately rhythm or metre processing failed to clearly demonstrate such a dichotomy (Peretz 1990, Peretz and Babai 1992, Liégois-Cauvel et al, 1999, Schuppert et al. 2000). But taking into account the data provided by Sakai in normal subjects (1999) and considering the pronounced local/global dichotomies in the melodic system (Peretz 1990), or in the visual modality (Heinze et al. 1994), the assumption of similar mechanisms during the processing of time structures seems to be justified.

Since in the present study subjects were required to do both, local and global level cognitive operations at the same time, effects on hemispheric lateralization are the result of a relative preponderance of the respective mode of processing. Furthermore, individual task-solving strategies such as the use of inner speech produce additional variability. It should be emphasised that in music rhythm and metre are linked together inseparably. The present study investigates "natural" conditions. At the moment it cannot be decided, to what extend rhythm processing or metre processing produced the cortical activation patterns and to what extend musical expertise influenced the use of the respective cognitive operations. To solve these questions, a further activation study including non musicians and investigating the two cognitive operations separately is in preparation.

Summary

In order to determine cortical structures involved in rhythm processing, slow-brain potentials (DC-potentials) were recorded from the scalp of 19 professional musicians during perception and imagination of rhythms. Subjects either listened to a simple or a complex rhythm or to a sustained tone. A fourth condition required the imagination of the previously presented simple rhythm. During acoustic stimulation, a bilateral increase in brain activity over frontal and parietal brain regions occurred, irrespective of whether the stimuli were temporally structured or not. Imagination of rhythms yielded a significant activation of bilateral posterior temporal regions. Group statistics revealed a tendency to left-hemispheric preponderance. The results are in favour of individually organised neuronal networks underlying the perception of time structures in professional musicians. The parietal activation is interpreted as a sign of a multimodal, - presumably visual - representation of rhythm.

Acknowledgements

This research was supported by grants from the Deutsche Forschungsgemeinschaft, SFB 370 /B8 and Al 269/4-2.

References

Altenmüller E. Hirnelektrische Korrelate der cerebralen Musikverarbeitung beim Menschen. Eur Arch Psychiatr Neurol Sci 235: 342-354 (1986)

Altenmüller E. Cortical DC-potentials as electrophysiological correlates of hemispheric dominance of higher cognitive functions. Intern J Neuroscience 47: 1-14 (1989)

Altenmüller E, Gerloff C. Psychophysiology and EEG. In: Niedermeyer E and Lopes da Silva F, (eds.) Electroencephalography. 4rd Ed.. Baltimore: Williams & Wilkins, pp 637-655 (1998)

Altenmüller E, Kriechbaum W, Helber U, Dichgans J, Petersen D. Cortical activation patterns during language processing. Clinical and linguistical aspects. Acta Neurochir 35: 12-21 (1993)

Altenmüller E, Gruhn W, Parlitz D, Kahrs J. Music learning produces changes in brain activation patterns: a longitudinal DC-EEG-Study. Int J Arts Medicine 5: 28-34 (1997)

Beisteiner R, Altenmüller E, Lang W, Lindinger G, Deecke L. Musicians processing music: measurement of brain potentials with EEG. Eur J Cogn Psychol 6. 311-327 (1994)

Brust J. Music and language: musical alexia and agraphia. Brain 103: 367-392 (1980)

Creutzfeldt O, Ojemann G. Neuronal activity in the human lateral temporal lobe. Activity during music. Exp Brain Res 77: 490-498 (1989)

David E, Finkenzeller P, Kallert S, Keidel WD. Akustischen Reizen zugeordnete Gleichspannungsänderungen am intakten Schädel des Menschen. Pflügers Arch 309: 362-367 (1969)

Dixon WJ and Mood AM. The statistical sign test. J Amer Statist Assoc 41: 557-566 (1946)

Frackowiak RSJ, Friston KJ, Frith CD, Dolan RJ, Mazziotta JC. Human Brain Function. Academic Press, London, (1997)

Fries W, Swihart AA. Disturbance of rhythm sense following right hemisphere damage. Neuropsychologia 28: 1317-1323 (1990)

Heinze HJ, Mangun GR, Burchert W, Hinrichs H, Scholz M, Münte TF, Gös A, Scherg M, Johannes S, Hundeshagen H, Gazzaniga MS, Hillyard SA: Combined spatial and temporal imaging of brain activity during visual selective attention in humans. Nature 372:543-546 (1994)

Henschen SE. Über Amusie. In: Klinische und anatomische Beiträge zur Pathologie des Gehirns, Vol 5, pp 137-213, Nordiska Bokhandeln, Stockholm (1920)

Homan RW, Herman J, Purdy P. Cerebral location of the international 10-20 system electrode placement. Electroenc clin Neurophysiol 66: 376-381 (1987)

Ingvar DH. "Memory of the future": an essay on the temporal organization of conscious awareness. Human Neurobiol. 4: 127-136 (1985)

Ivry R. Cerebellum involvement in the explicit representation of temporal information. Annals of the New York Academy of Sciences 682: 214-230 (1993)

Jones RB. Dynamic pattern structure in music: Recent theory and research. Perception and Psychophysics 41: 621-634 (1987)

Keidel WD. DC-Potentials in the auditory evoked response in man. Acta Otolaryng 71: 242-248 (1971).

Köhler W, Wegener J. Currents of human auditory cortex. J Cell Comparat Physiol 45: 25-54, (1955)

Kölsch S, Schröger E, Tervaniemi M. Superior pre-attentive auditory processing in musicians. Neuroreport 10: 1309-1313 (1999)

Lerdahl F, Jackendoff, R. A generative theory of tonal music. MIT-Press, Cambridge (1983)

Liégeois-Chauvel C, Peretz I, Babaï M, Laguitton V, Chauvel P. Contribution of different cortical areas in the temporal lobes to music processing. Brain 121: 1853-1867 (1998)

Lutzenberger W, Elbert T, Rockstroh B. A brief tutorial on the implications of volume conduction for the interpretation of the EEG. J Psychophysiol 1: 81-89 (1987)

Mavlov L. Amusia due to rhythm agnosia in a musician with left hemisphere damage: a non-auditory supramodal effect. Cortex 16: 331-338 (1980)

McCarthy G, Wood CC. Scalp distributions of event related potentials: an ambiguity associated with analysis of variance models. Electroenc clin Neurophysiol 62: 203-208 (1985)

Oldfield RC. The assessment and analysis of handedness: the Edinburgh handedness inventory. Neuropsychologia 9: 97-113 (1971)

Pantev C, Oostenveld R, Engelien A, Ross B, Roberts LE, Hoke M. Increased auditory cortical representation in musicians. Nature 392: 811-814 (1998)

Patel H, Price C, Baron JC, Wise R, Lambert J, Frackowiak SJ, Lechevalier B, Eustache F. The structural components of music perception. A functional anatomical study. Brain 120: 229-243 (1997)

Penhune VB, Zatorre RJ, Evans AC. Cerebellar contributions to motor timing: a PET study of auditory and visual rhythm reproduction. J Cognitive Neuroscience 10: 752-765 (1998)

Penhune VB, Zatorre RJ, Feindel WH. The role of auditory cortex in retention of rhythmic patterns as studied in patients with temporal lobe removals including Heschl's gyrus. Neuropsychologia 137: 315-331 (1999)

Peretz I. Processing of local and global musical information by unilateral brain-damaged patients. Brain 113: 1185-1205 (1990)

Peretz I. Babai M. The role of contour and intervals in the recognition of melody parts: evidence from cerebral asymmetries in musicians. Neuropsychologia 30: 277-292 (1992)

Roland PE, Friberg L. Localization of cortical areas activated by thinking. J Neurophysiol 5: 1219-1243 (1985)

Sakai K, Hikosaka O, Miyauchi S, Takino R, Tamada T, Iwata NK, Nielsen M. Neural representation of rhythm depends on its interval ratio. J Neurosci 19: 10074-10081 (1999)

Schlaug G, Jäncke L, Huang Y, Steinmetz H. In vivo evidence of structural brain asymmetry in Musicians. Science 267: 699-701 (1995)

Schuppert M, Münte TF, Wieringa BM, Altenmüller E. Receptive amusia: Evidence for cross-hemispheric neural networks underlying music processing strategies. Brain 123: 546-559 (2000)

Smith KC, Cuddy LL. Effects of metric and harmonic rhythm on the detection of pitch alterations in melodic sequences. J Exp Psychol 15: 457-471 (1989)

Sutula T, Cascino G, Cavazos J, Parada I, Ramirez L. Mossy fiber synaptic reorganization in the epileptic human temporal lobe. Ann Neurol 26: 321-330 (1989)

Tervaniemi M, Ilvonen T, Karma K, Alho K, Naatanen R. The musical brain: brain waves reveal the neurophysiological basis of musicality in human subjects. Neurosci Letters 226: 1-4 (1997)

Thomas C, Altenmüller E, Marckmann G, Kahrs J, Dichgans J. Language processing in aphasia: syndrome-specific lateralization patterns during recovery reflect cerebral plasticity in adults. Electroenc clin Neurophysiol 102: 86-97 (1997)

Uhl F, Goldenberg G, Lindinger G, Steiner M. Deecke L. Cerebral correlates of imagining faces, colours and a map. II. Negative cortical DC-potentials. Neuropsychologia 28: 81-93 (1990)

Uvstedt HJ. The method of examination in amusia. Acta Psychiat et Neurol 12: 447-455 (1937)

Vargha-Khadem F, Isaacs EB, Papaleloudi H, Polkey CE, Wilson T. Development of language in six hemispherectomized patients. Brain 114: 473-495 (1991)

Wertheim N, Botez M. Receptive amusia: a clinical analysis. Brain 84: 19-30 (1961)

T. Nakada (Ed.)
Integrated Human Brain Science: Theory, Method Application (Music)
© 2000 Elsevier Science B.V. All rights reserved

Chapter III.12

Interaction in Musical-Pitch Naming and Syllable Naming: An Experiment on a Stroop-like Effect in Hearing

Ken'ichi Miyazaki [1]

Department of Psychology, Faculty of Humanities, University of Niigata

Introduction

The present research investigates cognitive interactions in musical-pitch naming and syllable naming, a Stroop-like effect in hearing. The classical Stroop effect, originally reported by Stroop (1935), is represented by a cognitive interference phenomenon in naming the ink color of the printed color words. Typically, color naming is slower when the color is incongruent with the color designated by the word. On the contrary, reading words is easy and, usually, no interference occurs even when they are printed in wrong color. Since then, this phenomenon has stimulated cognitive psychologists interested in attentional processes to carry on a great number of experiments examining the interference in more detail (see MacLeod, for review, 1991).

Although experiments on the Stroop effect have focused primarily on the visual domain, there are several investigations that have found corresponding effects in the auditory domain; for example, interaction of pitch and word meaning (high or low) (Cohen & Martin, 1975; McClain, 1983; Walker & Smith, 1984), voice quality and speaker gender (Green & Barber, 1981, 1983), and ear of presentation and word meaning (left or right) (Pieters, 1981). In the auditory Stroop experiments that focused on the interaction of pitch and verbal meaning, only two values of pitch (high or low) were manipulated. Pitch could be characterized in a much more sophisticated manner; particularly in the musical context pitch is characterized by musical qualities (pitch classes or tone chromas) which are named by pitch alphabets (C, D, E, etc.) or conventionally used solfeggio syllables; for instance, tones of the C major scale are named as *do, re, mi, fa, so, la, si*. Thus, it is expected to find a similar interaction in the musical pitch domain more strongly corresponding to the classical Stroop interaction in the visual domain. Pitch class and syllables in this type of auditory Stroop effect correspond to color and words, respectively, in the visual Stroop effect.

The literature on the auditory Stroop effect is scarce on this issue. To our knowledge, there

---

[1] Correspondence: Ken'ichi Miyazaki, Department of Psychology, Faculty of Humanities, University of Niigata, Niigata 950-2181, Japan, e-mail: miyazaki@human.ge.niigata-u.ac.jp

is only one brief report (Zakay, Roziner, & Ben-Arzi, 1984) that has dealt with the interaction of sung pitch classes and pronounced syllables. In their experiment, subjects were selected for possession of absolute pitch and were required to identify pitch classes of vocal tones sung with syllables incongruent with pitch (pitch-word condition) or a fixed vowel (pitch condition). The subjects showed longer response times and more errors in the pitch-word condition than in the pitch condition. They interpreted these results as reflecting the two-stage model; pitch is recognized in the perceptual stage and then is associated with a corresponding pitch name in the response stage. However, the experiment of Zakay et al. (1984) is in a limited perspective in that it focuses on the nature of absolute pitch and deals with only the interfering effect of incongruent words on pitch naming.

From a more extensive perspective, it is assumed that listeners would be subject to interference in naming pitch of a vocal tone using the conventional solfeggio syllable when the tone is sung with a wrong syllable (e.g., the *do* note sung with a syllable *mi*); as a result, it may take longer time to name the incongruent tones, as Zakay et al. (1984) demonstrated. On the contrary, naming pitch of sung tones may be facilitated when pronounced syllables are congruent with sung pitch, resulting in shorter response times and/or fewer errors. On the other hand, this interaction is unlikely to be symmetric. Apparently, repeating (or shadowing) a pronounced syllable immediately after the vocal tone like a parrot is a much more automatic activity with virtually no attentional demands needed as compared with identifying a pitch-class name. It is, therefore, expected that the influence of sung pitch on syllable repetition should not occur in the pitch-syllable interaction. This expectation is in fact consistent with the common finding in the visual domain that usually no reverse Stroop effect is observed in the color-word interaction.

As to identifying musical pitch, absolute-pitch listeners are quite remarkable cases in that they are able to name pitch of isolated tones very accurately and quickly without recourse to any reference tone (Miyazaki, 1988). People with accurate absolute pitch seem to be highly practiced in converting musical pitch of single tones into verbal code (usually solfeggio syllables or pitch alphabets), and so possibly come to a state of automaticity in naming pitch. Indeed, absolute pitch listeners often tell that a pitch is always combined with its corresponding solfeggio syllable, and the piano tone sounds as if it is accompanied by a syllable corresponding to its pitch. Furthermore, Miyazaki (1995) demonstrated a reverse side of this remarkable ability. People with accurate absolute pitch had difficulty in the relative-pitch (pitch interval) identification task when a relative-pitch name and an absolute-pitch name were in conflict, presumably because the automatically accessed absolute-pitch name interfered with relative-pitch naming. This interference itself may be an example of the Stroop-like conflict within a pitch domain. The highly practiced pitch naming ability of absolute-pitch listeners may produce another more interesting conflict between a pitch name and a pronounced name. It is, then, hypothesized that their performance of repeating pronounced syllables should be subject to the influence of sung pitch of vocal tones, i.e., the reverse Stroop-like effect might occur for the absolute-pitch listeners, an exception to the typical pattern of the Stroop-like phenomenon.

The purpose of the present experiment is to investigate this possibility; how does a pronounced syllable of a vocal tone influence (interference and/or facilitate) on identification of sung pitch, and how does sung pitch influence on repeating a syllable? Furthermore, is there any difference in the pattern of the interaction between listeners with absolute pitch and those without absolute pitch?

Methods

Subjects

Forty-five undergraduate students who reported having normal hearing volunteered to participate in this experiment. All of them enrolled in an introductory psychology course at Niigata University and differed widely in musical experience. After the primary experiment was completed, the subjects performed an absolute-pitch identification test in which 60 different piano tones were presented randomly. According to the score, they were classified into three subgroups differing the accuracy of absolute-pitch identification; i.e., accurate AP subjects (AP1 group, $n = 10$) who scored higher than 90% correct, inaccurate AP subjects (AP2 group, $n = 13$) with scores from 50 to 85 % correct, and no-AP subjects (NAP group, $n = 22$) with scores lower than 45% correct.

Stimuli and Apparatus

Stimuli presented were recorded vocal tones sung by an experienced male or female singer. In recording, the singers repeatedly heard the major scale tones produced by a sampled piano-tone generator (Yamaha, MU-90) through headphones and concurrently sang the scale tones keeping pitch as precisely as possible. The pitch range of the scale was from C3 to B3 for the male singer and from C4 to B4 for the female singer. The scale-tone sequences were sung a sufficient number of times with each solfeggio syllable (*do, re, mi, fa, so, la,* and *si*). In each sequence, the singer sang the major scale tones with a fixed syllable independent of pitch. Additionally, the same major-scale tones sung in vowel /u:/ by the same singers and spoken solfeggio syllables in normal voice by other male and female speakers were recorded as stimuli for control conditions. All vocal sounds were recorded on a digital audio tape (DAT) recorder (SONY, DTC-2000ES) at a 48-kHz sampling rate through an electret condenser microphone (SONY, C-350), and then directly transferred to an Apple Power Macintosh 9600/300 computer via a sound card (lucid Technology).

The best samples selected from the recorded vocal sounds as test stimuli were normalized in amplitude and duration by editing and shaping the original waveforms using a waveform editing software (Bias, PEAK) on the Macintosh computer. Actually, the durations of the sung syllables were adjusted to approximately 400 to 550 ms and those of the spoken syllables were adjusted to 180 to 260 ms depending on the syllables pronounced.

Experimental stimuli were either congruent or incongruent in the pitch-syllable relationship. The congruent stimuli were sung with solfeggio-syllables that were congruent with their pitch, for instance, *do* for C, *re* for D, and so on. The incongruent stimuli were sung with syllables that were incongruent with their pitch. The incongruent syllables were those that were shifted upward in the major scale by 2 to 5 steps from the correct pitch syllables; for instance, in the incongruent condition, the C note (*do*, in the C major) was sung with an incongruent syllable *mi, fa, so,* or *la*. Stimuli for the control condition were always sung tones with a vowel /u:/ or spoken solfeggio-syllables in a normal speaking voice. Both experimental stimuli and control stimuli were produced by a male or a female singer/speaker. In total, the stimulus set included 56 incongruent stimuli (7 pitches x 4 syllables x 2 voices), 56 congruent stimuli, and 56 control stimuli (7 pitches or syllables x 2 voices x 4 replications); 7 pitches (syllables) in the latter two stimulus types were replicated 4 times each for appearing equally often to the incongruent stimuli.

At the beginning of each trial, a sampled piano tone generated from a tone generator (Yamaha MU-90) was played for the purpose of providing subjects with a reference for judging pitch of a test tone next to come. This tone had constant pitch of C4 and an approximate duration of 400 ms. Following this, a vocal test tone was presented with a stimulus onset asynchrony of 1 s.

The experiment was programmed on an Apple Macintosh G3 MT-266 computer using the PsyScope software developed at Carnegie Mellon University for designing psychological experiments (Cohen, MacWhinney, Flatt, & Provost, 1993). Stimuli were presented to subjects at comfortable listening levels over a pair of loudspeakers (SONY, SRS-Z1000) located in front of them. Subjects responded orally. Their vocal responses registered through a microphone (SONY C-350) triggered a voice-key of a PsyScope Button-Box. Time intervals from the onset of the test stimulus to the occurrence of the voice-triggered pulse were measured as response times by the computer in milliseconds.

Procedure

Subjects performed two different tasks in separate sessions. In the pitch-naming task, the subjects were required to name the pitch of the vocal test tone relative to the preceding tone as the reference (*do*). They were specifically instructed to focus on pitch of the vocal tone and to ignore its syllabic identity. On the contrary, in the syllable-naming task, the subjects' task was to repeat the syllable of the sung or spoken stimulus ignoring its pitch. No pitch reference was necessary for this task but the initial piano tone served simply as a cueing signal. Subjects were emphasized that they should respond as quickly and accurately as possible. There was a 3-s interval between the response occurrence and the onset of the reference tone of the next trial. Twenty practice trials were given prior to the experimental trials in each session. In each session, subjects performed 168 trials with the congruent, incongruent, or control stimulus presented in a random order with the restriction that two consecutive trials never contained the same pitch or the same syllable. Sessions of the pitch-naming task and the syllable-repetition task were given in a counterbalanced order with a rest period of an appropriate length between them.

After the main experiments completed, the subjects had the absolute-pitch test, in which 60 different pitches of chromatic scales were presented over 5 octaves from C2 to B6 with the standard A4 (440 Hz). The timbre was sampled piano tones. The subjects responded by pressing the corresponding key on the MIDI keyboard (Yamaha, CBX-K1) without regard to the octave position.

Results

Two types of measures, response times of correct responses and error rates, were obtained from the pitch task and the syllable task.

Response times

First, response time data of the two tasks were analyzed by a three-way analysis of variance with task (pitch and syllable) and congruency (congruent, incongruent, and control) as within-subjects factors and AP level (AP1, AP2, and NAP) as a between-subjects factor. The analysis revealed a significant main effect of task [$F(1, 42) = 131.8$, $MS = .072$, $p<.0001$], suggesting that, as a whole, response times in the syllable task were reliably shorter than those in the pitch task. All the interactions including task were significant, i.e., task by congruency [$F(2, 84) = 21.8$, $M = .003$, $p<.0001$], task by AP level [$F(2, 42) = 14.9$, $MS = .072$, $p<.0001$], and task by congruency by AP level [$F(4, 84) = 4.6$, $MS = .003$, $p<.005$]; hence the response time data were analyzed separately for each task.

Response times in the pitch task are shown in Figure 1. A two-way analysis of variance with congruency and AP level as factors revealed significant main effects for AP level [$F(2, 42) = 11.4$, $MS = .13$, $p<.0001$] and congruency [$F(2, 84) = 66.1$, $MS = .004$, $p<.0001$]; the interaction of AP level by congruency was not significant [$F(4, 84) = 1.4$, $MS = .004$, $p>.1$]. Subsequent pair-wise comparisons among AP levels using Tukey's *HSD* ($p<.05$) showed that the NAP group was reliably slower in pitch naming than the AP1 and AP2 groups, but the difference between the AP1 and AP2 groups was not reliable. Further comparisons among congruency conditions showed that response times in the incongruent condition were reliably longer than those in the congruent and control (vowel) conditions, but there was no reliable difference between the congruent and control conditions.

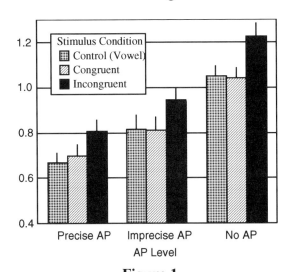 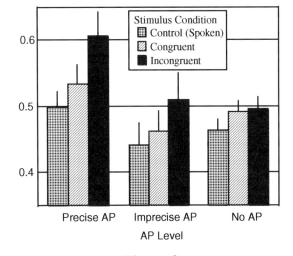

Figure 1
Averaged response times and standard errors in the pitch naming task.

Figure 2
Averaged response times and standard errors in the syllable repetition task.

Response times in the syllable task are shown in Figure 2. A two-way analysis of variance revealed that the main effect for congruency and the interaction of congruency with AP level were significant [$F(2, 84) = 64.9$, $MS = .0008$, $p<.0001$, and $F(4, 84) = 8.2$, $MS = .0008$, $p<.0001$, respectively], but the main effect for AP level was not significant [$F(2, 42) = 1.9$, $MS = 0.029$, $p>.15$]. Whereas the simple main effect of congruency was significant in all AP-level groups [$F(2, 84) = 54.5$, $p<.001$ for AP1, $F(2, 84) = 21.7$, $p<.001$ for AP2, and $F(2, 84) = 5.1$, $p<.01$ for NAP], the significant interaction of congruency by AP level suggested that the influence of congruency was different for AP-level groups. Hence, Tukey's *HSD* tests (p<.05) were carried out for pair-wise comparisons of congruency conditions for each AP level separately. This analysis showed that, for the AP1 group, the incongruent condition was significantly slower than the congruent condition, which was in turn slower than the control condition, for the AP2 group, the incongruent condition was reliably slower than both the congruent and control conditions but the latter two conditions were not reliably different, and for the NAP groups, there was no reliable difference between the congruent and incongruent conditions although these two conditions were reliably slower than the control condition. The simple main effect of AP level was not significant in the congruent and control conditions but significant in the incongruent condition. Tukey's *HSD* tests revealed that the AP1 subjects were slower in repeating syllables than the AP2 and NAP subjects.

Error rates

Figure 3 shows the percentage of error responses in the pitch task. A two-way analysis of variance revealed both main effects of AP level and congruency were significant [$F(2, 42) = 14.9$, $MS = 509.3$, and $F(2, 84) = 50.8$, $MS = 80.1$, $p<.0001$, respectively]. These main effects indicate that, in general, subjects showed the highest error rate in the incongruent condition, the lowest in the congruent condition, and the intermediate in the control condition, and that the AP1 subjects were most accurate, the AP2 subjects were intermediately accurate, and the NAP subjects were poorest in identifying musical pitch. A set of Tukey's *HSD post hoc* comparisons ($p<.05$) confirmed that all the differences among the subject groups and among the congruency conditions were reliable. However, there was a significant interaction of AP level by congruency [$F(4, 84) = 7.2$, $MS = 80.1$, $p<.0001$], suggesting that the effect of congruency contributed differently for the subject groups although the pattern of the effect was identical for all the groups. Subsequent analyses revealed that the simple main effect of AP level was significant for the incongruent condition [$F(2, 126) = 23.3$, $MS = 233.2$, $p<.001$] and for the control condition [$F(2, 126) = 13.0$, $MS = 233.2$, $p<.001$] and marginally significant for the congruent condition [$F(2, 126) = 2.8$, $MS = 233.2$, $p<.1$], and that the simple main effect of congruency was significant for the AP2 group [$F(2, 84) = 14.7$, $= 80.1$, $p<.001$] and for the NAP group [$F(2, 84) = 47.6$, $= 80.1$, $p<.001$] and marginally significant for the AP1 group [$F(2, 84) = 3.0$, $= 80.1$, $p<.1$]. For the syllable task, error data were not analyzed because only a negligible number of errors occurred

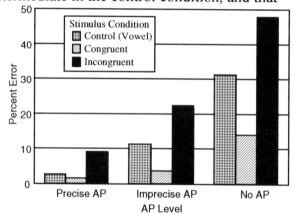

Figure 3
Averaged percentage of error responses in the pitch naming task.

Discussion

The present results clearly demonstrated that, for all subjects, performance of pitch naming of vocal tones was significantly influenced by syllables pronounced. In the pitch naming task, the performance was poorer and responses were slower when the vocal tones were sung with syllables incongruent with their pitch than when sung with congruent syllables or a neutral vowel. It appears to be considerably difficult for subjects to selectively attend to pitch of the vocal tone ignoring pronounced syllables when those syllables were incongruent with pitch to be named.

Subjects of the AP and the NAP groups may have used different strategies in performing the pitch-naming task. Subjects having no absolute pitch (NAP) are supposed to have identified pitch by the use of relative pitch comparing the test tone with the reference tone, whereas subjects with accurate absolute pitch (AP1) may have named pitch directly by the use of absolute pitch presumably without comparing with the reference tone. In spite of this difference in strategies, all the subject groups of different AP level were subject to interference with pitch naming from incongruent syllables, although it did not reach significance for error data of the AP1 group because they made relatively few errors in every condition probably for the benefit of accurate absolute pitch. This suggests that the incongruent syllables have an equivalent interfering effect with pitch naming based on both absolute pitch and relative pitch.

On the other hand, all the subject groups showed no significant difference in response times between the control (vowel) condition and the congruent condition, suggesting that pitch naming was not facilitated at least with respect to speed when the pronounced syllable was congruent with pitch. The lack of facilitation as opposed to considerable interference is consistent with the typical asymmetry of interference and facilitation reported in the color-word Stroop interaction that there is generally no facilitation by congruent words, or, if any, it is very small. On the contrary, a facilitation effect was actually found for response accuracy in the present pitch-naming task. The AP2 and the NAP subjects made significantly fewer errors in the congruent condition than in the control condition, although the AP1 subjects whose error rates were negligibly low did not. This facilitation in accuracy might be partially due to the fact that the AP2 and the NAP group contained subjects who made a number of errors and hence the difference in error rates between conditions was exaggerated for these groups. The apparent longer response times and more errors of the NAP group compared with the AP groups may reflect that the processing mode based on relative pitch is more demanding and requires a greater amount of cognitive resources than the absolute pitch mode. Additionally, this lower performance of the NAP group is supposed to be attributable to the difference in the amount of musical experience. Generally, absolute pitch is acquired through musical lessons particularly in piano in early childhood, and hence the better subjects performed in the absolute-pitch test, the more extensively they had been trained in music.

Apparently, the pattern of interference closely resembles to the visual Stroop phenomenon, interference between color and word. Various models have been so far proposed for explaining the Stroop effect, for example, the horse-race model based on the difference of processing speeds between color naming and word reading (Morton & Chambers, 1973), the automaticity model (Shiffrin & Schneider, 1977; Logan, 1978), and the parallel distributed processing model (Cohen, Dunbar, & McClelland, 1990). Similar explanations could be given to the Stroop-like phenomenon in musical pitch naming obtained in the present experiment.

From the perspective of automaticity focusing on the perceptual processing stage, the present results could be interpreted in the following way. In the pitch-naming task, subjects have to identify pitch of a sung tone and to name its corresponding pitch syllable without regard to the pronounced syllable. Identifying musical pitch is generally assumed to be a considerably demanding task requiring a lot of cognitive resources. Compared with pitch, the pronounced syllable is processed immediately, even almost automatically, and is recognized more quickly than the pitch name. When the syllable is incongruent with pitch of the sung tone, then, this incongruent syllable interferes with the decision process that has to produce the pitch name for responding, presumably resulting in longer response times and more errors. On the other hand, when the syllable is congruent with pitch, the congruent syllable now may facilitate processing, and possibly results in quicker responses and fewer errors. Alternative explanations might be raised from the perspective of the other models (see, MacLeod, 1991; Sugg & McDonald, 1994). To determine which model is preferable for accounting for the interference between pitch and syllables reported here, more detailed investigations are required.

If one can recognize pronounced syllables more automatically than sung pitch, it seems improbable that repeating a pronounced syllable (shadowing) is subject to interference when sung pitch is incongruent with the to-be-repeated syllable. In fact, it is commonly reported that there is no *reverse* Stroop effect in the color-word interaction, i.e., color naming does not interfere with word reading. In the present syllable task, as expected, the subjects without absolute pitch showed no interference in the reverse direction, i.e., there was no significant difference in response times in syllable repetition between in the congruent and incongruent conditions. By contrast, most importantly, the subjects with absolute pitch (both AP1 and AP2) showed significantly slower responses in the incongruent condition than in the congruent condition, demonstrating considerable interference with syllable repetition by incongruent pitch. This interference in the reverse direction could be accounted for also by the automaticity view assuming that, for the AP subjects, verbal labeling of musical pitch is a highly automatic process enough to have a potential of causing interference with another automatic syllable repetition process.

As compared with word reading of the original Stroop experiment that requires a conversion of letters into phonology, syllable repetition is assumed to be an even more automatic task that requires only phonological echoing. In this sense, syllable repetition represents a privileged loop (McLeod & Posner, 1984). It is noteworthy, therefore, that even the privileged syllable repetition was influenced by incongruent pitch in case of absolute pitch subjects. This indicates that verbal labeling of musical pitch is a highly automatic process people with absolute pitch have established.

Summary

An auditory analogue of the Stroop effect was demonstrated in an experiment in which a musical-pitch naming task and a syllable-naming task were used. In each trial, a reference piano tone with fixed pitch of C was followed by a vocal test tone sung by a male or female singer. Pitch of the sung note was selected from the C-major diatonic scale and was either congruent or incongruent with the pronounced syllable. Participants with absolute pitch and those without absolute pitch named vocally the musical pitch (pitch task) or repeated the syllable (syllable task). As expected from the visual Stroop effect, the incongruent syllable name interfered with pitch naming, resulting in more errors and longer response times compared with the congruent condition. More interestingly, participants with absolute pitch showed the reverse effect in the syllable task; the incongruent pitch interfered with syllable naming. In contrast, participants with no absolute pitch showed no such interference. This result is consistent with the assumption that absolute-pitch possessors have developed strong and mandatory associations between pitch and its note name.

References

Cohen G, Martin, M. Hemisphere differences in an auditory Stroop test. Perception & Psychophysics 1975; 17; 79-83.

Cohen JD, Macwhinney B, Flatt M, Provost J. PsyScope: An interactive graphic system for designing and controlling experiments in the psychology laboratory using Macintosh computers. Behavior Research Methods, Instruments and Computers 1993; 25; 257-71.

Cohen JD, Dunbar K, McClelland JL. On the control of automatic processes: A parallel distributed processing account of the Stroop effect. Psychological Review 1990; 97; 332-61.

Green E, Barber P. An auditory Stroop effect with judgments of speaker gender. Perception & Psychophysics 1981; 30; 459-66.

Green E, Barber P. Interference effects in an auditory Stroop task: Congruence and correspondence. Acta Psychologica 1983; 53; 183-94.

Logan GD. Attention in character-classification tasks: Evidence for the automaticity of component stages. Journal of Experimental Psychology: General 1978; 107; 32-63.

MacLeod CM. Half a century of research on the Stroop effect: An integrative review. Psychological Bulletin 1991; 109; 163-203.

McClain L. Stimulus-response compatibility affects auditory Stroop interference. Perception & Psychophysics; 1983; 33; 266-70.

McLeod P, Posner MI. Privileged loops from percept to act. In: Bouma H, Bouwhuis DG, editors. Attention and performance X: Control of language processes. Hillsdale, N.J. Lawrence Erlbaum, 1984; 55-66.

Miyazaki K. Musical pitch identification by absolute pitch possessors. Perception & Psychophysics 1988; 44; 501-12.

Miyazaki K. Perception of relative pitch with different references: Some absolute-pitch listeners can't tell musical interval names. Perception & Psychophysics 1995; 57; 962-70.

Morton J, Chambers SM. Selective attention to words and colours. Quarterly Journal of Experimental Psychology 1973; 25; 387-97.

Pieters JM. Ear asymmetry in an auditory spatial Stroop task as a function of handedness. Cortex 1981; 17; 369-79.

Shiffrin RM, Schneider W. Controlled and automatic human information processing: II. Perceptual learning, automatic attending, and a general theory. Psychological Review 1977; 84; 127-90

Stroop JR. Studies of interference in serial verbal reactions. Journal of Experimental Psychology 1935; 18; 643-62.

Sugg MJ, McDonald JE. Time course of inhibition in color-response and word-response versions of the Stroop task. Journal of Experimental Psychology: Human Perception & Performance; 1994; 20; 647-75.

Walker P, Smith S. Stroop interference based on the synaesthetic qualities of auditory pitch. Perception 1984; 13; 75-81.

Zakay D, Roziner I, Ben-Arzi S. On the nature of absolute pitch. Archives für Psychologie 1984; 136; 163-166.

Chapter III.13

Organization of Two-tone Facilitation in Neurons of the Primary Auditory Cortex in Rats

Tomohiro Donishi [1], Yasuhiko Tamai, and Akihisa Kimura

Department of Physiology, Wakayama Medical College

Introduction

In mammalian cerebral cortex, the primary auditory area (A1) has been found to be the location where a majority of neurons are narrowly tuned to frequency with short-latency responses (Merzenich et al. 1975; Phillips and Irvine 1981; Suga et al. 1983; Sally and Kelly 1988; Schwarz and Tomlinson 1990). The A1 is also characterized as well by its tonotopic organization, which is the species-specific ordered representation of sound frequency along the cortical surface. Therefore, a given pure tone can activate a group of neurons with spatially clustered distribution, although tuning sharpness varies among A1 neurons (Sutter and Schreiner 1995). Characteristic frequency (CF), known as the frequency at which the stimulus intensity necessary to generate a particular discharge rate is minimal, has been emphasized as one of the most significant functional properties of auditory neurons. However, whether the spacio-temporal activity pattern of total neurons in A1 is involved in auditory perception is still to be addressed by further studies on neuronal behavior to stimuli at frequencies far from the proper CF of the neuron. That is, when a given sound is presented, A1 neurons may be roughly divided into three groups; well-responding, not responding, and a marginal group between the former 2 ones. The neurons in the marginal group receive depolarizing input near threshold, providing the spacio-temporal activity pattern in A1. On the other hand, A1 has been found to be involved in auditory discrimination behavior in animals (Kelly and Whitfield 1971; Cranford et al. 1976; Neff 1977). Accordingly, to allow animals to make a fine auditory discrimination in which the difference between stimuli is physically very small, neurons in the marginal group may play an important role in the composition of neural circuits that enable animals to accomplish auditory-motor integration behaviors. However, well-responding group cannot encode the difference between stimuli since the neurons in this group respond with maximum firing rate (or maximum probability

---

[1] Correspondence: Tomohiro Donishi, Department of Physiology, Wakayama Medical College, Wakayama 641-0012, Japan, Tel and Fax: (81)-73-441-0622, e-mail: tdonishi@wakayama-med.ac.jp

of firing occurrence) to both stimuli.

In the present study we focused on the marginal group of auditory neurons and examined the possibility that this group is involved in representation of some aspect of auditory information, such as fine discrimination. We used two-tone complex as stimuli, and emphasized ratios between two tones since harmonic structure is generally observed in natural acoustic environment.

Methods

Surgical procedures

Experiments were carried out on 3 adult Wistar albino rats weighing 200-290g under chloral hydrate anesthesia (170mg/kg i.p.). Anesthesia was maintained with supplemental doses (80mg/kg i.p.) administered approximately every 50 min during expriment. The animal's head was placed in a stereotaxic apparatus with hollow ear bars and craniotomy was performed to expose the A1 (AP+3.5-6.5mm, ML5.0-8.0mm). The dura was removed and the vasculature pattern was photographed to allow precise location of the electrode penetrations. Warmed (37°C) saline was frequently applied to the surface of the brain to prevent drying and cooling throughout the experiment. To decrease brain pulsations, cerebrospinal fluid was drained from the foramen magnum.

Auditory stimulation

Pure tone and two-tone complex were used as auditory stimuli. The waveforms of the auditory stimuli were generated by a computer using data analyzing software (National Instruments, LabVIEW 4.0) and converted into analog voltage signals (12bit for ±10V, sampling frequency 80kHz) with analog output devices (National Instruments, NB-MIO-16 and NB-DMA2800). The analog signals were then transduced into sound by a sound stimulator (DIA Medical system, DPS-725) with a free field speaker (Pioneer, PT-R9) located 30cm in front of the head, corresponding to zero degree in azimuth and elevation. The frequency response of the speaker was flat to ±5dB between 1 and 50kHz. Waveforms of the two-tone complex were the sum of two of those pure tones with the same phase and amplitude. The analog output signals were not smoothed by a lowpass filter since the sampling frequency was higher than the audible range of rats and the sampling noises were essentially ignored. The duration of the stimuli was 20ms with 5-ms cosine ramp. The frequency and intensity of the pure tone were varied between 1-40kHz with 1-kHz step, and 60-80dB with 5-dB step, respectively. Stimulus repetition rate was ~0.2Hz.

Recording procedures

A tungsten-in-glass microelectrode (1.2-2.0MΩ) was inserted into the cortex normal to the cortical surface. Neural potentials were amplified 10,000-20,000 times and monitored on an oscilloscope and a loudspeaker. When unit discharges were encountered, their auditory responses were digitized (12bit for ±10V, sampling frequency 20kHz) with A/D converter (National Instruments, NB-MIO-16), and stored for the following off-line analysis. A single or small-cluster unit activities were detected and sorted for the post-stimulus time histogram (n=20). In order to elucidate the response properties of the marginal group, attention was paid to the excitatory threshold with respect to frequency at a given intensity. Thus, the frequency and intensity were selected at which the neuron exhibited response properties seen in the area just outside of the excitatory tuning curve. Two-tone complexes consisting of two pure tones were presented as well within the boarder area. The interval of frequencies of these tones was within 1 octave.

Data analysis

A single or small-cluster unit was classified as 'facilitatory' when its responses to a given two-tone complex were apparent while its component pure tone per se did not elicit prominent activity. For evaluation of the 'facilitatory' units, we used a post-stimulus time cumulative histogram: an accumulated graph of the post-stimulus time histogram of units over 20 tasks (cf. Figure 2D). In order to measure the facilitatory effect quantitatively, an index was defined as follows:

$$\text{facilitation index} = r_{1+2} - (r_1 + r_2)$$

where r_1 and r_2 are response magnitude to pure tone of a frequency f_1 and f_2 respectively, and r_{1+2} is the response to the two-tone complex consisting of frequencies f_1 and f_2. The response magnitude was the spike counts with significant high firing rate in the post-stimulus time cumulative histogram. The facilitation index could be quantitatively evaluated by the facilitatory effect.

Results

A total of 92 units were recorded at a depth of 580-1478μm from the cortical surface. Of these units, 78 (85%) exhibited transient burst in response to pure tones with a given range of frequency and intensity, and made it possible to measure latency of the unit responses (Figure 1). The latency of each unit was measured as the time between the stimulus onset and the beginning of alteration of the firing rate in the post-stimulus time cumulative histograms of the unit (cf. thin arrow in Figure 2D). The latency distribution histogram of these units demonstrates that 72 (92%) of them had latencies shorter than 18ms, which correspond to those of the A1 neurons. Further, most of units responded within a restricted frequency range of a given constant intensity, showing the existence of a border to their response area.

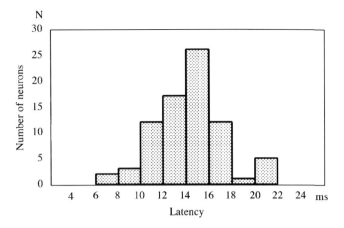

Figure 1

Latency distribution of units responsive to pure tones (n=78, bin width 2ms). Latency was measured as the time between the onset of the stimulation and the beginning of the steep alteration of the firing rate for each unit in the post-stimulus time cumulative histogram (cf. thin arrow in Figure 2). The majority (92%) of the units responded with short latencies (<18ms) corresponding to the A1 neurons.

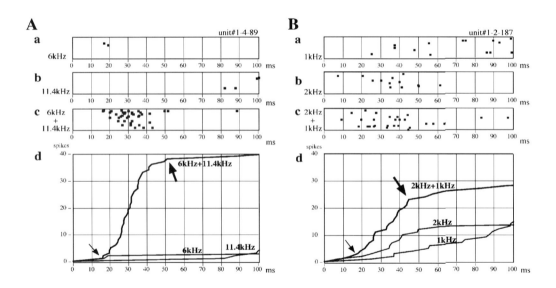

Figure 2

Two examples of units (A and B) demonstrating facilitatory effect to two-tone complex. The frequency used in Aab and Bab was within high and low frequency border in the response area, respectively. A: Each raster demonstrates the occurrence of neural discharges with respect to onset of the auditory stimuli in millisecond (n=20). Aab; Responses to pure tone of 6 and 11.4kHz, respectively. Ac; Responses to two-tone complex at 6kHz and 11.4kHz frequencies. Ad; Post-stimulus time cumulative histograms (n=20) for corresponding unit responses shown in Aabc. Note that the two-tone complex stimulation was very effective for this unit, while each pure tone was not. B: All data were treated in the same manner as in A, but at different frequencies. Note that change in firing rate is not obvious in the raw data shown in Babc, but apparent in the post-stimulus time cumulative histograms in Bd. Arrows indicate initiation (thin one) and end (thick one) of responses.

Facilitatory effect

Of the 78 units depicted in the previous section, 54 (69%) were found to show a facilitatory effect, which was characterized as responsiveness with robust spike bursts to the two-tone complex, while each component pure tone did not evoke significant responses. For example, a unit shown in the Figure 2A evoked virtually no response to pure tone per se at either 6kHz or 11.4kHz frequencies (Figure 2A a,b). However, obvious responses were observed when the two-tone complex at the same frequencies was presented in this unit (Figure 2A c). This effect was more obviously evaluated as alteration of firing rate in the post-stimulus time cumulative histogram (Figure 2A d). The pure tone response showed sparse spikes with low level accumulation in the post-stimulus time histogram, while the two-tone complex response consisted of several spikes with sharp accumulation in the post-stimulus time histogram. Another example of unit response is shown in the Figure 2B. This neuron seemed to fire sparsely in either pure tone or two-tone complex stimulation (Figure 2B a,b,c). However, the post-stimulus time cumulative histogram exhibited apparent facilitatory activities of the neuron in the two-tone complex stimulation as compared to that in the pure tone stimulation of the component frequency (Figure 2B d). Therefore, the facilitatory effect was expressed either by augmented spike occurrence or by increasing firing rate. Thus, it is possible that there might be two distinct synaptic activities underlying the facilitatory effect in a neuron or a small cluster of neurons.

Ratio between component frequencies

Figure 3 shows a facilitatory effect of one unit at high frequency boarder of the response area. Pure tone responses were recorded from frequency 15kHz to 32kHz in this preparation. Two-tone complex consisted of a fixed pure tone of 16kHz (arrow) and another higher frequency tone corresponding to the values shown in the abscissa of each plot (open circle in

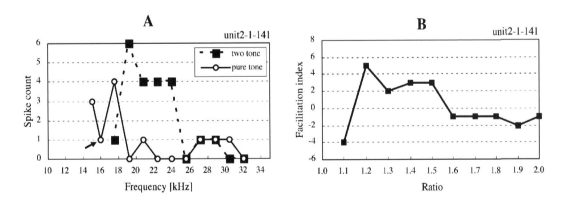

Figure 3
Facilitatory effect observed at high frequency border of the response area. A: Two curves indicating the spike count in response to pure tone and two-tone complex, respectively. The two-tone complex consisted of pure tones of 16kHz (arrow) and another frequency on X-axis value of each plot. Abscissa: stimulation frequency [kHz]. Ordinate: total spike count measured from the cumulative histogram (n=20) (cf. spike number between arrows in Figure 2). B: Facilitation index (see text) as function of the ratio of the frequencies of the two-tone complex. Note that the facilitatory effect occurred between the ratios 1.2 and 1.5.

Figure 3A). Spike count was largely increased in the two-tone complex stimulation (closed square and dashed line in Figure 3A). The extension of the excitatory response area in the two-tone complex stimulation was clearly evaluated in the facilitation index graph with respect to the frequency ratio (Figure 3B). Non facilitatory effect was indicated by an index value not greater than 0. The unit shown in Figure 3B was found to demonstrate the facilitatory effect at the ratio between 1.2 and 1.5. The relationship of the facilitation index with respect to the ratio can be considered as one of the characteristic properties of a given neuron. Thus, the ratio at which the facilitation index was maximum was defined as the best ratio. The majority of units indicating the facilitatory effect had the best ratio as shown in Figure 3B (ratio 1.2). To get further information on the facilitation effect of A1 cortex neurons, the unit responses were totally analyzed using the facilitation index and the best ratio. Figure 4 shows a distribution of the facilitation index with respect to the best ratio. There was no specific relationship between the facilitation index and the best ratio, indicating that any ratio in the two-tone complex stimulation facilitate neurons in the A1. That is, neurons in any area of the A1 cortex was responsive to the two-tone complex stimulation with low to high facilitatory activity.

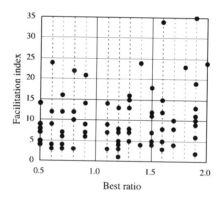

Figure 4
The distribution of the facilitation index with respect to the best ratio. The best ratio is defined as the ratio between component tone at which the facilitation index is maximum.

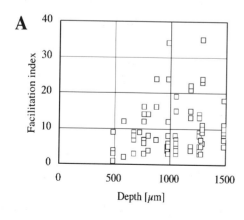 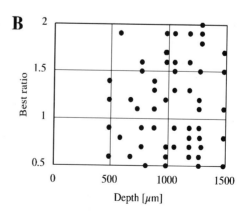

Figure 5
Distribution of the facilitation index and the best ratio with respect to the depth of neuron from the cortical surface. A: Relationship between the facilitation index and the depth. Note that the greater facilitation index is distributed over the deeper site in the cortex (>700μm). B: Relationship between the best ratio and the depth. No obvious deviation on the distribution is observed, indicating the uniformity of the best ratio distribution in any depth of the examined units.

Depth of recording site

Relationships between the facilitatory effect of neurons and the depth of the recording site were also examined with relation to the facilitation index and the best ratio. It was found that the facilitation index was greater only when the depth of the recording site from the cortical surface was over ~700μm as shown in Figure 5A. The difference of the distribution of the facilitation index with respect to the depth suggests the fine laminar organization in the A1 cortex. There was, however, no specific relationship between the best ratio and the depth as shown in Figure 5B.

Discussion

This study demonstrates that neurons in the auditory cortex elicit robust response following two-tone complex consisting of pure tones which are adjacent to its excitatory response area. The response pattern of the neurons to pure tones was in agreement with that of the A1 neurons in rats (Sally and Kelly 1988). There are many studies concerning the two-tone facilitation of neurons along auditory pathway as well as A1 (Ehret and Merzenich 1988; Jiang et al. 1996). However, most of them used the CF tone as one of components of two-tone complex and paid little attention to use two-tone pair of frequencies with both apart from CF but close to the boarder of response area as used in our study. It is possible that the facilitatory neurons observed in this study may encode specific ratio for a given single neuron, and that the A1 may process auditory information corresponding to the ratio between frequencies. In other words, the A1 neurons can detect an interval between two frequencies. In the natural environment in which animals survive, majority of sounds contain complex tones composed with different frequencies of harmonic overtone. Neurons facilitated by two-tone complex observed in this study may contribute to behavioral process to fine auditory discrimination. The auditory stimulation strategy and methods for data analysis used in this study can be a useful tool to elucidate further auditory information processing mechanism.

It has been known psychophysically that the perception of a pitch in the absence of spectral energy at that frequency in stimuli can occur on subjects in response to tone complex (Houtsma 1979). These results are thought to be partly due to the distortion product based on nonlinearity of the basilar membrane and the outer hair cells of the cochlea. In fact, the cubic difference tone ($2f_1-f_2$) was reported to be detected in the activity of auditory nerve fibers (Goldstein and Kiang 1968), and two-tone suppression effect was observed in the inner hair cells (Sellick and Russell 1979). The two-tone complex used in this study could also generate the cochlear distortion which could facilitatorily activate the cortical neurons. Since these phenomena must usually occur under natural environment, it is essential that the properties of the response of the auditory neurons to arbitrary sounds should be understood in conjunction with the distortion products. Thus, we conclude that the facilitatory effect of neurons with different frequency in the primary auditory cortex may be a part of neural circuits and may play an important role in discrimination behavior.

Summary

Since lesion of the auditory cortex has been resulted in a deficiency of auditory discrimination behaviors, we hypothesized that a group of neurons in the auditory cortex may contribute to the mechanism of fine discrimination of auditory signals. Discharges of 92 neurons in the primary auditory cortex of the albino rats were extracellularly recorded in response to a pure tone and two-tone complex under chloral hydrate anesthesia. The ratio of the component frequency was systematically varied. Most of neurons significantly responded to the two-tone complex consisting of two pure tones that solely elicited few responses. This facilitation effect by the two-tone complex exhibited nonmonotonic relationship with respect to the ratio of the component frequencies. The distribution of the ratio at which the facilitation effect was maximum was flat over any ratios used. We conclude that the facilitation effect of neurons in the primary auditory cortex may be a part of the neural circuitry and that it may play an important role in the accomplishment of discrimination behavior.

Acknowledgements

This work was supported by research funds from Wakayama Medical College. We would like to thank Dr. Bukasa Kalubi for his proofreading of the text.

References

Cranford, J. L., Igarashi, M. and Stramler, J. H. Effect of auditory neocortex ablation on pitch perception in the cat. J. Neurophysiol. 1976; 39: 143-152.

Ehret, G. and Merzenich, M. M. Complex sound analysis (frequency resolution, filtering and spectral integration) by single units of the inferior colliculus of the cat. Brain Res. Reviews 1988; 13: 139-163.

Goldstein, J. L. and Kiang, N. Y. S. Neural correlates of the aural combination tone $2f_1-f_2$. Proc. IEEE 1968; 56: 981-992.

Houtsma, A. J. M. Musical pitch of two-tone complexes and predictions by modern pitch theories. J. Acoust. Soc. Am. 1979; 66(1): 87-99.

Jiang, D., Palmer, A. R. and Winter, I. M. Frequency extent of two-tone facilitation in onset units in the ventral cochlear nucleus. J. Neurophysiol. 1996; 75: 380-395.

Kelly, J. B. and Whitfield, I. C. Effects of auditory cortical lesions on discriminations of rising and falling frequency-modulated tones. J. Neurophysiol. 1971; 34: 802-816.

Neff, W. D. The brain and hearing: auditory discriminations affected by brain lesions. Ann. Otol. 1977; 86: 500-506.

Phillips, D. P. and Irvine, D. R. F. Responses of single neurons in physiologically defined primary auditory cortex (AI) of the cat: frequency tuning and responses to intensity. J. Neurophysiol. 1981; 45: 48-58.

Sally, S. L. and Kelly, J. B. Organization of auditory cortex in the albino rat: sound frequency. J. Neurophysiol. 1988; 59: 1627-1638.

Schwarz, D. W. F. and Tomlinson, R. W. W. Spectral response patterns of auditory cortex neurons to harmonic complex tones in alert monkey (Macaca mulatta). J. Neurophysiol. 1990; 64: 282-298.

Sellick, P. M. and Rusell, I. J. Two tone suppression in cochlear hair cells. Hear. Res. 1979; 1: 227-236.

Suga, N., O'neill, W. E., Kujirai, K. and Manabe, T. Specificity of combination-sensitive neurons for processing of complex biosonar signals in auditory cortex of the mustached bat. J. Neurophysiol. 1983; 49: 1573-1626.

Sutter, M. L. and Schreiner, C. E. Topography of intensity tuning in cat primary auditory cortex: single-neuron versus multiple-neuron recordings. J. Neurophysiol. 1995; 73: 190-204.

T. Nakada (Ed.)
Integrated Human Brain Science: Theory, Method Application (Music)
© 2000 Elsevier Science B.V. All rights reserved

Chapter III.14

Music Therapy in Parkinson's Disease: Improvement of Parkinsonian Gait and Depression with Rhythmic Auditory Stimulation

Naoko Ito, Akito Hayashi [1], Weiju Lin, Norio Ohkoshi, Masahiko Watanabe, and Shin'ichi Shoji

Department of Neurology, Institute of Clinical Medicine, University of Tsukuba

Introduction

Gait disturbance is one of the most frequent and intractable motor disturbances in Parkinson's disease. The main treatments for gait disturbance are antiparkinsonian drugs, stereotaxic surgery, and the electrical stimulation of the basal ganglia. In order to maintain the quality of life of patients, the role of rehabilitation is also very important.

Abnormalities of both time estimation and time reproduction have been reported in cases of Parkinson's disease, and locomotor deficits associated with Parkinson's disease, particularly repetitive movements such as tapping and gait, may reflect a disturbance of internal rhythm formation. Numerous reports have described Parkinson's-related disturbances of the rhythm, time estimation, or tempo of repetitive movements. Time estimation, i.e., the 'internal clock', has been reported to be abnormally slow in PD (Pastor et al., 1992). Disturbances of rhythm formation in patients with Parkinson's disease have been reported based on the study of repetitive finger-tapping responses to periodic signals (Nakamura et al., 1978).

It is known that gait training with rhythmic sound or music is effective for the treatment of parkinsonian gait (Thaut et al., 1996; Enzensberger et al., 1996). However, there have been no trials investigating rhythmic auditory stimulation alone without gait training in Parkinson's disease. The aim of this study was to determine the potential efficacy of rhythmic auditory stimulation without gait training for treating for the gait disturbance in patients with Parkinson's disease, and the effect on depression in the disorder was studied.

---

[1] Correspondence: Akito Hayashi, Department of Neurology, Institute of Clinical Medicine, University of Tsukuba, 1-1-1 Tennoudai, Tsukuba, 305-8575, Japan, Tel and Fax: (81)-298-53-3224, e-mail: akito@md.tsukuba.ac.jp

Materials and Methods

Subjects

The subjects were 25 outpatients (10 men and 15 women; mean age ± SD, 70.0 ± 7.7 years) with idiopathic Parkinson's disease and gait disturbance. The degree of Hoehn and Yahr stage was 3 ± 0.5 (mean ± SD). The duration of the illness was 7.0 ± 5.4 years (mean ± SD; table 1). All patients were asked to adhere a consistency of the medication and rehabilitation regimen as before.

In addition, we examined the gait of 130 healthy senescent volunteers (96 men and 34 women; mean age ± SD, 71.1 ± 5.4 years) as a control. Informed consent was obtained from all participants.

Number:	25 (male/female=10/15)			
age:	70.0 ± 7.7 y.o.	BL:	154 ± 11 cm	
duration:	7.0 ± 5.4 y.	BW:	49.1 ± 10.8 kg	
H&Y stage:	3 ± 0.5 deg.			
Medication:	levodopa:	3.7 ± 1.5 tab.		#1
	dopamine agonist:	3.0 ± 2.8		#2
	trihexyphenidyl:	2.2 ± 2.7		#3
	doroxydopa:	1.5 ± 2.0		#4

1 tab.= 100 mg in #1 & #4, 2 mg in #3,

1 tab #2.= 2.5 mg in bromocriptine, 250 µg in pergolide

Table 1
Patient profile

Rhythmic auditory stimulation and music

The frequency of rhythmic auditory stimulation was set at 2 Hz, since this has been reported as the normal- gait frequency of healthy gait of healthy individuals of between 60 and 79 years of age (Öberg et al., 1993). A metronome was used to produce a rhythmic auditory stimulus, and a computer was used to embed the stimulus in several styles of music; subjects could choose their favorite from among classical, Japanese- popular, or children's music.

Experimental procedures

The patient's task was simply to listen to the audio tape at home, without gait training, for at least one hour a day for three to four weeks. Each patient was asked to keep a diary in to ascertain that the task was surely performed daily.

Assessment

Before and after the task, we measured the gait speed, stride length, and step cadence of the fast gait of patients across a 20 m expanse with one turn under a no-external-cue condition (subjects were simply asked to walk as fast as they could) by means of videotaping. Two additional questionnaires were also administered, one measuring the depressive state of subjects by a self-rating depression scale (SDS; Zung, 1965) before and after the task, and one investigating gait and mental condition, etc., by self-assessment after the task.

Statistics

We measured gait speed, step cadence, and stride length of the fast gait of both parkinsonian patients and healthy controls, SDS and self-assessment questionnaires were scored. Results are represented as means ± standard deviations.
Improvement rate (percent of change) was calculated as

$$\text{Improvement rate (\% rate)} = \frac{A-B}{B}$$

where A is the value after the task, and B is the value before the task. A mean improvement rate in all subjects was calculated from an each improvement rate not from mean values pre and post- task. Statistical comparisons were done by paired t test; $p < 0.05$ was considered to indicate statistical significance.

Results

Comparison of gait between parkinsonian patients and healthy controls

Parkinsonian gait in patients older than 60 years (22 patients; mean age, 71.8 years) was different from that in controls (130 people; mean age, 71.1 years). The speed was slower, the stride length was shorter, i.e., 'petit pas' in French, and the frequency of cadence was slower in parkinsonian patients (Table 2 and Figure 1).

Changes of Gait Parameters

After the task, the patients significantly improved their gait speed (pre: 50.0 m/min; post: 58.2 m/min; $p < 0.0001$; Figure 2A, Table 3) and stride length (pre: 41.7 cm/step; post: 47.2 cm/step; $p < 0.001$; Figure 2C, Table 3).

	control	PD patient
subjects (male/female)	130 (96/34)	22 (9/13)
age (years)	71.1 ± 5.4	71.8 ± 4.7
gait speed (m/min.)	86.4 ± 12.2	51.2 ± 16.9
step length (cm)	67.7 ± 8.7	42.4 ± 12.3
cadance (steps/min)	128.0 ± 11.7	117.6 ± 16.2

Table 2
Comparison of gait between PD patients and controls. Gait speed was examined under condition of walking as fast as possible (fast gait).

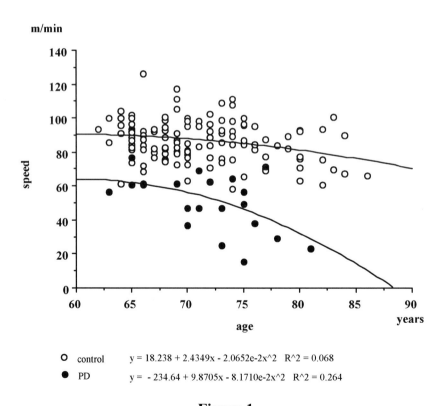

○ control $\quad y = 18.238 + 2.4349x - 2.0652e{-}2x^2 \quad R^2 = 0.068$
● PD $\quad y = -234.64 + 9.8705x - 8.1710e{-}2x^2 \quad R^2 = 0.264$

Figure 1
Each gait speed was plotted (open circles; age-matched controls; closed circles, PD patients). There was a tendency that the gait speed in PD patients became much slower as getting older compared with that in the same age control.

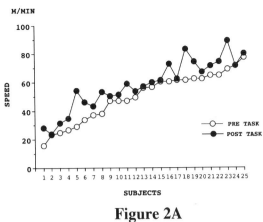

Figure 2A

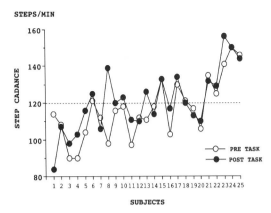

Figure 2B

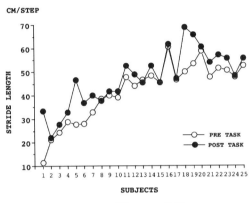

Figure 2C

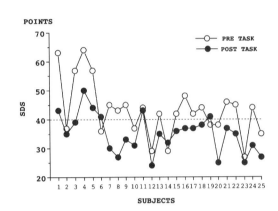

Figure 2D

Figure 2

Gait parameters and SDS. Gait speed (A), Step cadence (B), Stride length (C), SDS (D). Each data of B, C, and D was in the order equal to A (the order of subjects in gait speed is slow to fast). In C, there was a positive correlation between gait speed and stride length.

	pre-task	post-task	change (%)
gait speed (m/min)	50.0 ± 17.1	58.2 ± 17.3***	20.7 ± 22.6
stride length (cm)	41.7 ± 12.3	47.2 ± 11.7**	20.0 ± 40.0
cadance (steps/min)	117.0 ± 16.1	120.8 ± 16.6	4.2 ± 11.6
SDS (point)	43.1 ± 9.5	35.0 ± 6.7***	17.2 ± 13.9

***p<0.0001, **p<0.001, *p<0.05

Table 3

Changes in gait and depression scale after the task. Improvement rate is the absolute value of change (%) in SDS.

Nineteen out of 25 patients (76 %) showed an improvement of over 5 % in gait speed, and 14 of 25 patients (56 %) showed an improvement of over 10 % in gait speed. Although the step cadence was changed from 117.0 to 120.8 steps/min, but this change was not statistically significant (Figure 2B, Table 3).

Self-rating depression scale (SDS)

Before the task, the mean SDS score was 43.1 points. Nine patients scored below 40 points (a score under 40 points indicates an absence of depression), 12 patients (48 %) showed a depressive tendency (40 to 49 points) and 4 patients (16 %) were in a depressive state (over 50 points). After the task, the mean SDS score decreased to 35.0 points, and this decrease represented a significant improvement in SDS (p<0.0001, Figure 2D, Table 3). Twelve patients who scored above 40 before the task were within the normal range (< 40) after the task.

Self-assessment questionnaires after the task

Fourteen patients reported sensing a post- task improvement in gait, and 11 patients felt no change (Figure 3). There were discrepancies between the subjective feelings and objective measurement values. For example, some patients who sensed an improvement in gait did not show an improvement in gait speed, while some patients who did not sense any changes improved their gait speed. With regard to depression, 24 out of 25 patients felt that their feelings had lifted.

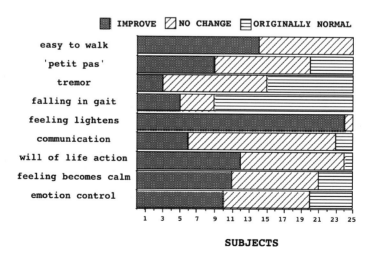

Figure 3
Self-assessment questionnaires after the task. Note that 14 patients reported sensing a post- task improvement in gait and that 24 out of 25 patients felt that their feelings had lifted.

Discussion

In this study, it is a most important finding that only rhythmic auditory stimulation without gait training had an effect on the gait disturbance in Parkinson's disease.

In our study, gait speed was improved by a rate of 20.7 %, from 50.0 m/min to 58.2 m/min after only rhythmic stimulation without gait training. Thaut et al. reported that the gait speed in Parkinson's disease was improved by a rate of 24.1 %, from 48.7 m/min to 58.3 m/min after 3 weeks of gait training with rhythmic auditory stimulation (Thaut, 1996). Based on the results that the improvement rates were not so changed in two different methods, it was suggested that even only rhythmic auditory stimulation for parkinsonian gait could play an important role in the efficacy. Moreover, since both these effects were observed in gait testing under a no-external-cue condition, it is considered that the influence of rhythmic auditory stimulation on gait may remain after the stimulation. This indicates the possible effect of rhythmic entrainment mechanisms. Thaut et al. reported that the effect of rhythmic auditory stimulation with gait training was maintained for about a month, based on the results of follow-up tests (Thaut et al., 1999).

In Parkinson's disease, it is known that time estimation, i.e., 'internal clock' or 'internally-generated rhythms', is abnormally slow (Pastor et al., 1992; Lange et al., 1995). Thus parkinsonian patients cannot smoothly perform repetitive movements, in which rhythm is essential, such as tapping (Nakamura et al., 1978) or walking. With regard to the mechanism by which rhythmic auditory stimulation improves the performance of such movements in the absence of gait training, external stimulation may stabilize destabilized processes of internal rhythmic formation (Freeman et al., 1993; Thaut et al., 1996). In other words, rhythmic auditory stimulation may have made it easier for parkinsonian patients to reproduce the internal rhythmic generation of gait, thereby allowing them to walk faster and more smoothly.

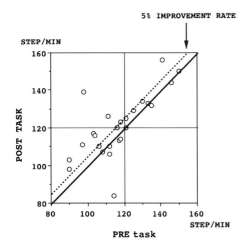

Figure 4

Correlation of pre and post-step cadence. Seven of 16 patients whose cadences were under 120 steps/min showed an increase in cadence.

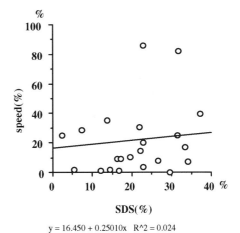

Figure 5

Changes in gait and depression scale after the task. Improvement rate is the absolute value of change (%) in SDS.

In our study, the frequency of rhythmic stimulation was set at 2 Hz (120 per min). Thaut set the frequency of rhythmic stimulation to be 10 % faster than the cadence of the subjects steps (Thaut et al., 1996), and Enzenberger set it at 1.5 Hz (96 per min) without specifying their rationale (Enzensberger et al., 1996). It has been reported, that the mean step cadence in the normal gait of healthy people (60 to 79 years) is 119.0 steps/min (Öberg et al., 1993). In our study of healthy senescent volunteers, the step cadence was 128.0 steps/min (2.13 Hz) in the fast gait (Table 2). Because the rhythm frequency of 120 per min is close to an average gait rhythm among healthy older individuals, it was chosen as the appropriate frequency of rhythmic stimulation for gait. Figure 4 compares the pre and post-step cadences. Seven of 16 patients whose cadences were under 120 steps/min showed an increase in cadence, whereas only 1 of 9 patient whose cadences were over 120 steps/min. Based on the these findings, the 2 Hz rhythmic stimulation may have an important contribution to the observed increase in cadence.

Parkinson's disease is commonly complicated by the existence of comorbid depression. In our study, there were 16 patients (56 %) who showed a tendency toward depression. An obvious improvement of depressive state was observed after the task. Sixteen patients who had a depressive tendency or a depressive state all improved and 13 patients among them became within normal range. With regard to the reason why their mental states alleviated, there could be a possibility that they could relax by listening music. We investigated the relationship between gait and depression improvements, but there was no significant positive correlation between these parameters (p=0.47, r=0.153, Figure 5). Thus the gait improvement could not be attributed to the mental improvement, not vice versa. According to Cummings's review, depression in Parkinson's disease may be caused by the involvement of frontal dopaminergic projections rather than by such PD- associated disabilities as gait disturbance (Cummings, 1992). The discrepancy between gait and mental improvements observed here supports that depression may not be caused by gait disturbance in Parkinson's disease.

In conclusion, our study indicates that music embedded by rhythmic auditory stimulation without gait training is very effective for treating the gait disturbance and depression associated with Parkinson's disease. The technique demonstrated here is thought to be very simple and efficient. Therefore, we recommend the use of this method for patients with Parkinson's disease and reconfirmed the usefulness of music therapy in Parkinson's disease.

Summary

We studied the effect of rhythmic auditory stimulation without gait training on parkinsonian gait. The subjects were 25 outpatients with idiopathic Parkinson's disease and gait disturbance. The patient's task was simply to listen to rhythmic auditory stimulation (2Hz) embedded in familiar styles of music at home, without gait training, for at least one hour a day for three to four weeks. After the task, patients' gait speed and stride length were significantly improved ($p<0.0001$, $p<0.001$, respectively). The step cadence was not significantly changed, but there was a tendency that the cadence rhythm under 2 Hz approached the rhythm of auditory stimulation. The self-rating depression scale was also significantly improved ($p<0.05$). The self-assessment of patients regarding several symptoms after the task was also described. In conclusion, our study indicates that music embedded by rhythmic auditory stimulation without gait training is very effective for treating the gait disturbance and depression associated with Parkinson's disease. With regard to the mechanism by which rhythmic auditory stimulation improves the performance of such movements in the absence of gait training, external stimulation may have made it easier for parkinsonian patients to reproduce the internal rhythmic generation of gait, thereby allowing them to walk faster and more smoothly.

References

Cummings JL. Depression and Parkinson's disease: a review. Am J Psychiatry. 1992; 146:443-454.

Enzensberger W, Fischer PA. Metronome in Parkinson's disease. Lancet 1996; 347: 1337.

Freeman JS, Cody FWJ, Shady W. The influence of external timing cues upon the rhythm of voluntary movements in Parkinson's disease. J Neuro Neurosurg Psychiatry 1993; 56: 1078-1084.

Keith RA, Granger CV, Hamilton BB, Sherwin FS. The functional independence measure: A new tool for rehabilitation. Adv Clin Rehabili 1987; 1: 6-18.

Lange KW, Tucha O, Steup A, Gsell W, Naumann M. Subjective time estimation in Parkinson's disease. J Neural Transm Suppl 1995;46:433-438.

McIntosh GC, Brown SH, Rice RR, Thaut MH. Rhythmic auditory-motor facilitation of gait patterns with Parkinson's disease. J Neuro Neurosurg Psychiatry 1997;62: 22-26.

Nakamura R, Nagasaki H, Narabayashi H. Disturbances of rhythm formation in patients with Parkinson's disease: part I. Characteristics of tapping response to the periodic signals. Percept Mot Skills 1978;46(1):63-75.

Öberg T, Karsznia A, Öberg K. Basic gait parameters: reference data for normal subjects, 10 79 years of age. J Rehabil Res Dev. 1993; 30:210-223.

Pastor MA, Artieda J, Jahanshahi M, Obeso JA. Time estimation and reproduction is abnormal in Parkinson's disease. Brain 1992; 115: 211-225.

Thaut MH, McIntosh GC, Rice RR, Miller RA, Rathbun J, and Brault JM. Rhythmic auditory stimulation in gait training for Parkinson's disease patients. Mov Disord 1996; 11: 193-200.

Thaut MH, Kenyon GP, Schauer ML, McIntosh GC. The connection between rhythmicity and brain function. IEEE Eng Med Biol Mag 1999; 18(2):101-108.

Zung, WWK. A self rating depression scale. Arch Gen Psychiatry 1965; 12: 63-70.

Chapter III.15

A Beat Tracking Model by Recurrent Neural Network

Ken'ichi Ohya [1]

Department of Electronics and Computer Science,
Nagano National College of Technology

Introduction

Artificial neural networks are thought to simulate certain primitive models of human brain function. In computer music research, Large has used neural networks for simulating beat tracking system[1] and Page, self-organizing neural networks for perception of musical sequences[7].

There are many models or architectures of neural networks. The recurrent neural network is one of the most complex among them. The recurrent neural network is a neural network that has feedback connections recurrently to each neuron.

It is known that the recurrent neural network can produce some kinds of rhythm patterns. Indeed, those patterns have been thought to have some relation to rhythm generators in human, especially central pattern generators (CPGs)[2,3,11].

Recurrent neural networks composed of continuous-time, continuous-variable neuron model can be trained to learn spatiotemporal patterns[8,9] and even chaotic dynamics[10]. Adaptive nonlinear pair oscillators with local connections, APOLONN, is one of the architectures that has been applied for speech synthesis[10].

Since the APOLONN can also learn complex dynamics, it is possible to use it as a sound generator by using a teacher signal of a waveform of an acoustic instrument including natural fluctuations of amplitudes and periodicities, on the architecture. A waveform of a piano tone, known as a mixture of attack noise, simple vibrations and their fluctuations, was used for a teacher signal, and good results of synthesized sound were obtained[5]. Fluctuations of the output are quite natural, because of the architecture's features; that does not memorize fluctuations of the original data, but does chaotic or nonlinear dynamics behind the original.

---

[1] Correspondence: Ken'ichi Ohya, Department of Electronics and Computer Science, Nagano National College of Technology, Nagano 381-8550, Japan, Tel and Fax: (81)-26-295-7085, e-mail: ohya@ei.nagano-nct.ac.jp

And it was shown that it's not easy to learn and synthesize of the sound of the violin [6].

Recurrent Neural Network

As a model of neuron, continuous-time, continuous-variable neuron model is used. One pair of the neurons, that is fully connected, can generate very complex dynamics pattern depending on the values of their weight connections, and can be considered as a kind of nonlinear oscillator. Even one pair of recurrent neural networks can be used as a rhythm perception model [4].

Equations of dynamics of the output of this kind of neurons are given as

$$\tau_i \frac{du_i}{dt} = -u_i + f\left(\sum_{j=1}^{n} W_{ij} u_j\right) + I_i$$

where $u_i(t)$ is the i-th unit output at a time t, τ_i a time delay constant, $f(x)$ a sigmoid function, I_i an external input of the i-th unit, W_{ij} a connection weight from the j-th unit to the i-th unit.

Some architectures of recurrent neural networks can be trained to learn spatiotemporal pattern [8,9] and even chaotic dynamics [10].

Adaptive nonlinear pair oscillators with local connections (APOLONN), is one of the architectures that was applied for speech synthesis [10].

The architecture of APOLONN consists of many pairs of oscillators (Figure 1).

A pair of oscillator is locally connected with its neighboring pairs, and all neurons are connected to one neuron; the output neuron.

Each pair of oscillator generates various kinds of complex patterns depending on their

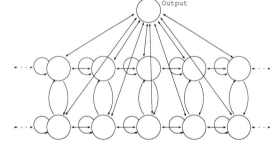

Figure 1
APOLONN

parameters, such as time delay constant τ_i or their weight connections. Since each pair is locally connected to neighboring pairs, oscillation of one pair is not independent from other neurons. The total system can produce further complex nonlinear patterns that are rich in frequencies.

Simulations and Results

The architecture of APOLONN with sufficient number of neurons has capability to learn arbitrary any dynamics, any waveform including fluctuations of amplitude and periodicities, of an acoustic musical instrument. (See Ref. 10 for learning algorithm details.)

I used APOLONN as a beat tracking system, and tried to seek possibilies of beat tracking.

In this simulation, the number of pairs of oscillators was 20, the time delay constant τ_i of each pair was set to slightly different from the neighboring pair, the ratio of the τ_i between two neighboring pairs was 0.9.

Firstly, I tried APOLONN to learn some beat impulses. Figure 2 shows my desired results. My intention is following: the system can learn some impulses, and after learning, the system can produce beat impulses periodically.

But the results came up to my expectations. APOLONN tried to learn the spatiotemporal pattern of steep impulse, but it seemed very very difficult to learn that very steep curve. APOLONN with more neurons also failed.

And then, when I tried APOLONN to learn only one beat impulse. Figure 3, periodic waveform was obtained Figure 4.

Then I tried APOLONN this time to learn this obtained periodic waveform. The result is shown as Figure 5.

This Figure 5 shows that it is not easy for APOLONN to learn steep impulse waveform, but it shows a sign of learning triangle beat impulse waveform.

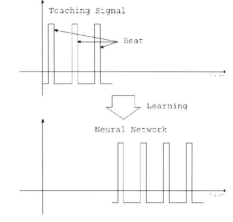

Figure 2

desired simulation result

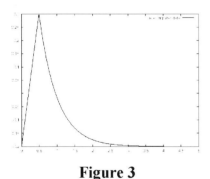

Figure 3

teaching signals: one impulse

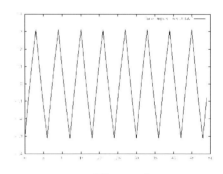

Figure 4

after learning: one impulse

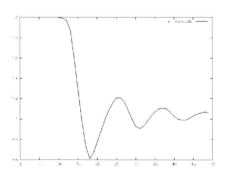

Figure 5

after learning: relearn obtained waveform

Summary

In this paper, a beat tracking model using recurrent neural network is presented. As an individual neuron model, continuous-time continuous-value neuron model is adopted. Therefore the total system is described as many differential equations, the number of equations is the same as that of neurons. As an architecture of the beat tracking model, APOLONN is adopted. Very steep beat impulses were fed into that model. When a triangle beat impulses were fed, the system can produce weak periodic wave.

Acknowledgement

The author would like to thank Yasuaki Nakano for his helpful comments.

References

1. Large, E.W. & Kolen, J.F. (1993). "A Dynamical Model of the Perception of Metrical Structure". Presented at Society for Music Perception and Cognition. Philadelphia, June.
2. Matsuoka K., "Sustained oscillations generated by mutually inhibiting neurons with adaptation," *Biol Cybern*, vol.52, pp.367-376, 1985.
3. Matsuoka K., "Mechanics of frequency and pattern control in the neural rhythm generators," *Biol Cybern*, vol.56, pp.345-353, 1987.
4. Ohya, Ken'ichi. (1994). "A Rhythm Perception Model by Neural Rhythm Generators". *Proceedings of the 1994 International Computer Music Conference*, pp.129-130.
5. Ohya, Ken'ichi. (1995). "A Sound Synthesis by Recurrent Neural Network". *Proceedings of the 1995 International Computer Music Conference*, pp.420-423.
6. Ohya, Ken'ichi. (1998). "Sound Variations by Recurrent Neural Network Synthesis". *Proceedings of the 1998 International Computer Music Conference*, pp.280-283.
7. Page, M.P.A. (1993). "Modelling Aspects of Music Perception Using Self-organizing Neural Networks". Unpublished doctoral dissertation, University of Cardiff.
8. Pearlmutter, B. A. (1989). "Learning state space trajectories in the recurrent neural network". *Neural Computation*, **1** (2), pp.263-269.
9. Sato, M. (1990). "A learning algorithm to teach spatio\-temporal patterns to recurrent neural networks". *Biological Cybernetics*, **62**, pp.259-263.
10. Sato, M., Joe, K., & Hirahara T. (1990). "APOLONN brings us to the real world: Learning nonlinear dynamics and fluctuations in nature". *Proceedings of the International Joint Conference on Neural Networks*, San Diego, **I**, pp.581-587.
11. Taga G., Yamaguchi Y., Shimizu H., "Self-organized control of bipedal locomotion by neural oscillators in unpredictable environment", *Biol Cybern*, vol.65, pp.147-159, 1991.

T. Nakada (Ed.)
Integrated Human Brain Science: Theory, Method Application (Music)
© 2000 Elsevier Science B.V. All rights reserved

Chapter III.16

Increased Activation of Supplementary Motor Area During a Hand Motor Task Tuned to Auditory Rhythm

Masanobu Saito [b], Takashi Kujirai [a b 1], Noboru Saito [b], Kayoko Kujirai [c], Giito Izuta [d], and Takashi Yamaguchi [d]

[a]*Department of Neurology,* [b]*Division of Higher Brain Function, National Yonezawa Hospital Yamagata University,* [c]*School of Medicine,* [d]*School of Engineering*

Introduction

Neuroimaging technique has been applicable to the sophisticated detection for activated areas in the brain. Especially with the advent of positron emission tomography (PET), single-photon emission tomography (SPECT) and the quantitative examination of regional cerebral blood flow (rCBF), a sensitive detection and characterization of regional changes in brain metabolism have been provided even when subjects are performing specific tasks. Therefore, such a functional imaging allows us to examine patterns of cortical or basal ganglia activation, as evidenced by regional blood flow, associated with specific performance of different motor tasks. Among other functional imagings, non-metabolizing xenon computed tomography (Xe-CT) is of a particular benefit for the investigation of appropriate metabolism in the brain because xenon is easily transmissible to the neuronal membrane providing little delay of trapping time of the tracer and also the technique is quite safe to bodies. Xe-CT scanning is so applicable in some hand motor tasks requiring the speed and precise time correlated to the movement execution. For example, rhythm which is one of main factors of music is beneficial for our daily behavior, but is requisite to predict as well as to memorize the time strictly. Non-metabolizing Xe-CT scanning would help us draw conclusions over the possible role in some motor tasks. Then how does rhythm contribute to play a beneficial role to the usage of motor circuits in the brain? In order to verify which parts

---

[1] Correspondence: Takashi Kujirai, Department of Neurology, National Yonezawa Hospital, 26100-1, Misawa, Yonezawa, Yamagata 992-1202, Japan, Tel: (81)-238-223-210, Fax: (81)-238-226-691
Abbreviations: Xe-CT= xenon-computed tomography, SMA= supplementary motor area, rCBF= regional cerebral blood flow

of brain areas are activated during a tuned motor task to rhythmic auditory cues, we studied 5 normal volunteers who gave written informed consent using Xe-CT scanning.

Methods

All the volunteers were neurologically free and showed no abnormality in taking auditory rhythm. They were on a bed, and were asked to listen to a triple time with a metronome. They were also required to push a switch button just at the time of the third tick sound after learning the rhythm precisely (task A), or to push it in the self-paced way randomly (task B). During the study, every volunteer was covered adhesively over their mouths with a mask and were compelled to inhale non-metabolizing Xe-gas for 3 minutes and to expel it for 4 minutes. 4 different slice levels of helical CT-scanning for the region of interest (ROI) were determined for the monitoring the regional cerebral blood flows (rCBFs), in which are available to trace areas such as the supplementary motor area (SMA), sensorimotor cortex, premotor cortex, basal ganglia and cerebellar nuclei.

Two levels are parallel to the surface of the frontal lobe and the others are along to the orbito-meatus (OM) line. The width of the scanning for each slice is 1 cm. During the study, 7 scannings were performed for each slice level. The size of tracing xenon is 0.625 mm x 0.625 mm in a pixel. According to the experimental design as shown in the fig.1, rCBF, κ (flow rate constant) and λ (tissue partition coefficient) were evaluated for each level of CT-scanning. Here, among these factors, the following formula holds; $CBF = \lambda \cdot \kappa / m$ (m: equilibrium constant). The subtraction between two tasks (: task A – task B) on the rCBF imaging were carried out for each CT scanning level. Finally, the change in different areas among two tasks in terms of rCBF is evaluated for statistical analysis.

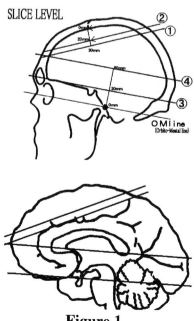

Figure 1

This figure explains the actual level for slicing the brain. 1 or 2 level is parallel to the surface of the frontal lobe. 3 or 4 is also parallel to the orbito-meatal line which is determined by the anatomical connections between orbitus and meatus. The distances from the basal line are shown in this figure. These lines are reflected in the brain as shown at the bottom.

Results

Fig.2 shows an example of the slice level covering supplementary motor area in a subject who is a right handed man aged 22. The degree of rCBF is associated the colour suggesting that the highest value of rCBF is white and the lowest is black. The colour image on the left is by the hand motor task tuned to auditory rhythm (Task A) and on the left is by the self initiated task (Task B). Fig.3 on the left shows the subtraction from Task A to Task B, detecting the activation of the bilateral supplementary motor areas as well as premotor cortexes, in which the contralateral sides are preferentially rather activated. The size of the hot area is rather small. On the right, the real helical CT scanning at the same slice level as the Xe- image on the left. CT scanning is also available to have a clue of the anatomical cortical areas by following specific cortical gyrus. Focusing of different regions of interest (ROIs), the change between the two tasks are evaluated. All data were summarized as shown in the fig.4. At almost all the levels of CT scanning, rCBFs of task A were elevated except for the ipsilateral putaminal and caudate areas, which is the striatum by a general term ($p < 0.05$). The rCBF after the subtraction suggests the decrease in value. In contrast, especially at the contralateral of SMA, task A showed significant augmentation of rCBF ($p < 0.01$). At the SMA, here is the evaluation for all three factors implying the change mode of the brain metabolism, which are λ, κ and rCBF. Premotor cortexes shows the only change in λ-factor on the ipsilateral side to the finger movement ($p < 0.05$). In contrast, other areas such as sensorimotor cortexes, basal ganglia nuclei and cerebellar nuclei remained almost unchanged.

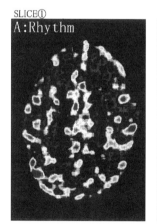
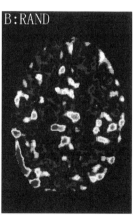
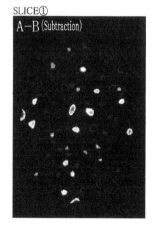

Figure 2
This is an example of the image by Xe-CT scanning from one subject aged 22 years old. On the left, in the task tuned to auditory rhythm, around the midline areas great activation is demonstrated. In contrast, on the right, the areas are poorly activated in the self-paced way randomly.

Figure 3
This is the image after the subtraction method from Task A to Task B. In this method, the subtraction implies probably the only activation by the rhythm. Note that the spot area in the midline preferentially on the left side is greatly activated. The size is rather small, but it stands out against its background of the brain. On the right, the real scanning of the brain is shown in the same level as evaluated for the rCBF on the left.

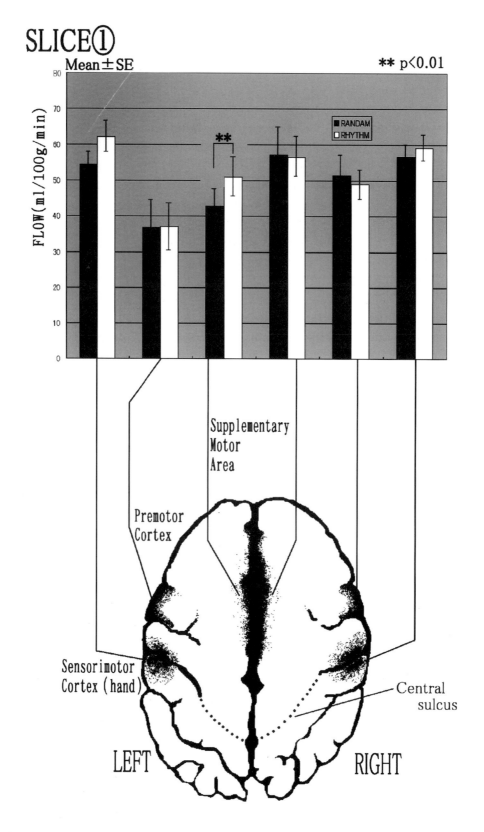

Figure 4
A variety of brain areas are determined for the evaluation of rCBF. Note that only supplementary motor area has significant difference between the two tasks, in terms of rCBF. In short, rCBF in the tuned motor task is significantly greater than that in the random task ($p < 0.01$).

Discussion

Functional neuroimaging like Xe-CT scanning has provided the clue to the precise activated position in the brain, which is considerably to correlated to the tasks. Xe-CT scanning showed a great activation of the rostral midline area, contralateral to the side of the tuned hand motor task, even after performing the subtraction method. Then, where is the activated area in terms of anatomical structure? In general, there are few approaches to assure us that the rostral midline area is the precise position of the SMA in man. In fact, in patients with the involvement of SMA, most researchers rely on the place by specific clinical signs such as supplementary seizures. In normal volunteers, however, it is not the case. Nevertheless, with increased understanding of neuroanatomical findings by functional neuroimaging associated with sequential or complex finger movements has come refinements in the concept that SMA is definitely located at the rostral midline area in man. In addition, it has been acceptable that SMA has also somatotopical organization like the motor strip. The activation using Xe-CT scanning, therefore, is probably located at SMA.

The hot area related to the tuned hand motor task, however, is located in the midline, relatively far, rostral to the primary motor cortex. Luders et al described that the somatotopy of the SMA is quite similar to the body image lying with the head placed anteriorly. Then, the hand area is probably at rather rostral position than the area in front of M1. Again, the hot region of hand area in the contralateral SMA by the subtraction method is also rather small in size (Fig.3). This finding is reasonable in the respect of the fact that most parts of the SMA are considered to show the trunk in the body.

Taken together, the findings as we demonstrated would give rise to the following questions. First, how does the hand motor task tuned to auditory rhythm activate SMA? Second, does the tuned task prefer to activate SMA rather than do premotor cortex? The study using Xe-CT scanning is insufficient for the adequate answers to these questions. We have also demonstrated here that movement-related cortical potentials (MRCPs) are augmented at the periods of premovement from -500 to 0 ms. In contrast, if subjects are required to push a button immediately after noticing the third phone, these components are greatly inhibited (Kujirai T et al, 1999). Some later components of MRCP are considered to be derived from the contralateral SMA. Then, how is the mechanism of the activation brought about during the tuned task? We speculate that the mechanism of the augmentation is due to the effect based on a variety of physiological events such as the attention and the memory of the rhythm. Little prediction, for instance, definitely is lack of appropriate motor planning. In this respect, Passingham (Passingham, 1996) also describes that the supplementary motor area is activated when external rhythmic tones are predicted. If so, then why do external phones fail to activate the premotor cortex rather than SMA? Unfortunately, our proper explanation is not adequate yet. A little speculation onto the mechanism at present is like this. Once subjects learn the external rhythm, their movements may be carried out using procedural or working memory of the tuned task. With the procedure like inertial load, the attention and/or the prediction of the time may activate the contralateral SMA, not utilizing the motor circuits through the premotor cortex. Finally, the decrease of rCBF in the ipsilateral striatum is quite a puzzle. In fact, one of the reasons may be based on the subtraction method mathematically. The contralateral striatum is relatively activated higher than the other, which seems to be theoretically compatible with the general idea.

Conclusions

Increased activation of SMA is demonstrated using Xe-CT scanning. Auditory rhythm can assist the execution of motor programming especially by activating SMA. The attention and the prediction of the time are probably requisite for the activation.

Summary

Neuroimaging technique provides with detailed cortical activation associated with some sensorimotor tasks. Rhythm, one of main factors of music, is of great benefit for some daily behavior. To investigate which cortical areas are activated during a tuned motor task, we studied 5 subjects who gave written informed consent, using Xenon-contrast CT scanning. They were required to push a button, tuning to the third phone of a melody of the triple time with a metronome, or being self-paced randomly. The melody contained three tone bursts every 500 ms for each. The subtraction of Xe-CT scanning between the two tasks demonstrated the only activation ($p<0.01$) in the contralateral supplementary motor area (SMA) in terms of regional cerebral blood flow (rCBF). In contrast, the rCBFs in the ipsilateral caudate and putamen were decreased ($p<0.05$). These results suggest that rhythm may assist the execution of motor programming especially by activating SMA.

Acknowledgements

The authors should like to thank all of the volunteers for the study. This work was supported by the Research Grant for Longevity Sciences (11-C) from the Ministry of Health and Welfare.

References

Deeke L, Scheid P, Kornhuber HH. Distribution of readiness potential, pre-motion positivity, and motor potential of the human cerebral cortex preceding voluntary finger movements. Exp Brain Res 1969;7:158-168.

Ikeda A, Luders HO, Burgess RC, Shibasaki H. Movement-related potentials recorded from supplementary motor area and primary motor area: role of supplementary motor area in voluntary movements. Brain 1992;115:1017-1043.

Jahansahi M, Jenkins IH Brown RG, Marsden CD, Passingham RE, BrooksDJ. Self-initiated versus externally triggered movements. I. An investigation using measurement of regional cerebral blood flow with PET and movement-related potentials in normal and Parkinson's disease subjects. Brain 1995;118:913-933.

Passingham RE. Functional specialization of the supplementary motor area in monkeys and humans. In: Luders HO, editors Supplementary motor area. Advances in Neurology vol.70.

New York: Raven Press; 1996 p.105-116.

Tanji J, Shima K. Role for supplementary motor area cells in planning several movements ahead. Nature 1994;371:413-416.

Chapter III.17

Somatosensory Gating During a Hand Motor Task Tuned to Auditory Rhythm

Hiroyuki Watanabe [b], Takashi Kujirai [a b d 1], Masanobu Saito [b], Noboru Saito [b], Kayoko Kujirai [c], and Shoogo Ueno [d]

[a] *Department of Neurology,* [b] *Division of Higher Brain Function, National Yonezawa Hospital*
Yamagata University, [c] *School of Medicine*
[d] *University of Tokyo, Department of Biomedical Engineering*

Introduction

Peripheral sensory information is probably unnecessary for the movement of the target muscle. For example, the execution of some motor tasks impairs the ability to perceive the sensation or even the stimulation of the moving part, and by increasing the sensory threshold some components of somatosensory evoked potentials (SEPs) are suppressed, which is termed as gating phenomenon. The modulation, however, sometimes assumes in the attitude of the increment of the components. Based on such a specific modulation of upward sensory transmission, the movement is finally accomplished in the way the motor plan is made. This gating effect occurs at numerous sites along the sensory pathways, including the dorsal column nuclei, the thalamus and the somatosensory cortex. Even when the instruction of the movement is given in advance, motor cortical reflexes by way of peripheral sensory transmission is greatly inhibited. Evarts et al (Evarts E and Tanji J, 1974) speculated that the modulation is probably originated in the precentral areas rather than the primary motor cortex. Here, it is obvious that our daily behavior is, to some extent, more easily carried out while tuning to rhythm, which is one of main factors of music. Such an external rhythmic cues or instructions would modulate somatosensory information in moving parts. Regarding this speculation, we have demonstrated the activation of supplementary motor area (SMA) and the augmentation of movement-related cortical potentials in some components around NS',

---

[1] Correspondence: Takashi Kujirai, Department of Neurology, National Yonezawa Hospital, 26100-1, Misawa, Yonezawa, Yamagata 992-1202, Japan, Tel: (81)-238-223-210, Fax: (81)-238-226-691
Abbreviation: SEP= somatosensory evoked potential

which are probably derived from SMA. If so, then does sensory gating occur during a tuned motor task as well? And does this effect take place at a limited time? In order to verify when and how somatosensory gating occurs during the motor task tuned to auditory rhythm, we studied 7 normal volunteers who gave written informed consent using somatosensory evoked potentials (SEPs).

Methods

All of the volunteers for the study were the laboratory stuff aged from 22 to 54 years old. They were right handed and were on a bed comfortably during the study. They were required to listen to the three rhythmic tone bursts every 500 ms for each, which is a triple time. Focusing on the third phone, they were asked to push it just at the time of the third tick sound after predicting the time precisely. Somatosensory evoked potentials (SEPs) were recorded following electrical stimulation of the right median nerve at the wrist. The SEP was recorded from two electrodes on the scalp referred to the earlobes on both sides. One electrode was over the somatosensory area (2cm posterior to C3) and the other was 3 cm anterior to C3. Recordings were made with a band-pass of 3Hz to 3kHz. The stimulation at the wrist occurred at various intervals from the initial phones from 0 to 1500 ms. The mean average of SEP was analyzed after 100 trials of the motor task. SEP for C3. Recordings were made with a band-pass of 3Hz to 3kHz. The stimulation at the wrist occurred at various intervals from the initial phones from 0 to 1500 ms. The mean average of SEP was analyzed after 100 trials of the motor task. SEP for the control was obtained from the recordings at 0 ms because the motor set or the attention to the phone is almost all the time prepared during the study. The short latency of SEP comprises several components such as N20, N20-P25, P25-N33, N33-P40 and P40-N65 in the parietal recording and N18, N18-P22, N30-P40 and P40-N65 in the frontal recording. The amplitudes in peak-to-peak for each component were evaluated and the data were analyzed for statistical analysis.

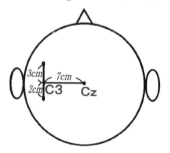

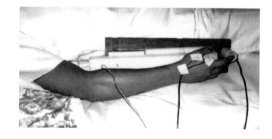

Figure 1

The top: The recording sites for somatosensory evoked potentials are shown in this figure. All volunteers are right handed, and two electrodes are placed on the scalp as follows; one is 2 cm posterior to C3 and the other 3 cm anterior to C3.

The bottom: This photograph shows the maneuver for the hand motor task tuning to a triple tone bursts every 500 ms. A subject is required to push a button at the time of the third phone after predicting the time. Surface EMG is also monitored to see whether the task is carried out appropriately.

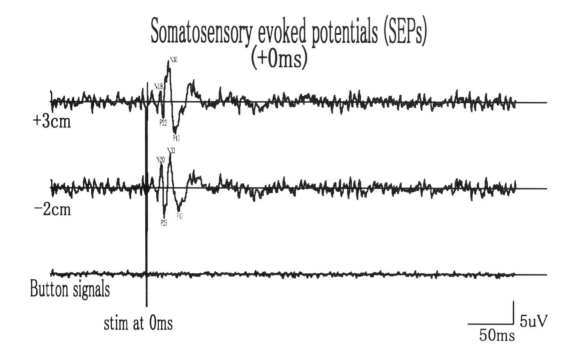

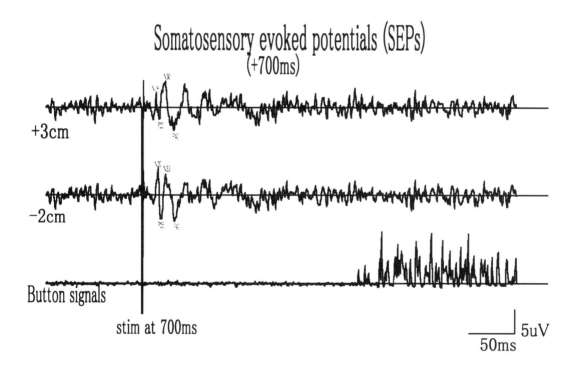

Figure 2

This figure shows examples of somatosensory evoked potentials (SEPs) recorded from one subject. The top: SEPs at 0 ms shows well-defined components in the frontal and parietal recordings. The bottom: At 700 ms from the initial sound, some components such as parietal p25-N33 are apparently inhibited in amplitude. Switch signals are obtained 150 or 200 ms later from the stimulus time. The inhibition takes place even before the actual movement.

The superimposition of SEPs
(+0ms～,+700ms～,n=7)

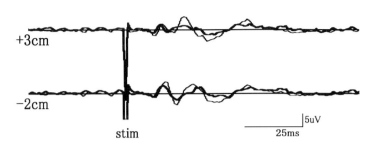

Figure 3
This figure shows the grand average of SEPs among 7 volunteers. Two grand averages are superimposed. The intervals from the initial sound are 0 and 700 ms here. Note that some components such as the frontal P22-N30 and parietal P25-N33 are greatly inhibited.

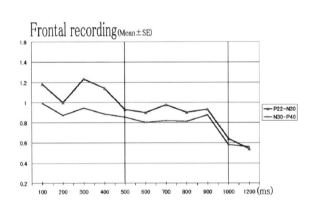

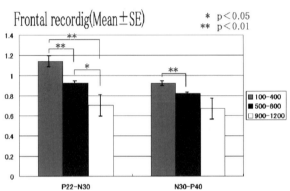

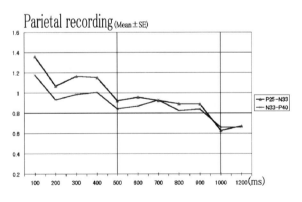

Figure 4
The time course of the modulatory effect on the SEP is shown in this figure. The frontal P22-N30 as well as parietal P25-N33 is augmented from 0 to 500 ms, and then its modulatory behavior becomes gradual inhibition toward the actual movement.

Figure 5
From Fig.4, the time course is divided into 3 blocks; the first block is from 100 to 400 ms, the second from 500 to 800 ms and the third from 900 to 1200 ms. Then, for each block, the mean average of ratio is evaluated respectively. Note that the frontal P22-N30 and parietal P25-N33 in the first block are significantly suppressed even before the movement.

Results

Fig.1 shows an example of SEPs from the parietal and frontal recordings. Even before the actual movement of pushing a button, for example, SEPs at 700ms from the initial phone are suppressed in amplitude by 10% or so. The modulation of SEPs in the parietal P25-N33 component is the augmentation from 100 to 500 m. The magnitude gradually gets lesser and after around 500 ms the modulation becomes the inhibition. During the actual movement around 1000 ms the SEPs in all components are greatly inhibited (Fig.2). In view of blocks of intervals such as from 100 to 400 ms, parietal P25-N33 component has significant difference among the other blocks ($p<0.05$). Some other components also are statistically significant (Fig.3).

Discussion

The present results show that a hand motor task tuned to auditory rhythm can affect the size of several components of the SEP produced by stimulation of the contralateral median nerve at the wrist. On both recordings, parietal and frontal components are gradually inhibited as toward the time of the third phone. For instance, the parietal P25-N33 component is augmented before 500 ms from the initial phone, and from around 500 ms is gradually suppressed up to the time of the movement. By monitoring the sEMG, even around 500 or 600 ms from the initial sound, the actual movement by pushing a button is not still undertaken with physiological silence similar to the rest situation. If so, then how is the several components of SEP inhibited? The volleys from the primary motor cortex might have landed onto the structures in which the sensory gating can take place. This possibility seems to be unlikely because the movement is achieved 400 or 500 ms later, which implies that the time is rather earlier for the beginning of the gating.

The speculation from this study alone is quite limited. However, subjects probably know or can predict the precise time of pushing a button after learning the rhythm. With respect to the premovement or preparation of the movement, in fact, Kujirai et al (Kujirai et al,1999) demonstrated that some later components of movement-related cortical potentials (MRCPs) are augmented around from -500 ms. In this case, -500 ms in the MRCPs is almost equal to the interval of 500 ms here. In addition, we have shown the activation of supplementary motor area in the hand motor task while tuning to auditory rhythm. Taken together, the modulation from excitation to inhibition in the parietal P25-N33 component may be due to some events in the supplementary motor area activated by taking the auditory rhythm. Presumably, through a series of the studies associated with auditory rhythm, activation of the SMA is a prominent finding. As speculated by Evarts et al (Evarts E and Tanji J, 1974), SMA plays a crucial role to the somatosensory gating. In brief, around 500 ms or so from the initial phone, as long as volunteers are tuning to the rhythm, they are subject to have the somatosensory gating by activating SMA.

Conclusions

While tuning to auditory rhythm, somatosensory gating is carried out probably by activating the supplementary motor area. Using the gating rhythmic motor command is finally performed for the execution especially at the premovement periods.

Summary

Motor output is capable of modulating of upward sensory transmissions. The modulation of sensory information, so-called somatosensory gating, is prominent in some motor tasks. In contrast, motor task is probably facilitated in accordance with rhythmic external cues. To investigate when and how sensory modulation occurs during a beneficial motor task tuned to auditory rhythm, we studied 7 subjects who gave written informed consent. They were required to push a button, tuning to the third phone of a melody of the triple time. The melody contained three tone bursts every 500 ms for each. Somatosensory evoked potentials (SEPs) by stimulating the contralateral median nerve were recorded during the task. The parietal components of SEPs such as N20-P25 were gradually decreased in amplitude ($p < 0.05$) from around the second phone. This finding suggests that rhythmic motor command may participate in modulating sensory information for the execution even before the actual movement.

Acknowledgments

The authors should like to thank all of the volunteers for the study. This work was supported by the Research Grant for Longevity Sciences (11-C) from the Ministry of Health and Welfare.

References

Evarts EVE and Tanji J. Gating of motor cortex reflexes by prior instruction. Brain Res 1974;71:479-494.

Kujirai T, Sato M, Rothwell JC and Cohen LG. The effect of transcranial magnetic stimulation on median nerve somatosensory evoked potentials. Electroencephalogr Clin Neurophysiol 1993;89:227-234.

Kujirai T, Watanabe H, Saito M, Saito N, Tuburaya K, Kujirai K and et al. Motor programming as a hand motor task tuned to auditory rhythm. Abstract. ISNM '99.

Saito M, Kujirai1 T, Watanabe H, Saito N, Tuburaya K, Kujirai K and et al. Increased activation of supplementary motor area during a hand motor task tuned to auditory rhythm. Abstract. ISNM '99.

T. Nakada (Ed.)
Integrated Human Brain Science: Theory, Method Application (Music)
© 2000 Elsevier Science B.V. All rights reserved

Chapter III.18

Motor Programming as a Hand Motor Task Tuned to Auditory Rhythm

Takashi Kujirai [a,b,e,1], Hiroyuki Watanabe [b], Kayoko Kujirai [c], Giito Izuta [d], Takashi Yamaguchi [d], and Shoogo Ueno [e]

[a]*Department of Neurology,* [b]*Division of Higher Brain Function, National Yonezawa Hospital*
Yamagata University, [c]*School of Medicine,* [d]*School of Engineering*
[e]*University of Tokyo, Department of Biomedical Engineering*

Introduction

Rhythm is one of main factors of music. It has probably a beneficial effect on some specific behavior such as bradykinesia in patients with Parkinson's disease. Bradykinesia or slowness is the difficulty in initiating and executing a motor plan, and particularly difficult for patients are sequential finger motor acts, confined to distal muscles. Once the stimulus to move from the environment is perceived, however, surprisingly they would find themselves performing motor acts more easily. Especially while tuning to rhythm by auditory or visual cues, they would seem to have no impairment in performing a motor plan. According to Marsden (Marsden, 1982), movement execution requires the selection of appropriate motor programs from the memory store. In addition, the activation of supplementary motor area, which is probably correlated to movement-related cortical potentials, are also plays a crucial role to executing motor program for discrete or sequential movements. We have demonstrated the activation of the contralateral supplementary motor area during a tuned hand motor task using Xe-CT-scanning (Saito *et al*, 1999) If so, then movement-related cortical potentials (MRCPs) may be augmented according to the motor programming while tuning to rhythm. In order to verify whether MRCPs are changed in amplitude during tuned motor tasks or whether motor evoked potentials (MEPs) by transcranial magnetic stimulation are varied, we

---

[1] Correspondence: Takashi Kujirai, Department of Neurology, National Yonezawa Hospital, 26100-1, Misawa, Yonezawa, Yamagata 992-1202, Japan, Tel: (81)-238-223-210, Fax: (81)-238-226-691
Abbreviations: MRCP= movement-related cortical potential, MEP= motor evoked potential FDI= the first dorsal interosseus muscle

studied 7 normal volunteers who gave written informed consent according to the local ethical committee approval at National Yonezawa Hospital.

Methods

Seven normal volunteers participated in the following study. All the volunteers for the study were the laboratory stuff aged from 22 to 54 years old (mean age 34.5 years old). They were seated in a comfortable reclining chair or on a bed during the study. They were also required to push a switch-button according to the experimental design (Fig.1). The study is divided into 7 different sessions for recording movement-related cortical potentials (MRCPs). Each session was finished in 10 minutes. They took a rest for about 10 minutes to avoid a mental fatigue for the next session. One session (the 3H task in the Fig.1), for example, they were listening to the three rhythmic tone bursts every 500 ms for each, which is a triple time. Focusing on the third phone whose frequency was 1.5 kHz equal to about one octave above the base phone (1 kHz), they were asked to push the switch just at the time of the third phone after predicting it precisely. This session was also performed for the study of transcranial magnetic stimulation.

Two experiments as follows were carried out.

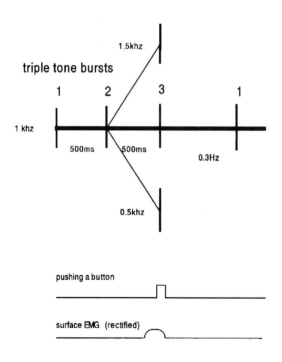

MRCPs for 7 different motor tasks

1) pushing a button with randomly self-decided pace without any cues (SFP)
2-4) pushing a button tuning to a triple time with the frequency of the third phone
    2) 1.5 KHz (3H)
    3) 1.0 KHz (3N)
    4) 0.5 KHz (3L)
5) pushing a button immediately after noticing the third phone (3Hr)
6) pushing a button tuning to memorized triple time in mind without any cues (IMG)
7) pushing a button immediately after noticing the different frequency of the phone, which occurs at 1 Hz (RAND)

The task for MEPs

pushing a button tuning to a triple time higher frequency of the third phone (1.5 KHz)

Figure 1
The study was carried out according to the paradigm as shown in the figure. All subjects were asked to listen to the triple tone bursts (1 kHz, 500 ms for each, for instance). Focusing on the third phone, they were also required to push a button after predicting the exact time precisely. The surface electromyography (sEMG) was monitored in the first dorsal interosseus muscle during the study. The onset of the rectified sEMG was triggered for the average of backgrounds of electroencephalography on the scalp by the international 10-20 system. 7 different paradigms were performed for the evaluation of the change of movement-related cortical potentials (MRCPs). Each abbreviation such as 3H reflects the study respectively.

1) Movement-related cortical potentials (MRCPs)

The electroencephalogram (EEG) was recorded from silver/silver chloride surface electrodes, (time constant 3000 ms) with recording electrodes placed at 16 positions such as C3, Cz and Pz according to 10-20 international system, referenced bilaterally to electrodes on both earlobes, during the study. The electromyogram (EMG) in their first interosseus muscles and the push-button signal were also recorded. The onset of the rectified EMG was triggered for back averaging of EEG. About 100 sweeps were averaged in each condition for each subject. After identifying every component of MRCP, their latencies and amplitudes were evaluated for statistical analysis among different motor tasks. Grand average of the MRCPs of 7 subjects for each session was also obtained.

2) Motor evoked potentials (MEPs)

At the same state as mentioned in the experiment-1, transcranial magnetic stimuli over the contralateral hand motor area were given using a Magstim 200 stimulator connected to a figure-of-eight coil (outer loop diameters of 7 cm; maximum output rating of 2.4 Tesla). The target muscle was the right first dorsal interosseus muscle (FDI). The surface electromyogram was recorded from the FDI using the belly-tendon method. MEPs were recorded during push-button tasks or during non-push-button task as a control. When the button is pushed, the switch signal was also monitored for the detection of a delay from the third phone. Peak-to-peak amplitudes of MEPs were evaluated for statistical analysis among different motor tasks.

Results

Grand averages of MRCPs were obtained for the evaluation of the mean amplitudes for the each component. The onset of the rectified EMG in the target muscle was 0 ms for the trigger of back averaging of EEG. The each component was easily detected in the grand averages. The latencies at +50 ms, 0, -200, -350 and -700 ms were determined for the statistical analysis among different sessions. As shown in the upper row in the Fig.2, for example, the two MRCPs in the SFP and 3H tasks are superimposed, and the significant difference between them are shown as * $p < 0.05$ or ** $p < 0.01$. In the SFP task the amplitude at 0 ms for C3 is -0.54±0.33 uV (Mean±SE), and the mean amplitude at the same position in the 3H task is -1.65±0.37 uV (Mean±SE) with the statistical difference ($p < 0.05$). The MRCP at -350 ms for C3, Cz and Pz in the 3H task is also significantly greater in amplitude than that in the SFP. Statistical difference holds among other sessions in terms of the amplitudes. In contrast, intriguingly enough, the amplitude remained virtually unchanged at 0 ms, irrespective of the frequency of the third phone (Fig.3). Again, at 0 ms for C3, the amplitude of the grand average in IMG is larger than that in the IMG, which failed to be significant (the lower row in the Fig.2). In the task that subjects are required to push a button immediately after noticing the third phone, the amplitude at -200 ms for is significantly smaller that that in the 3H task ($p < 0.05$, Fig.2). The degree of the amplitude is much prominent in the RAND task ($p < 0.01$).

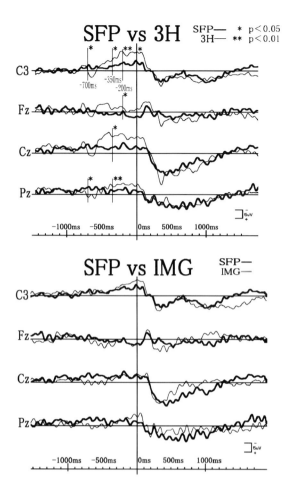
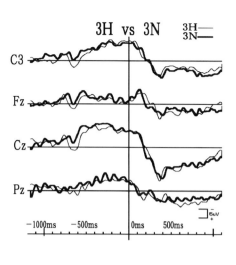
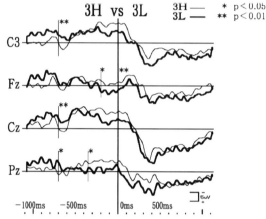

Figure 2
This figure is one of examples for the task-1 (3H), in which volunteers were asked to push a button in their self-paced ways randomly or to push it tuning to the third phone (about one octave higher or higher frequency than the others by 0.5 kHz=1.5kHz) of the auditory triple time. Here, 4 different recording sites were shown for the evaluation; C3, Fz, Cz and Pz. The two waves for the tasks are superimposed. The vertical line implies the onset of the sEMG in the FDI muscle, which is 0 ms here for the MRCPs. Note that MRCPs from around -600 to +100 ms at C3, Cz, and Pz in the task tuned to auditory rhythm are larger in amplitude than that in the self-paced way (SFP).

Figure 3
A variety of combinations of MRCPs are shown in this figure. All recordings are superimposed. The vertical line indicates the onset of sEMG which is 0 ms for the potentials. The combinations are as follows; the self-paced task (SFP) versus the image task (IMG), SFP versus the tuned task to the rhythm with higher frequency of the third phone (3H). Note that a prominent augmentation in the task of 3H compared with the SFP task. Even in the image task, slight increment is shown at C3 recording which fails to have the statistical significance.

Fig.4 shows the results of the transcranial magnetic stimulation study. At 700 ms after the initial phone, for example, MEPs are virtually unchanged, but the facilitation in amplitude gradually takes place from around 700 ms even before the surface electromyogram discharge suggesting the actual movement. The delay of the switch signal from the onset of sEMG is almost 50 ms for each trial. At 1000 ms, the actual movements sometimes, however, produce the prominent facilitation of MEPs.

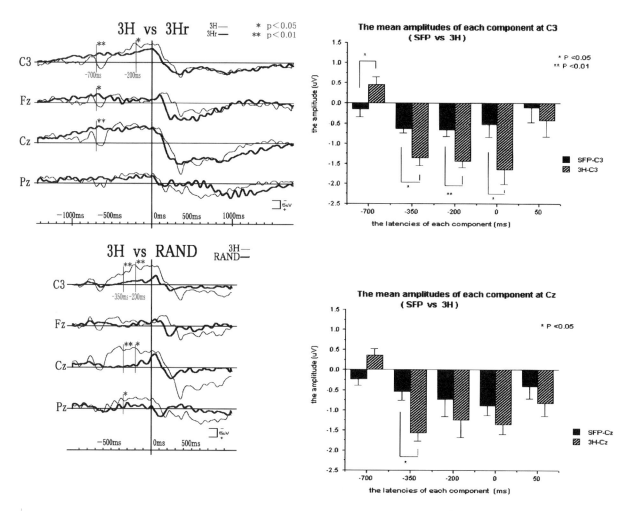

Figure 4

This figure shows the following superimposed MRCPs; 3H versus the tuned task to the rhythm with the same frequency of the third phone (3N), 3H versus the tuned task with lower frequency of the third phone (3L), 3H versus the IMG, 3H versus the reaction task to the higher frequency of the third phone (3Hr) and 3H versus the reaction task to the randomized phone (RAND). Note that MRCPs are augmented irrespective of the frequency of the third phone in comparison with that in the SFP task.

Figure 5

As shown in the Fig.2, the MRCPs are changed in amplitude, especially at + 50, 0, -200. -350 and -700 ms. The amplitudes for these components were evaluated for each subject. Note that tuned tasks to auditory rhythm show prominent augmentation of some components of MRCPs, irrespective of the frequencies, with the statistical significance ($p < 0.05$).

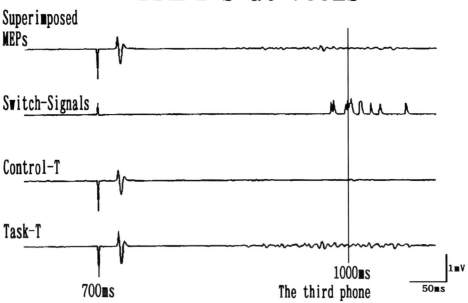

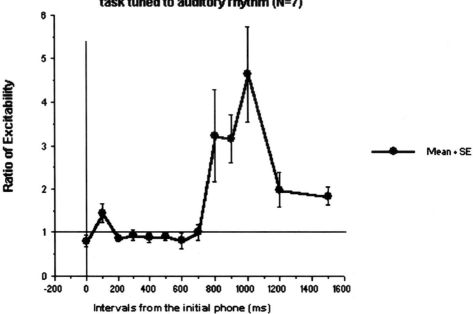

Figure 6

The top: This figure shows an example of the change of motor evoked potentials during a motor task. Transcranial magnetic stimuli during the task were delivered at 700 ms from the initial phone. The mean average of MEPs of ten trials for pushing a button tuned to the auditory rhythm or for the rest state was made after the study. The upper row is the average of all trials for each recording. The lower row is the example is the mean averages of pushing a button task and of the rest state. The upper trace is sEMG in the FDI muscle and the lower the signals for pushing a button for all trials. The onset of sEMG proceeded the signal of the switch of the button by nearly 50 ms. Note that the mean average of MEP for pushing a button is larger than that for the rest state.

The bottom: MEPs were changed in amplitude at various intervals from the initial phone. Around 700 ms from the initial phone, MEPs remained virtually unchanged. Note that MEPs were greatly facilitated after the intervals of 800 ms. Around 1000 ms, MEPs were prominent in terms of the facilitation.

Discussion

Movement-related cortical potentials (MRCPs) have been extensively recorded since the original reports by Kornhuber and Deecke (Deek L *et al*, 1967). A better wave of MRCP is divided into several components, two of which are easily recognized as the early symmetric part (Bereitschaftspotential; BP) and the late asymmetric part of the premovement slow negativity (negative slope; NS'). Likewise, the components of MRCPs in the study are distinguished NS' from BP.

The study here shows several unique findings of MRCPs with respect to the auditory rhythm, which is one of main factors of music. First, with the tuning to a triple time, MRCPs around from -350 ms to 0 ms are larger in amplitude in the contralateral sensorimotor cortex and frontal midline areas. In fact, the peak amplitude reflecting the later components occurring around the latencies from -200 ms to the time of movement onset is greatly lateralized to C3 area. Second, even in the image task of the rhythm, MRCPs at 0 ms for C3 shows a little larger than that in the self paced way (SFP). In addition to the image task, the MRCPs in the reaction-time task to the third phone remains still larger in amplitude around the latencies of the later components than that in the SFP. In contrast, when subjects were required to do the same task immediately after the phone different from the background sound (1 kHz), they would not be able to predict the time of the phone. Then, as it turns out, the same components of MRCPs are greatly inhibited probably because motor tasks were performed in the situation that the motor plan had been insufficient to carry out. Third, the later components like NS' are scarcely under the influence by the frequency of the third phone. Taken together, the amplitude of the components are probably associated with the time, especially the predicted time of forthcoming sounds by focusing on them. The memory of the auditory rhythm here must be quite necessary for the execution of the hand motor tasks during tuning to external rhythmic cues, even in the image without the actual movement. In brief, the tuning to rhythm using the attention and the memory is of an appropriate usage of motor programming or of a beneficial effect on motor tasks.

With regard to the actual motor tasks, the facilitation of MEP is apparent after 700 ms from the initial phone (Fig.4) to 1500 ms. Concerning the fact that the later components of MRCPs are lateralized to C3, it is quite interesting that the time course of the facilitation of MEPs is similar to the MRCP at C3. The steeper slope of the MRCP before the execution of the movement may imply the flow or the command of the movement. In fact, transcranial magnetic stimulation technique provides the investigation of the excitability of human motor cortex (Kujirai T *et al*, 1993). The facilitation of MEPs probably occurs around periods after the decision of the execution of the movement. The facilitation suggests, more or less, the volleys from the primary motor cortex. Then, where is the origin for the components during the period such as NS', just before the final motor command? Unfortunately, however, our knowledge regarding the precise origin of the each component of MRCP is unsatisfactory. Recent studies using magnetoencephalogram or neuro-imaging technique such as positron emission topography (PET) and functional MRI (fMRI) have been the most accessible to the precise origin. As postulated by Passingham (Passingham, 1996), the supplementary motor area is activated when tones are predicted, which is corresponding to the results using Xe-CT scanning we have demonstrated here (Saito *et al*, 1999).

Conclusions

The tuning to auditory rhythm probably using the attention and the prediction is of a specific effect on the enhancement of some components of movement-related cortical potentials (MRCPs) and also is of a great usage of motor programming.

Summary

Rhythm, one of main factors of music, is of great benefit for our daily behavior. Some motor tasks are probably easily performed when tuned to external rhythmic cues, especially in patients with Parkinson's disease. To investigate how motor cortex prepares for motor execution during a tuned motor task, we studied 7 subjects who gave written informed consent. They were required to push a button, tuning to the third phone of a melody of the triple time. The melody contained three rhythmic tone bursts every 500 ms for each. Movement-related cortical potentials (MRCPs) or motor evoked potentials (MEPs) in the first dorsal interosseus muscle (FDI) were recorded during the task. The amplitude of MRCP was greater in the tuned task than in the self-paced task ($p<0.05$). The MEPs were also facilitated at around 1000 ms. These results indicate that rhythm may contribute to improving the usage of cortical command.

Acknowledgments

The authors should like to thank all of the volunteers for the study. This work was supported by the Research Grant for Longevity Sciences (11-C) from the Ministry of Health and Welfare.

References

Deeke L, Scheid P, Kornhuber HH. Distribution of readiness potential, pre-motion positivity, and motor potential of the human cerebral cortex preceding voluntary finger movements. Exp Brain Res 1969;7:158-168.

Kujirai T, Caramia MD, Rothwell JC, Day BL and Thompson PD and et al. Corticocortical inhibition in human motor cortex. J Physiol (Lond) 1993;471:501-519.

Marsden CD. The mysterious motor function of the basal ganglia: the Robert Wartenberg lecture. Neurology 1982;32:514-539.

Saito M, Kujirai1 T, Watanabe H, Saito N, Tuburaya K, Kujirai K and et al. Increased activation of supplementary motor area during a hand motor task tuned to auditory rhythm. Abstract. ISNM '99.

Passingham RE. Functional specialization of the supplementary motor area in monkeys and humans. In: Luders HO, editors Supplementary motor area. Advances in Neurology vol.70. New York: Raven Press;1996 p.105-116.

Chapter III.19

Motor Imagery Tuned to Auditory Rhythm Activates the Motor Cortex

Kayoko Kujirai[a,1], Takashi Kujirai[b], Takeo Kato[c], and Makoto Tominaga[a]

[a] *Dept. of Laboratory Medicine and* [c] *3rd Dept. of Internal Medicine, Yamagata University, School of Medicine*
[b] *Dept. of Neurology, National Yonezawa Hospital*

Introduction

In order to accomplish a movement task, motor programming is crucial to translate the inertial decision into the movement execution. With the advent of a variety of neuroimaging techniques such as single-photon emission tomography (SPECT), positron emission tomography (PET) and functional magnetic resonance imaging (fMRI), human inertial experiences have been investigated by measuring the regional activities associated with their mental tasks. However, these maneuvers fail to demonstrate temporal changes of activation in a specific time. In this respect, as described by Kosslyn SM et al (Kosslyn SM et al, 1999), transcranial magnetic stimulation technique is quite available to investigate the relationship between some mental tasks and the times, and also to compensate for the obscurity because it is capable of studying cortical excitability in almost real time (Kujirai T et al, 1993). With regard to the findings of the mental tasks, focusing on the motor imagery, we demonstrate the effect on the motor cortical excitability by using transcranial magnetic stimulation technique.

Methods

Subjects and experimental design

Six right-handed healthy subjects (aged from 21 to 63 years old) participated in the study.

---

[1] Correspondence: Kayoko Kujirai, Department of Laboratory Medicine, Yamagata University School of Medicine, Yamagata, Yamagata 990-9585, Japan, Tel: (81)-236-285-406, Fax: (81)-236-285-409
Abbreviations: MEP= motor evoked potential, FDI= the first dorsal interosseus muscle.

All gave informed consent and the procedures had the approval of the local ethical committee. Subjects were seated in a reclining chair with their eyes closed, and were asked to listen to a melody of the triple sound. The melody contained three rhythmic tone bursts (1 kHz, 100 dB, plateau 50 ms) every 500 ms. Then, subjects were required to image the abduction of the index finger at the third sound of the melody (at 1000 ms from the first sound) (Fig.1). Surface electromyogram (EMG) was recorded from the first dorsal interosseous (FDI) muscles of the right hand using silver/silver-chloride surface electrodes by a belly-tendon method. Electrodes were also placed on forearm flexor (FF) muscles to check electrical silence during the experiment. The EMG signals were amplified and filtered (bandwidth 20-3 kHz).

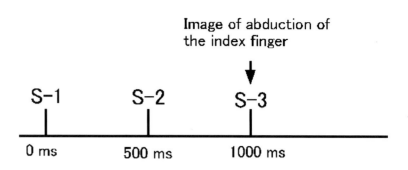

Figure 1
The melody contained three rhythmic tone bursts (1 kHz, 100 dB, plateau 50 ms) every 500 ms. Subjects were required to image abduction of the index finger tuning to the third sound of the melody (at 1000 ms after from the first sound). Transcranial magnetic stimulation (TMS) was given using a Mag Stim 200 stimulator. Motor evoked potentials (MEPs) were recorded at various intervals from the first sound (Time=0 ms). H-reflex was also measured at each interval.

Transcranial magnetic stimulation

Transcranial magnetic stimulation (TMS) was given using a Mag Stim 200 stimulator connected to a figure-of-eight coil with external diameters of 9 cm and a peak magnetic field of 2.2 Tesla. The stimulating coil was placed over the left hemisphere at the optimal site for eliciting motor evoked potentials (MEPs) in the FDI. The stimulating intensity was used to obtain MEPs having a peak-to-peak amplitude of about 500 μV. MEPs were recorded at various intervals from the first sound (Time=0 ms). The peak-to-peak amplitude of MEPs with motor imagery (Image + T) and that of MEPs without image (T) were evaluated after 10 trials at each time.

H-reflex recordings

In four subjects, the effect of motor imagery on the size of H-reflexes was studied. They were asked to image the flexion of the right index finger at around the third sound, and H-reflexes were recorded from the forearm flexor (FF) muscles with stimuli of 1-ms duration to the median nerve at the elbow. The peak-to-peak amplitude of the mean average of responses was measured at each time.

Statistics

A paired t-test was carried out to evaluate the difference between the two conditions. The null hypothesis was rejected at the 0.05 level.

Results

Figure-2 shows an example of motor evoked potentials (MEPs) recorded from FDI. Transcranial magnetic stimulation was given at 1100 ms from the initial sound. Motor imagery increased the amplitude of MEPs.

Figure-3 shows the time course of MEPs. The time of 0 ms implies the onset of the initial auditory rhythm. The ordinate in the graph shows the ratio of the excitability for each time. From around the interval of 800 ms, transcranial magnetic stimulation study demonstrated a gradual augmentation of MEPs. The peak excitation of MEPs was observed at 1000 or 1100 ms in most subjects except for one person, in whom the peak was shown at 1300ms. The mean excitation ratio of MEPs, for example, was 2.64 ± 0.37 (M$\pm$SE) at the interval of 1100 ms. Motor imagery significantly increased the amplitude of MEPs at 1000 and 1100 ms ($p<0.05$).

An example of H-reflex study at 1100 ms was shown in Figure-4. In the H-reflex study, the excitability of the reflex arc in the spinal level remained virtually unchanged throughout the time course from 0 up to 1500 ms (Figure-5).

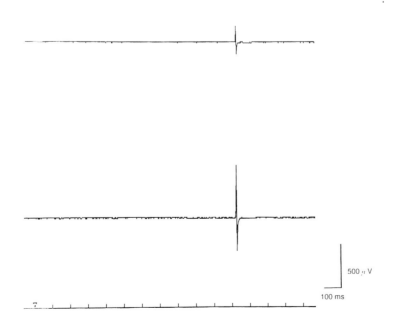

Figure 2
Transcranial magnetic stimulation (TMS) was given using a Mag Stim 200 stimulator connected to a figure-of-eight coil at 1100 ms after the initial sound. The stimulating coil was placed over the left hemisphere. Surface EMG responses were recorded from the first dorsal interosseous (FDI) muscles of the right hand using silver/silver-chloride surface electrodes by a belly-tendon method. Motor imagery increased the amplitude of MEPs.; Top: MEP without image of abduction of the index finger., Bottom: MEP with the image.

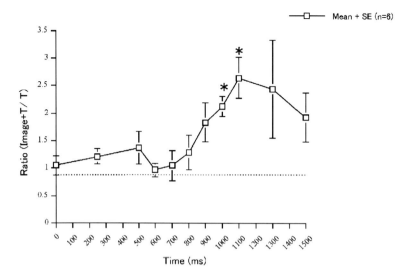

Figure 3
Mean ratio of the peak-to-peak amplitudes of MEP with motor imagery (Image +T) and that of MEPs without image (T) obtained from six subjects. MEPs were recorded at various intervals from the initial sound (Time =0 ms). The stimulating intensity was used to obtain MEPs having a peak-to-peak amplitude of about 500 μ V. The amplitude of MEP was evaluated after 10 trials at each time. The facilitatory effect was observed significantly at 1000 and 1100 ms from the initial sound (*$p<0.05$).

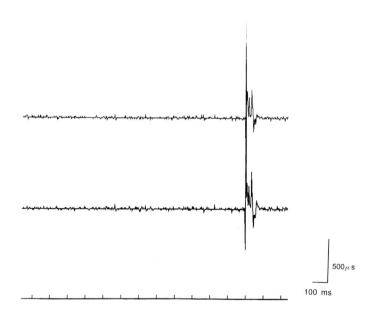

Figure 4
H-reflex study of motor imagery. Subjects were asked to image the flexion of the right index finger at around the third sound. H-reflexes were recorded from the flexor forearm muscles with stimuli of 1-ms duration to the median nerve at the elbow at various intervals from the first sound. An example is shown at 1100 ms from the initial sound. No significant effect was observed on the size of H-reflex by motor imagery.
Top: H-reflex without the image, Bottom: H-reflex with the image.

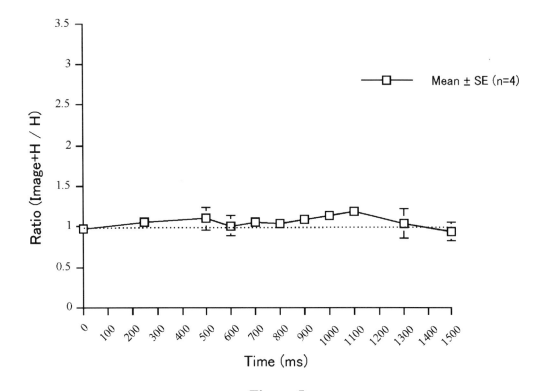

Figure 5
Mean ratio of the sizes of H-reflex (Image +H/ H) from 4 subjects. The peak-to-peak amplitude of the mean average of responses was measured at each time. Motor imagery did not affect the size of H-reflex significantly.

Discussion

The study demonstrated the prominent dissociation of the excitability between the H-reflexes and MEPs by transcranial magnetic stimuli during motor imagery. Here, these results would give rise to a few questions about the excitation of MEPs, probably associated with auditory rhythm, which is one of main factors of music.

First, where is the origin of the excitation of MEP derived from during a tuned hand motor task? The volleys induced by transcranial magnetic stimulation can go down onto the α-motoneurons at the velocity that suggests contribution of the most largest fibers in the cotricospinal tracts. Then the excitation probably occurred at the spinal level or the cortical level. Maybe other anatomical structures below the spinal level are also involved. However, the last possibility is rather unlikely because any movement by way of the final common pathway was not detected. The study may be quite limited regarding to the precise origin. However, in consideration of the simple anatomic structure of the pyramidal tract, the dissociation strongly supports the concepts that the motor cortex is activated by the tuned imagery task.

Second, then, does the excitability depend on the imagery or on the tune to auditory rhythm? This question is difficult to resolve in the limited study performed here. However, we also demonstrated great activation of the supplementary motor area during a tuned hand motor task when the activation was compared to that of during a self-paced task (Saito M et al,

1999). In addition, movement-related cortical potential (MRCP) for the premovement is more augmented in a tuned task than that of in a self-paced task (Kujirai T et al, 1999). Taken together, by tuning to rhythmic sound, the final command of the movement after the motor programming is probably ready to convey the signals onto the pyramidal cells in the primary motor cortex (M1), like the volleys by magnetic stimulation, which are supposed to go down through the pyramidal tract. In brief, auditory rhythm contributes to the usage of the final common command appropriately and the imagery is capable of the activation of M1 after the decision of the movement execution, even in mind without the real execution.

Conclusions

A motor imagery tuning to auditory rhythm can facilitate motor cortical excitability. Taking auditory rhythm probably has beneficial effects on cortical motor programming in a daily behavior.

Summary

Rhythm is of great usage for movement execution. Then, does motor imagery activate the motor cortex even without actual movement? To investigate cortical motor excitability during the imagery of the abduction of the index finger, we studied 6 subjects who gave written informed consent. They were required to have the imagery in mind, tuning to the third phone of a melody of the triple time. The melody contained three rhythmic tone bursts every 500 ms for each. Motor evoked potentials (MEPs) were recorded in the first dorsal interosseus muscle (FDI) at various intervals from the initial sound. At 1000 and 1100 ms from the initial phone, MEPs were greatly facilitated ($p<0.05$). In contrast, H reflex study showed no statistical differences. These findings suggest that the facilitation is of cortical origin. In conclusion, motor imagery tuned to auditory rhythm participates in cortical motor programming.

Acknowledgment

The authors should like to thank all of the volunteers for the study. This work was supported by the Research Grant for Longevity Sciences (11-C) from the Ministry of Health and Welfare.

References

Kosslyn SM, Pascual-Leone A, Felician O, Camposano S, Keenan JP, Thompson WL and et al. The role of area 17 in visual imagery : convergent evidence from PET and rTMS.

Science 1999; 284:167-170.

Kujirai T, Caramia MD, Rothwell JC, Day BL and Thompson PD and et al. Corticocortical inhibition in human motor cortex. J Physiol (Lond) 1993; 471:501-19.

Kujirai T, Watanabe H, Saito M, Saito N, Tuburaya K, Kujirai K and et al. Motor programming as a hand motor task tuned to auditory rhythm. Abstract. ISNM 1999.

Saito M, Kujirai T, Watanabe H, Saito N, Tuburaya K, Kujirai K and et al. Increased activation of supplementary motor area during a hand motor task tuned to auditory rhythm. Abstract. ISNM 1999

Chapter III.20

Effect of Different Music Contents on EEG Activity and Mood States

Kumi Naruse[1] and Haruo Sakuma

Graduate School of Human Culture, Nara Women's University

Introduction

Music has an effect on the auditory sense, as well as on the function of mind and body such as cognition, task efficiency, and emotions (Snyder M. & Chlan, L., 1999). Today, our society is full of stress, and people participate in a variety of activities for pleasure and stress reduction. Listening to music is one of them.

Music is primarily composed of melody, rhythm, and harmony, as well as high-low pitch and long-short duration of the individual sounds. These components combine in countless ways to affect the listener, giving it different properties. Recently, in the field of psychophysiology, as an attempt to explain the biological substrate of affection or emotion, such as a positive affective state or anxiety, a large amount of research has been conducted concerning brain activity, especially regarding frontal brain asymmetry (Ahern et al., 1985; Benca et al., 1999; Davidson et al., 1990; Hatfield et al., 1987; Petruzzello et al., 1994; Wheeler et al., 1993). However, few studies have been conducted to explain the effect of music on brain activity. The purpose of the present study was to examine the influence of music and its different properties on brain activity and the mood state of listeners. The hypotheses are as follows:
1. Different musical properties can cause mood state changes in the listener.
2. Mood state changes caused by music reflect the alpha power asymmetry of brain activity.

Methods

Subjects

The subjects were 7 female students with a mean age of 24.9yr. (age range 21 to 33 yr.)

---

[1] Correspondence: Kumi Naruse, Graduate School of Human Culture, Nara Women's University, Kita-Uoya-Nishi Machi, 630-8506 Nara, Japan, Tel & Fax: (81)-742-71-2588

Presented Music

Pre-research was conducted on 4 music pieces using the Affective Value Scale of Music (AVSM; Taniguchi., 1995). These were "Pomp and Circumstance No.1" (Elger), "Radetzky-Marsch" (J.Straus), "The Swan of Tuonela" (Sibelius), and "Adagio d' Albinoni" (Albinoni). The AVSM parameters consisted of 5 items, elation, friendliness, excitement, lilt, and solemnity. Sixty-eight female junior-college students cooperated in the AVSM pre-research study. "Radetzky" (MH; Music-High) and "Adagio" (ML; Music-Low) were chosen, because of their high scores for elation, excitement, and lilt (Table 1). These musical selections were played on a tape recorder in the forward direction for the main part of our study.

Table 1 The mean scores of the affective value of music (pre-research)

Music title	Elation	Friendliness	Excitement	Lilt	Solemnity
Pomp and circumstance	15.85	8.09	15.84	9.68	13.79
Radetzky Marsch	18.35	10.07	15.21	13.57	11.97
The swan of tuonela	5.82	12.31	8.5	5.65	14.31
Adagio d'Albinoni	4.79	11.72	9.81	5.44	15.9

EEG analyzing

Twelve-channel EEGs data (Fz, F3, F4, Cz, C3, C4, Pz, P3, P4, T3, T4, Oz), using averaged ear lobes as a reference, were recorded (Nihon Kohden Neurofax 4514). The time constant was 0.3sec. The obtained spectral data were compressed into four frequency bands, and the alpha (8-13Hz) and beta (13-30Hz) band data are reported here. Laterality coefficients were calculated using the expression $LC = (R-L)/(R+L)$.

Evaluation of Mood

The mood of the subjects was measured by the Multiple Mood Scale (MMS; Terasaki.1992) which consisted of 8 items, depression (anxiety), liveliness, well being, friendliness, concentration, startle, hostility and boredom. The rating was based on 5-point scales.

Procedure

The experiment was carried out in a sealed room, with the temperature kept below 25c and background noise below 45db. Subjects sat on a chair, and were asked to keep their eyes closed and maintain a restful attitude. EEG activity during this rest time was recorded for 5min. and then the subjects answered the MMS questionnaire. Next, in the music session, the subjects were asked to listen to the two chosen music pieces with closed eyes while their EEG activity was recorded. After 4min, they answered the MMS and AVSM questionnaires. The procedure was the same for both music pieces, MH and ML, however, the listening order was at random. A second session was designed with the subjects sitting for the same length of time, but without hearing any music presentation. This was called the no-music (NM) session. After 4min, the subjects answered the MMS questionnaire. The music session and no-music session were carried out on different days.

Results

Affective value of presented music

Figure 1 shows the mean scores of the affective value of the two musical pieces. The score of MH (Radetzky) were significantly higher than ML (Adagio) for "elation" ($p<.01$), "excitement" ($p<.01$), and "lilt" ($p<.05$).

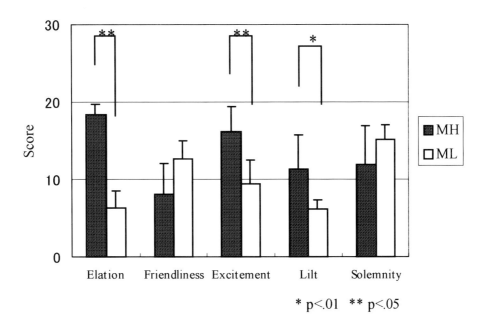

* $p<.01$ ** $p<.05$

Figure 1
The mean scores of the affective value of MH and ML

Mood state

Figure 2 shows the mean scores of the mood states after listening to MH and ML, as well as the NM session. The scores were calculated by subtracting the music session score from the pre-music score. The "concentration" and "friendliness" scores decreased in all sessions. The "liveliness" score increased after MH, while it decreased after ML and NM. The differences were significant ($p<.01$). The "boredom" score after listening to MH was lower than ML and NM, and there was a significant difference between MH and NM ($p<.05$).

EEG activity

Figure 3 shows the rate of percentage amplitude of the alpha band. The alpha band amplitude increased during the music session for both MH and ML, in comparison to the pre-music rest period. However, it decreased in the frontal area (Cz, C3) and parietal area (P4) during the NM session. The beta band amplitude did not significantly change.

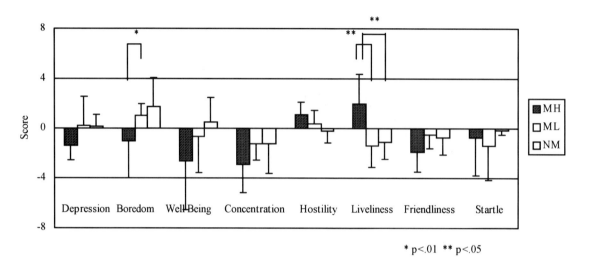

* $p<.01$ ** $p<.05$

Figure 2
The mean scores of mood states

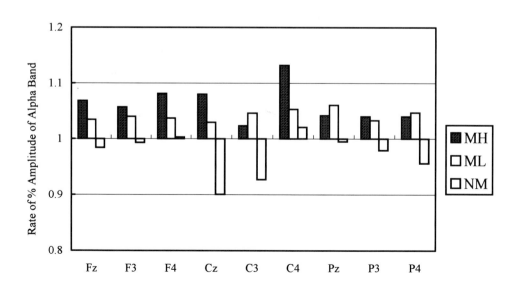

Figure 3
The rate of percentage amplitude of the alpha band in 3 conditions (MH, ML and NM)

Laterality coefficient in the frontal area

Figure 4-1,2,3 show the change of laterality coefficient in the frontal area. The laterality coefficient score for both music sessions indicated that the alpha band amplitude was more evident in the right hemisphere while the beta band was more evident in the left hemisphere. This tendency was especially observed in phases 2 and 3 (90～150s after the start) during the MH session. In phase 4, this tendency was reversed for both music sessions. However, in regards to the no-music session, the score of laterality coefficient indicated that the alpha band amplitude was more evident in the left hemisphere and the beta band amplitude was more evident in the right hemisphere, throughout the session.

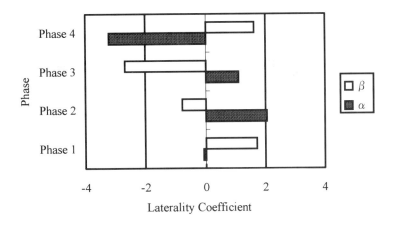

Figure 4-1
Laterality coefficient in frontal area during MH (Phase1: 60～90s Phase2: 90～120s Phase3: 120～150s Phase4: 150～180s)

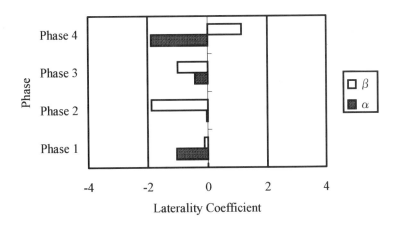

Figure 4-2
Laterality coefficient in frontal area during ML (Phase1: 60～90s Phase2: 90～120s Phase3: 120～150s Phase4: 150～180s)

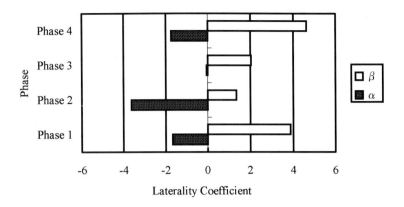

Figure 4-3
Laterality coefficient in frontal area during NM (Phase1: 60～90s Phase2: 90～120s Phase3: 120～150s Phase4: 150～180s

Discussion

1. Different musical properties can cause mood state changes in the listener.

It is known that music possesses two types of affective value, stimulating and calm. The results of the AVSM scores indicate that the affective value of MH (Radetzky) was more stimulating than ML (adagio). After listening to MH, the "liveliness" score increased while "boredom" decreased. This result suggests that the different properties cause different mood states in the listener. However, the relationship between the affective value of the music and the mood states of the listener requires more research and consideration of the various components of music.

2. Mood state changes caused by music reflect the alpha power asymmetry of brain activity.

Generally, increased alpha band power activity is thought to reflect a decreased activity in the corresponding region of the underlying cerebral cortex (Benca et al., 1999). Moreover, the alpha power asymmetry in the frontal area has an effect on affection and emotions. It is known that an alpha band increase in the right hemisphere is related to a positive affective-state and a decrease in the left brain is related to a negative affective-state (Ahern et al., 1985; Benca et al., 1999; Davidson et al., 1990; Petruzzello., 1994; Wheeler., 1993). Concerning the effect of music on brain activity, Iwaki et al. (1996) reported that regardless of music type, subjects in a high arousal state were calmed, while subjects with low arousal were aroused while listening to music. These results suggest that music has a modulation effect on arousal. Iwaki et al. (1997) reported further that the frontal interhemispheric coherence values of the alpha-2 band increased during the stimulating session, while the coherence values did not change during the calm music.

In this study, the alpha band amplitude increased in both music sessions. This result demonstrates that listening to music has an effect on increasing the alpha band, but that different music properties have no effect. The value of laterality coefficient in the frontal area showed the alpha band increase and the beta band decrease in the right hemisphere during the MH session. It suggests that a positive affective-state occurred with more stimulating music and that the different mood states, which resulted from listening to music, had an effect on the alpha power asymmetry.

In a previous study regarding physical activity, the alpha band amplitude increased in the right hemisphere after exercise and the hypothesis of cerebral lateralization was proposed (Hatfield & Landers, 1987) It is common to do physical activity while listening to music. Copeland & Franks (1991) reported that subjects who ran while listening to soft and slow music tended to extend their exercise time and underestimate the value of Rating of Perceived Exertion (RPE). We hope to investigate the effect of music during more dynamic activity in the future.

Summary

This experiment was to examine the influence of music on EEG activity and mood state. Tv ive-channel EEGs, averaged ear lobes as reference, were recorded on 7 female students during periods of pre-music rest and 4-minutes three music conditions ("No", "Radetzky-Marsch", and "Adagio d'Albinoni"). Affective Value Scale of Music (AVSM) and Multiple Mood Scale Questionnaire (MMSQ) were also administered and related to EEG power spectrum changes induced by three music conditions. The amplitude of alpha band increased in the two music conditions, "Radetzky-Marsch" and "Adagio d'Albinoni", in comparison with pre-music rest period. According to AVSM, positive feeling and vigor scores after listening to "Radetzky-Marsch" also significantly increased, but didn't increased in "Adagio d'Albinoni". Laterality coefficients in the frontal area during the music, "Radetzky-Marsch", were showed to be associated with concentration. These results were discussed with the viewpoint of hemispheric specialization for mood state.

References

Ahern,G.L., Schwrtz,G.E. Differential lateralization for positive and negative emotion in the human brain: EEG spectral analysis. Neuropsychologia 1985; 23:745-755.

Benca,R.M., Obermeyer,W.H., Larson,C.L., Yun,B., Dolski,I., Kleist,K.D., Weber,S.M., Davidson,R.J. EEG alpha power and alpha power asymmetry in sleep and wakefulness. Psychophysiol 1999; 36:430-436.

Copeland,B.L., Franks,B.D.:Effects of types and intensities of background music on treadmill endurance. J Sport Med Physical Fitness 1991; 31:100-103.

Davidson,R.J., Ekman,P., Saron,C.D., Senulis,J.A., Friesen,W.V. Approach-withdrawal and cerebral asymmetry: Emotional expression and brain physiology Ⅰ. J. Pers. Soc. Psychol 1990; 58:330-341.

Hatfield,B.D., Landers,D.M. Psychophysiology in exercise and sport research: an overview. Exercise, Sports Sci.Rev., 1987; 15:351-386.

Iwaki,T., Hayashi,M. & Hori,T. Study of arousal modulation effects of different affectional music on EEG activity. Jap J EEG and EMG 1996; 24:30-37.

Iwaki,T., Hayashi,M. & Hori,T. Changes in alpha band EEG activity in the frontal area after stimulation with music of different affective content. Percept Mot Skills 1997; 84: 515-526.

Petruzzello, S.J., Landers, D.M. State anxiety reduction and exercise: does hemispheric activation reflect such changes? Med. Sci. Sports Exerc 1994; 26:1028-1035.

Snyder M. Chlan, L. Music Therapy Annu Rev Nurs Res 1999; 17:3-25.

Taniguchi, T. Construction of an affective value scale of music and examination of relations between the scale and a multiple mood scale. Jap J Psychol 1995; 65: 463-470.

Terasaki, M. & Kishimoto, Y. & Koga, A. Construction of a multiple mood scale. The Jap J Psychol 1992; 62: 350-356.

Wheeler,R.E., Davidoson,R.J., Tomarken,A.J. Frontal brain asymmetry and emotion reactivity: A biological substrate of affective style. Psychophysiol, 1993; 30:82-89.

Chapter III.21

Strong Rightward Asymmetry of the Planum Parietale Associated with the Ability of Absolute Pitch

Kota Katanoda[a,1], Kohki Yoshikawa[b], and Morihiro Sugishita[a]

[a] *Department of Cognitive Neuroscience, Faculty of Medicine, University of Tokyo*
[b] *Department of Radiology, Institute of Medical Science, University of Tokyo*

Introduction

Music is a complex cognitive function unique to human. An immense literature has accumulated on the hemispheric lateralization of musical processes in the brain. The present study focused on an ability rare but useful to musicians, absolute pitch (AP). It enables its possessors to identify or produce a given musical pitch without using an external reference pitch. Three groups of normal right-handed subjects participated in this study: musicians with AP, musicians without AP, and non-musicians. High-resolution in vivo magnetic resonance (MR) morphometry of the planum temporale (PT) and the planum parietale (PP) was performed, and left-right asymmetry of the PT, PP, and the sum of these two (PT+PP), was compared among three subject groups. These two regions were of interest, as they were both found to participate in auditory processings (Binder et al., 1997) and strong leftward asymmetry of the PT was reported to be associated with musical ability in a recent study by Schlaug et al (1995).

Methods

Subjects

Thirty musicians and 24 non-musicians participated in this study after giving written informed consent. Edinburgh Handedness Inventory laterality quotients ranged from 60 to 100 (mean 90.8). Subject characteristics were summarized in Table 1.

---

[1] Correspondence: Kota Katanoda, Department of Cognitive Neuroscience, Faculty of Medicine, University of Tokyo 113-0011, Japan, Tel: (81)-3-5841-3574

Table 1 Subject characteristics and performance on pitch identification test

Group	Number	Age [years old]	Handedness	Duration of musical training [years]	Onset of musical training [years old]	AP test mean error [semits]
Musicians with AP	16	22.0 (1.9)	90.4 (13.2)	17 (3.2)	4.8 (2.6)	0.10 (0.16)
Musicians without AP	14	22.1 (2.3)	86.1 (7.7)	13.3 (6.2)	8.5 (5.7)	1.74 (1.04)
Non-musicians	24	24.7 (4.4)	95.2 (9.0)	0.0 (0.0)	-	-

AP: absolute pitch.; Numbers in parentheses are standard deviations.
Handedness is expressed as Edinburgh Handedness Inventory laterality quotients

Absolute pitch test

All musicians underwent a pitch identification test. Twenty simple piano tones ranging from $F_3^\#$ to E_6 (middle C is C_4) were presented binaurally to the subjects. Subjects were asked to identify the pitch and octave of each piano tone by writing on a musical staff. Subjects whose average error scores were within 0.6 semitone of the target were regarded as AP possessors.

Image acquisition

Imaging was performed on a 1.5 Tesla General Electric Signa Horizon system. T1-weighted images were acquired as a set of 124 contiguous axial or coronal slices using a spoiled-gradient-echo sequence (TR = 33 ms, TE = 3 ms, flip angle = 30, 256×256 matrix, 1.3 mm thick, pixel dimensions: 0.94×0.94 mm or 0.70×0.70 mm).

Anatomical definitions and measurements

MR morphometry was performed using the same method and definitions as described by Steinmetz et al (1991). The length of the PT and PP was determined respectively on each sagittal slice by tracing a cursor manually. By summing up length measurements on single slices multiplied by the slice thickness, an estimate of the total surface area of each region was provided. Left-right asymmetry of each region was expressed as the asymmetry coefficients d, which were calculated by the equation $d = (R - L) / [0.5 (R + L)]$ with R and L being the surface area of each region.

Statistical analyses

One-way multiple analysis of variance (MANOVA) was performed using dPT and dPP as dependent variables and subject group as an independent factor with three levels: musicians with and without AP, and non-musicians. Subsequently, univariate one-way analyses of variance (ANOVA) with subject group as an independent factor were performed for dPT, dPP, and d (PT+PP), respectively. Two orthogonal contrasts and one additional contrast were tested: (1) musicians with AP versus musicians without AP, (2) musicians versus non-musicians, and (3) musicians with AP versus non-musicians (corrected for multiple comparisons using the Scheffe's procedure).

Results

Table 2 shows mean areas and asymmetry coefficients of the PT, PP, and PT+PP. A one-way MANOVA yielded a significant overall effect of subject group (F=4.17; df =4, 100; P<0.005). Subsequent univariate one-way ANOVAs were carried out for dPT, dPP and d(PT+PP) respectively, which revealed the following results. The PT was similarly left-lateralized in three subject groups, and unlike the study by Schlaug et al. (1995), no significant difference was observed among them (P > 0.90, corrected). In contrast, the PP was differentially right-lateralized in three subject groups. The PP was significantly more lateralized to the right in musicians with AP than in musicians without AP (F=6.18; df =2, 51; P<0.005, corrected). This region was also significantly more right-lateralized in musicians with AP than in non-musicians (F=7.26, df =2, 51; P<0.005, corrected). With respect to the total area of the two regions (PT+PP), no significant difference was observed among three subject groups (P>0.20, corrected). Figure 1 shows sagittal MR images of the brain typical for each subject group.

Discussion

Leftward asymmetry of the PT in right handers has been accepted as an anatomical evidence of left hemisphere dominance for language related auditory processings. AP enables the retrieval of an arbitrary association between the pitch of a tone and its verbal label, and such verbal coding can be interpreted as an analogy to the identification of speech sounds. In our data, however, no significant difference was observed among three subject groups in the asymmetry of the PT, suggesting no association between the ability of AP and leftward asymmetry of this language-related region.

Auditory association cortex is distributed in the PP as well as in the PT (Witelson and Kigar, 1992). A functional neuroimaging study has demonstrated that the PP, right more extensively than the left, is involved in a kind of tone processings (Binder et al, 1997). Our result of strong rightward asymmetry of the PP in AP possessors anatomically suggests that the right PP bears closer relation to the ability of AP than the left. This finding forms a sharp contrast to the fact that the left PT, which is involved in auditory verbal processes, is larger than the right. Interestingly, the degree of the AP-related rightward PP asymmetry was significantly greater than that of the language-related PT asymmetry (a paired t-test against the null hypothesis that the sum of dPT and dPP is zero, t =3.39; df = 5; P<0.005), although the total area of these two regions (PT+PP) did not show rightward asymmetry because of relatively smaller areas of the PP than the PT (Table 2). Regarding the acquisition of AP, there is a notion called the early-learning theory, which states that AP can be acquired only during a limited early period of development. In our data, the onset age of musical training in musicians with AP showed a tendency to be earlier than that in musicians without AP, although not significant (4.8 years and 8.5 years old respectively, P=0.05, Table 1). Our anatomical findings can thus be given the following interpretation: the right PP is associated with some absolute aspects of pitch processings and in AP possessors auditory associate cortex of this region extensively developed due to early exposures to absolute pitch processings, which may have resulted in the relatively strong rightward asymmetry of this region.

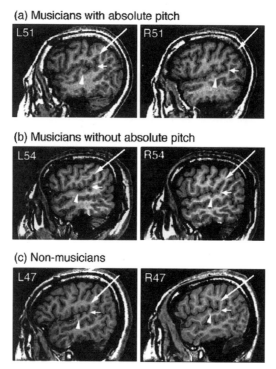

Figure 1
Sagittal magnetic resonance images of the brain typical for each of the three subject groups. (a) Musicians with absolute pitch (AP). (b) musicians without AP. (c) non-musicians. The hemisphere and the slice location are given on the upper left of each image (L: left, R: right, The numbers indicate the distance in mm from the midline.). Arrow heads indicate the anterior border of the planum temporale (PT). Short arrows indicate the posterior border of the PT, which is identical to the anterior border of the planum parietale (PP). Long arrows indicate the posterior border of the PP.

Table 2 Mean areas [mm^2] and asymmetry coefficients of the peri-Sylvian regions

Regions	PT			PP			PT+PP		
Group	Left	Right	dPT	Left	Right	dPP	Left	Right	d(PT+PP)
Musicians with AP	982 (322)	677 (188)	-0.33 (0.42)	193 (149)	436 (155)	0.84* (0.62)	1175 (322)	1114 (172)	-0.03 (0.28)
Musicians without AP	1038 (216)	734 (191)	-0.34 (0.36)	298 (134)	341 (195)	0.05 (0.52)	1336 (291)	1075 (214)	-0.21 (0.25)
Non-musicians	965 (266)	693 (208)	-0.33 (0.36)	287 (147)	321 (201)	0.09 (0.65)	1252 (276)	1014 (250)	-0.21 (0.30)

AP: absolute pitch.; Numbers in parentheses are standard deviations.
Asymmetry coefficient d = (R - L) / [0.5 (R + L)] ; Positive and negative values of d indicate rightward and leftward asymmetry respectively.; *Significantly larger compared to musicians without AP (ANOVA, P<0.005, corrected for multiple comparisons).

Summary

Absolute pitch (AP) is the ability to identify or produce a given musical pitch without the use of an external reference pitch. Using magnetic resonance morphometry, we examined the relation between the presence of this ability and the anatomical asymmetry of two peri-Sylvian cortical regions: one is the planum temporale, the posterior supratemporal area behind the first transverse gyrus of Heschl, and the other is the planum parietale, the posterior wall of the posterior ascending ramus of the Sylvian fissure. The planum temporale, which is involved in auditory verbal processes on the left hemisphere, was similarly left-lateralized in musicians with or without AP. In contrast, the planum parietale was significantly more right-lateralized in musicians with AP than musicians without AP. This result suggests that the planum parietale, right more closely than left, is associated with absolute aspects of pitch processings.

Acknowledgements

We thank S. Shigeno for her invaluable assistance in the preparation for our pitch identification test, and the students of Tokyo National University of Fine Arts and Music for their kind participation in our study. This work was supported by a grant-in-aid (09490006) from the Japanese Ministry of Education, Science, Sports and Culture and a grant (L-2-12) from the Japan Society for the Promotion of Science.

References

Binder JR, Frost JA, Hmmeke TA, Cox RW, Rao SM, Prieto T. Human brain language areas identified by functional magnetic resonance imaging. J Neurosci 1997; 17: 353-362.

Miyazaki K. Musical pitch identification by absolute pitch possessors. Percept & Psychophysics 1988; 44: 501-512.

Schlaug G, Jancke L, Huang, Y, Steinmetz H. In vivo evidence of structural brain asymmetry in musicians. Science 1995; 267: 699-701.

Steinmetz H, Volkmann J, Jancke L, Freund HJ. Anatomical left-right asymmetry of language-related temporal cortex is different in left- and right-handers. Ann Neurol 1991; 29: 315-319.

Witelson SF, Kigar DL. Sylvian fissure morphology and asymmetry in men and women: bilateral differences in relation to handedness in men. Comp Neurol 1992; 323: 326-340.

Index

absolute pitch (AP), 351, 416, 421, 487
adiabatic condition, 80, 82
adiabatic full passage (AFP), 80, 81
adiabatic rapid passage, 81, 86
aliasing, 165
allophonic variation, 294
alpha band, 484, 485
amusia, 352
anatomical realization, 5, 19
APOLONN, 445
aquaporin 4, 9
astrocyte, 5, 8
    assembly, 8
    electron dense layer, 9
    matrix, 8
    network, 8
    processes, 8
attentional trace hypothesis, 390
auditory association cortex, 489
auditory attention, 390, 395
auditory cortex
    primary, 326, 347, 348, 350, 425
    secondary, 347
auditory discrimination, 425
auditory grouping, 292, 295
auditory illusions, 279
auditory nerve, 361
auditory processing, 14, 291
    pre-attentive, 291, 292, 332
auditory stream segregation, 292

B_0, 49
B_0 inhomogeneity, 50, 51
Backus–Gilbert theory, 155, 167, 169
ballistocardiogram, 103, 104, 106
Beer–Lambert law, 240, 271
beta band, 485
binding, 17, 24, 27
blind separation, 124, 132
blind source separation (BSS), 92
blood oxygenation level dependent (BOLD), 43
BOLD contrast, 44, 50, 51
BOLD phenomenon, 46
Bolzmann constant, 66
boundary element method (BEM), 113, 155
Brillouin equation, 65
Broca's aphasia, 339

carotid artery, 251, 252, 253, 254, 262
carotid endarterectomy (CEA), 254, 255, 256, 257
cellular automata, 11, 12, 14, 16
central limit theorem, 158
central pattern generators (CPGs), 445
cerebral blood flow (CBF), 251, 260, 347, 349, 351, 352
    regional (rCBF), 241, 258, 259, 261, 451
cerebral peduncle, 230
cerebrovascular disorders, 252
characteristic frequency (CF), 425

chromatic charge, 305
chromophore, 239, 240, 271
coarse graining, 13, 14, 22
coherence, 32, 33, 34
Coherent Infomax, 24
contextual disambiguation, 24, 26, 31, 36
contextual field (CF), 24, 26, 34, 35, 185
convection
    Bénard (type of), 10, 11, 12, 21
    free, 10, 11
    Marangoni, 12
    thermal, 11
convection cells, 11, 12
    Marangoni (hexagonal), 12, 14, 15
convective flow, 4, 6, 10, 11
copper, 250, 251
coronal cuts, 65
correlation length, 13, 14
cortical
    columns, 6, 15
    cortex, 6
Cramer–Rao lower bound, 173, 180
cytochrome oxidase, 239, 249, 250, 251, 254, 256, 257

decimation, 13, 22
depression, 442
detection probability curve (DPC), 172
diamagnetic, 69
dielectric resonance, 72
diffusion anisotropy, 221, 233
diffusion characteristic function, 221, 222
diffusion tensor, 222, 223, 227, 228
director musices, 300, 307, 308, 312
dopamine, 267
Doppler effect, 283
dorsolateral frontal cortex, 351
dorsolateral prefrontal cortex (DLPF), 267, 269, 270
duration contrast articulation, 301
duration contrast, 301, 303
dynamic grouping, 24, 25, 26, 30, 34, 36

Edinburgh Handedness Inventory, 401, 487
eigenvalue images, 223, 225
eigenvalue, 96, 159, 220, 221, 227, 233, 234
eigenvector, 96, 159, 221, 227, 233, 234
electroencephalography (EEG), 27, 32, 101, 109, 193, 243, 252, 263, 465, 480
electric source imaging (ESI), 153, 154, 155, 156, 168, 185
electrocorticography (ECoG), 124, 176
electron-dense layer and dendritic ramification (ELDER), 9, 10, 11, 12, 15, 16

ellipsoid, 221
emotion, 352, 479
emotional communication, 368
emphasis rules, 301
entropy-vortex wave, 6, 7, 10, 15, 16, 17
epilepsy, 103, 124
equation
    Betchov's intrinsic, 22
    Biot–Savart, 216
    Boussinesq, 4
    Euler's, 6
    Navier–Stocks, 14
    non-linear Schrödinger (NLS), 22
    of continuity, 219
    Sarvas, 218
equivalent current dipole (ECD), 157
Euler angles, 111
event related potential (ERP), 123, 127, 193, 389, 150, 176, 185, 325, 370
    auditory, 291
evoked potential
    motor (MEP), 463, 472
    somatosensory (SEP), 175, 252, 254, 256, 257, 457
    visual (VEP), 32, 175, 178
extracellular space
    compartmentalization, 8
    conventional, 9
    dry, 8, 11

fast spin echo, 50
ferromagnetic susceptibility, 50
ferromagnetic, 69
Fick's law, 219
fictitious activation, 51, 58
finite element method, 113
Fisher information matrix (FIM), 173
Frish–Hasslacher–Pomeau (FHP) lattice, 14
FSE, 56
full width at half maximum (FWHM), 48
functional magnetic resonance imaging (fMRI), 72, 91, 101, 109, 243, 326, 370
    high field (HF-fMRI), 337

gait, 437, 442
gamma activity, 34, 157, 186
Gamma-band, 28, 29, 36
gestalt grouping, 31
gliosis, 225, 231
glutamate, 26
    GLU, 27
gyrus
    fusiform, 347
    parahippocampal, 352

Haas effect, 282
hand motor task, 453, 461, 463
harmonic charge, 301, 304
harmonics, 280
harmony, 374, 376, 378
hemispheric specialization, 17, 18
high resolution NMR, 55
high-density EEG, 31, 147
high-field system, 65
H-reflex, 472

independent component analysis (ICA), 91, 92, 124
   algorithm, 124, 144
   decomposition, 126, 127, 128, 129, 130, 136, 137
   training, 126
independent component–cross correlation–sequential epoch (ICS) analysis, 91, 95
infomax, 132, 133, 134, 135, 138, 139
interslice partial volume phenomenon, 64
intraparietal, 61

Kanji, 341
kinetic viscosity, 4, 15
k-space, 54

lambda chart analysis, 221, 223, 224
lambda chart, 221, 222
language, 18, 337, 366, 367, 368, 369
Laplacian
   EEG, 176
   maps, 176, 177
   transfer function, 178
lattice-gas shell, 12, 15, 16, 17
localized-induction approximation, 17
logical power supply, 15

magnetic resonance angiograpy (MRA), 55
magnetic shimming, 49, 54
magnetic susceptibility, 69
magnetoencephalography (MEG), 109, 201, 294, 326
   evoked by visual stimulation, 210, 213
   high spatial resolution, 202
   radial, 207, 208
Maxwell's equation, 57
medial lemniscus, 230
melodic charge, 301, 303, 307, 308, 309, 312
melodic processing, 346
   precuneus, 353
   subcallosal cingulate, 353
   superior temporal (STG), 347, 350
metre, 399

mismatch negativity (MMN), 291, 292, 293, 294, 325, 326, 329, 331, 396
   generators, 326, 327
modified boxcar, 62
molecular layer, 10
molecular motion, 219
Monte Carlo simulation, 167, 180
mood, 480, 484
motion correction, 58, 60
motion ghost, 63
motor imagery, 472
movement-related cortical potentials (MRCPs), 453, 461, 463, 465, 469, 476
moyamoya disease, 263
MRI, 32
   vertical (system), 71
Multiple Mood Scale (MMS), 480
music literacy, 341
music performance, 299, 300, 315
   horizontal programming, 317
   vertical programming, 319
musical phrase, 194, 196
myoglobin, 239

N400, 376, 378
n-back task, 269
near-infrared spectroscopy (NIRS), 239, 249,
   fMRI guided, 267
   functional, 242
   imaging, 239
negative displacement (Nd), 389, 395
Nelder–Mead algorithms, 111
NMDA, 27, 35
nuclear magnetic resonance (NMR), 47
Nyquist
   criterion, 163, 183
   frequency, 175, 183
   inter-electrode distance, 175
Nyquist artifact, 62
Nyquist ghost, 57, 58, 61, 62, 63

offset independent adiabaticity, 83
optical topography, 244, 245
oxygen oversupply hypothesis, 46

P3, 393
P600, 374, 376, 378
paramagnetic susceptibility, 50, 65
paramagnetic, 69
Parkinson's disease, 267, 435, 463, 470
partial volume phenomenon, 53
performance grammar, 307, 309
phoneme, 294, 327, 366
pictorial mathematics, 237

piecewise parabolic method, 4
  free convection, 21
  PPM, 21
pitch, 279, 281, 282, 294, 301, 310, 346, 348, 357, 359, 360, 361, 362, 392, 415, 416
  discrimination, 328, 329
  processing, 331
pixel misalignment, 58
planum parietale (PP), 487, 491
planum temporale (PT), 400, 487, 491
positron emission tomography, (PET), 32, 326, 349
precedence effect, 282
primary motor cortex, 457, 461, 469
primary visual cortex, 14, 243
principal component analysis (PCA), 23, 93, 127, 129, 130
principal eigenvector imaging ($PE_{vec}I$), 227, 234
principal quantum, 14
Promax rotation, 127, 130

quadrature detection, 54

recurrent neural network, 445
red nucleus, 229
relative pitch, 416
re-normalization, 13
Reynolds number, 9, 15
rhythm, 282, 399, 410, 449, 453, 457, 462, 469, 470
rhythmic auditory simulation, 436, 442
ripple model, 14
room temperature shim coils, 49

scalp Laplacian, 150, 151, 178
second order tensor, 220, 228, 234
selective attention, 24, 25, 35, 36
selective excitation, 48
self-organization, 4, 10, 11, 14, 15, 18, 19, 21
self-rating depression scale (SDS), 440
semantic expectancy, 371, 376
sequential epoch analysis (SEA), 91, 94
simple tone, 194, 196
simulacra, 279, 281
single photon emission computed tomography (SPECT), 259, 260, 261
singular value decomposition (SVD), 93
slab, 55
slice definition, 56
solitons, 17, 22
sound patterns, 293
sound-pattern traces, 293, 295
sound traces
  long term exploratory, 293
  short term exploratory, 292, 295
spatial filters (SFs), 202, 206, 209, 216
spatial signal-to-noise ratio (SNR) transform, 151, 152, 158, 186
spatial spectrum analysis, 161
spherical harmonics, 161, 162, 164
spin density, 51
spin-lattice and spin-spin relaxation time, 51, 52
statistical parametric mapping (SPM), 92, 166
stereotactic surgery, 59
Stroop effect, 415, 421
  auditory, 415
  visual, 415
structural magnetic resonance imaging (sMRI), 110, 120
substantia nigra, 230, 267
sulcus
  inferior temporal, 341
  transverse occipital, 341
supplementary motor area, (SMA), 350, 453, 457, 461, 463, 469
supratemporal plane, 292
susceptibility effects, 50

$T_2{}^*$, 44
$T2^*$ contrast, 50, 51, 53
$T2^*$ weighted imaging, 46
tensor analysis
  diffusion, 233
  full (FTA), 223, 227, 234, 236
  single ellipsoid diffusion, 228
three dimensional anisotropy contrast magnetic resonance axonography (3DAC MRX), 227, 235, 236
tone, 280, 281, 282, 301
transcranial Doppler (TCD), 245, 253, 254, 256
transcranial magnetic stimulation (TMS), 469, 472
two-dimensional Fourier transform (2DFT), 48

universality, 13

Varimax rotation, 127, 130
virtual sphere, 4, 6, 12
visual processing, 13
vortex model, 3, 10, 16
vortex-entropy wave, 11
voxel volume, 59

voxel, 53

Wada test, 342
Wald test, 181
Wallerian degeneration, 223, 230
weak susceptibility effects, 46, 51, 52, 53
Wechsler Adult Intelligence Scale (WAIS), 269
Wernicke's aphasia, 339, 342
white noise, 194, 196
working memory, 18, 267, 346, 348, 351

Xenon computed tomography (Xe-CT), 449, 463